Adaptive Inverse Control

IEEE Press
445 Hoes Lane
Piscataway, NJ 08854

Adaptive Inverse Control

A Signal Processing Approach

Reissue Edition

Bernard Widrow
Department of Electrical Engineering
Stanford University

Eugene Walach
IBM Israel Ltd.
Science & Technology

IEEE PRESS SERIES ON POWER ENGINEERING
Mohamed E. El-Hawary, *Series Editor*

IEEE
IEEE PRESS

BICENTENNIAL 1807 WILEY 2007 BICENTENNIAL

WILEY-INTERSCIENCE
A John Wiley & Sons, Inc., Publication

Published by John Wiley & Sons, Inc., Hoboken, New Jersey.
Published simultaneously in Canada.

Wiley Bicentennial Logo: Richard J. Pacifico

Library of Congress Cataloging-in-Publication Data is available.

ISBN 978-0-470-22609-4

10 9 8 7 6 5 4 3 2 1

FSC
www.fsc.org
MIX
Paper from responsible sources
FSC® C013604

We dedicate this work to our families over the generations. They helped us and inspired us.

I would like to dedicate this book to my grandsons Jeffrey and Adam Sklarin, to their parents Rick and Debbie, to my daughter Leslie, to my wife Ronna Lee, and to the memory of my parents Moe and Ida Widrow.

Bernard Widrow

I would like to dedicate this book to my son Elad, to my daughter Algith, to my wife Rina, to my mother Sarah and to the memory of my father Benjamin Walach.

Eugene Walach

Contents

Preface

In this book, methods of adaptive signal processing are borrowed from the field of digital signal processing to solve problems in dynamic systems control. Adaptive filters, whose design and behavioral characteristics are well known in the signal processing world, can be used to control plant dynamics and to minimize the effects of plant disturbance. Plant dynamic control and plant disturbance control are treated herein as two separate problems. Optimal least squares methods are developed for these problems, methods that do not interfere with each other. Thus, dynamic control and disturbance cancelling can be optimized without one process compromising the other. Better control performance is the result. This is not always the case with existing control techniques.

Inverse control of plant dynamics involves feed-forward compensation, driving the plant with a filter whose transfer function is the inverse of that of the plant. Inverse compensation is well known in signal processing and communications.

Every MODEM in the world uses adaptive filters for channel equalization. Similar techniques are described here for plant dynamic control. Inverse control is feed-forward control. The same precision of feedback that is obtained with existing control techniques is also obtained with adaptive feed-forward control since feedback is incorporated in the adaptive algorithm for obtaining the parameters of the feed-forward compensator.

Inverse control can be used effectively with minimum phase and non-minimum phase plants. It cannot work with unstable plants, however. They must first be stabilized with conventional feedback, of any design that simply achieves stability. Then the plant and stabilizing feedback can be treated as an equivalent stable plant that can be controlled in the usual way with adaptive inverse control. Model reference control can be readily incorporated into adaptive inverse control.

Adaptive noise cancelling techniques are described that allow optimal reduction of plant disturbance, in the least squares sense. Adaptive noise cancelling does not affect inverse control of plant dynamics. Inverse control of plant dynamics does not affect adaptive disturbance cancelling. If initial feedback is needed to provide plant stabilization, the design of the stabilizer has no effect on the optimality of the adaptive disturbance canceller.

The designs of the adaptive inverse controller and of the adaptive disturbance canceller are quite simple once the control engineer gains a mastery of adaptive signal processing. This book provides an introductory presentation of this subject with enough detail to do system design. The mathematics is simple and indeed the whole concept is simple and easy to implement, especially when compared with the complexity of current control methods.

Adaptive inverse control is not only simple, but it affords new control capabilities that can often be superior to those of conventional systems. Many practical examples and applications are shown in the text.

Another feature of adaptive inverse control is that the same methods can be applied to adaptive control of nonlinear plants. This is surprising because nonlinear plants do not have transfer functions. But approximate inverses are possible. Experimental results with nonlinear plants have shown great promise. Optimality cannot be proven yet, but excellent results have been obtained. This is a very promising subject for research. The whole area of nonlinear adaptive filtering is a fascinating research field that already shows great results and great promise.

This book was originally published under the title *Adaptive Inverse Control*. We are grateful to IEEE Press and John Wiley, Inc. for bringing it back into print. We are also grateful to colleagues Gene Franklin, Karl Johan Åström, Jose Cruz, Brian Anderson, Paul Werbos, and Shmuel Merhav for their early comments, suggestions, and feedback. We are grateful to former Stanford students Steve Piche, Michel Bilello, Gregory Plett, and Ming-Chang Liu who confirmed the results with experiments and who assisted with preparation of the drawings and final manuscript.

BERNARD WIDROW
Stanford, California

EUGENE WALACH
Haifa, Israel

Chapter 1

The Adaptive Inverse Control Concept

1.0 INTRODUCTION

Adaptive filtering techniques have been successfully applied to adaptive antenna systems [1–20]; to communications problems such as channel equalization [21–30] and echo cancelation in long-distance telephony [31–39]; to interference canceling [40–46]; to spectral estimation [47–57]; to speech analysis and synthesis [58–60]; and to many other signal processing problems. It is the purpose of this book to show how adaptive filtering algorithms can be used to achieve adaptive control of unknown and possibly time varying systems.

The system to be controlled, usually called the "plant," may be noisy, that is, subject to disturbances, and for the most part it may be unknown in character.[1] The plant and its internal disturbances may be time variable in an unknown way. In some cases, the plant might even be unstable. Adaptive control systems for such plants would be advantageous over fixed systems since the parameters of adaptive systems can be adjusted or tailored to the unknown and varying requirements of the plant to be controlled. Adaptivity finds a natural area of application in the control field [88].

In the past two decades or so, many hundreds of papers have been published on adaptive control systems in the *Transactions* of the IEEE Control Systems Society, in *Automatica*, in the IFAC (International Federation for Automatic Control) journals and conference proceedings, and elsewhere. At the same time, a very large number of papers on adaptive signal processing and adaptive array processing have appeared in the *Transactions* of the IEEE Signal Processing Society, Antennas and Propagation Society, Communications Society, Circuits and Systems Society, Aerospace and Electronics Society, the *Proceedings of the IEEE*, and elsewhere. Many books have been published on these subjects. The two schools of thought, adaptive controls and adaptive signal processing, have developed almost independently. The control theorists have by and large studied adaptive control using

[1]Some prior knowledge of the character of the plant and its internal disturbances will be needed in order to establish proper control. For example, at least a rough idea of the transient response time of the plant would be required in order to model it adaptively. Some idea of how rapidly the plant characteristics change for plants that vary over time would be needed. Some knowledge of the plant disturbance would be useful, such as disturbance power level at the plant output. Detailed knowledge of the plant and its disturbances would not be required however.

state variable feedback coupled through variable parameter networks to regulate unknown plants and to control their disturbances. The signal processing people have been working on problems that for the most part involve adapting weights of transversal filters by gradient methods and employing the resultant adaptive filters to systems without feedback (except for feedback in the adaptive process itself). The signal processing people have found a great number of practical applications for their work, and so have the adaptive control people.

The goal of this book is not to bridge the gap between these two schools of thought but to attack certain problems in adaptive control from an alternative point of view using the methodology of adaptive signal processing. The result is what we call "adaptive inverse control."

We begin with a discussion of direct modeling (or identifying) the characteristics of the unknown plant using simple adaptive filtering methods. Then we show how similar methods, with some modification but in a different configuration, can be used for inverse modeling (or equalization or deconvolution). Inverse plant models can be used to control plant dynamics. Next we show how both direct and inverse models can be used in the same adaptive process to minimize the effects of plant disturbance. In this development, we assume that the plant is completely controllable and observable, that it can (in a quasistatic sense) be represented in terms of an input-output transfer function (albeit an unknown one), and that the plant is stable (if unstable, someone has previously applied stabilization feedback). The plant may be either minimum-phase or nonminimum-phase.

The basic ideas of adaptive inverse control have been under development at Stanford University over the course of many years. The earliest related work is described in a paper by Widrow on blood pressure regulation [61]. Subsequent work is reported in several papers that were presented at Asilomar conferences [62, 63]. A Ph.D. dissertation by Shmuel Schaffer was concerned with model-reference adaptive inverse control [64]. A tutorial on the work is given by Widrow and Stearns [65]. The first paper on adaptive inverse control including adaptive plant disturbance canceling was presented by Widrow and Walach in 1983 at the First IFAC Workshop in Control and Signal Processing in San Francisco [66]. The second presentation was by Widrow in 1986 in a keynote talk at the Second IFAC Workshop on Adaptive Systems in Control and Signal Processing, University of Lund, Sweden [67]. There have been almost no other publications on inverse control and disturbance canceling until recently. Several recent publications in the neural network literature have appeared concerning nonlinear adaptive inverse control [95, 96, 97].

1.1 INVERSE CONTROL

A conventional control system like the one illustrated in Fig. 1.1 uses feedback, sensing the response of the plant to be controlled, comparing this response to a desired response, and using the difference to excite an actuator or controller whose output drives the plant input to cause the plant output to follow the desired response more closely.

The system of Fig. 1.1 has unity feedback and is often called a *follow-up* system since the objective is that the plant output follow the input signal or the *command input*. Any difference between the plant output and the command input signal is an *error signal* sensed by the controller which amplifies and filters it to drive the plant to reduce the error.

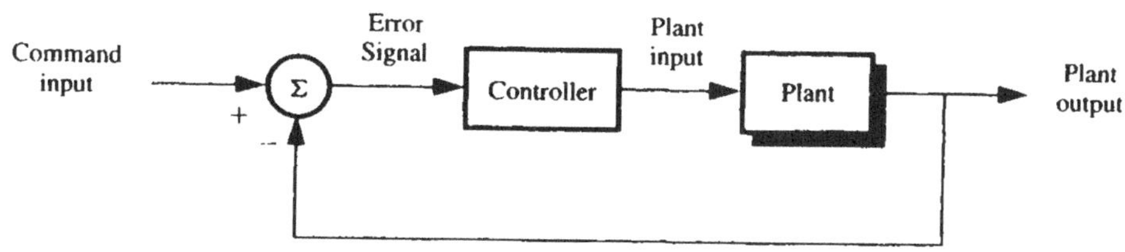

Figure 1.1 A conventional feedback control system.

The use of feedback must be done in a careful way to prevent instability and to achieve satisfactory dynamic response. When the plant characteristics are time variable or nonstationary, it is sometimes necessary to design the controller to vary with the plant. A common objective in doing this would be to minimize the mean square of the error. But achieving this objective is generally difficult. If one knew the plant characteristics versus time, one might be able to determine the best controller versus time. Not knowing the plant, an identification process could be used to estimate plant characteristics over time, and these characteristics could be used to determine the controller over time. Another idea would be to parametize the controller and vary the parameters to directly minimize mean square error. The difficulty with this approach is that, regardless of how the controller is parametized, the mean square error versus the parameter values would be a function not having a unique extremum and one that could easily become infinite if the controller parameters were pushed beyond the brink of stability.

The objective of the present work is to take an alternative look at the subject of adaptive control. The approach to be developed, adaptive inverse control, in some sense involves open-loop control and it is quite different from the feedback-control approach in Fig. 1.1. We attempt to develop a form of adaptive control that is simple, robust, and precise. With some knowledge of the subject of adaptive filtering, adaptive inverse control is easy to understand and use in practice.

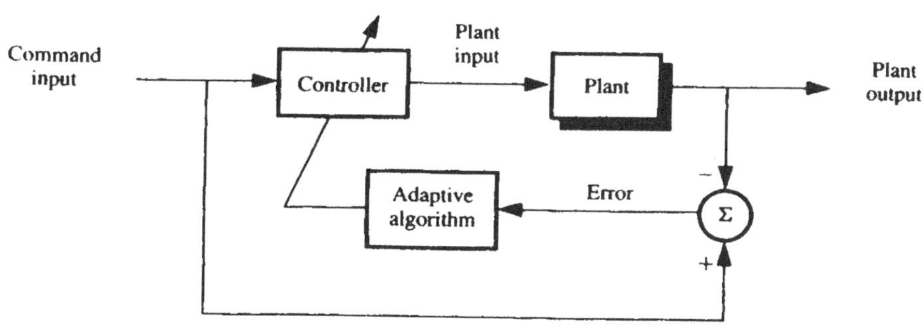

Figure 1.2 Basic concept of adaptive inverse control.

The basic idea of adaptive inverse control is to drive the plant with a signal from a controller whose transfer function is the inverse of that of the plant itself. The idea is illustrated with the system of Fig. 1.2. The objective of this system is to cause the plant output to follow the command input. Since the plant is generally unknown, it is necessary to adapt or to adjust the parameters of the controller in order to create a true plant inverse. An error signal, the difference between the plant output and the command input, is used by an adaptive algorithm to adjust the controller's parameters to minimize the mean square of this error.

Referring to Fig. 1.2, the controller in this diagram can be thought of as a filter having an input and an output. This controller has adjustable parameters, and adjustability is indicated by the arrow going through the box. Control of the adjustable parameters is done by means of an "adaptive algorithm," which is driven by the error signal. A usual objective for the adaptive algorithm would be minimization of mean square error, the error being the difference between the plant output and the command input.

Comparing the system of Fig. 1.1 with that of Fig. 1.2, *minimizing mean square error is done in the first case by using the error signal directly in a feedback process for forming the plant input signal, whereas in the second case the error signal is used in a feedback process to control the parameters of the controller and is not fed back directly to the plant input. The first case is feedback control and the second case is feedforward control. In both cases, feedback is used to ensure precise system responses.*

If adapting the controller of Fig. 1.2 were to make the error small, the controller would have become an inverse of the plant. The cascade of the controller and plant would thereby have a combined transfer function matching a gain of unity.

We assume that the plant is linear and that it varies slowly so that it is quasistatically stationary. We assume that the controller has converged and that it too is linear and quasistatically stationary. The dynamic characteristics of both the plant and controller may be represented by transfer functions, and the transfer function of the controller would be the reciprocal of that of the plant.

If the plant has internal delay, the inverse controller may have difficulty in overcoming it. The controller would need to be a predictor. Furthermore, if the plant is nonminimum-phase (transfer function zeros in the right half of the s-plane or outside the unit circle in the z-plane), then the inverse controller would want to have poles in the right half of the s-plane or outside the unit circle in the z-plane. Such an inverse would normally be unstable. Means for overcoming these difficulties are developed in Chapter 5.

The system of Fig. 1.2 illustrates the basic concept of adaptive inverse control. How the adaptive algorithm works to control the parameters of the controller is a matter to be developed below.

Sometimes it is desired that the plant output track not the command input itself but a delayed or smoothed version of the command input. The system designer would generally know the smoothing characteristic to be used. A smoothing model can be readily incorporated into the adaptive inverse control concept. How this may be done is illustrated in Fig. 1.3. The smoothing model is generally designated as the *reference model* in the control theory field. Thus, the system of Fig. 1.3 may be called a *model-reference* adaptive inverse control system. The model-reference idea is due to Whitaker, and an early reference is [89]. A recent reference is [88].

The reference model is chosen to have the same dynamic response that the designer would like for the entire system. Referring to Fig. 1.3, it is evident that this result would be obtained by once again adapting the controller to cause the mean square error to be low. In this case, the cascade of the controller and the plant after convergence would have a dynamic response like that of the reference model. The product of the controller and plant transfer functions would closely approximate the transfer function of the reference model.

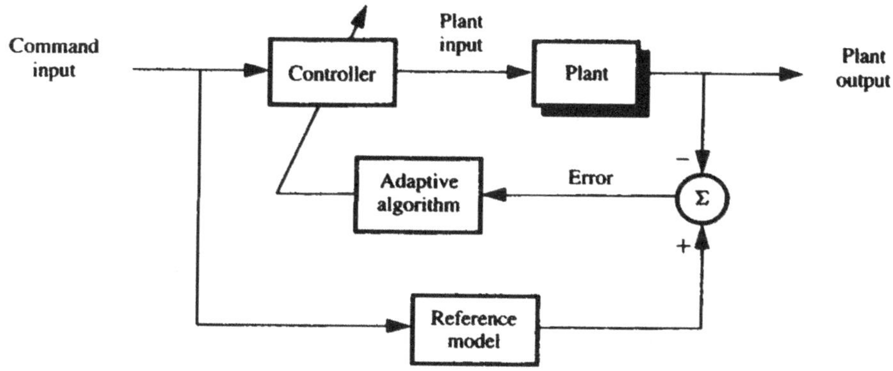

Figure 1.3 Model-reference adaptive inverse control.

Plant noise and disturbance present a problem for the adaptive inverse control approach. Lack of feedback from the plant output back to the plant input permits internal plant noise and disturbance to exist unchecked at the plant output. Various signal processing methods for noise canceling have been developed [40–46], and with some modification they have been applied to the cancelation of plant noise and disturbance. The basic scheme is shown in Fig. 1.4. Use is made of both a plant model and a plant inverse model. The plant model has the same transfer function as the plant, while the plant inverse model has a transfer function which is the reciprocal of that of the plant. How these models can be obtained will be described below. For the present argument, we assume that these models exist.

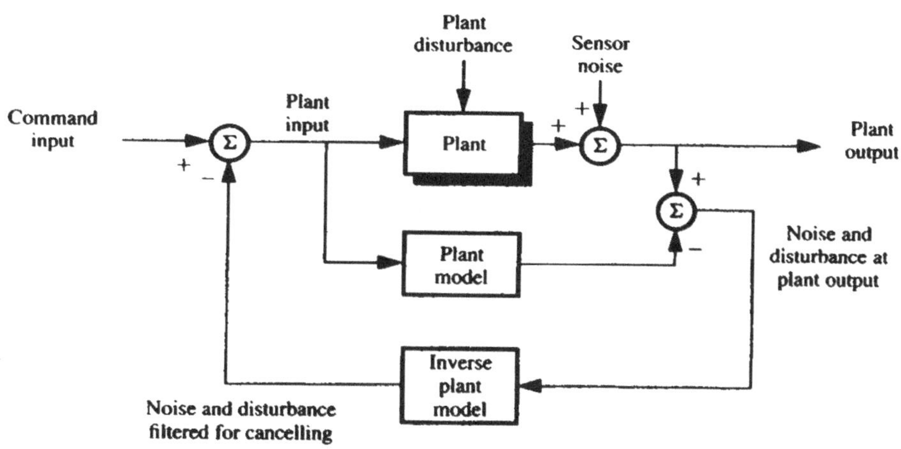

Figure 1.4 Canceling plant noise and disturbance.

In the control literature, plant disturbance is often represented as an additive noise at the plant input. Sensing the plant output is done with a detector or sensor that may be noisy. Sensor noise is often represented as an additive noise at the plant output. In the system of Fig. 1.4, the plant noise and disturbance are separated from the plant's dynamic output response. The plant input drives both the plant and its model (which is free of noise and disturbance). The difference between the plant output and the plant model output is the plant noise and disturbance as they appear at the plant output. The sum of the plant output noise and disturbance is used to drive the inverse plant model to generate filtered noise and disturbance for subtraction from the plant input. The ultimate effect is to cancel noise and disturbance at the plant output [67].

With almost perfect direct and inverse models, one can show that the transfer function from the point of injection of the plant sensor noise to the plant output point is close to zero. This implies that the sum of the plant noise and plant disturbance will be highly attenuated at the plant output. One can show furthermore that the dynamic response of the plant is essentially unchanged even while the plant's noise and disturbance are canceled by the disturbance canceling feedback. It is interesting to note that with perfect direct and inverse models, the system of Fig. 1.4 will be an unusual feedback system. The feedforward link of the major loop will have zero gain when the dynamics of the plant model perfectly balances that of the plant itself.

An adaptive inverse control system including the plant noise and disturbance canceling features of Fig. 1.4 and the model-reference control features of Fig. 1.3 are shown in Fig. 1.5. In a practical system of this type, separate adaptive processes would be needed to obtain the plant model, the inverse plant model, and the controller.

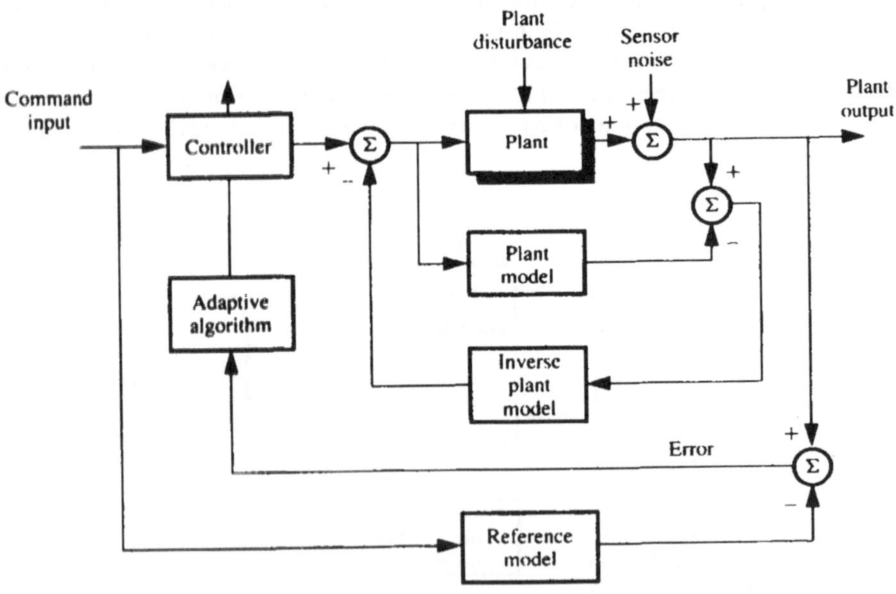

Figure 1.5 Model-reference adaptive inverse control system with plant noise and disturbance canceling.

With regard to ordinary control systems, like the one shown in Fig. 1.1, control of plant noise and disturbance is done with feedback. But by incorporating feedback for noise and disturbance control, the plant dynamics are inevitably altered. A compromise is generally required in the design process to obtain good dynamic response and good noise and disturbance control all at the same time.

Our approach is a different one in that it involves separate adaptive processes for (a) the control of plant dynamics and (b) for the control of plant noise and disturbance. By handling the problems in this way, the corresponding adaptive subsystems are relatively simple, easy to analyze, and easy to optimize.

The system of Fig. 1.5 represents our approach to adaptive control in an overall way. In the remainder of this text, we develop a variety of practical adaptive algorithms for implementation of the various subsystems of Fig. 1.5, and we analyze their performance and study how they interact in the overall control context. Simulation results are presented to verify theory and to demonstrate workability.

1.2 SAMPLE APPLICATIONS OF ADAPTIVE INVERSE CONTROL

The objective of this section is to illustrate the application of adaptive inverse control to a variety of control problems. These problems are in some sense classic in nature, and they will be discussed from many points of view in the chapters to follow. Here we see without much explanation results of control by adaptive inverse control techniques applied to a set of exemplary problems.

1.2.1 Dynamic Control of a Minimum-Phase Plant

The plant to be controlled has the transfer function

$$\frac{s+0.5}{(s+1)(s-1)}. \tag{1.1}$$

This plant is minimum-phase and is unstable. The first step is to stabilize it with feedback. A root-locus diagram is shown in Fig. 1.6. It is clear from this diagram that the plant can be stabilized by making use of the simple unity feedback system of Fig. 1.7, by setting the loop gain within the stable range $\infty > k > 2$. The loop gain was set to $k = 4$ for this control experiment. The closed loop transfer function is minimum-phase and has two poles in the left half of the s-plane.

The plant and its stabilization are continuous (analog) systems. The adaptive inverse control part, as it would be in the real world, is discrete (digital). A diagram of the complete system is shown in Fig. 1.8, including the necessary analog-to-digital conversion (ADC) and digital-to-analog conversion (DAC) components. The command input is sampled and is fed to both the inverse controller and the reference model. The controller output is converted to analog form, using a zero-order hold, to drive the plant and its stabilization loop. The error signal used to adapt the inverse controller is discrete. This is the difference between the reference model output and the sampled plant output. The system of Fig. 1.8 is modeled after that of Fig. 1.3.

Referring to Fig. 1.8, we define the *discretized equivalent plant* to be the discrete transfer function from the digital input to the DAC, to the sampled output of the plant. This includes the plant and its stabilization loop, shown in Fig. 1.7, and the DAC, and the sampler at the plant output. The impulse response of the discretized equivalent plant is shown in Fig. 1.9. The chosen sampling rate was 10 samples per second.

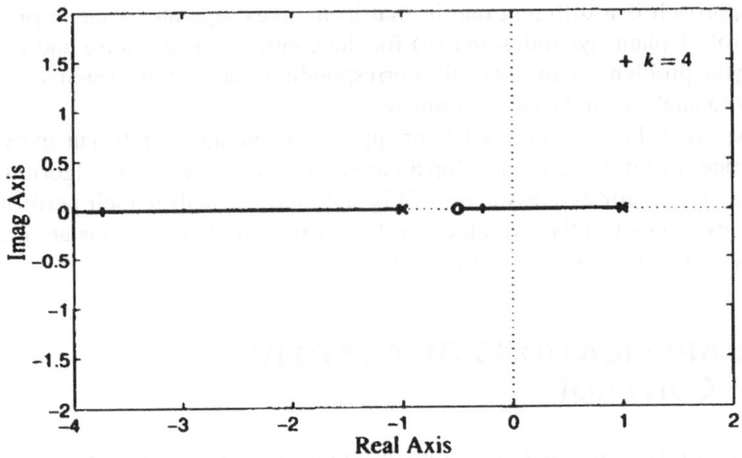

Figure 1.6 Root-locus of minimum-phase plant with proportional feedback.

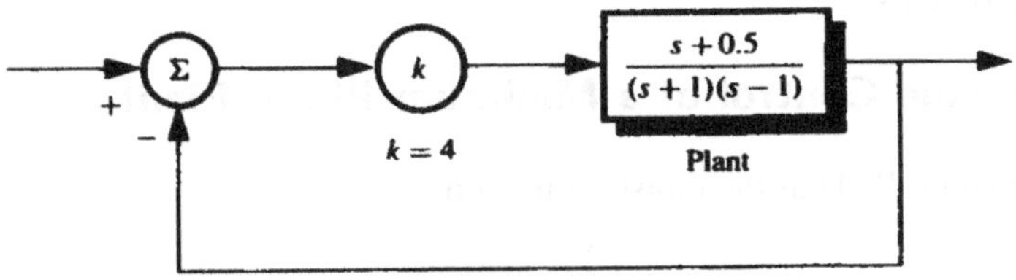

Figure 1.7 Minimum-phase plant stabilized with proportional feedback, $k = 4$.

The reference model chosen for this experiment was a one-pole digital filter with a one-second time constant. Its impulse response is shown in Fig. 1.10. The objective was to cause the overall system response, from the samples of the command input to the samples of the plant output, to be as close to the response of the reference model as possible in the least squares sense. The sampled command input was a first-order Markov process, generated by filtering white noise with a one-pole one-second time constant digital filter. The controller was allowed to have 100 weights, and the theoretically optimal impulse response for that many weights is shown in Fig. 1.11. The learned impulse response of the inverse controller is shown in Fig. 1.12. Notice the similarity that it has to the optimal impulse response.

It is useful to compare the plant output to the reference model output. At the beginning, the inverse controller is learning and not yet performing well. The plant output and the reference model output are shown over the first 200 samples in Fig. 1.13. They do not track. It is too early in the process.

The effects of learning can be observed in Fig. 1.14. The plant output begins to track properly after about 3,000 samples. This corresponds to about 300 seconds in real time. The entire sequence has 100,000 samples, corresponding to 10,000 seconds in real time. A comparison of the two outputs over the last 200 samples of the sequence is shown in Fig. 1.15.

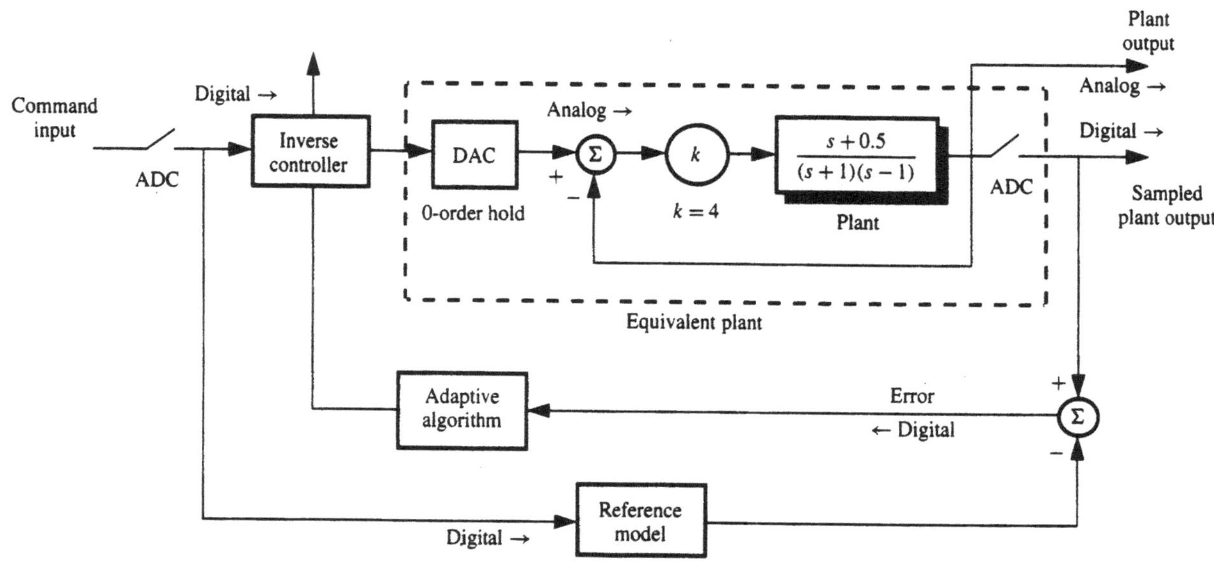

Figure 1.8 Model-reference adaptive inverse control of a stabilized minimum-phase plant.

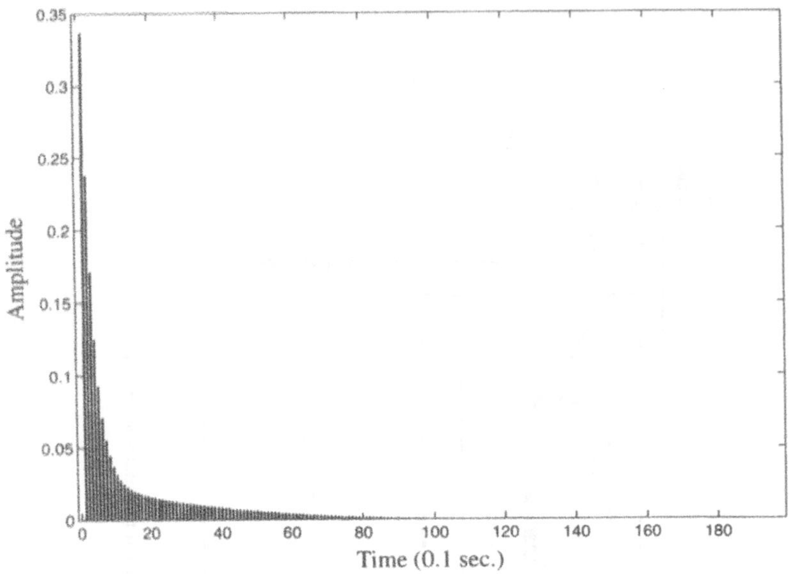

Figure 1.9 Impulse response of discretized equivalent minimum-phase plant.

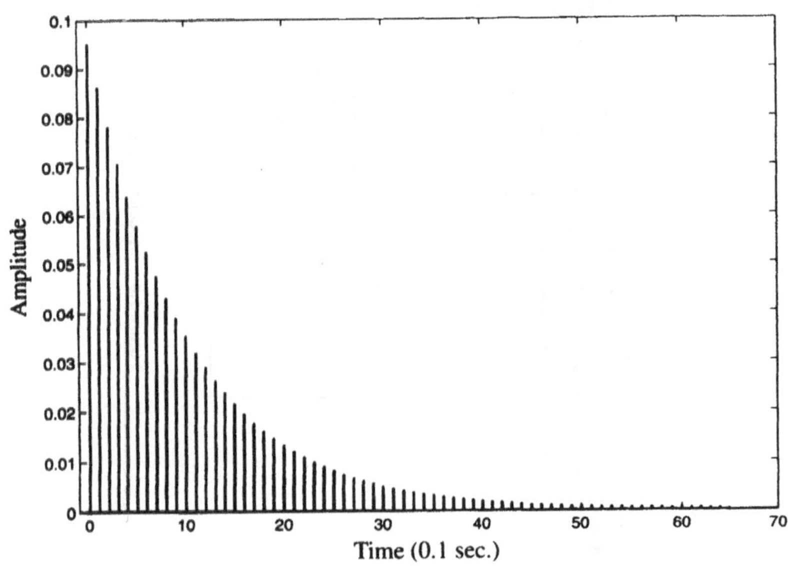

Figure 1.10 Impulse response of reference model.

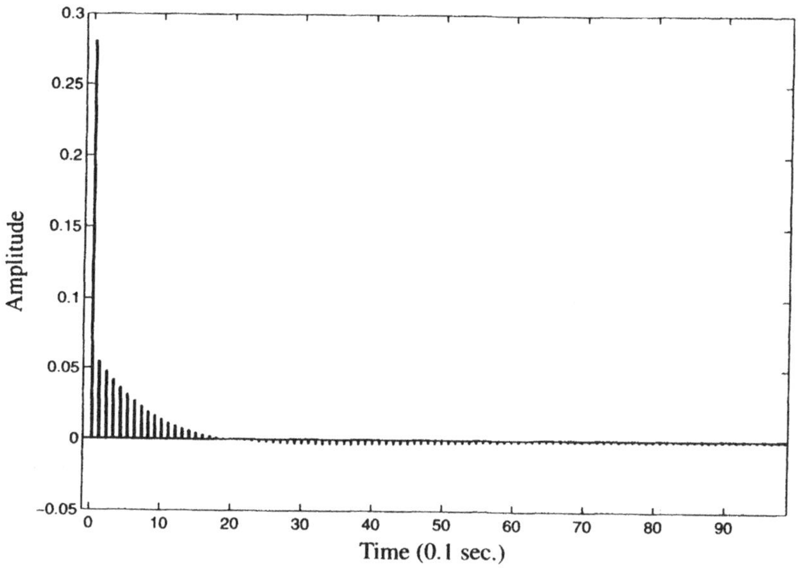

Figure 1.11 Impulse response of optimal 100-weight inverse controller.

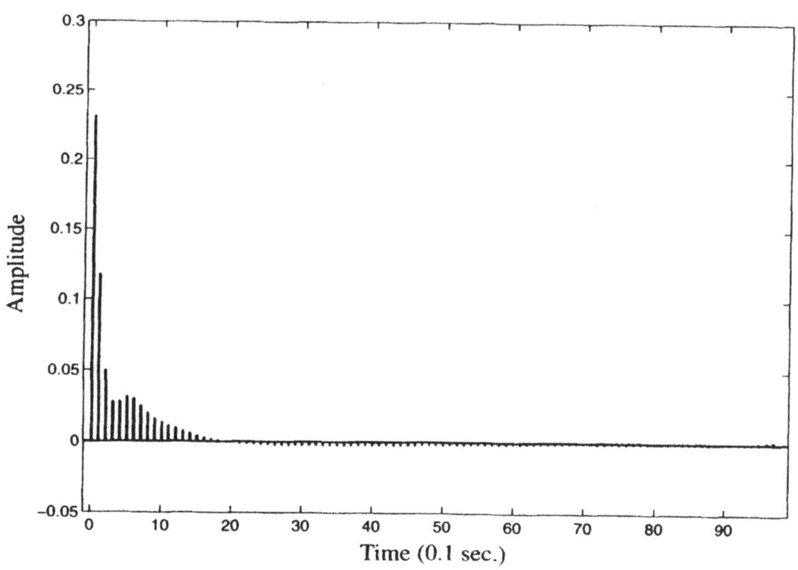

Figure 1.12 Impulse response of the adaptive controller after 100,000 learning iterations.

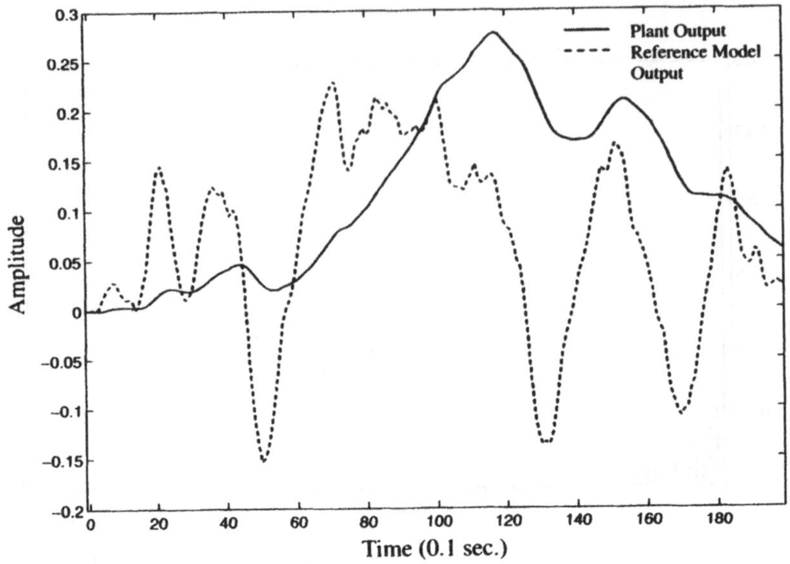

Figure 1.13 Plant output and reference model output during first 200 samples with minimum-phase plant.

Tracking can be seen to be essentially perfect. Model-reference inverse control does indeed work in this case.

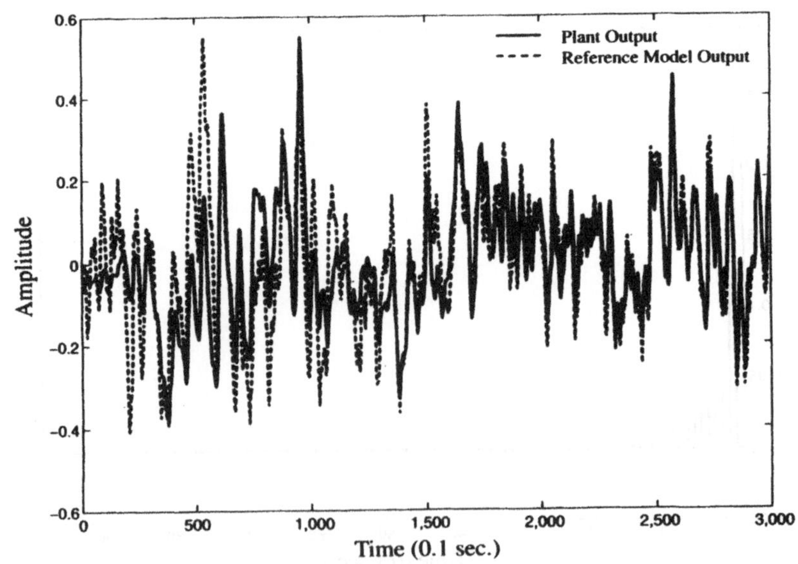

Figure 1.14 Plant output and reference model output over sequence of 3,000 samples, demonstrating controller learning.

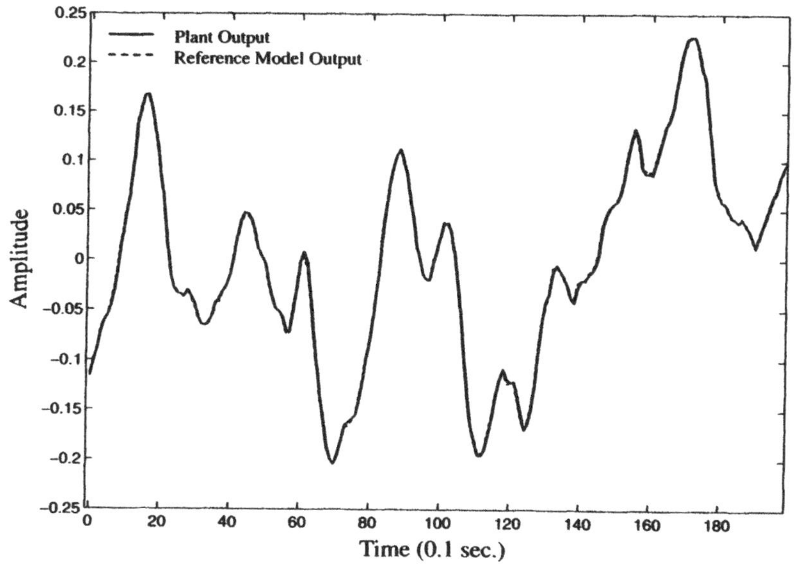

Figure 1.15 Plant output and reference model output during last 200 samples of sequence of 100,000 samples.

1.2.2 Dynamic Control of a Nonminimum-Phase Plant

For this experiment, a more challenging one, the plant to be controlled has the transfer function

$$\frac{(s-0.5)}{(s+1)(s-1)}. \tag{1.2}$$

Since this plant is unstable, the first step is once again to stabilize it with feedback. This plant cannot be stabilized with simple proportional feedback, like that of Fig. 1.7, but it can be stabilized by using feedback with a compensating network. The compensating network used in this experiment had the transfer function

$$\frac{(s+1)}{(s+7)(s-2)}. \tag{1.3}$$

A root-locus plot is shown in Fig. 1.16, and the stabilization feedback diagram is shown in Fig. 1.17. The compensating network has a zero at $s = -1$, and poles at $s = 2$ and -7. The stable range for the loop gain is $\infty > k > 20$. For this experiment, k was chosen to be $k = 24$. The result is a stabilized plant that remains nonminimum-phase.

1.2.3 Canceling Disturbance in the Minimum-Phase Plant

The stabilized plant was incorporated into an adaptive inverse control system like the one shown in Fig. 1.8. The sampling rate was chosen to be 10 samples per second. The reference model was a one-pole digital filter with a one-second time constant, but in this case the reference model included an eight-second delay. The impulse response of the reference

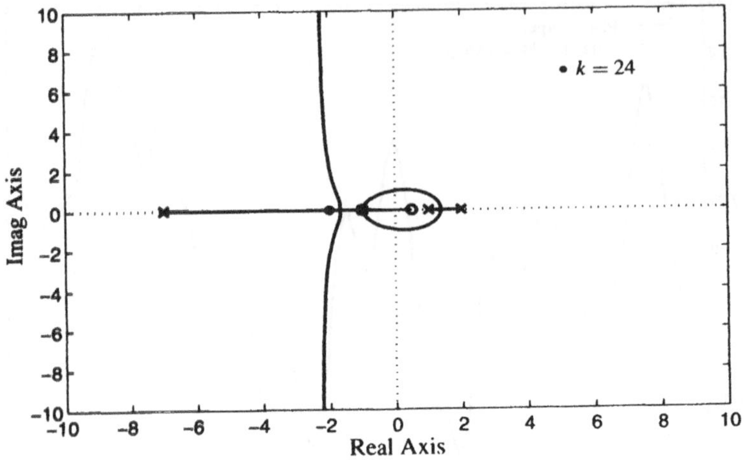

Figure 1.16 Root-locus plot for nonminimum-phase plant with compensating network.

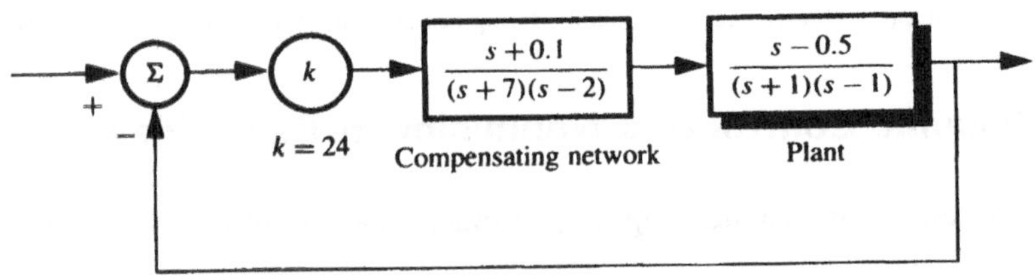

Figure 1.17 Nonminimum-phase plant with compensating network stabilization. $k = 24$.

model including the delay is shown in Fig. 1.18. The delay is necessary for inverse control of a nonminimum-phase plant. This will be explained in Chapter 5.

The impulse response of the discretized stabilized plant, which includes the plant and its stabilizer (shown in Fig. 1.17), and the DAC and the sampler at the plant output, are shown in Fig. 1.19. This is the impulse response from the input to the DAC to the sampled output of the plant.

The impulse response of the converged inverse controller is shown in Fig. 1.20. This controller has 150 weights. It was obtained from a learning algorithm. The form and shape of this impulse response could not be inferred intuitively.

To observe and demonstrate workability, we compare the plant output with the reference model output when the entire system is driven by a command input. For this experiment, the command input was a first-order Markov process generated by applying white noise to a one-pole digital filter with a one-second time constant. Figure 1.21 shows plots of plant output and reference model output for the first 200 samples (the first 20 seconds) of a training sequence. Figure 1.22 shows 1,000 samples of this sequence. Learning can be observed with good tracking after about 500 samples, corresponding to 50 seconds of operation. The last 200 samples of the 20,000-sample sequence are shown in Fig. 1.23. The

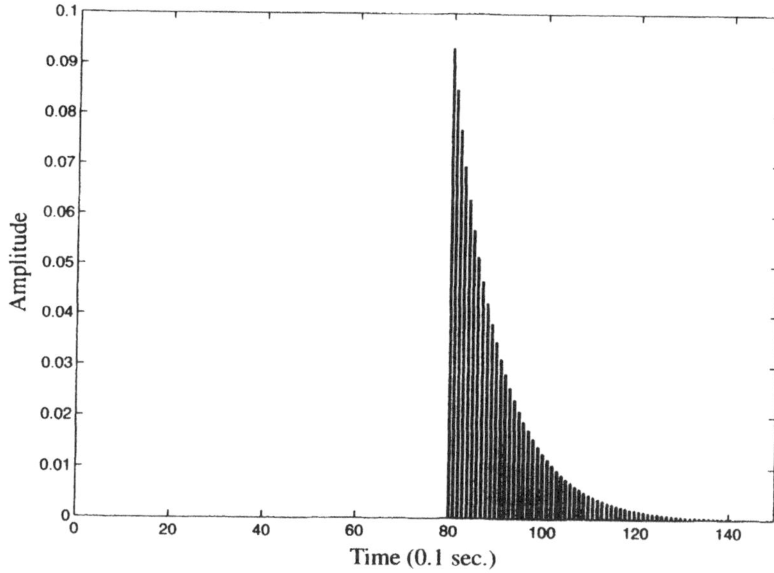

Figure 1.18 Impulse response of reference model, including a delay of 8 seconds.

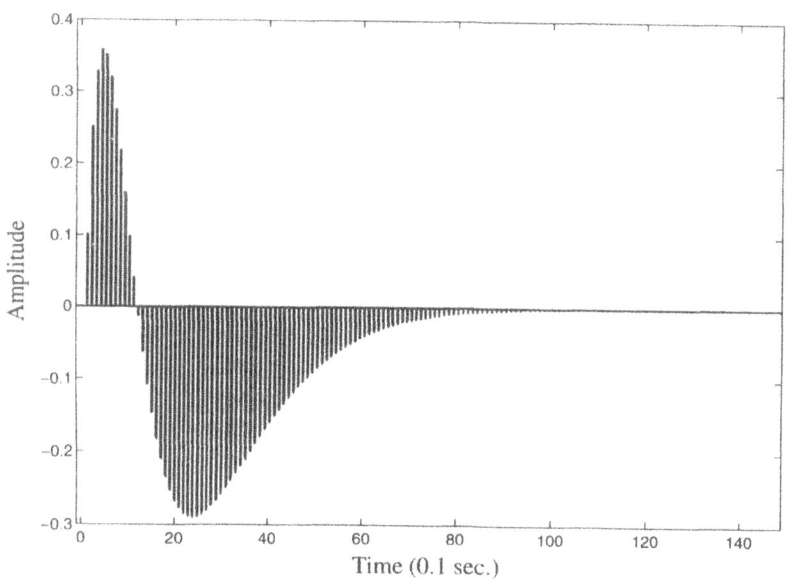

Figure 1.19 Impulse response of discretized stabilized nonminimum-phase plant with $k = 24$.

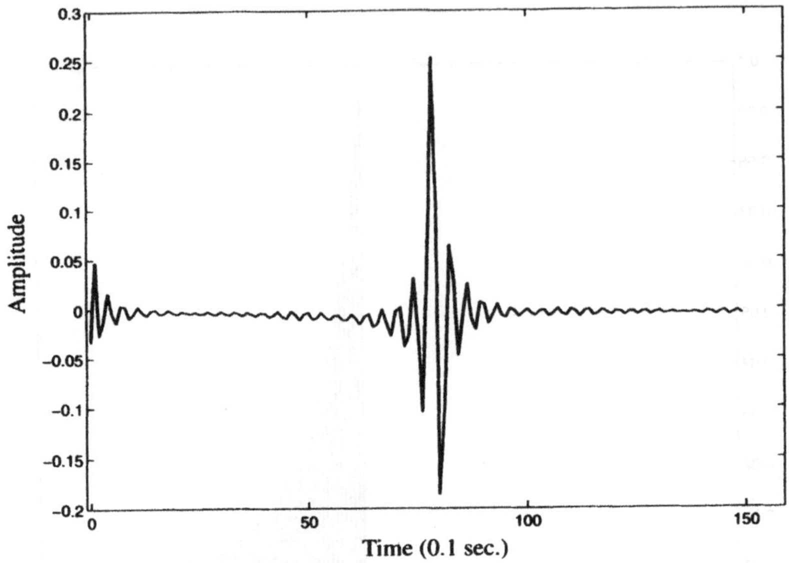

Figure 1.20 Converged impulse response of adaptive inverse controller for nonminimum-phase plant.

tracking is excellent, demonstrating precise control of a nonminimum-phase plant by means of adaptive inverse control, once learning has taken place.

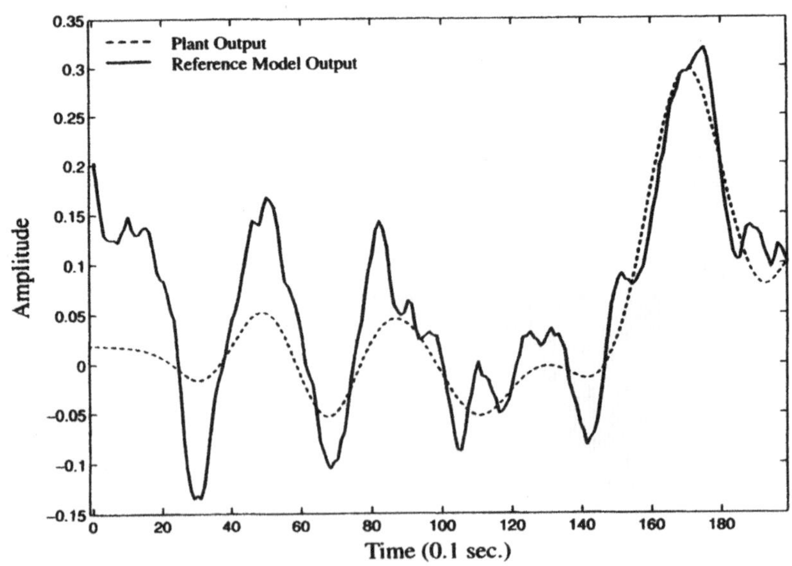

Figure 1.21 Plant output and reference model output during first 200 samples with nonminimum-phase plant.

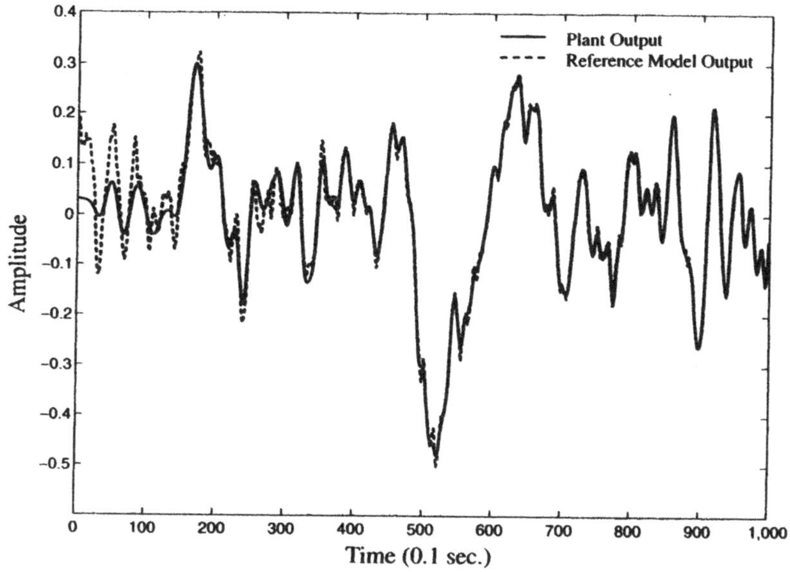

Figure 1.22 Plant output and reference model output during sequence of 1,000 samples, demonstrating learning with a nonminimum-phase plant.

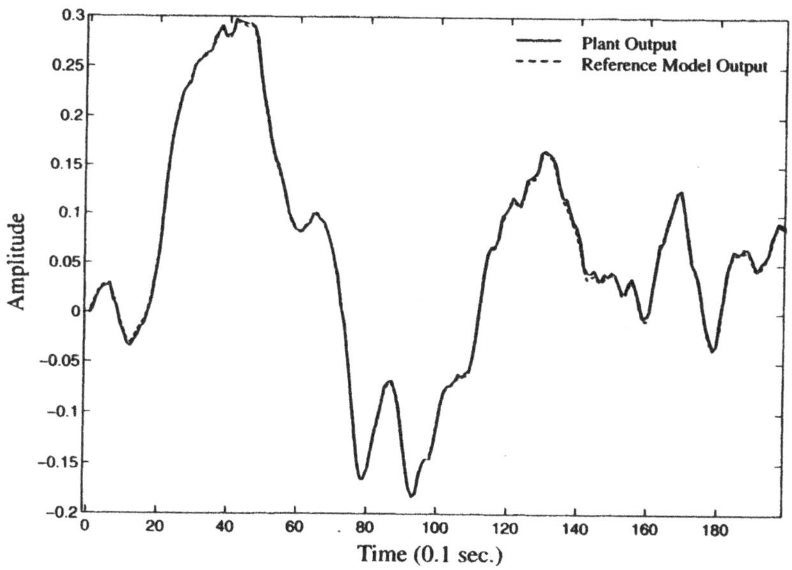

Figure 1.23 Plant output and reference model output during last 200 samples of training sequence with nonminimum-phase plant.

1.2.4 Canceling Disturbance in the Minimum-Phase Plant

Since both plant disturbance and sensor noise show up at the plant output, it is convenient for our purposes to lump them together and to simply call the combined effect plant disturbance, represented as an additive disturbance at the plant output. We shall do this throughout the book.

The minimum-phase plant described by (1.1) was stabilized in accord with the block diagram of Fig. 1.7. Noise is now injected into the plant output to simulate plant disturbance. An adaptive noise canceler fashioned after the one of Fig. 1.4 is used in this experiment to minimize the disturbance appearing at the plant output. The stabilized plant and its noise canceler are integrated into a model-reference adaptive inverse control system like the one diagrammed in Fig. 1.5. The resulting system is shown in Fig. 1.24.

Referring to Fig. 1.24, an adaptive model was made of the equivalent plant, and the converged impulse response is shown in Fig. 1.25. An adaptive inverse model of the equivalent plant was made for use in the disturbance canceler, and its converged impulse response is shown in Fig. 1.26.

The inverse controller was adapted, and its converged impulse response is shown in Fig. 1.27. It is useful to note that although this controller was adapted to control a noisy equivalent plant, and the learning process took place in the presence of plant disturbance, the converged dynamics of the controller were essentially unaffected by the plant disturbance. The controller impulse response of Fig. 1.12 was adapted for the same equivalent plant, but without plant disturbance and without an adaptive disturbance canceler. Comparison of Figs. 1.12 and 1.27 shows that both impulse responses are almost the same. The separation of the functions of *dynamic control* and *plant disturbance control* is a subject to be discussed in detail in Chapters 8 and 9. This separation is a characteristic of the adaptive inverse control approach.

The effectiveness of the adaptive disturbance canceler can be assessed from an inspection of Fig. 1.28. The plant output disturbance (which is computed as the difference between the plant output and what the plant output would be if there were no disturbance) is squared at each sample time and plotted in Fig. 1.28. There is no averaging in this plot, as the squared values are plotted over time. The disturbance canceling feedback loop was open until the five-thousandth sample in the time sequence. During this epoch, there was more than enough time for adaptive modeling of the equivalent plant and adaptive inverse modeling of this equivalent plant. Both models were required by the disturbance canceler. Then the loop was closed and the immediate noise reduction became apparent. It will be shown in Chapter 8 that this form of disturbance canceler is optimal in the least squares sense even when the plant is under dynamic control.

The response of the overall system is exemplified by the plots of Figs. 1.29–1.32. Figure 1.29 shows a comparison of the outputs of the reference model and the plant at the beginning of the learning sequence. The controller has not yet converged, and this causes overall error. Also, the disturbance canceler has not been turned on yet. Plant output disturbance contributes further to the overall error. Figure 1.30 shows the entire sequence of 10,000 error samples. The controller has learned its function after the first several hundred samples. One can see the results of convergence of the controller at the beginning of the plot. Error persists, however, because of the plant disturbance. The disturbance canceler is turned on at 5,000 samples, and then the error is reduced to its lowest possible level. In Fig. 1.31, the

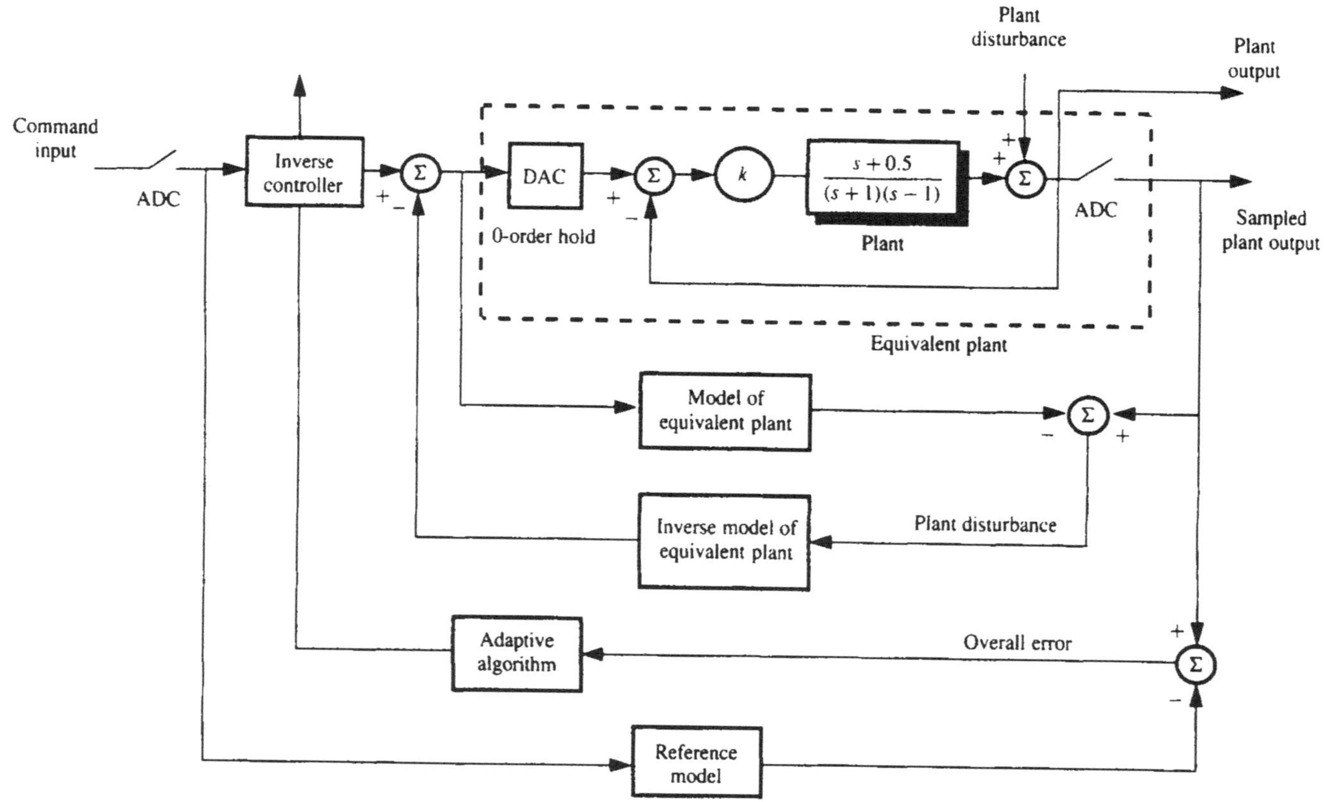

Figure 1.24 Model-reference inverse control system for minimum-phase plant with adaptive plant disturbance canceler.

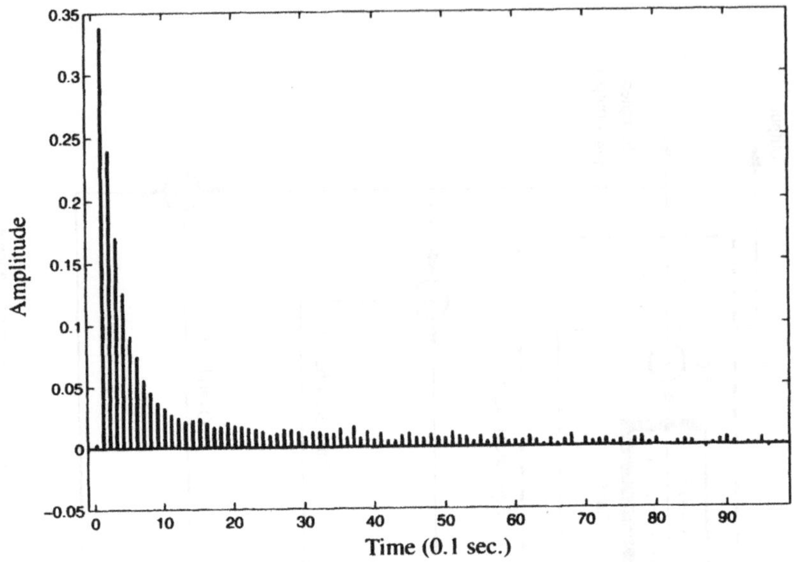

Figure 1.25 Impulse response of converged adaptive model of equivalent plant (minimum-phase).

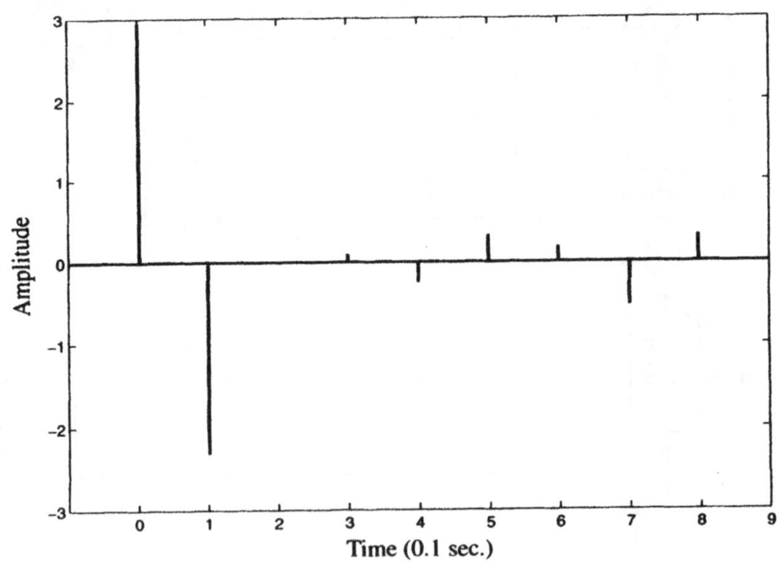

Figure 1.26 Converged impulse response of adaptive inverse model of equivalent plant (minimum-phase), for use in plant disturbance canceler.

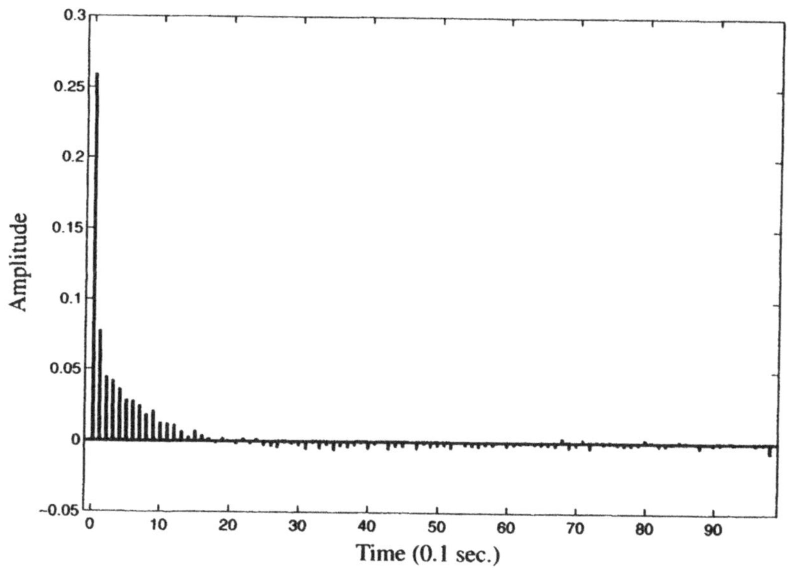

Figure 1.27 Convergence of model-reference inverse controller for equivalent plant (minimum-phase). Plant disturbance was present during learning process.

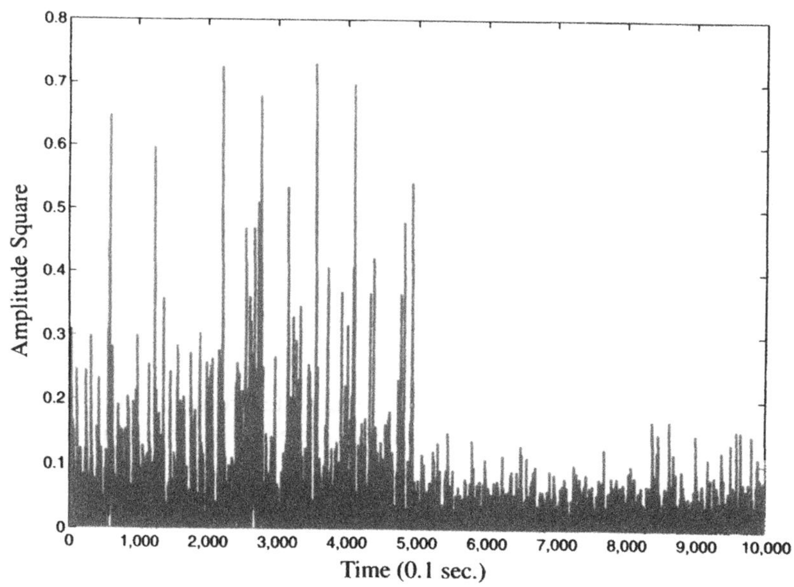

Figure 1.28 Square of plant output disturbance versus time. The disturbance canceler was turned on at the five-thousandth sample.

square of this error is graphed over time, without averaging. Figure 1.32 shows the model-reference output and the plant output over the last 200 samples in the sequence. Tracking between them is almost perfect, except for the small residual of uncanceled plant disturbance. The controller has long since converged and provides the proper dynamic control over the plant, and the disturbance canceler is turned on and doing an optimal job of canceling plant disturbance. The plant is under control!

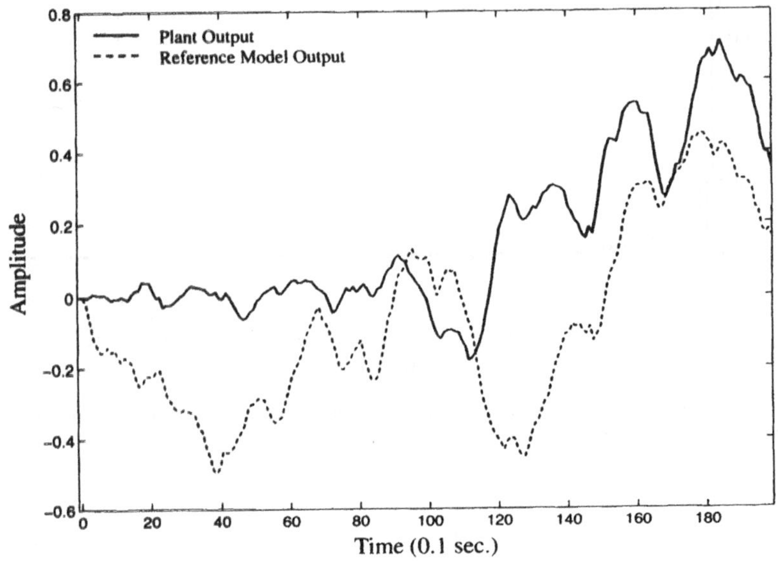

Figure 1.29 Comparison of reference model output with plant output at the beginning of the learning sequence. Plant disturbance is present.

.3 AN OUTLINE OR ROAD MAP FOR THIS BOOK

Section 1.1 has provided an overview of the idea of adaptive inverse control. This approach treats dynamic control of the plant as a separate problem from that of control of plant disturbance:

- Dynamic control is effected to cause the overall system response from command input signal to plant output signal to match that of a selected reference model.
- Plant disturbance is controlled to minimize the mean square of the plant output noise and disturbance.

Treating these as two separate problems is a very effective approach since the processes of dynamic control and noise and disturbance control can be optimized without one compromising the other.

The basic block diagrams of Section 1.2 are highly simplified. There are many details that need to be considered before one can successfully apply adaptive inverse control to

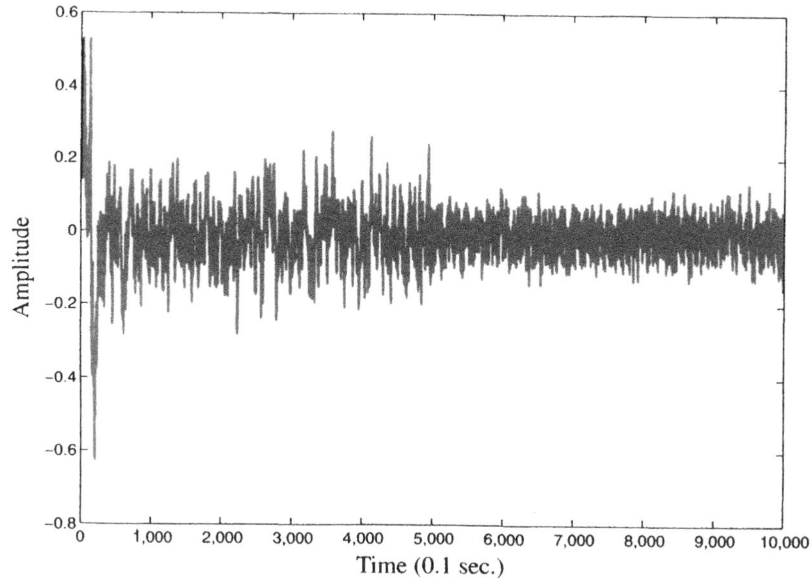

Figure 1.30 Entire learning sequence of 10,000 samples. Plot of error, equal to difference between reference model output and plant output. Disturbance canceler turned on at 5,000 samples.

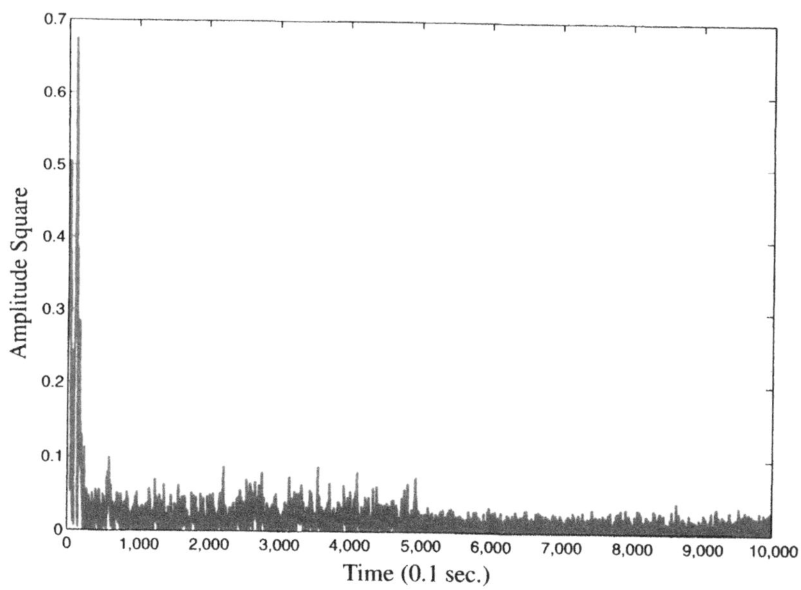

Figure 1.31 Square of overall system error plotted without averaging, over entire 10,000-sample learning sequence. Disturbance canceler turned on at 5,000 samples.

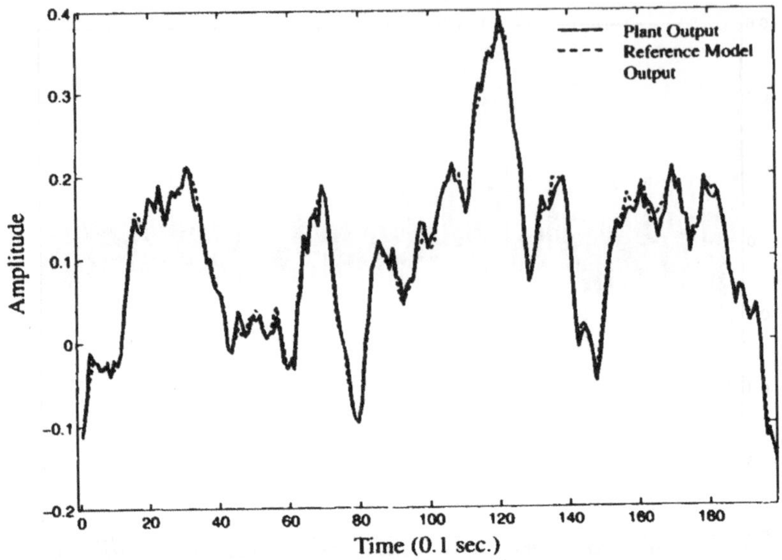

Figure 1.32 Comparison of reference model output with plant output over last 200 samples of training sequence. Controller has long ago converged, and plant disturbance is being canceled.

practical problems. Our goal is to explain some of the many different techniques that are available for the practical application of adaptive inverse control. The theory behind the various techniques is also derived and explained.

This book contains 12 chapters and 8 appendices. Since we expect that the subject will be new to most readers, we thought it appropriate to include in this introductory chapter a "road map" of the book. We are addressing both signal processing engineers and control engineers. The methodology taught here, the mathematical techniques and the adaptive signal processing techniques, will be more familiar to signal processing people. On the other hand, the problems that are being discussed are more familiar to control people.

We next describe the book and explain how it is organized.

Chapter 1: The Adaptive Inverse Control Concept

Chapter 1 introduces the idea of adaptive control and explains the need for it. It describes the more usual approaches to adaptive control and outlines the adaptive inverse control idea and its history. This chapter explains in an overall way how adaptive inverse control manages to control plant dynamics, and separately, plant noise or plant disturbances. It briefly reviews the subject matter of each of the chapters of this book. The objective is to aid the control engineer and the signal processing engineer in gaining an understanding and a perspective on adaptive inverse control and its applications.

Chapter 2: Wiener Filters

Wiener filters are best linear least squares filters which are very useful for prediction, estimation, interpolation, signal and noise filtering, and so on. To design them, prior knowledge of the appropriate statistical properties of the input signal(s) is required. The theory of Wiener filtering is also very useful in determining the asymptotic behavior of adaptive filters. For this reason, we present a brief chapter on the subject.

This chapter is a review of Wiener theory from the discrete point of view. In discrete form, Wiener theory is appropriate for analyzing digital adaptive filters. Causal and noncausal Wiener filters are studied. Both forms are used to analyze the behavior of adaptive filters as inverse controllers and plant disturbance cancelers. A very simple approach to causal Wiener filtering was devised by Bode and Shannon [68]. Shannon-Bode theory is explained in detail. This theory is extremely useful in the development of adaptive inverse control.

Chapter 3: Adaptive LMS Filters

This chapter reviews the theory of adaptive digital filtering, which is essential to the development of adaptive inverse control. This subject is discussed in many papers [1–60], among them several by B. Widrow and co-authors. Adaptive filters are discussed at length in several textbooks such as Widrow and Stearns [65], Haykin [69, 70] Cowan and Grant [71], and Treichler and colleagues [72].

The chapter begins with the idea of an adaptive filter, a tapped delay line with variable coefficients or tap weights driven by the LMS algorithm of Widrow and Hoff [73–77]. This is a gradient algorithm based on the method of steepest descent. The use of this algorithm to adjust the weights to minimize mean square error is described in the context of several practical applications. An important application is that of plant identification, necessary for adaptive control. Wiener theory is used to describe asymptotic adaptive behavior. The speed of convergence and the effects of gradient noise (caused by obtaining gradients with finite amounts of input-signal data) are analyzed. Fast adaptation causes noisy weights which, in excess, can result in poor performance.

The efficiency of adaptive algorithms is discussed. Two algorithms operating with the same level of noise in their weights can be compared. The one that converges faster is the more efficient. It is shown that when the input of an adaptive filter is stationary and the eigenvalues are equal or close in value, the LMS algorithm performs at a level of efficiency close to a theoretical maximum. When the eigenvalues are highly disparate, an algorithm similar to LMS but based on Newton's method would be needed to approach the theoretical maximum in efficiency. For nonstationary inputs, the conventional steepest-descent LMS algorithm performs with close to maximum efficiency [77].

This chapter concludes with a description of adaptive filtering as applied to adaptive noise canceling. The principles of adaptive noise canceling are derived. Experimental results are presented illustrating the effectiveness of noise canceling techniques to problems in adult electrocardiography (removing 60-Hz interference) and to fetal electrocardiography (removing interference from the maternal heart to reveal tiny signals from the fetal heart).

Chapter 4: Adaptive Modeling

The idea of LMS adaptive filtering for plant identification was discussed in Chapter 3. This chapter analyzes the sources of error in adaptive plant modeling, such as the effects of (a) weight noise, (b) inadequate input signal or dither signal, (c) plant noise, and so forth. Expressions are obtained for the covariance of the error or the difference between the plant's impulse response and the adaptive model's impulse response. These expressions are used in subsequent chapters to analyze the performance of adaptive inverse control systems.

Chapter 5: Inverse Plant Modeling

Direct adaptive modeling or plant identification was discussed in detail in Chapter 4. Chapter 5 discusses inverse adaptive modeling. The direct model has a transfer function similar to that of the plant being modeled. The inverse model has a transfer function like the reciprocal of the plant transfer function.

Forming a stable inverse transfer function is easy when the plant is minimum-phase. But doing this for a nonminimum-phase plant requires a two-sided Wiener impulse response which, to be causal, needs to be delayed and truncated. A theoretically optimal solution is obtained with the help of Shannon-Bode theory. The theory shows how approximate inverses can be formed for nonminimum-phase plants. Although such inverses are delayed, they can be very effective.

Model-reference inverses can be formed in a similar way. The result is that the cascade of the plant and its inverse develops an impulse response equal to that of a selected *reference model*. The inverse filter becomes the controller for the plant. The entire control system then has a dynamic response like that of the reference model. This is model-reference adaptive inverse control.

If the plant being controlled is subject to disturbance, the plant inverse may be obtained by adapting the inverse filter against a direct model of the plant instead of against the plant itself. Gradient noise in the weights of both the direct model and the inverse model affect the accuracy of the inverse model. Expressions are obtained for the variance of the error at the plant output due to noise in the inverse filter's weights. This knowledge is very important to the design of the adaptive controller.

Chapter 6: Adaptive Inverse Control

Schemes for adaptive inverse control were first presented in Chapter 1, but as shown there, they lack sufficient detail to be used in practice.

Practical model-reference control systems are shown in Chapter 6 for both disturbance-free plants and for plants with enough disturbance to interfere with the inverse modeling process. To alleviate the effects of plant disturbance, direct modeling is done first, and the inverse modeling process is done offline using a disturbance-free plant model to form the inverse. The idea of a dither signal, used in the direct modeling process, is introduced here.

If the dither is strong, good direct and inverse modeling is the result. But the dither adds to the plant output disturbance. A theory is developed to optimize the power of the dither to minimize the summed effects of the control error and plant disturbance at the plant output.

Adaptive control of blood pressure in experimental animals is reported. Successful experiments were performed regulating a dog's blood pressure in real time. The animal was in an induced state of shock. Under computer control, life-sustaining administration of therapeutic drugs took place. Dynamic control of blood pressure was accomplished by adaptive inverse control.

Chapter 7: Other Configurations for Adaptive Inverse Control

Two other algorithms for finding the inverse controller when plant disturbance is present are described in Chapter 7. These algorithms and variations on them are called *filtered-X* and *filtered-ϵ* LMS algorithms. Their advantages and disadvantages relative to each other and to the algorithms of Chapter 6 are discussed in detail. Conditions for stability are derived. Learning rates are obtained, as are expressions for weight noise variance. All of this can be used to predict the performance of adaptive inverse control systems.

A practical problem, canceling noise that leaks through earphones in a high-noise environment, illustrates application of both the filtered-X algorithm and the filtered-ϵ algorithm. This signal processing problem is very much like a control problem.

Chapter 8: Plant Disturbance Canceling

In the previous chapters a variety of techniques were developed for achieving precise dynamic control, even for plants subject to disturbance. These methods control plant dynamics but do nothing to control, reduce, or eliminate plant disturbance, however. An optimal adaptive scheme for reducing or eliminating plant disturbance is described and analyzed in this chapter. Its learning rate is determined. Shannon-Bode theory [68] is used to prove optimality and to derive expressions for the power of the residual plant output disturbance after adaptive cancelation. It is proven for linear systems that no other method can reduce the mean square of the plant output disturbance to a lower level. Expressions are derived for overall system output mean square error for simultaneous application of adaptive inverse control and adaptive plant disturbance canceling.

A simulation of an adaptive disturbance canceling system is reported. Application of this system to real-time aircraft ride control is illustrated. Fast control of the airplane's ailerons could reduce the vertical component of disturbance due to random updrafts and downdrafts.

Chapter 9: System Integration

An entire control system consisting of an inverse controller, a plant to be controlled, and an adaptive disturbance canceler for the plant is described and analyzed in Chapter 9. A surprising result develops from the analysis. If the plant model $\widehat{P}(z)$, which is used to obtain both the controller $\widehat{C}(z)$ and the noise-canceling feedback filter $Q(z)$, does not perfectly represent the plant $P(z)$, the errors in $\widehat{P}(z)$ and the resulting errors in $\widehat{C}(z)$ and $Q(z)$ combine in such a way that the overall system transfer function from command input to plant output is unaffected by the errors in $\widehat{P}(z)$. In other words, small errors in $\widehat{P}(z)$ cause small errors in the design of the controller $\widehat{C}(z)$. In addition, small errors in $\widehat{P}(z)$ cause small changes in the transfer function of the plant $P(z)$ in feedback connection with its disturbance canceler. The effects of these two transfer function errors cancel, leaving the overall system transfer function unaffected by small errors in the plant model $\widehat{P}(z)$.

The effect of small errors in $\widehat{P}(z)$ upon the disturbance canceler's ability to cancel plant disturbance is small. This result was obtained in Chapter 8. The effectiveness of the adaptive disturbance canceler is therefore not significantly impaired by small errors in $\widehat{P}(z)$.

Once the plant model $\widehat{P}(z)$ has been obtained, the rest of the system falls into place. Offline processes can be used to compute $Q(z)$ for the plant disturbance canceler, and to compute the controller $\widehat{C}(z)$. $\widehat{P}(z)$ is also used in the plant disturbance canceler.

The determination of $\widehat{P}(z)$ is not critical, however. Small errors in $\widehat{P}(z)$ only slightly degrade the performance of the disturbance canceler and have no effect on the precision of the dynamic control of the plant. This is an important result. Although the plant is being controlled by an open-loop controller, feedback in the adaptive algorithms and interactions among them result in precise dynamic control and close to optimal disturbance canceling.

Chapter 10: Multiple-Input Multiple-Output (MIMO) Adaptive Inverse Control Systems

MIMO systems are used when a physical plant to be controlled has multiple actuators, all of which have interacting effects on the process, and multiple sensors. Modeling and controlling MIMO systems are much more complicated than doing the same with SISO (single-input single-output systems).

This chapter begins with a review of digital signal processing for MIMO systems. The review section introduces notation used throughout the chapter. MIMO systems may be represented with block diagrams and flow graphs. Each signal line carries a bundle of signals, that is, a signal vector. Transfer functions are matrices. The rules of matrix algebra apply, and transfer functions are not commutable. This has a considerable effect on the design of adaptive algorithms to do plant modeling, inverse modeling, and plant disturbance canceling, all of which are described for MIMO systems.

Two methods for devising MIMO inverse controllers are presented in Chapter 10. One is an algebraic technique, and the other is based on the filtered-ϵ algorithm. Methods for adaptive plant disturbance canceling are also described.

Integrated MIMO systems incorporating inverse dynamic control and plant disturbance canceling in a single control system are shown. There are many ways to do this, several of which are illustrated in Chapter 10.

An application is described concerning the problem of reducing noise in a volume of space by using several loudspeakers and several sensing microphones in a MIMO controller configuration. Control of noise in an airplane cabin is illustrated here as an example. These are signal processing problems which are mathematically identical to control problems.

Chapter 11: Nonlinear Adaptive Inverse Control

The purpose of this chapter is to show how to do adaptive "inverse control" with nonlinear plants of both the SISO and MIMO types. Although nonlinear dynamic plants generally do not have inverses, techniques like inverse control can still be used.

Many of the rules of MIMO systems apply to nonlinear systems, such as noncommutability of filtering operations. An additional rule for nonlinear systems is that plant behavior should be modeled only with input signal character and power level corresponding to those of the actual plant input. Scaling and linearity do not work.

Inverse control of nonlinear plants and plant noise and disturbance canceling require the use of nonlinear adaptive filters. This chapter shows how to make nonlinear filters as tapped delay lines with Volterra networks [93], [94] connecting the tap signals to the variable weights, or with neural networks connected to the tap signals. Both of these methods process input signals nonlinearly to make output signals. A Volterra filter may be adapted by the LMS algorithm. A neural network filter would generally be adapted by the *backpropagation* algorithm of Werbos [78], and Rumelhart and colleagues [79], [80]. Backpropagation is the most widely used training algorithm for neural networks worldwide. It is an outgrowth of and a substantial generalization of the LMS algorithm.

How to do nonlinear plant modeling, with and without dither, is described in this chapter. When using dither, the superposed natural plant command signals, which could be nonstationary and which sometimes could be larger or sometimes smaller than the dither, mix nonlinearly in the plant. Simple direct modeling is not so simple. Means of dealing with this interaction are explained.

Once the plant model has been obtained, an inverse controller can be devised. Doing this with the filtered-ϵ LMS algorithm is illustrated. Care is exercised to ensure that all adaptive filters during training have input signals that have the right amplitude levels and the right signal characteristics.

Nonlinear plant noise canceling is demonstrated next. Online and offline processes for obtaining $\widehat{P}$, $\widehat{P}^{-1}$, Q, and $\widehat{C}$ are shown. The chapter ends with a block diagram for an integrated nonlinear control system which could be SISO or MIMO. The system incorporates both plant disturbance canceling and nonlinear inverse dynamic control.

Chapter 12: Pleasant Surprises

This chapter summarizes the principal theoretical results of adaptive inverse control. So many of these results were unexpected and worked out so nicely that we called the chapter "Pleasant Surprises."

Appendix A: Stability and Misadjustment of the LMS Adaptive Filter

This appendix summarizes learning theory for adaptive transversal filters based on the LMS adaptive algorithm. Key issues are learning rate, stability, and effects of noise in the weights (misadjustment). Most of this work has been published elsewhere by Widrow and colleagues [73–77]. More precise stability conditions and formulas for misadjustment has been reported by Horowitz and Senne [81]. Simplified derivations are given here that provide substantially the same results as those of Horowitz and Senne.

The work presented in this appendix is used throughout the book and is brought together here for the convenience of the reader.

Appendix B: Comparative Analyses of Dither Modeling Schemes A, B, and C

Basically, there are four ways of doing plant modeling which are illustrated in this book. The first method uses the natural plant input signals encountered during normal system operation to do the modeling. The second method, scheme A, adds a dither signal to the natural plant input to augment this signal and cause the combination to be "persistently exciting" [82]. The objective is to ensure that all frequency components are present in the modeling signal. The third method, scheme B, addresses an issue that could arise from nonstationarity of the natural plant input signal or if this signal has a highly nonuniform spectral character. Modeling might in some cases be done better with the dither alone, without the natural signal. With scheme B, the input signal to the adaptive modeling filter is pure dither. Carrying the idea one step further, the fourth method, scheme C, removes the natural plant output component (due to the natural plant input) from the "desired response" or training signal of the adaptive modeling filter. Scheme C uses only dither for the input and for the desired response of the adaptive plant model.

All four of these methods are used throughout the book. They have advantages and disadvantages relative to each other. This appendix develops ranges of stable operating conditions, learning rates, and misadjustment for all of the methods.

Appendix C: A Comparison of the Self-Tuning Regulator of Åström and Wittenmark with the Techniques of Adaptive Inverse Control

Some of the finest work in adaptive control has been done by Åström and Wittenmark. Their self-tuning regulator [82–88] is known worldwide for its simplicity, elegance, and utility. It can be used to control plant disturbance and plant dynamics, but it does not control them independently like adaptive inverse control. The purpose of Appendix C is to compare these two approaches to adaptive control in order to appraise their relative strengths and weaknesses.

The self-tuning regulator can be represented as a system consisting of a plant, a feedback controller, and a feedforward controller. The feedback controller can be designed to minimize plant output noise and disturbance power. But the feedback changes the dynamics. Compensation can be made for this, and at the same time, the overall response can be made to match a model response $M(z)$ by properly choosing the feedforward input controller.

It is shown in this appendix that an optimal feedback controller for a self-tuning regulator can always be chosen, and the resulting feedback portion of the system will always be stable. The requirements on the feedforward controller may turn out to make it unstable, however. (This is unstable in the sense of having poles outside the unit circle in the z-plane.) When this happens, one can design a stable feedforward controller that can approximately realize the required transfer function with a delayed response. The self-tuning regulator is indeed a general methodology.

Adaptive inverse control can always realize the optimal plant noise canceler without experiencing instability due to feedback. Furthermore, the ideal inverse controller will be stable, with its poles inside the unit circle if the plant is minimum-phase. This ideal controller can be realized approximately by a transversal filter without delay. A delayed response is only required when the plant is nonminimum-phase. Adaptive inverse control is a general methodology.

Adaptive inverse control and the self-tuning regulator are two completely different approaches to the problem of adaptive control of plant dynamics and plant disturbance. Their methods of adaptation differ, their adaptive filters are of different structure, their errors in dynamic response are different, but their abilities to cancel plant disturbance should be equal. Which approach is better, more robust, faster to converge, easier to implement, easier to design and debug, and easier to understand? The answers are probably problem dependent.

Appendix D: Adaptive Inverse Control for Unstable Linear SISO Plants

If the plant to be controlled is unstable, it is impossible to control it with adaptive inverse control. The reason for this is that a feedforward controller, even though it is adaptive, will leave the plant unstable.

The first step in the utilization of adaptive inverse control for an unstable plant is stabilization with feedback. This appendix shows that the choice of stabilization feedback is not critical as long as the unstable plant is stabilized. The plant together with its feedback stabilizer should be treated as a unit, as a stable *equivalent plant*. The equivalent plant can be outfitted with an adaptive disturbance canceler and with an inverse controller, just like an ordinary stable plant.

If two different feedback filters can stabilize the plant, it is shown in this appendix that the minimal plant output disturbance (after adaptive disturbance cancelation) is the same for both stabilizing filters. Since no one stabilizer does better than any other, the choice of stabilizer is not critical. All one needs is some form of feedback filter that will stabilize the plant, and the adaptive disturbance canceler will deliver optimal performance.

The same is true for the design of the inverse controller. The choice of feedback stabilization is immaterial. One can always design an inverse controller for the stabilized plant. If the inverse controller needs delay for its proper realization, the required delay will not depend on the choice of the stabilization filter. Also, the length of the inverse controller will not depend on this choice.

The conclusion of Appendix D is that plant instability is no impediment to the use of adaptive inverse control. One needs to know only enough about the plant to design a feedback stabilizer for it. The design could be accomplished by experimentation. The design is not at all critical as long as the plant is stabilized. Once stabilized, the plant and its stabilizer should be treated as a stable equivalent plant, and adaptive inverse control should be applied in the usual way. Although a wide variety of stabilizer designs would generally be acceptable, forming any one design would not always be a straightforward process and indeed may require some effort, particularly if the unstable plant is nonlinear and/or MIMO.

Appendix E: Orthogonalizing Adaptive Algorithms: RLS, DFT/LMS, and DCT/LMS

This appendix was prepared by Dr. Francçise Beaufays, based on her Ph.D. dissertation research in the Department of Electrical Engineering at Stanford University. Extreme eigenvalue spread of the autocorrelation matrix of the input to an adaptive filter causes convergence problems for the LMS algorithm. To speed up convergence and maintain stability, she proposes a new adaptive algorithm, the DCT/LMS.

She begins with a brief discussion of the RLS (recursive least squares) algorithm. This algorithm is most often very effective, but it is complicated and has its own stability problems.

The next algorithm discussed is the DFT/LMS. This algorithm first Fourier transforms the signals at the taps of a filter's tapped delay line. The purpose is to achieve orthogonalization, which the DFT does well, except for the phenomenon of *leakage*. Once the tap signals are orthogonalized (approximately) and power normalized, they are weighted and summed to produce the filter output. The weights are adapted by conventional LMS. Although adaptation is done simply with steepest descent, the behavior is similar to adaptation with Newton's method.

Finally, the DCT/LMS algorithm is discussed. Structurally, this algorithm is very similar to DFT/LMS, with the DCT (digital cosine transform) substituted for the DFT. Ms. Beaufays shows that the DCT/LMS algorithm performs significantly better than the DFT/LMS algorithm when the adaptive filter input is first-order Markov, a type of signal that is common in signal processing and controls. This type of signal comes from applying white noise to a one-pole filter.

The DFT/LMS and DCT/LMS algorithms are easy to implement and are computationally robust. All of their parts, DFT or DCT, power normalization (like AGC in a radio or TV), and LMS algorithm are "bulletproof." These algorithms should enjoy greater acceptance and application in the future.

Appendix F: A MIMO Application: An Adaptive Noise Canceling System Used for Beam Control at the Stanford Linear Accelerator Center

This appendix was prepared by Dr. Tom Himel, a research physicist at the Stanford Linear Accelerator Center (SLAC). It is based on his experience with control of the beam of a two-mile long linear accelerator on the Stanford campus whose output drives positrons and electrons in opposite directions along the arcs of a circular collider. To achieve collisions suitable for physics research, the electron and positron beams must be controlled in position to within several microns of each other after each travels distances of about three miles.

Control of the beam is critical. The U.S. Department of Energy, the sponsor, spends millions of dollars a year operating the accelerator and has spent many hundreds of millions building it over the past 25 years. Physicists from all over the world depend on this machine for their research.

Until recently, the beam was controlled by many servo loops positioned along it, with manual control of set points. Intercoupling between stages has always been a problem for overall control. The beam output of one stage is the beam input for the next. Adaptive techniques are now used to obtain precise local beam positioning and at the same time, to obtain decoupling between stages.

The corrective methodology is based on adaptive noise canceling. The adaptive disturbance canceler at each stage is an eight-input, eight-output MIMO system.

The adaptive system (implemented in software) has been installed and working for more than a year, 24 hours a day, 7 days a week. The result is much better accelerator operation. It is automatic, without the need of human operator intervention, and the frequency of collision events has increased. This is an operational system, no longer a laboratory demonstration.

Appendix G: Thirty Years of Adaptive Neural Networks: Perceptron, Madaline, and Backpropagation

This appendix is a reprint of a paper by Widrow and Lehr that was published in the September 1990 issue of the *Proceedings of the IEEE*. For the reader who is familiar with the LMS algorithm and adaptive filters, this paper introduces the subject of neural networks. It shows how the LMS algorithm can be extended to form the *backpropagation* algorithm, the most widely used learning procedure for neural networks.

Appendix H: Neural Control Systems

This appendix shows how to construct nonlinear adaptive inverse control systems by making use of neural networks and the backpropagation algorithm. It also describes applications of neural control systems.

One such application is the truck backer-upper. A simulated trailer truck is steered by a neural controller while backing to a loading platform. After many backing runs from many different initial conditions, the controller learns to steer the truck by making many mistakes and learning what not to do. Once learning is complete, the truck can be placed in almost any initial state, states not previously encountered, and the controller steers the truck while driving backward to the loading dock. The truck backer-upper is an example of how a nonlinear controller can learn and, in a real sense, design itself.

Principles of inverse neural control similar to those incorporated in the truck backing system have been used to control the functioning of electric furnaces and for chemical process control. These applications are described in some detail and are not laboratory exercises. They are industrial applications of high commercial significance and are examples of what is currently being called "intelligent control systems."

This completes the road map of the book. We now invite you to travel down the road with us. We wish you a productive journey.

Bibliography for Chapter 1

[1] B. WIDROW, K.M. DUVALL, R.P. GOOCH, and W.C. NEWMAN, "Signal cancellation phenomena in adaptive antennas: Causes and cures," *IEEE Trans. Antennas Propag.*, Vol. AP-30 (May 1982), pp. 469–478.

[2] C.W. JIM, "A comparison of two LMS constrained optimal array structures," *Proc. IEEE*, Vol. 65 (December 1977), pp. 1730–1731.

[3] L.J. GRIFFITHS and C.W. JIM, "An alternative approach to linearly constrained adaptive beamforming," *IEEE Trans. Antenna Propag.*, Vol. AP-30 (January 1982), pp. 27–34.

[4] R.P. GOOCH, "Adaptive pole-zero array processing," in *Proc. 16th Asilomar Conf. Circuits Syst. Comput.*, Santa Clara, CA, November 1982.

[5] P. HOWELLS, "Intermediate frequency side-lobe canceller," U.S. Patent 3202 990, August 24, 1965.

[6] Special issue on adaptive antennas, *IEEE Trans. Antennas Propag.*, Vol. AP-24, No. 5 (September 1976).

[7] W.F. GABRIEL, "Adaptive arrays—An introduction," *Proc. IEEE*, Vol. 64 (February 1976), pp. 239–272.

[8] B. WIDROW, "Adaptive antenna systems," *Proc. IEEE*, Vol. 55, No. 12 (December 1967), pp. 2143–2159.

[9] L.J. GRIFFITHS, "A simple adaptive algorithm for real-time processing in antenna arrays," *Proc. IEEE*, Vol. 57, No. 10 (October 1969), pp. 1696–1704.

[10] O.L. FROST, III, "An algorithm for linearly constrained adaptive array processing," *Proc. IEEE*, Vol. 60, No. 8 (August 1972), pp. 926–935.

[11] C.L. ZAHM, "Applications of adaptive arrays to suppress strong jammers in the presence of weak signals," *IEEE Trans. Aerosp. Electron. Systems*, Vol. AES-9 (1973), pp. 260–271.

[12] R.T. COMPTON, JR., R.J. HUFF, W.G. SWARNER, and A.A. KSIENSKI, "Adaptive arrays for communciation systems: An overview of research at the Ohio State University," *IEEE Trans. Antennas Propag.*, Vol. AP-24, No. 5 (September 1976), pp. 599–606.

[13] R.T. COMPTON, JR., "An experimental four-element adaptive array," *IEEE Trans. Antennas Propag.*, Vol. AP-24, No. 5 (September 1976), pp. 697–706.

[14] L.E. BRENNAN, AND I.S. REED, "Convergence rate in adaptive arrays," Technology Service Corp., Rep. TSC-PD-177-4, January 13, 1978.

[15] W.D. WHITE, "Adaptive cascade networks for deep nulling," *IEEE Trans. Antennas Propag.*, Vol. AP-26, No. 3 (May 1978), pp. 396–402.

[16] S.P. APPLEBAUM, "Adaptive arrays," *IEEE Trans. Antennas Propag.*, Vol. AP-24 (September 1976), pp. 585–598.

[17] E. WALACH, "On superresolution effects in maximum-likelihood antenna arrays," *IEEE Trans. Antennas Propag.*, Vol. AP-32, No. 3 (March 1984), pp. 259–263.

[18] L.J. GRIFFITHS, "A comparison of multidimensional Wiener and maximum-likelihood filters for antenna arrays," *Proc. IEEE (Letters)*, Vol. 55 (November 1967), pp. 2045–2047.

[19] N.K. JABLON, "Effect of element errors on half-power beamwidth of the Capon adaptive beamformer," *IEEE Trans. on Circuits and Systems*, Vol. CAS-34, No. 7 (July 1987), pp. 743–751.

[20] N.K. JABLON, "Steady state analysis of the generalized sidelobe canceller by adaptive noise cancelling techniques," *IEEE Trans. Antenna Propag.*, Vol. AP-34 (March 1986), pp. 330–337.

[21] R.W. LUCKY, "Automatic equalization for digital communication," *Bell Syst. Tech.*, Vol. 44 (April 1965), pp. 547-588.

[22] R.W. LUCKY, "Techniques for adaptive equalization of digital communication systems," *Bell Syst. Tech.*, Vol. 45 (February 1966), pp. 255–286.

[23] A. GERSHO, "Adaptive equalization of highly dispersive channels for data transmission," *Bell Syst. Tech.*, Vol. 48 (January 1969), pp. 55–70.

[24] M.M. SONDHI, "An adaptive echo canceller," *Bell Syst. Tech.*, Vol. 46 (March 1967), pp. 497–511.

[25] J.G. PROAKIS, *Digital communications* (New York: McGraw-Hill, 1983).

[26] R.W. LUCKY, J. SALZ, and E.J. WELDON, JR., *Principles of data communication* (New York: McGraw-Hill, 1968).

[27] R.D. GITLIN, J.F. HAYES, and S.B. WEINSTEIN, *Data communications principles* (New York: Plenum Press, 1992).

[28] S. QURESHI, "Adaptive equalization—a comprehensive review," *IEEE Communications Magazine* (March 1982), pp. 9–16.

[29] J.G. PROAKIS, "Adaptive digital filters for equalization of telephone channels," *IEEE Trans. Audio Electroacoust.*, Vol. AU-18, No. 2 (June 1970), pp. 484–497.

[30] J.G. PROAKIS and J.H. MILLER, "An adaptive receiver for digital signaling through channels with intersymbol interference," *IEEE Trans. Info. Theory*, Vol. IT-15, No. 4 (July 1969), pp. 484–497.

[31] S.J. CAMPANELLA, H.G. SUYDERHOUD, and M. ONUFRY, "Analysis of an adaptive impulse response echo canceller," *Comsat Tech. Rev.*, Vol. 2, No. 1 (Spring 1972), pp. 1–37.

[32] D.L. DUTTWEILER, "A twelve-channel digital echo canceller," *IEEE Trans. Communications*, Vol. COM-26 (May 1978), pp. 647–653.

[33] D.L. DUTTWEILER, and Y.S. CHEN, "Performance and features of a single chip FLSI echo canceller," in *Proc. NTC 79*, Washington, DC, November 17–19, 1979.

[34] M.M. SONDHI and D.A. BERKLEY, "Silencing echoes on the telephone network," *Proc. IEEE*, Vol. 68 (August 1980), p. 948.

[35] F.K. BECKER and H.R. RUDIN, "Application of automatic transversal filters to the problem of echo suppression," *Bell Syst. Tech.*, Vol. 45 (December 1966), pp. 1847–1850.

[36] M.M. SONDHI, and A.J. PRESTI, "A self-adaptive echo canceller," *Bell Syst. Tech.*, Vol. 45 (December 1966), pp. 1851–1854.

[37] D.D. FALCONER, K.H. MUELLER, and S.B. WEINSTEIN, "Simultaneous two-way data transmission over a two-wire circuit," in *Proc. NTC*, Dallas, December 1976.

[38] T.R. ROSENBURGER and E.J. THOMAS, "Performance of an adaptive echo canceller operating in a noisy, linear time-invariant environment," *Bell Syst. Tech.*, Vol. 50 (March 1971), p. 785.

[39] N. DEMYTKO, and L.K. MACKECHNIE, "A high speed digital adaptive echo canceller," *Aust. Telecommun. Rev.*, Vol. 7 (1973), pp. 20–27.

[40] B.WIDROW, *et al.*, "Adaptive noise cancelling: principles and applications," *Proc. IEEE*, Vol. 63, No. 12 (December 1975), pp. 1692–1716.

[41] J. GLOVER, "Adaptive noise cancelling of sinusoidal interference," Ph.D. diss., Stanford Electronics Lab., Stanford University, Stanford, CA (December 1975).

[42] J. GLOVER, "Adaptive noise cancelling applied to sinusoidal interferences," *IEEE Trans. Acoust., Speech, Signal Processing*, Vol. ASSP-25, No. 6 (December 1977), pp. 484–491.

[43] M.R. SAMBUR, "Adaptive noise cancelling for speech signals," *IEEE Trans. Acoust., Speech, Signal Processing*, Vol. ASSP-26 (October 1978), p. 419.

[44] M.J. SHENSHA, "Non-Wiener solutions of the adaptive noise canceller with a noisy reference," *IEEE Trans. Acoust., Speech, Signal Processing*, Vol. ASSP-28 (August 1980), p. 468.

[45] L.J. GRIFFITHS, "An adaptive lattice structure for noise-cancelling applications," in *Proc. ICASSP-78*, p. 87.

[46] L.J. GRIFFITHS, and R.S. MEDAUGH, "Convergence properties of an adaptive noise cancelling lattice structure," *Proc. 1978 IEEE Conf. Decision and Control*, San Diego, CA (January 1979), pp. 1357–1361.

[47] L.J. GRIFFITHS, "Rapid measurement of instantaneous frequency," *IEEE Trans. Acoust., Speech, Signal Processing*, Vol. ASSP-23 (April 1975), pp. 209–222.

[48] J.P. BURG, "Maximum entropy spectral analysis," presented at the 37th Annual Meeting, Soc. Exploration Geophysicists, Oklahoma City, OK, 1967.

[49] J.R. ZEIDLER, *et al.*, "Adaptive enhancement of multiple sinusoids in uncorrelated noise," *IEEE Trans. Acoust., Speech, Signal Processing*, Vol. ASSP-26 (June 1978), p. 240.

[50] D.R. MORGAN, "Response of a delayed-constrained adaptive linear predictor filter to a sinusoid in white noise," in *Proc. 1981 ICASSP*, p. 271.

[51] J.A. EDWARDS, and M.M. FITELSON, "Notes on maximum-entropy processing," *IEEE Trans. Info. Theory*, Vol. IT-19 (March 1973), p. 232.

[52] R.T. LACOSS, "Data adaptive spectral analysis methods," *Geophysics* (August 1971), p. 661.

[53] A. VAN DEN BOS, "Alternative interpretation of maximum entropy spectral analysis," *IEEE Trans. on Info. Theory*, Vol. IT-17 (July 1971), p. 493.

[54] A. POPOULIS, "Maximum entropy and spectral estimation: A review," *IEEE Trans. Acoust., Speech, Signal Processing*, Vol. ASSP-29 (December 1981), p. 1176.

[55] W.F. GABRIEL, "Spectral analysis and adaptive array superresolution techniques," *Proc. IEEE*, Vol. 68 (June 1980).

[56] B. FRIEDLANDER, "Lattice methods for spectral estimation," *Proc. IEEE*, Vol. 70, No. 9 (September 1982), pp. 990–1017.

[57] B. WIDROW, and S.D. STEARNS, *Adaptive signal processing* (Englewood Cliffs, NJ: Prentice Hall, 1985), chap. 12.

[58] M. MORF, A. VIEIRA, and D.T. LEE, "Ladder forms for identification and speech processing," *Proc. 1977 IEEE Conf. Decision and Control*, New Orleans, LA (December 1977), pp. 1074–1078.

[59] J.I. MAKHOUL and L.K. COSELL, "Adaptive lattice analysis of speech," *IEEE Trans. Acoust., Speech, Signal Processing*, Vol. ASSP-29 (June 1981), p. 654.

[60] F. ITAKURA and S. SAITO, "Digital filtering techniques for speech analysis and synthesis," in *Proc. 7th Int. Conf. Acoust.*, Vol. 3, Paper 25C-1, 1971, p. 261.

[61] B. WIDROW, "Adaptive model control applied to real-time blood-pressure regulation," in *Pattern Recognition and Machine Learning; proceedings*, ed. K.S. Fu (New York: Plenum Press, 1971), pp. 310–324.

[62] B. WIDROW, J. MCCOOL, and B. MEDOFF, "Adaptive control by inverse modeling," in *Conf. Rec. of 12th Asilomar Conference on Circuits, Systems and Computers*, Santa Clara, CA, November 1978, pp. 90–94.

[63] B. WIDROW, D. SHUR, and S. SHAFFER, "On adaptive inverse control," in *Conf. Rec. of 15th Asilomar Conference on Circuits, Systems and Computers*, Santa Clara, CA, November 1981, pp. 185–189.

[64] S. SHAFFER, "Adaptive inverse-model control," Ph.D. diss., Stanford University, August 1982.

[65] B. WIDROW, and S.D. STEARNS, *Adaptive signal processing* (Englewood Cliffs, NJ: Prentice Hall, 1985).

[66] E. WALACH and B. WIDROW, "Adaptive signal processing for adaptive control," in *IFAC Workshop on Adaptive Systems in Control and Signal Processing*, San Francisco, CA, 1983.

[67] B. WIDROW, "Adaptive inverse control," in *Second IFAC Workshop on Adaptive Systems in Control and Signal Processing*, Lund, Sweden, July 1986.

[68] H.W. BODE, and C.E. SHANNON, "A simplified derivation of linear least square smoothing and prediction theory," *Proc. IRE*, Vol. 38 (April 1950), pp. 417–425.

[69] S. HAYKIN, *Introduction to adaptive filters* (New York: Macmillan, 1984).

[70] S. HAYKIN, *Adaptive filter theory* (Englewood Cliffs, NJ: Prentice Hall, 1986).

[71] C.F.N. COWAN, and P.M. GRANT, *Adaptive filters* (Englewood Cliffs, NJ: Prentice Hall, 1985).

[72] J.R. TREICHLER, C.R. JOHNSON, JR., and M.G. LARIMORE, *Theory and design of adaptive filters* (New York: John Wiley, 1987).

[73] B. WIDROW, and M.E. HOFF, "Adaptive switching circuits," in *IRE WESCON Conv. Rec.*, Pt. 4, 1960, pp. 96–104.

[74] B. WIDROW, and J.M. MCCOOL, "A comparison of adaptive algorithms based on the methods of steepest descent and random search," *IEEE Trans. Antennas Propag.*, Vol. AP-24, No. 5 (September 1967), pp. 615–637.

[75] B. WIDROW, J.M. MCCOOL, M.G. LARIMORE, and C.R. JOHNSON, Jr., "Stationary and nonstationary learning characteristics of the LMS adaptive filter," *Proc. IEEE*, Vol. 64, No. 8 (August 1976), pp. 1151–1162.

[76] B. WIDROW, "Adaptive filters," in *Aspects of Network and Systems Theory*, ed. R.E. Kalman and N. De Claris (New York: Holt, Rinehart and Winston, 1970), pp. 563–587.

[77] B. WIDROW, and E. WALACH, "On the statistical efficiency of the LMS algorithm with nonstationary inputs," *IEEE Trans. Info. Theory—Special Issue on Adaptive Filtering*, Vol. 30, No. 2, Part 1 (March 1984), pp. 211–221.

[78] P. WERBOS, *Beyond Regression: New Tools for Prediction and Analysis in the Behavioral Sciences*, Ph.D. diss., Harvard University, August 1974.

[79] D.E. RUMELHART, and J.L. MCCLELLAND, *Parallel Distributed Processing*, Vols. I and II (Cambridge, MA: M.I.T. Press, 1986).

[80] B. WIDROW and M.A. LEHR, "30 years of adaptive neural networks: perceptron, madaline, and backpropagation," *Proceedings of the IEEE*, Vol. 78, No. 9 (September 1990), pp. 1415–1442.

[81] L.L. HOROWITZ, and K.D. SENNE, "Performance advantage of complex LMS for controlling narrow-band adaptive arrays," *IEEE Trans. on Circuits and Systems*, Vol. CAS-28, No. 6 (June 1981), pp. 562–576.

[82] L. LJUNG, *System identification: Theory for the user* (Englewood Cliffs, NJ: Prentice Hall, 1987).

[83] K. ÅSTRÖM, *Introduction to stochastic control* (New York: Academic Press, 1970).

[84] K. ÅSTRÖM, and B. WITTENMARK, "On self-tuning regulators," *Automatica*, Vol. 9, No. 2 (1973).

[85] K. ÅSTRÖM, and B. WITTENMARK, "Analysis of a self-tuning regulator for nonminimum-phase systems," in *IFAC Symp. on Stochastic Control*, Budapest, Hungary, 1974.

[86] K. ÅSTRÖM, U. BORISSON, L. LJUNG, and B. WITTENMARK, "Theory and application of self tuning regulators," *Automatica*, Vol. 13 (1977), pp. 457–476.

[87] K. ÅSTRÖM, and B. WITTENMARK, "Self-tuning controllers based on pole-zero placement," *Proc. IEEE*, Vol. 127, Pt. D, No. 3 (May 1980), pp. 120–130.

[88] K. ÅSTRÖM, and B. WITTENMARK, *Adaptive control* (Reading, Mass: Addison-Wesley, 1989).

[89] P.V. OSBURN, H.P. WHITAKER, and A. KEZER, "New developments in the design of adaptive control systems," Paper No. 61-39, Institute of Aeronautical Sciences, February 1961.

[90] G.F. FRANKLIN, J.D. POWELL, and A. EMAMI-NAEINI, *Feedback control of dynamic systems* (Reading, Mass.: Addison-Wesley, 1986).

[91] G.F. FRANKLIN, J.D. POWELL, and M.L. WORKMAN, *Digital control of dynamic systems*, 2nd ed. (Reading, Mass.: Addison-Wesley, 1990).

[92] K. OGATA, *Modern control engineering*, 2nd ed. (Englewood Cliffs: Prentice Hall, 1990).

[93] D.F. SPECHT, "Generation of polynomial discriminant functions for pattern recognition," *IEEE Trans. on Electronic Computers*, Vol. EC-16, No. 3 (June 1967), pp. 308–319.

[94] V.J. MATHEWS, "Adaptive polynomial filters," *IEEE Signal Processing Magazine*, Vol. 8, No. 3 (July 1991), pp. 10–26.

[95] D. PSALTIS, A. SIDERIS, and A.A. YAMAMURA, "A multilayered neural network controller," *IEEE Control Systems Magazine*, Vol. 8 (April 1988), pp. 17–21.

[96] K.J. HUNT, and D. SBARBARO, "Neural networks for nonlinear internal model control," *IEEE Proceedings-D*, Vol. 138, No. 5 (September 1991), pp. 431–438.

[97] K.J. HUNT, D. SBARBARO, R. ZBIKOWSKI, and P.J. GANTHROP, "Neural networks for control systems—a survey," *Automatica*, Vol. 28, No. 6 (November 1992), pp. 1083–1112.

Chapter 2

Wiener Filters

2.0 INTRODUCTION

Wiener filters are best linear least squares filters which are used for prediction, estimation, interpolation, signal and noise filtering, and so forth. To design them, prior knowledge of the appropriate statistical properties of the input signal(s) is required. The problem is that this prior knowledge is often not available. Adaptive filters are used instead, making use of input data to learn the required statistics. Wiener filter theory is important to us however, because the adaptive filters used here converge asymptotically (in the mean) on Wiener solutions. Understanding Wiener filters is therefore necessary for the understanding of adaptive filters. Wiener filter theory and adaptive filter theory are fundamental to adaptive inverse control.

The idea of best linear least squares filtering was introduced by Norbert Wiener in 1949 [1]. The purpose of this chapter is to explain how Wiener filters work and how they can be designed, given the statistical properties of the input signals. Simple forms of Wiener filters can be made optimal without regard to causality. More complicated Wiener filters can be designed, using the Shannon-Bode [2] approach, to be causal and optimal in the least squares sense. A nice perspective on this is given by Kailath [3].

The discussion of Wiener filters will be made with regard to discrete-time *digital* fii-ters rather than analog filters. The reason for this is that the modern implementation of Wiener filters and adaptive filters is digital almost everywhere. It should be noted however, that Wiener's original work was analog, dealing with continuous rather than discrete signals and systems.

In the discussion to follow, Section 2.1 will describe correlation functions, their transforms, and relations between input and output signals of linear discrete filters driven by stochastic inputs. In Section 2.2, the two-sided (noncausal) Wiener filter will be derived. In Section 2.3, the Shannon-Bode approach to causal Wiener filter design will be given.

2.1 DIGITAL FILTERS, CORRELATION FUNCTIONS, z-TRANSFORMS

Figure 2.1 shows a block diagram of a simple causal linear digital filter. The boxes labeled "z^{-1}" represent unit delays. The circles labeled "h_0," "h_1," and so on are weights or gains, multiplying coefficients with no delay. The output of the filter is g_k, where k is

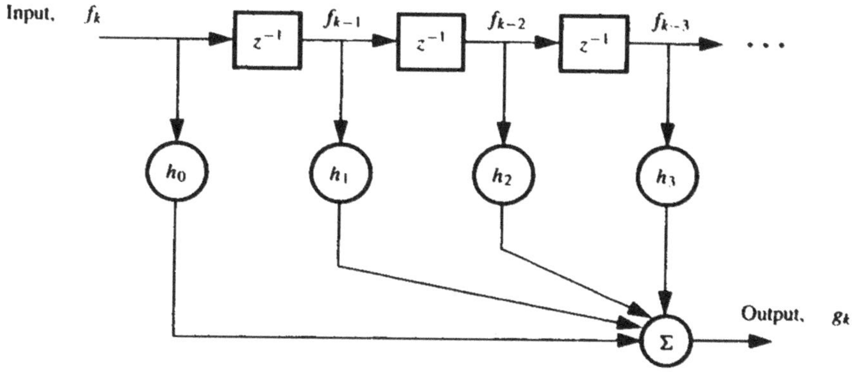

Figure 2.1 A causal linear digital filter.

the time index. The output is a weighted sum of the present input f_k, and its delayed versions f_{k-1}, f_{k-2}, and so on. The filter of Fig. 2.1 is often called a tapped delay line filter or transversal filter. The impulse response of the filter is a string of impulses whose values are h_0, h_1, and so on. The output g_k can be expressed algebraically as

$$\begin{aligned} g_k &= f_k h_0 + f_{k-1} h_1 + f_{k-2} h_2 + \dots \\ &= \sum_{l=0}^{\infty} f_{k-l} h_l. \end{aligned} \tag{2.1}$$

The above summation is a convolution of the input signal f_k with the impulse response h_k, which can be represented symbolically as

$$g_k = f_k * h_k = \sum_{l=0}^{\infty} f_{k-l} h_l. \tag{2.2}$$

The z-transform of the input f_k is defined as

$$F(z) \triangleq \sum_{k=-\infty}^{\infty} f_k z^{-k} \tag{2.3}$$

The z-transform of the output signal g_k can be obtained from the definition in Eq. (2.3) and the convolution in Eq. (2.2). Accordingly,

$$\begin{aligned} G(z) &\triangleq \sum_{k=-\infty}^{\infty} g_k z^{-k} \\ &= \sum_{k=-\infty}^{\infty} z^{-k} \sum_{l=0}^{\infty} f_{k-l} h_l \\ &= \sum_{k=-\infty}^{\infty} \sum_{l=0}^{\infty} z^{-k} f_{k-l} h_l. \end{aligned} \tag{2.4}$$

We now assume that the order of summation can be changed,[1] so that

$$G(z) = \sum_{l=0}^{\infty} \sum_{k=-\infty}^{\infty} z^{-l} h_l z^{-k+l} f_{k-l}. \tag{2.5}$$

Summing over k for any given value of l yields

$$\begin{aligned} G(z) &= \sum_{l=0}^{\infty} z^{-l} h_l \sum_{k=-\infty}^{\infty} z^{-k+l} f_{k-l} \\ &= \sum_{l=0}^{\infty} z^{-l} h_l \cdot F(z). \end{aligned} \tag{2.6}$$

Summing over l gives

$$G(z) = H(z) \cdot F(z). \tag{2.7}$$

Convolution in the time domain therefore corresponds to multiplication in the transform domain. The z-transform of the impulse response is $H(z)$, and this is called the transfer function of the filter.

A digital filter having a two-sided (noncausal) impulse response is shown in Fig. 2.2. The branches labeled "z" represent unit time advances. The branches labeled "z^{-1}" represent unit time delays, as before. In a real-time sense, the advances cannot exist physically, since they must be perfect predictors. The delays, of course, can exist. The filter of Fig. 2.2 cannot be physically built for real-time operation. The response of such a filter can be approximated, however, in a delayed and truncated form.

The output signal g_k of Fig. 2.2 can be expressed as a convolution of the input with the impulse response:

$$g_k = f_k * h_k = \sum_{l=-\infty}^{\infty} f_{k-l} h_l. \tag{2.8}$$

The derivation is similar to (2.1) and (2.2). The z-transform of this output signal can be written as

$$G(z) = H(z) \cdot F(z). \tag{2.9}$$

The derivation is similar to (2.6) and (2.7).

Assume now that the input signal f_k is stochastic, stationary, and ergodic. We define the autocorrelation function of f_k as

$$\phi_{ff}(m) \triangleq E[f_k \cdot f_{k+m}], \tag{2.10}$$

where the symbol $E[\cdot]$ represents expectation. We multiply the time sequence f_k by itself, lagged by m time delays, then we average. Since time average and ensemble average are equal (f_k is ergodic by assumption) the autocorrelation function can be written as a time average:

$$\phi_{ff}(m) = \lim_{N\to\infty} \frac{1}{2N+1} \sum_{l=-N}^{N} f_{k-l} \cdot f_{k-l+m}. \tag{2.11}$$

[1]The order of summation in (2.4) is interchangeable as long as the series expansion for $F(z)$, given by (2.3), and the corresponding expansion for $G(z)$ have a common region of absolute convergence in the z-domain.

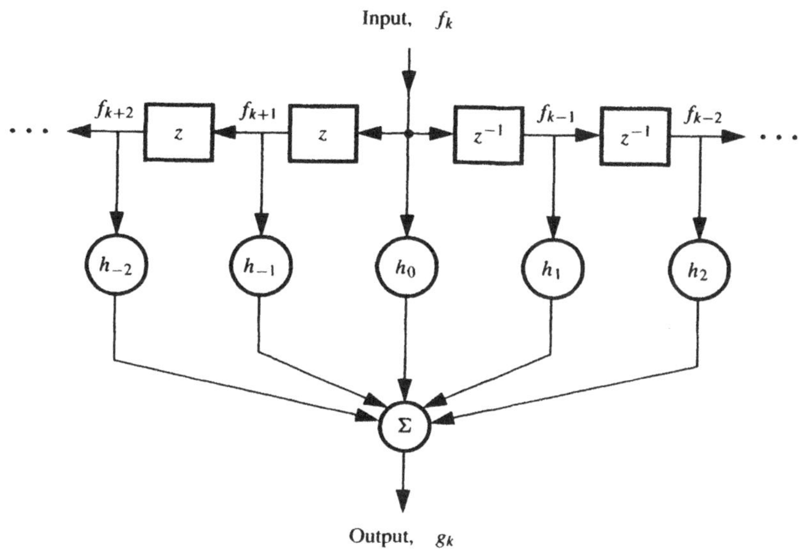

Figure 2.2 A noncausal two-sided digital filter.

The autocorrelation function can be approximately calculated from real data using (2.11) by making N large, but finite.

Given a digital filter like that of Fig. 2.1 or Fig. 2.2, the crosscorrelation function between the input f_k and the output g_k is defined as

$$\phi_{fg}(m) \triangleq E[f_k \cdot g_{k+m}] = \phi_{gf}(-m). \tag{2.12}$$

Using (2.2), this can be expressed as

$$\begin{aligned} \phi_{fg}(m) &= E\left[f_k \sum_{l=0}^{\infty} f_{k-l+m} h_l \right] \\ &= E\left[\sum_{l=0}^{\infty} h_l f_k f_{k-l+m} \right]. \end{aligned} \tag{2.13}$$

Since h_l is fixed for all l, and since f_k is stochastic, the expectation can be written as

$$\begin{aligned} \phi_{fg}(m) &= \sum_{l=0}^{\infty} h_l E[f_k f_{k-l+m}] \\ &= \sum_{l=0}^{\infty} h_l \phi_{ff}(m-l) \\ &= h_m * \phi_{ff}(m). \end{aligned} \tag{2.14}$$

Therefore, the crosscorrelation between input and output of a linear digital filter is the convolution of the input autocorrelation function with the impulse response. The same result is obtained for causal or noncausal filters.

Taking the z-transform of (2.14), we have

$$\Phi_{fg}(z) = H(z) \cdot \Phi_{ff}(z). \tag{2.15}$$

The autocorrelation function of the output of the digital filter is

$$\begin{aligned}
\phi_{gg}(m) &= E[g_k \cdot g_{k+m}] \\
&= E\left[\sum_{l=0}^{\infty} f_{k-l} h_l \sum_{\mu=0}^{\infty} f_{k-\mu+m} h_\mu\right] \\
&= \sum_{l=0}^{\infty}\sum_{\mu=0}^{\infty} h_l h_\mu E[f_{k-l} f_{k-\mu+m}] \\
&= \sum_{l=0}^{\infty}\sum_{\mu=0}^{\infty} h_l h_\mu \phi_{ff}(m+l-\mu) \\
&= \sum_{l=0}^{\infty} h_l \sum_{\mu=0}^{\infty} h_\mu \phi_{ff}(m+l-\mu).
\end{aligned} \tag{2.16}$$

Making use of (2.14) and (2.16), we obtain

$$\phi_{gg}(m) = \sum_{l=0}^{\infty} h_l \phi_{fg}(m+l). \tag{2.17}$$

Equation (2.17) is similar to a convolution. To place it in such form, define a time-reversed impulse response as

$$\tilde{h}_l \triangleq h_{-l}, \quad \text{or} \quad \tilde{h}_{-l} \triangleq h_l. \tag{2.18}$$

Accordingly,

$$\begin{aligned}
\phi_{gg}(m) &= \sum_{l=0}^{\infty} \tilde{h}_{-l} \phi_{fg}(m+l) \\
&= \sum_{l=0}^{-\infty} \tilde{h}_l \phi_{fg}(m-l).
\end{aligned} \tag{2.19}$$

This is a convolution, and can be written as

$$\begin{aligned}
\phi_{gg}(m) &= \tilde{h}_m * \phi_{fg}(m) \\
&= h_{-m} * \phi_{fg}(m).
\end{aligned} \tag{2.20}$$

Making use of (2.14), (2.20) becomes

$$\phi_{gg}(m) = h_{-m} * h_m * \phi_{ff}(m). \tag{2.21}$$

This is a double convolution. This derivation was done assuming that the impulse response was causal. The same result can be obtained with a noncausal or with a two-sided impulse response.

It is useful to have the z-transform of (2.21). It can be written as

$$\Phi_{gg}(z) = H(z^{-1}) \cdot H(z) \cdot \Phi_{ff}(z). \tag{2.22}$$

It should be noted that the z-transform of the time-reversed impulse response is equal to the z-transform of the impulse response with opposite exponent on z,

$$\sum_{l=-\infty}^{\infty} h_{-l} z^{-l} = \sum_{l=-\infty}^{\infty} h_l z^l = H(z^{-1}). \tag{2.23}$$

2.2 TWO-SIDED (UNCONSTRAINED) WIENER FILTERS

Two-sided Wiener filters are easiest to design because their impulse responses are unconstrained. We begin with this design. Refer to Fig. 2.3.

The digital filter has an input signal and it produces an output signal. This filter will be a Wiener filter if its impulse response is chosen to minimize mean square error. The error is defined as the difference between the filter output and the desired response:

$$\epsilon_k = d_k - g_k. \tag{2.24}$$

When working with Wiener filters, the desired response signal generally exists only conceptually. The statistical properties of the imagined desired response signal and the statistical relationship of this signal to the filter input signal are assumed to be known by the filter designer. The situation is quite different when dealing with adaptive filters. Here the desired response exists as an actual signal which must be available as an input to the real-time adaptive algorithm in order to achieve learning and adaptation. The Wiener filter does not learn. Its design is fixed, based on a priori statistical knowledge.

The impulse response of the Wiener filter is obtained by finding an expression for mean square error and minimizing this with respect to the impulse response. Squaring both sides of (2.24) and using (2.8), we get

$$\begin{aligned} \epsilon_k^2 &= d_k^2 + g_k^2 - 2d_k g_k \\ &= d_k^2 + \sum_{l=-\infty}^{\infty} \sum_{\mu=-\infty}^{\infty} h_l h_\mu f_{k-l} f_{k-\mu} - 2 \sum_{l=-\infty}^{\infty} h_l f_{k-l} d_k. \end{aligned} \tag{2.25}$$

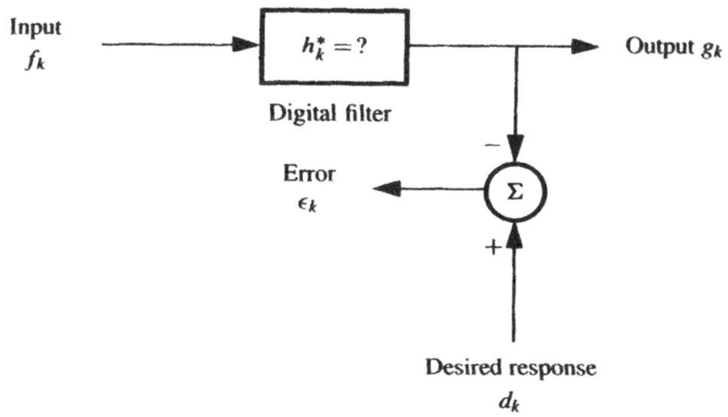

Figure 2.3 The Wiener filter.

Taking expectation of both sides yields an expression for mean square error (MSE):

$$E[\epsilon_k^2] = E[d_k^2] + \sum_{l=-\infty}^{\infty}\sum_{\mu=-\infty}^{\infty} h_l h_\mu E[f_{k-l} f_{k-\mu}] - 2\sum_{l=-\infty}^{\infty} h_l E[f_{k-l} d_k] \tag{2.26}$$

$$= \phi_{dd}(0) + \sum_{l=-\infty}^{\infty}\sum_{\mu=-\infty}^{\infty} h_l h_\mu \phi_{ff}(l-\mu) - 2\sum_{l=-\infty}^{\infty} h_l \phi_{fd}(l).$$

Taking partial derivative of the MSE with respect to the jth impulse of the filter impulse response,

$$\frac{\partial E[\epsilon_k^2]}{\partial h_j} = 0 + 2\sum_{l=-\infty}^{\infty} h_l \phi_{ff}(j-l) - 2\phi_{fd}(j). \tag{2.27}$$

We must set this derivative to zero for all values of j. The result is the Wiener impulse response, h_k^*, determined by

$$\sum_{l=-\infty}^{\infty} h_l^* \phi_{ff}(j-l) = \phi_{fd}(j), \quad \text{all } j. \tag{2.28}$$

This is the Wiener-Hopf equation, and it is in the form of a convolution

$$h_k^* * \phi_{ff}(k) = \phi_{fd}(k). \tag{2.29}$$

Taking z-transforms of both sides yields

$$H^*(z)\cdot\Phi_{ff}(z) = \Phi_{fd}(z), \text{ or} \tag{2.30}$$

$$H^*(z) = \frac{\Phi_{fd}(z)}{\Phi_{ff}(z)}.$$

The transfer function of the Wiener filter is $H^*(z)$ and is easily obtained from the z-transforms of the autocorrelation function of the input signal and the crosscorrelation function between the input and the desired response. The Wiener impulse response can be obtained by inverse z-transformation of $H^*(z)$.

Since the mean square error is minimized by use of the Wiener filter, an expression for minimum MSE can be obtained by combining the Wiener-Hopf equation (2.28) with the general expression for MSE given by (2.26):

$$\left(E[\epsilon_k^2]\right)_{\min} = \phi_{dd}(0) + \sum_{l=-\infty}^{\infty} h_l^* \phi_{fd}(l) - 2\sum_{l=-\infty}^{\infty} h_l^* \phi_{fd}(l) \tag{2.31}$$

$$= \phi_{dd}(0) - \sum_{l=-\infty}^{\infty} h_l^* \phi_{fd}(l).$$

A very useful fact regarding Wiener filters, that the crosscorrelation between the error and the input signal to each of the tap gains or weights is zero, can be demonstrated as follows. In general, this crosscorrelation is

$$E[f_{k-m}\epsilon_k] = E[f_k\epsilon_{k+m}] = E[f_k(d_{k+m} - g_{k+m})]$$

$$= E\left[f_k d_{k+m} - \sum_{l=-\infty}^{\infty} h_l f_{k-l+m} f_k\right] \tag{2.32}$$

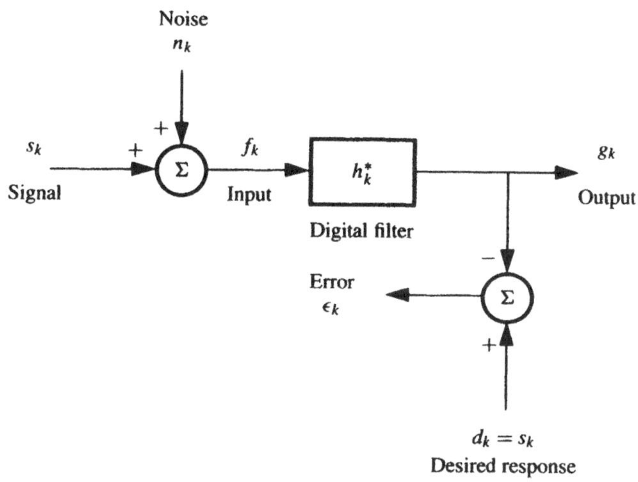

Figure 2.4 A Wiener noise filter.

$$= \phi_{fd}(m) - \sum_{l+-\infty}^{\infty} h_l \phi_{ff}(m-l).$$

If the impulse response is optimized in accord with the Wiener-Hopf equation, the result is

$$E[f_k \epsilon_{k+m}] = \phi_{fd}(m) - \sum_{l=-\infty}^{\infty} h_l^* \phi_{ff}(m-l) = 0. \tag{2.33}$$

We will make use of this fact about Wiener filters in subsequent chapters.

The following example illustrates how Wiener filter theory may be used. Refer to Fig. 2.4. A signal s_k is corrupted by additive noise n_k. Let their sum f_k be the input to a Wiener filter. The purpose of the filter is to eliminate the noise and reproduce the signal as best possible in the least squares sense. So the Wiener filter input is

$$f_k = s_k + n_k. \tag{2.34}$$

The output of the Wiener filter is

$$g_k = f_k * h_k^*, \tag{2.35}$$

and the desired response is

$$d_k = s_k. \tag{2.36}$$

For the present example, we are given that the autocorrelation function for the signal s_k is

$$\phi_{ss}(m) = \frac{10}{27}\left(\frac{1}{2}\right)^{|m|}, \quad \text{all } m, \tag{2.37}$$

and the autocorrelation function of the noise n_k is

$$\phi_{nn}(m) = \frac{2}{3}\delta(m), \tag{2.38}$$

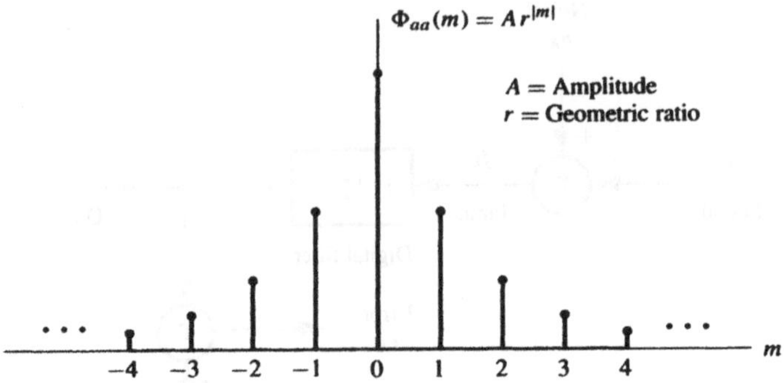

Figure 2.5 A geometric autocorrelation function.

where $\delta(\ )$ is the Kronecker delta function. The noise is thus "white," uncorrelated over time, with a mean square value of 2/3, and a mean value of zero. The signal is correlated over time, has a mean square value of 10/27, and a mean value of zero. The signal and noise are assumed to be uncorrelated with each other.

The transfer function of the Wiener filter is given by (2.30). We need to obtain $\Phi_{fd}(z)$ and $\Phi_{ff}(z)$ to get this function. We can first find $\phi_{fd}(m)$ and $\phi_{ff}(m)$, and then z-transform them. Since the signal and noise are uncorrelated with each other,

$$\begin{aligned}\phi_{ff}(m) &= \phi_{ss}(m) + \phi_{nn}(m) \\ &= \frac{10}{27}\left(\frac{1}{2}\right)^{|m|} + \frac{2}{3}\delta(m).\end{aligned} \tag{2.39}$$

The z-transform of this equation is

$$\Phi_{ff}(z) = \Phi_{ss}(z) + \Phi_{nn}(z). \tag{2.40}$$

The z-transform of the noise autocorrelation is

$$\Phi_{nn}(z) = \frac{2}{3}. \tag{2.41}$$

The z-transform of the signal autocorrelation is

$$\Phi_{ss}(z) = \sum_{m=-\infty}^{\infty}\left[\frac{10}{27}\left(\frac{1}{2}\right)^{|m|}\right]z^{-m}. \tag{2.42}$$

To evaluate this z-transform, examine the geometric autocorrelation function pictured in Fig. 2.5, given by

$$\phi_{aa}(m) = Ar^{|m|},$$

where A is a scale factor, and r is a geometric ratio less than one.

In accord with the definition of the z-transform (2.3), the transform of the autocorrelation function of Fig. 2.5 can be expressed as

$$\begin{aligned}\sum_{m=-\infty}^{\infty}[Ar^{|m|}]z^{-m} &= A[1 + rz^{-1} + r^2z^{-2} + r^3z^{-3} + \ldots] \\ &\quad + A[rz + r^2z^2 + r^3z^3 + \ldots].\end{aligned} \tag{2.43}$$

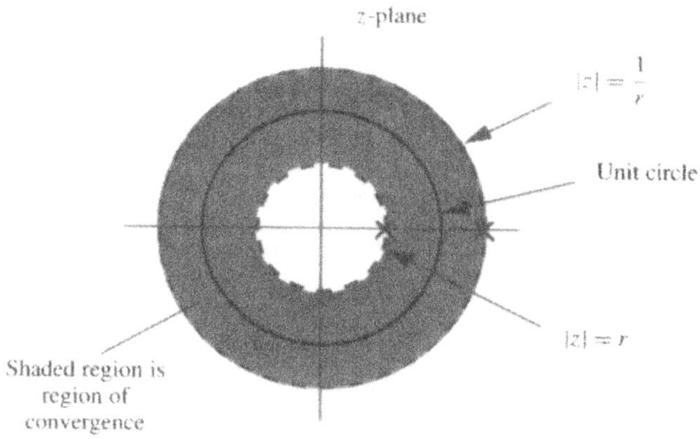

Figure 2.6 Region of convergence for geometric autocorrelation function.

The two sums are geometric and converge for certain values of z. The first series converges absolutely where

$$|rz^{-1}| < 1, \quad \text{or } |z| > r. \tag{2.44}$$

The second series converges absolutely where

$$|rz| < 1, \quad \text{or } |z| < \frac{1}{r}. \tag{2.45}$$

Therefore, the z-transform (2.23) exists where

$$\frac{1}{r} > |z| > r. \tag{2.46}$$

Under this condition,

$$\begin{aligned} \sum_{m=-\infty}^{\infty} [Ar^{|m|}]z^{-m} &= \frac{A}{1-rz^{-1}} + \frac{Azr}{1-rz} \\ &= \frac{A(1-r^2)}{(1-rz^{-1})(1-rz)}. \end{aligned} \tag{2.47}$$

This function has two poles in the z-plane, as shown in Fig. 2.6. The figure shows the region of convergence of the transform.

Making use of (2.47), the transform (2.42) can be evaluated as

$$\Phi_{ss}(z) = \frac{5/18}{(1-\frac{1}{2}z^{-1})(1-\frac{1}{2}z)}. \tag{2.48}$$

Combining this with (2.41), we obtain

$$\begin{aligned} \Phi_{ff}(z) &= \frac{5/18}{(1-\frac{1}{2}z^{-1})(1-\frac{1}{2}z)} + \frac{2}{3} \\ &= \frac{20-6z-6z^{-1}}{18(1-\frac{1}{2}z^{-1})(1-\frac{1}{2}z)}. \end{aligned} \tag{2.49}$$

Next we seek $\phi_{fd}(k)$. This can be written as

$$\begin{aligned}\phi_{fd}(m) &= E[f_k \cdot d_{k+m}] \\ &= E[(s_k + n_k)d_{k+m}].\end{aligned} \tag{2.50}$$

The desired response is equal to the signal, so that

$$\begin{aligned}\phi_{fd}(m) &= E[(s_k + n_k)s_{k+m}] \\ &= E[s_k \cdot s_{k+m}] \\ &= \phi_{ss}(m).\end{aligned} \tag{2.51}$$

We have used the fact that the signal and noise are uncorrelated with each other. Accordingly,

$$\Phi_{fd}(z) = \Phi_{ss}(z). \tag{2.52}$$

We can now get the transfer function of the Wiener filter,

$$H^*(z) = \frac{\Phi_{fd}(z)}{\Phi_{ff}(z)} = \frac{\Phi_{ss}(z)}{\Phi_{ff}(z)}. \tag{2.53}$$

Using (2.49) and (2.48),

$$H^*(z) = \frac{5}{20 - 6z - 6z^{-1}} = \frac{5/18}{(1 - \frac{1}{3}z^{-1})(1 - \frac{1}{3}z)}. \tag{2.54}$$

This transform resembles (2.47). By inspection, we can inverse transform (2.54) to obtain

$$h_k^* = \frac{5}{16}\left(\frac{1}{3}\right)^{|k|}, \quad \text{all } k. \tag{2.55}$$

This is the Wiener impulse response.

The minimum MSE can be computed using (2.31):

$$\begin{aligned}\left(E[\epsilon_k^2]\right)_{\min} &= \phi_{dd}(0) - \sum_{l=-\infty}^{\infty} h_l^* \phi_{fd}(l) \\ &= \frac{10}{27} - \sum_{l=-\infty}^{\infty} \frac{5}{16}\left(\frac{1}{3}\right)^{|l|}\left(\frac{10}{27}\right)\left(\frac{1}{2}\right)^{|l|} \\ &= \frac{10}{27} - \sum_{l=-\infty}^{\infty} \frac{25}{216}\left(\frac{1}{6}\right)^{|l|} \\ &= \frac{5}{24}.\end{aligned} \tag{2.56}$$

The ratio of minimum MSE to the mean square of the desired response is

$$\frac{5/24}{10/27} = 56.25\%. \tag{2.57}$$

Although this is a high ratio, the problem is a difficult one and this is the best that can be done with a linear Wiener filter.

2.3 SHANNON-BODE REALIZATION OF CAUSAL WIENER FILTERS

The Wiener-Hopf equation for the unconstrained Wiener filter is given by Eq. (2.28). This was obtained by differentiating the MSE with respect to all of the impulses in the impulse response and setting the resulting derivatives to zero. Our interest now focuses on the realization of causal Wiener filters. Their impulse responses are constrained to be zero for negative time. The optimal causal impulse response has zero response for negative time and has zero derivatives of MSE with respect to impulse response for all times equal to and greater than zero. Accordingly, the causal Wiener-Hopf equation is

$$\sum_{l=0}^{\infty} h^*_{l_{\text{causal}}} \phi_{ff}(j-l) = \phi_{fd}(j), \quad j \geq 0 \tag{2.58}$$

$$h^*_{j_{\text{causal}}} = 0, \quad j < 0. \tag{2.59}$$

This is not a simple convolution like the Wiener-Hopf equation for the unconstrained case, and special methods will be needed to find useful solutions. The approach developed by Shannon and Bode [2] will be used.

We begin with a simple case. Let the filter input be white with zero-mean and unit variance, so that

$$\phi_{ff}(m) = \delta(m). \tag{2.60}$$

For this input, the causal Wiener-Hopf equations (2.58)–(2.59) become

$$\left\{ \begin{array}{rcl} \sum_{l=0}^{\infty} h^*_{l_{\text{causal}}} \phi_{ff}(j-l) = h^*_{j_{\text{causal}}} & = & \phi_{fd}(j), \quad j \geq 0 \\ h^*_{j_{\text{causal}}} & = & 0, \quad j < 0 \end{array} \right\}. \tag{2.61}$$

With the same white input, but without the causality constraint, the Wiener-Hopf equation would be

$$\sum_{l=-\infty}^{\infty} h^*_l \phi_{ff}(j-l) = h^*_j = \phi_{fd}(j), \quad \text{all } j. \tag{2.62}$$

From this we may conclude that when the input to the Wiener filter is white, the optimal solution with a causality constraint is the same as the optimal solution without constraint, except that with the causality constraint the impulse response is set to zero for negative time. With a white input, the causal solution is easy to obtain, and it is key to Shannon-Bode. You find the unconstrained two-sided Wiener solution and lop off the noncausal part in the time domain. But you don't do this if the input is not white because, in that case, (2.61) and (2.62) do not hold.

Usually, the input to the Wiener filter is not white. Accordingly the first step in the Shannon-Bode realization involves whitening the input signal. A whitening filter can always be designed to do this, using a priori knowledge of the input autocorrelation function or its z-transform.

Assume that the z-transform of the autocorrelation function is a rational function of z that can be written as the ratio of a numerator polynomial in z to a denominator polynomial in z. Factor both the numerator and denominator polynomials. The autocorrelation function is symmetrical, and its z-transform has the following symmetry:

$$\phi_{ff}(m) = \phi_{ff}(-m), \tag{2.63}$$
$$\Phi_{ff}(z) = \Phi_{ff}(z^{-1}).$$

Accordingly, there must be symmetry in the numerator and denominator factors:

$$\Phi_{ff}(z) = A\frac{(1-az^{-1})(1-az)(1-bz^{-2})(1-bz^{2})\ldots}{(1-\alpha z^{-1})(1-\alpha z)(1-\beta z^{-2})(1-\beta z^{-2})\ldots}. \tag{2.64}$$

With no loss in generality,[2] assume that all the parameters $a,\ b,\ c,\ \ldots,\ \alpha,\ \beta,\ \ldots$ have magnitudes less than one. If $\Phi_{ff}(z)$ is factored as

$$\Phi_{ff}(z) = \Phi^{+}_{ff}(z)\cdot\Phi^{-}_{ff}(z), \tag{2.65}$$

we have

$$\Phi^{+}_{ff}(z) = \sqrt{A}\frac{(1-az^{-1})(1-bz^{-2})\ldots}{(1-\alpha z^{-1})(1-\beta z^{-2})\ldots} \tag{2.66}$$

$$\Phi^{-}_{ff}(z) = \sqrt{A}\frac{(1-az)(1-bz^{-2})\ldots}{(1-\alpha z)(1-\beta z^{-2})\ldots}. \tag{2.67}$$

All poles and zeros of $\Phi^{+}_{ff}(z)$ will be inside the unit circle in the z-plane. All poles and zeros of $\Phi^{-}_{ff}(z)$ will be outside the unit circle in the z-plane. Furthermore,

$$\Phi^{+}_{ff}(z) = \Phi^{-}_{ff}(z^{-1}),\quad \Phi^{+}_{ff}(z^{-1}) = \Phi^{-}_{ff}(z). \tag{2.68}$$

The whitening filter can now be designed. Let it have the transfer function

$$H_{\text{whitening}}(z) = \frac{1}{\Phi^{+}_{ff}(z)}. \tag{2.69}$$

To verify its whitening properties, let its input be f_k with autocorrelation function $\phi_{ff}(m)$, and let its output be i_k with autocorrelation function $\phi_{ii}(m)$. Using (2.22), the transform of the output autocorrelation function is

$$\begin{aligned}\Phi_{ii}(z) &= H_{\text{whitening}}(z^{-1})\cdot H_{\text{whitening}}(z)\cdot\Phi_{ff}(z)\\ &= \frac{1}{\Phi^{+}_{ff}(z^{-1})}\cdot\frac{1}{\Phi^{+}_{ff}(z)}\cdot\Phi_{ff}(z)\\ &= \frac{1}{\Phi^{-}_{ff}(z)}\cdot\frac{1}{\Phi^{+}_{ff}(z)}\cdot\Phi_{ff}(z)\\ &= 1.\end{aligned} \tag{2.70}$$

The autocorrelation function of the output is

$$\phi_{ii}(m) = \delta(m). \tag{2.71}$$

[2]The only case not included involves zeros of $\Phi_{ff}(z)$ exactly on the unit circle. This is a special case and it requires special treatment. These zeros are assumed to be somewhere off the unit circle, and they are placed back on it by a limiting process.

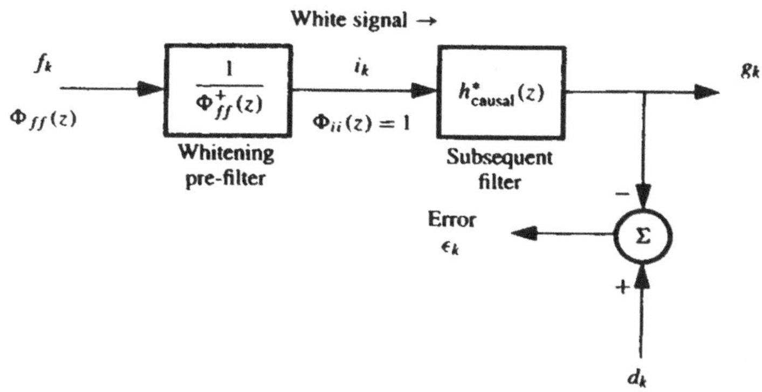

Figure 2.7 Shannon-Bode realization of causal Wiener filter.

Therefore the output is white with unit mean square.

It is important to note that the whitening filter (2.69) is both causal and stable, since the zeros of $\Phi^+_{ff}(z)$ are all inside the unit circle in the z-plane. Furthermore, using this whitening filter does nothing irreversible to the signal. The inverse of the whitening filter has a transfer function equal to $\Phi^+_{ff}(z)$, which is also stable and causal because $\Phi^+_{ff}(z)$ has all of its poles inside the unit circle. The whitening filter can be readily utilized since it is causal, stable, and invertible with a causal, stable filter.

The Shannon-Bode realization is sketched in Fig. 2.7. The input signal is f_k. The output of the causal Wiener filter is g_k. The desired response is d_k. The error is ϵ_k.

The input f_k is first applied to a whitening prefilter, in accord with (2.69). This does nothing irreversible to the input signal, it merely simplifies the mathematics. If the subsequent filtering operation does not benefit from the whitening filter, it could eliminate it immediately by inverse filtering.

The subsequent filter is easily designed, since its input is white. We represent its transfer function by $Y^*_{\text{causal}}(z)$ and design it without regard to causality, then lop off the noncausal part of its impulse response

Without regard to causality, the best subsequent filter would have the transfer function

$$Y^*(z) = \frac{\Phi_{fd}(z)}{\Phi_{ff}(z)} \cdot \Phi^+_{ff}(z) = \frac{\Phi_{fd}(z)}{\Phi^-_{ff}(z)} \tag{2.72}$$

To obtain this result, use is made of the unconstrained Wiener solution (2.30), taking into account that the input is whitened by (2.69).

The subsequent filter is made causal by inverse transforming (2.72) into the time domain and lopping off the noncausal part. The causal part is then z-transformed to yield a transfer function. This operation cannot be done with z-transforms alone. A special notation has been devised to represent taking the causal part:

$$Y^*_{\text{causal}}(z) = [Y^*(z)]_+ = \left[\frac{\Phi_{fd}(z)}{\Phi^-_{ff}(z)}\right]_+ . \tag{2.73}$$

The Shannon-Bode realization of the causal Wiener filter can now be formulated. The parts are evident from Fig. 2.7. The transfer function of the causal Wiener filter is

$$H^*_{\text{causal}}(z) = \frac{1}{\Phi^+_{ff}(z)}\left[\frac{\Phi_{fd}(z)}{\Phi^-_{ff}(z)}\right]_+. \tag{2.74}$$

An example helps to illustrate how this formula is used. We will rework the above noise filtering example, only in this case the Wiener filter will be designed to be causal.

The first step is to factor $\Phi_{ff}(z)$ in accord with (2.65), (2.66), and (2.67). $\Phi_{ff}(z)$ for this example is given by (2.49):

$$\Phi_{ff}(z) = \frac{20 - 6z - 6z^{-1}}{18(1 - \frac{1}{2}z)(1 - \frac{1}{2}z^{-1})} = \frac{(1 - \frac{1}{3}z)(1 - \frac{1}{3}z^{-1})}{(1 - \frac{1}{2}z)(1 - \frac{1}{2}z^{-1})}. \tag{2.75}$$

Therefore,

$$\Phi^+_{ff} = \frac{(1 - \frac{1}{3}z^{-1})}{(1 - \frac{1}{2}z^{-1})}, \quad \Phi^-_{ff} = \frac{(1 - \frac{1}{3}z)}{(1 - \frac{1}{2}z)}. \tag{2.76}$$

$\Phi_{fd}(z)$ is given by (2.52) and (2.48). From this and (2.76) we obtain

$$\begin{aligned} Y^*(z) = \frac{\Phi_{fd}(z)}{\Phi^-_{ff}(z)} &= \frac{5/18}{(1 - \frac{1}{2}z^{-1})(1 - \frac{1}{3}z)} \\ &= \left[\frac{\frac{1}{6}z^{-1}}{1 - \frac{1}{2}z^{-1}} + \frac{\frac{1}{3}}{1 - \frac{1}{3}z}\right]. \end{aligned} \tag{2.77}$$

The filter $Y^*(z)$ is generally two-sided in the time domain. It must be stable. In order for (2.77) to be stable, its two terms must correspond to time domain components as follows:

$$\frac{\frac{1}{6}z^{-1}}{1 - \frac{1}{2}z^{-1}} \rightarrow \text{(right-handed time function)}.$$

$$\frac{\frac{1}{3}}{1 - \frac{1}{3}z} \rightarrow \text{(left-handed time function)}.$$

Any other interpretation would cause instability. The time function corresponding to the first term is sketched in Fig. 2.8(a). The time function for the second term is sketched in Fig. 2.8(b). The time function for the sum of these terms is shown in Fig. 2.8(c). The causal part of the sum is sketched in Fig. 2.8(d). Taking the z-transform of this time function yields

$$\begin{aligned} Y^*_{\text{causal}}(z) &= \frac{1}{3} + \frac{1}{6}z^{-1} + \frac{1}{12}z^{-2} + \dots \\ &= \left(\frac{\frac{1}{3}}{1 - \frac{1}{2}z^{-1}}\right). \end{aligned} \tag{2.78}$$

Including the whitening filter, the transfer function of the causal Wiener filter is

$$H^*_{\text{causal}}(z) = \left(\frac{1 - \frac{1}{2}z^{-1}}{1 - \frac{1}{3}z^{-1}}\right)\left(\frac{\frac{1}{3}}{1 - \frac{1}{2}z^{-1}}\right) \tag{2.79}$$

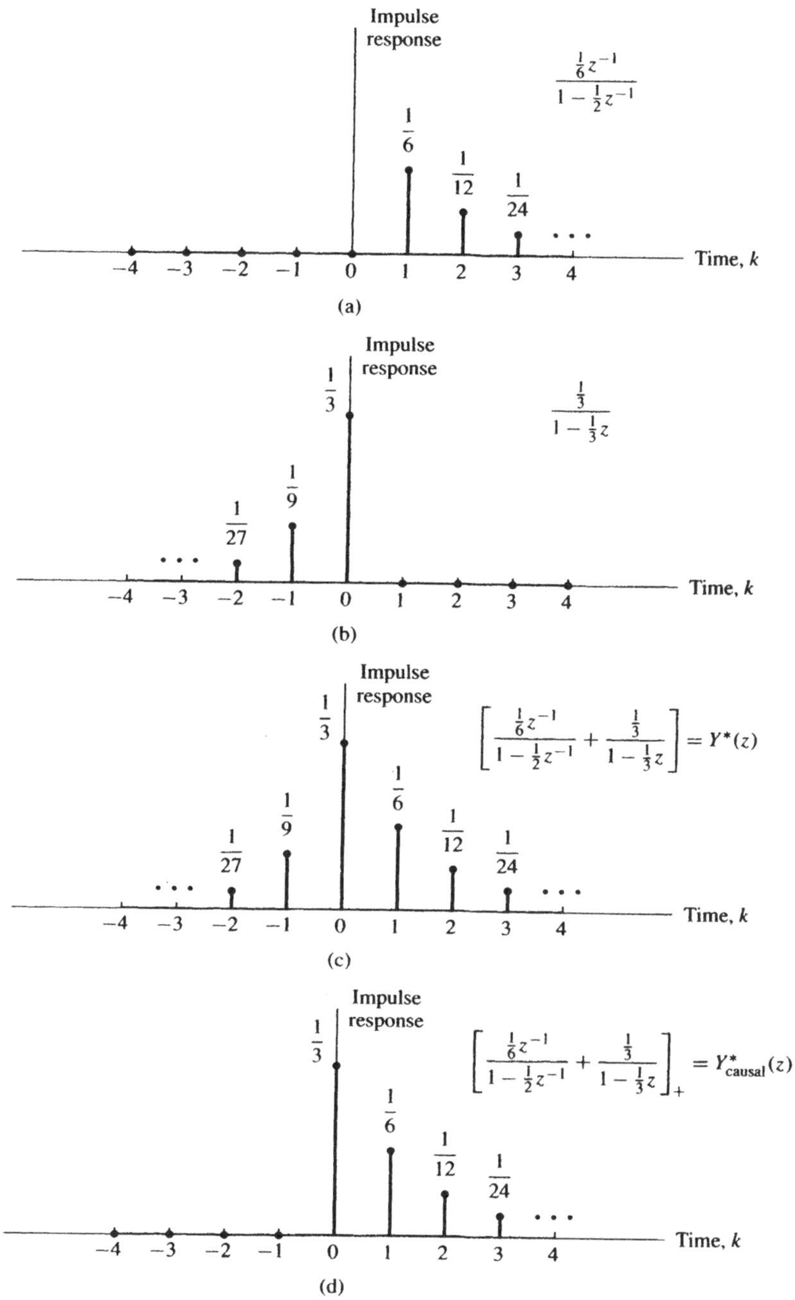

Figure 2.8 Correspondence between time functions and z-transforms for noise filtering example.

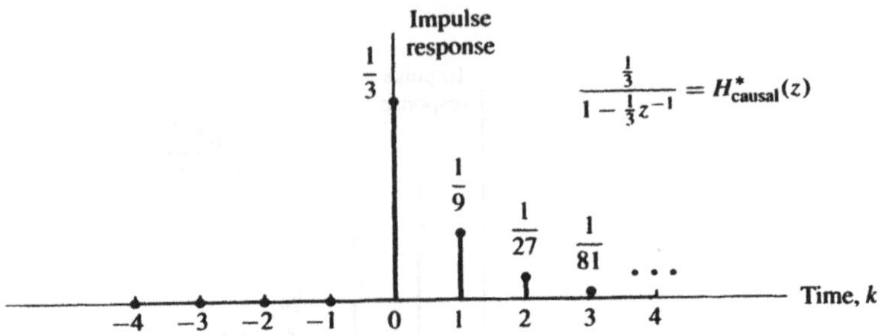

Figure 2.9 Causal Wiener filter for noise filtering example.

$$= \frac{\frac{1}{3}}{1-\frac{1}{3}z^{-1}}.$$

This is quite a different result from (2.55), the transfer function of the unconstrained Wiener filter. The impulse response corresponding to the transfer function (2.79) is sketched in Fig. 2.9.

It is expected that the causal Wiener filter would not perform as well as the unconstrained noncausal Wiener filter. To verify this, we need an expression for minimum mean square error of the causal Wiener filter.

The general equation for mean square error is (2.26). The minimum MSE of the causal Wiener filter is obtained by substituting the causal Wiener-Hopf Eq. (2.58) into Eq. (2.26). The result is

$$\left(E[\epsilon_k^2]\right)_{\text{min causal}} = \phi_{dd}(0) - \sum_{l=0}^{\infty} h^*_{l_{\text{causal}}} \phi_{fd}(l). \tag{2.80}$$

Equation (2.80) can now be used to find the minimum MSE. For the above example,

$$\phi_{dd}(m) = \phi_{ss}(m), \text{ and} \tag{2.81}$$

$$\phi_{fd}(m) = \phi_{ss}(m) = \frac{10}{27}\left(\frac{1}{2}\right)^{|m|}.$$

From (2.79), the causal Wiener impulse response is

$$h^*_{k_{\text{causal}}} = \frac{1}{3}\left(\frac{1}{3}\right)^k, \quad k \geq 0. \tag{2.82}$$

Accordingly,

$$\left(E[\epsilon_k^2]\right)_{\text{min causal}} = \frac{10}{27} - \sum_{l=0}^{\infty} \frac{10}{3(27)}\left(\frac{1}{6}\right)^l = \frac{6}{27}. \tag{2.83}$$

The ratio of the minimum MSE to the mean square of the desired response is

$$\frac{6/27}{10/27} = 60\%. \tag{2.84}$$

This is a poorer result than 56.25% obtained with the unconstrained Wiener filter. Causality did cost some performance.

2.4 SUMMARY

The principal results obtained in this chapter are the following:

- The unconstrained Wiener filter (two-sided) is determined by a convolution known as the Wiener-Hopf equation:

$$\sum_{l=-\infty}^{\infty} h_l^* \phi_{ff}(j-l) = \phi_{fd}(j), \quad \text{all } l. \tag{2.28}$$

- The transfer function of the two-sided Wiener filter is

$$H^*(z) = \frac{\Phi_{fd}(z)}{\Phi_{ff}(z)}. \tag{2.30}$$

- The minimum MSE of this filter is

$$\left(E[\epsilon_k^2]\right)_{\text{min}} = \phi_{dd}(0) - \sum_{l=-\infty}^{\infty} h_l^* \phi_{fd}(l). \tag{2.31}$$

- The causal Wiener filter is determined by the causal form of the Wiener-Hopf equation, which is not a simple convolution:

$$\sum_{l=0}^{\infty} h_{l_{\text{causal}}}^* \phi_{ff}(j-l) = \phi_{fd}(j), \quad j \geq 0 \tag{2.58}$$

$$h_{j_{\text{causal}}}^* = 0, \quad j < 0. \tag{2.59}$$

- The transfer function of the causal Wiener filter is

$$H_{\text{causal}}^*(z) = \frac{1}{\Phi_{ff}^+(z)} \left[\frac{\Phi_{fd}(z)}{\Phi_{ff}^-(z)}\right]_+ . \tag{2.74}$$

- The minimum MSE for this filter is

$$\left(E[\epsilon_k^2]\right)_{\text{min causal}} = \phi_{dd}(0) - \sum_{l=0}^{\infty} h_{l_{\text{causal}}}^* \phi_{fd}(l). \tag{2.80}$$

- It has been established that for any Wiener filter, causal or two-sided, the crosscorrelation between the error ϵ_k and the signal input to any of the tap gains or weights f_{k-m} is zero.

Bibliography for Chapter 2

[1] N. WIENER, *Extrapolation, interpolation, and smoothing of stationary time series with engineering applications* (New York: John-Wiley, 1949).

[2] H.W. BODE and C.E. SHANNON, "A simplified derivation of linear least square smoothing and prediction theory," in *Proc. IRE*, Vol. 38 (April 1950), pp. 417–425.

[3] T. KAILATH, *Lectures on Wiener and Kalman filtering* (New York: Springer-Verlag, 1981).

Chapter 3

Adaptive LMS Filters

3.0 INTRODUCTION

The theory of adaptive filtering is fundamental to adaptive inverse control. Adaptive filters are used for plant modeling, for plant inverse modeling, and to do plant disturbance canceling. At every step of the way, adaptive filtering is present. It is important to think of the adaptive filter as a building block, having an input signal, having an output signal, and having a special input signal called the "error" which is used in the learning process. This building block can be combined with other building blocks to make adaptive inverse control systems.

The purpose of this chapter is to present a brief overview of the theory of adaptive digital filtering. We will describe several applications for adaptive filters, and will discuss stability, rate of convergence, and effects of noise in the impulse response.[1] We will derive relationships between speed of adaptation and performance of adaptive systems. In general, faster adaptation leads to more noisy adaptive processes. When the input environment of an adaptive system is statistically nonstationary, best performance is obtained by a compromise between fast adaptation (necessary to track variations in input statistics) and slow adaptation (necessary to contain the noise in the adaptive process). A number of these issues will be studied both analytically and by computer simulation. The context of this study will be restricted to adaptive digital filters driven by the LMS adaptation algorithm of Widrow and Hoff [1]–[5]. This algorithm and algorithms similar to it have been used for many years in a wide variety of practical applications [6].

We are reviewing a *statistical theory of adaptation*. Stability and rate of convergence are analyzed first; then gradient noise and its effects upon performance are assessed. The concept of "misadjustment" is defined and used to establish design criteria for a sample problem, an adaptive predictor. Finally, we consider an application of adaptive filtering to adaptive noise canceling. The principles of adaptive noise canceling are derived and some experimental results are presented. These include utilization of the noise canceling techniques to improve results of adult and fetal electrocardiography.

[1] Since the parameters of the adaptive filter are data dependent and time variable during adaptation, the adaptive filter does not have an impulse response defined in the usual way. An instantaneous impulse response can be defined as the impulse response that would result if adaptation were suddenly stopped and the parameters were fixed at instantaneous values.

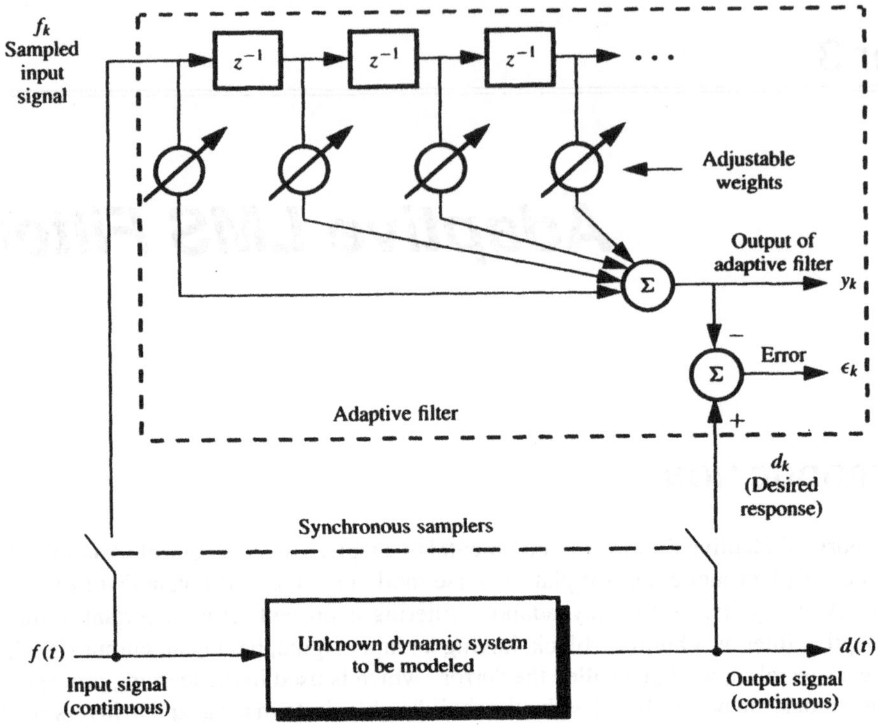

Figure 3.1 Modeling an unknown system by a discrete adaptive filter.

3.1 AN ADAPTIVE FILTER

The form of adaptive filter to be considered here comprises a tapped delay line, variable weights (variable gains) whose input signals are the signals at the delay-line taps, a summer to add the weighted signals, and an adaptation process that automatically seeks an optimal impulse response by adjusting the weights. Figure 3.1 illustrates the adaptive filter as it would be used in modeling an unknown dynamic system. This filter is causal and has a finite impulse response (FIR).

In addition to the usual input signals, another input signal, the *desired response*, must be supplied during the adaptation process to generate the error signal. In Fig. 3.1, the same input is applied to the adaptive filter as to the unknown system to be modeled. The output of the unknown system provides the desired response for the adaptive filter. In other applications, considerable ingenuity may be required to obtain a suitable desired response (or some form of equivalent) for an adaptive process.

The adaptive filter of Fig. 3.1 is a discrete-time or *digital* filter. The unknown system to be modeled is a continuous-time or *analog* filter. The inputs to the adaptive filter are therefore sampled versions of the input and output signals of the unknown system. The weights of the adaptive filter are adjusted by an automatic algorithm to minimize mean square error. Adjustability is indicated by the arrows through the weights. When the weights converge and the error becomes small, the impulse response of the adaptive filter will closely match a sampled version of the impulse response of the unknown system. Then, both filters with the same input (or sampled version) will produce essentially the same output (or sampled version).

3.2 THE PERFORMANCE SURFACE

An analysis of the adaptive filter can be developed by considering first the *adaptive linear combiner* of Fig. 3.2, a subsystem of the adaptive filter of Fig. 3.1, comprising its most significant part. This combinational system can be connected to the elements of a phased array antenna to make an adaptive antenna [7], or to a quantizer to form a single artificial *neuron*, that is, an adaptive threshold element (*Adaline* [1], [8] or threshold logic unit (TLU) [9]) for use in neural networks and in adaptive logic and pattern-recognition systems. It can also be used as the adaptive portion of certain learning control systems [10], [11]; as a key portion of adaptive filters for channel equalization [12]; for adaptive noise canceling [13]; or for adaptive systems identification [14]–[23].

In Fig. 3.2, a set of n input signals is weighted and summed to form an output signal. The inputs occur simultaneously and discretely in time. The kth input signal vector is

$$\mathbf{X}_k = [x_{1k}, x_{2k}, \ldots, x_{lk}, \ldots, x_{nk}]^T. \tag{3.1}$$

The set of weights is designated by the vector

$$\mathbf{W}^T = [w_1, w_2, \ldots, w_l, \ldots, w_n]. \tag{3.2}$$

For the present analysis, let the weights remain fixed. The kth output signal will be

$$y_k = \sum_{l=1}^{n} w_l x_{lk} = \mathbf{W}^T \mathbf{X}_k = \mathbf{X}_k^T \mathbf{W}. \tag{3.3}$$

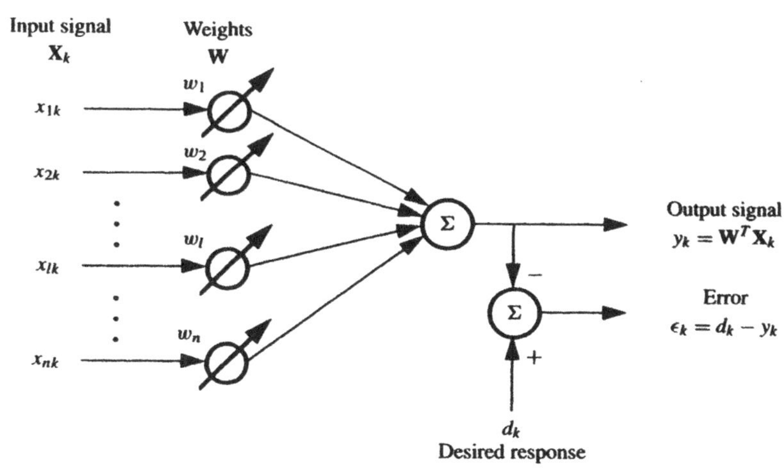

Figure 3.2 Adaptive linear combiner.

The input signals and desired response are assumed to be stationary ergodic processes. Denoting the desired response as d_k, the error at the kth sampling time is

$$\epsilon_k = d_k - y_k = d_k - \mathbf{W}^T\mathbf{X}_k = d_k - \mathbf{X}_k^T\mathbf{W}. \tag{3.4}$$

The square of this error is

$$\epsilon_k^2 = d_k^2 - 2d_k\mathbf{X}_k^T\mathbf{W} + \mathbf{W}^T\mathbf{X}_k\mathbf{X}_k^T\mathbf{W}. \tag{3.5}$$

The mean square error ξ, the expected value of ϵ_k^2, is

$$\begin{aligned} \text{MSE} = \xi \triangleq E[\epsilon_k^2] &= E[d_k^2] - 2E[d_k\mathbf{X}_k^T]\mathbf{W} + \mathbf{W}^T E[\mathbf{X}_k\mathbf{X}_k^T]\mathbf{W} \\ &= E[d_k^2] - 2\mathbf{P}^T\mathbf{W} + \mathbf{W}^T\mathbf{R}\mathbf{W}, \end{aligned} \tag{3.6}$$

where the crosscorrelation vector between the input signals and the desired response is defined as

$$E[d_k\mathbf{X}_k] = E\begin{bmatrix} d_k x_{1k} \\ d_k x_{2k} \\ \vdots \\ d_k x_{nk} \end{bmatrix} \triangleq \mathbf{P}, \tag{3.7}$$

and where the symmetric and positive definite (or positive semidefinite) input correlation matrix $\mathbf{R}$ of the $\mathbf{X}$-input signal is defined as

$$E[\mathbf{X}_k\mathbf{X}_k^T] = E\begin{bmatrix} x_{1k}x_{1k} & x_{1k}x_{2k} & \cdots \\ x_{2k}x_{1k} & x_{2k}x_{2k} & \cdots \\ \vdots & \vdots & \\ & & x_{nk}x_{nk} \end{bmatrix} \triangleq \mathbf{R}. \tag{3.8}$$

It may be observed from (3.6) that the mean square error (MSE) performance function is a quadratic function of the weights, a *bowl-shaped* surface; the adaptive process will be continuously adjusting the weights, seeking the bottom of the bowl. This may be accomplished by steepest descent methods [24], [25] discussed below.

3.3 THE GRADIENT AND THE WIENER SOLUTION

The method of steepest descent uses gradients of the performance surface in seeking its minimum. The gradient at any point on the performance surface may be obtained by differentiating the MSE function, Eq. (3.6), with respect to the weight vector. The gradient vector is

$$\nabla \triangleq \begin{Bmatrix} \frac{\partial E[\epsilon_k^2]}{\partial w_1} \\ \vdots \\ \frac{\partial E[\epsilon_k^2]}{\partial w_n} \end{Bmatrix} = -2\mathbf{P} + 2\mathbf{R}\mathbf{W}. \tag{3.9}$$

Set the gradient to zero to find the optimal weight vector $\mathbf{W}^*$:

$$\mathbf{W}^* = \mathbf{R}^{-1}\mathbf{P}. \tag{3.10}$$

We assume that $\mathbf{R}$ is positive definite and that $\mathbf{R}^{-1}$ exists. Equation (3.10) then is the Wiener-Hopf equation in matrix form.

Equation (3.10) gives the Wiener solution for a digital filter whose impulse response has finite length (FIR, finite impulse response). Generally, this filter would be causal. However (3.10) could be made to apply just as well to noncausal finite impulse response filters.

Equation (3.10) can be rewritten as

$$\mathbf{R}\mathbf{W}^* = \mathbf{P}. \tag{3.11}$$

This relation has similarity to (2.29), the Wiener-Hopf equation for unconstrained filters whose impulse responses could extend infinitely in both directions over time comprising two-sided IIR, infinite impulse responses. It is not as easy to compare (3.11) with (2.58), the Wiener-Hopf equation for causal filters. The various forms of the Wiener-Hopf equation, although similar, are especially devised to meet the constraints, or lack thereof, imposed on the Wiener impulse response.

The minimum MSE for the finite impulse response case is obtained from (3.10) and (3.6):

$$\xi_{\min} = E[d_k^2] - \mathbf{P}^T\mathbf{W}^*. \tag{3.12}$$

Substituting (3.10) and (3.12) into (3.6) yields a useful formula for MSE:

$$\xi = \xi_{\min} + (\mathbf{W} - \mathbf{W}^*)^T\mathbf{R}(\mathbf{W} - \mathbf{W}^*). \tag{3.13}$$

Define $\mathbf{V}$ as the difference between $\mathbf{W}$ and the Wiener solution $\mathbf{W}^*$:

$$\mathbf{V} \triangleq (\mathbf{W} - \mathbf{W}^*). \tag{3.14}$$

Therefore,

$$\xi = \xi_{\min} + \mathbf{V}^T\mathbf{R}\mathbf{V}. \tag{3.15}$$

Differentiation of (3.15) yields another form for the gradient:

$$\nabla = 2\mathbf{R}\mathbf{V}. \tag{3.16}$$

The input correlation matrix, being symmetric and positive definite or positive semidefinite, may be represented in normal form as

$$\mathbf{R} = \mathbf{Q}\mathbf{\Lambda}\mathbf{Q}^{-1} = \mathbf{Q}\mathbf{\Lambda}\mathbf{Q}^T, \tag{3.17}$$

where $\mathbf{Q}$ is the eigenvector matrix, the orthonormal modal matrix of $\mathbf{R}$, and $\mathbf{\Lambda}$ is its diagonal matrix of eigenvalues:

$$\mathbf{\Lambda} = \text{diag}[\lambda_1, \lambda_2, \cdots, \lambda_p, \cdots, \lambda_n]. \tag{3.18}$$

Equation (3.15) may be reexpressed as

$$\xi = \xi_{\min} + \mathbf{V}^T\mathbf{Q}\mathbf{\Lambda}\mathbf{Q}^{-1}\mathbf{V}. \tag{3.19}$$

Define a transformed version of $\mathbf{V}$ as

$$\mathbf{V}' \triangleq \mathbf{Q}^{-1}\mathbf{V} \text{ and } \mathbf{V} = \mathbf{Q}\mathbf{V}'. \tag{3.20}$$

Accordingly, Eq. (3.15) may be put in normal form as

$$\xi = \xi_{\min} + \mathbf{V}'^T\mathbf{\Lambda}\mathbf{V}'. \tag{3.21}$$

This expression has only square terms in the primed coordinates, no crossterms. The primed coordinates are therefore the principal axes of the quadratic surface. Transformation (3.20) may also be applied to the weight vector,

$$\mathbf{W}' = \mathbf{Q}^{-1}\mathbf{W} \text{ and } \mathbf{W} = \mathbf{Q}\mathbf{W}'. \tag{3.22}$$

.4 THE METHOD OF STEEPEST DESCENT

The method of steepest descent makes each change in the weight vector proportional to the negative of the gradient vector:

$$\mathbf{W}_{k+1} = \mathbf{W}_k + \mu(-\nabla_k). \tag{3.23}$$

The scalar parameter μ is a convergence factor that controls stability and rate of adaptation. The gradient at the kth iteration is ∇_k. Using (3.14), (3.16), (3.17), and (3.20), Eq. (3.23) becomes

$$\mathbf{V}'_{k+1} - (\mathbf{I} - 2\mu\mathbf{\Lambda})\mathbf{V}'_k = 0. \tag{3.24}$$

This homogeneous vector difference equation is uncoupled. It has a simple geometric solution in the primed coordinates [4] which can be expressed as

$$\mathbf{V}'_k = (\mathbf{I} - 2\mu\mathbf{\Lambda})^k\mathbf{V}'_0, \tag{3.25}$$

where $\mathbf{V}'_0$ is an initial condition

$$\mathbf{V}'_0 = \mathbf{W}'_0 - \mathbf{W}^{*\prime}. \tag{3.26}$$

For stability of (3.25), it is necessary that

$$1/\lambda_{\max} > \mu > 0, \tag{3.27}$$

where $\lambda_{\max}$ is the largest eigenvalue of $\mathbf{R}$. From (3.25), we see that transients in the primed coordinates will be geometric; the geometric ratio for the pth coordinate is

$$r_p = (1 - 2\mu\lambda_p). \tag{3.28}$$

Note that the pth eigenvalue is λ_p.

An exponential envelope can be fitted to a geometric sequence. If the basic unit of time is considered to be the iteration cycle, time constant τ_p can be determined as follows:

$$r_p = \exp\left(-\frac{1}{\tau_p}\right) = 1 - \frac{1}{\tau_p} + \frac{1}{2!\tau_p^2} - \cdots. \tag{3.29}$$

The case of general interest is slow adaptation, that is, large τ_p. Therefore,

$$r_p = (1 - 2\mu\lambda_p) \simeq 1 - \frac{1}{\tau_p} \tag{3.30}$$

or

$$\tau_p \simeq \frac{1}{2\mu\lambda_p}. \tag{3.31}$$

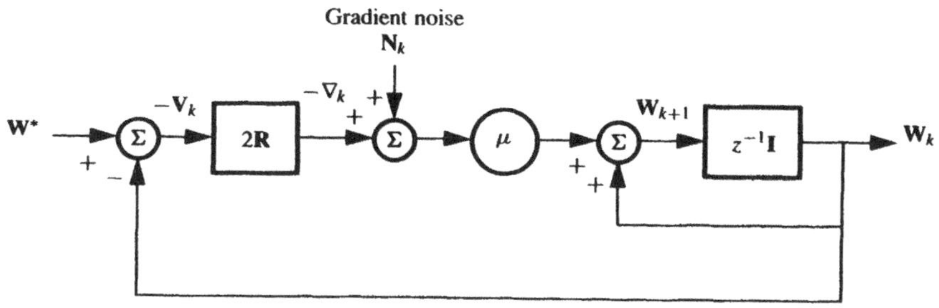

Figure 3.3 Feedback model of steepest descent.

Equation (3.31) gives the time constant of the pth mode for slow adaptation, with small μ. The time constant is expressed in number of iteration cycles.

Steepest descent can be regarded as a feedback process where the gradient plays the role of vector error signal. The process, if stable, tends to bring the gradient to zero. Figure 3.3 shows a feedback model[2] for a stationary quadratic MSE surface being searched by the method of steepest descent. The model is equivalent to the following set of relations:

$$\begin{aligned} \mathbf{W}_k &= \mathbf{W}_{k+1}|_{\text{delayed one iteration}} \\ \mathbf{W}_{k+1} &= \mathbf{W}_k + \mu(-\nabla_k + \mathbf{N}_k) \\ \nabla_k &= 2\mathbf{R}(\mathbf{W}_k - \mathbf{W}^*) = 2\mathbf{R}\mathbf{V}. \end{aligned} \tag{3.32}$$

Notice that this model has an input not mentioned earlier, *gradient noise* $\mathbf{N}_k$, which exists because gradients estimated at each iteration cycle with finite amounts of input data are imperfect or noisy.

3.5 THE LMS ALGORITHM

The LMS algorithm is an implementation of steepest descent using measured or estimated gradients:

$$\mathbf{W}_{k+1} = \mathbf{W}_k + \mu(-\widehat{\nabla}_k). \tag{3.33}$$

The estimate of the true gradient is $\widehat{\nabla} = \nabla + \mathbf{N}_k$, equal to the true gradient plus gradient noise.

The error ϵ_k of the adaptive linear combiner of Fig. 3.2 is given by (3.4). A crude gradient estimate may be obtained by squaring the single value of ϵ_k and differentiating it as if it were the mean square error:

$$\widehat{\nabla}_k = \begin{Bmatrix} \frac{\partial \epsilon_k^2}{\partial w_1} \\ \vdots \\ \frac{\partial \epsilon_k^2}{\partial w_n} \end{Bmatrix} = 2\epsilon_k \begin{Bmatrix} \frac{\partial \epsilon_k}{\partial w_1} \\ \vdots \\ \frac{\partial \epsilon_k}{\partial w_n} \end{Bmatrix} = -2\epsilon_k \mathbf{X}_k. \tag{3.34}$$

[2]This has been called performance feedback [1], [26].

Substituting (3.34) into (3.23) yields the LMS algorithm:

$$\mathbf{W}_{k+1} = \mathbf{W}_k + 2\mu\epsilon_k\mathbf{X}_k. \tag{3.35}$$

Since a new gradient estimate is obtained with each data sample, an adaptive iteration is effected with the arrival of each sample.

The gradient estimate of (3.34) may be computed in a practical system without squaring, averaging, or differentiation, and it is elegant in its simplicity and efficiency. All components of the gradient vector are obtained from a single data sample, without perturbation of the weight vector. Each gradient vector is an *instantaneous gradient*. Since these gradient estimates are obtained without averaging, they contain large noise components. The noise, however, becomes averaged and attenuated by the adaptive process, which acts as a low-pass filter in this respect.

It is important to note that for a fixed value of **W**, the LMS gradient estimate is unbiased. This can be demonstrated in the following way. Using (3.34) together with (3.4), we get

$$\begin{aligned} E[\widehat{\nabla}_k] = -2E[\epsilon_k\mathbf{X}_k] &= -2E[d_k\mathbf{X}_k - \mathbf{X}_k\mathbf{X}_k^T\mathbf{W}] \\ &= -2(\mathbf{P} - \mathbf{R}\mathbf{W}). \end{aligned} \tag{3.36}$$

From (3.9), the formula for the true gradient, we obtain

$$\nabla = -2(\mathbf{P} - \mathbf{R}\mathbf{W}). \tag{3.37}$$

Therefore,

$$E[\widehat{\nabla}] = \nabla. \tag{3.38}$$

Although these gradient estimates are noisy, many small steps taken in the directions of the negative instantaneous gradients will, on average, go in the correct direction for steepest descent.

Many convergence proofs for LMS algorithms have appeared in the literature over the years [27], [28], [29], [30], [31]. Each one demonstrates convergence under a given set of assumptions. Equation (3.27) turns out to be sufficient for convergence of the weight vector in the mean, but not sufficient for convergence of the variance of the weight vector. A stronger condition for convergence of mean and variance is needed, namely,

$$\frac{1}{\operatorname{tr}\mathbf{R}} > \mu > 0. \tag{3.39}$$

This condition for LMS stability is derived and discussed in [32]. A brief derivation of (3.39) is also included in Appendix A. The trace of **R**, equal to the sum of the mean squares of all the signals going into the weights, would be known, at least approximately. When the linear combiner is connected to the tapped delay line of an adaptive filter, the sum of mean squares equals the power level of the signal going into the adaptive filter multiplied by the number of weights. One would generally have some idea of the power level of the signal going into a filter, making it easy to apply (3.39).

3.6 THE LEARNING CURVE AND ITS TIME CONSTANTS

During adaptation, the error ϵ_k is nonstationary as the weight vector adapts toward $\mathbf{W}^*$. The MSE can be defined only on the basis of ensemble averages. From (3.21), we obtain

$$\xi_k = \xi_{\min} + \mathbf{V}_k'^T \mathbf{\Lambda} \mathbf{V}_k'. \tag{3.40}$$

Imagine an ensemble of adaptive processes, each having individual stationary ergodic inputs drawn from the same statistical population, with all initial weight vectors equal. The MSE ξ_k is a function of iteration number k obtained by averaging over the ensemble at iteration k.

Using (3.25), but assuming no noise in the weight vector, Eq. (3.40) becomes

$$\begin{aligned}\xi_k &= \xi_{\min} + \mathbf{V}_0'^T (\mathbf{I} - 2\mu\mathbf{\Lambda})^k \mathbf{\Lambda} (\mathbf{I} - 2\mu\mathbf{\Lambda})^k \mathbf{V}_0' \\ &= \xi_{\min} + \mathbf{V}_0^T (\mathbf{I} - 2\mu\mathbf{R})^k \mathbf{R} (\mathbf{I} - 2\mu\mathbf{R})^k \mathbf{V}_0. \end{aligned} \tag{3.41}$$

When the adaptive process is convergent, it is clear from (3.40) that

$$\lim_{k\to\infty} \xi_k = \xi_{\min} \tag{3.42}$$

and that the geometric decay in ξ_k going from ξ_0 to $\xi_{\min}$ will, for the pth mode, have a geometric ratio of r_p^2 and a time constant

$$\tau_{p_{\text{mse}}} \triangleq \frac{1}{2}\tau_p = \frac{1}{4\mu\lambda_p}. \tag{3.43}$$

The result obtained by plotting MSE against number of iterations is called the *learning curve*. It has a number of modes equal to the number of distinct eigenvalues of $\mathbf{R}$. Due to noise in the weight vector, actual practice will show ξ_k to be higher than indicated by (3.41). The learning curve shows the reduction of MSE resulting from repeated application of the adaptation algorithm.

3.7 GRADIENT AND WEIGHT-VECTOR NOISE

Gradient noise will affect the adaptive process both during initial transients and in steady state. The latter condition is of particular interest here.

Assume that the weight vector is close to the Wiener solution. Assume, as before, that $\mathbf{X}_k$ and d_k are stationary and ergodic and that $\mathbf{X}_k$ is uncorrelated over time, that is,

$$E[\mathbf{X}_k^T \mathbf{X}_{k+l}] = 0, \quad l \neq 0. \tag{3.44}$$

The LMS algorithm uses an unbiased gradient estimate

$$\hat{\nabla} = -2\epsilon_k \mathbf{X}_k = \nabla_k - \mathbf{N}_k, \tag{3.45}$$

where ∇_k is the true gradient and $\mathbf{N}_k$ is a zero-mean gradient estimation noise vector, defined above. When $\mathbf{W}_k = \mathbf{W}^*$, the true gradient is zero. But the gradient would be estimated according to (3.34), and this would be equal to pure gradient noise:

$$\mathbf{N}_k = 2\epsilon_k \mathbf{X}_k. \tag{3.46}$$

According to Wiener filter theory, when $\mathbf{W}_k = \mathbf{W}^*$, ϵ_k and $\mathbf{X}_k$ are uncorrelated. If they are assumed to be zero-mean Gaussian, ϵ_k and $\mathbf{X}_k$ are statistically independent. As such, the covariance of $\mathbf{N}_k$ is

$$\begin{aligned}\text{cov}[\mathbf{N}_k] &= E[\mathbf{N}_k\mathbf{N}_k^T] = 4E[\epsilon_k^2\mathbf{X}_k\mathbf{X}_k^T] \\ &= 4E[\epsilon_k^2]E[\mathbf{X}_k\mathbf{X}_k^T] \\ &= 4E[\epsilon_k^2]\mathbf{R}.\end{aligned} \tag{3.47}$$

When $\mathbf{W}_k = \mathbf{W}^*$, $E[\epsilon_k^2] = \xi_{\min}$. Accordingly,

$$\text{cov}[\mathbf{N}_k] = 4\xi_{\min}\mathbf{R}. \tag{3.48}$$

As long as $\mathbf{W}_k \simeq \mathbf{W}^*$, we conclude that the gradient noise covariance is given approximately by (3.48) and that this noise is stationary and uncorrelated over time. This conclusion is based on (3.44) and (3.46) and on the Gaussian assumption.

Projecting the gradient noise into the primed coordinates,

$$\mathbf{N}'_k = \mathbf{Q}^{-1}\mathbf{N}_k. \tag{3.49}$$

The covariance of $\mathbf{N}'_k$ becomes

$$\begin{aligned}\text{cov}[\mathbf{N}'_k] = E[\mathbf{N}'_k\mathbf{N}'^T_k] = E[\mathbf{Q}^{-1}\mathbf{N}_k\mathbf{N}_k^T\mathbf{Q}] &= \mathbf{Q}^{-1}\text{cov}[\mathbf{N}_k]\mathbf{Q} \\ &= 4\xi_{\min}\mathbf{Q}^{-1}\mathbf{R}\mathbf{Q} \\ &= 4\xi_{\min}\mathbf{\Lambda}.\end{aligned} \tag{3.50}$$

Although the components of $\mathbf{N}_k$ are correlated with each other, those of $\mathbf{N}'_k$ are mutually uncorrelated and can, therefore, be handled more easily.

Gradient noise propagates and causes noise in the weight vector. Accounting for gradient noise, the LMS algorithm can be expressed conveniently in the primed coordinates as

$$\mathbf{W}'_{k+1} = \mathbf{W}'_k + \mu(-\widehat{\nabla}'_k) = \mathbf{W}'_k + \mu(-\nabla'_k + \mathbf{N}'_k). \tag{3.51}$$

This equation can be written in terms of $\mathbf{V}'_k$ in the following way:

$$\mathbf{V}'_{k+1} = \mathbf{V}'_k + \mu(-2\mathbf{\Lambda}\mathbf{V}'_k + \mathbf{N}'_k). \tag{3.52}$$

Note once again that, since the components of $\mathbf{N}'_k$ are mutually uncorrelated and since (3.52) is diagonalized, the components of noise in $\mathbf{V}'_k$ will also be mutually uncorrelated.

Near the minimum point of the error surface in steady state after adaptive transients have died out, the mean of $\mathbf{V}'_k$ is zero, and the covariance of the weight-vector noise may be obtained as follows. Postmultiplying both sides of (3.52) by their transposes and taking expected values yields

$$\begin{aligned}E[\mathbf{V}'_{k+1}\mathbf{V}'^T_{k+1}] = {} & E[(\mathbf{I}-2\mu\mathbf{\Lambda})\mathbf{V}'_k\mathbf{V}'^T_k(\mathbf{I}-2\mu\mathbf{\Lambda})] + \mu^2E[\mathbf{N}'_k\mathbf{N}'^T_k] \\ & +\mu E[\mathbf{N}'_k\mathbf{V}'^T_k(\mathbf{I}-2\mu\mathbf{\Lambda})] + \mu E[(\mathbf{I}-2\mu\mathbf{\Lambda})\mathbf{V}'_k\mathbf{N}'^T_k].\end{aligned} \tag{3.53}$$

It has been assumed that the input vector $\mathbf{X}_k$ is uncorrelated over time; the gradient noise $\mathbf{N}_k$ is accordingly uncorrelated with the weight vector $\mathbf{W}_k$, and therefore $\mathbf{N}'_k$ and $\mathbf{V}'_k$ are uncorrelated. Equation (3.53) can thus be expressed as

$$\begin{aligned}E[\mathbf{V}'_{k+1}\mathbf{V}'^T_{k+1}] = {} & (\mathbf{I}-2\mu\mathbf{\Lambda})E[\mathbf{V}'_k\mathbf{V}'^T_k](\mathbf{I}-2\mu\mathbf{\Lambda}) \\ & +\mu^2E[\mathbf{N}'_k\mathbf{N}'^T_k].\end{aligned} \tag{3.54}$$

Furthermore, if $\mathbf{V}'_k$ is stationary, the covariance of $\mathbf{V}'_{k+1}$ is equal to the covariance of $\mathbf{V}'_k$, which may be expressed as

$$\text{cov}[\mathbf{V}'_k] = (\mathbf{I} - 2\mu\mathbf{\Lambda})\text{cov}[\mathbf{V}'_k](\mathbf{I} - 2\mu\mathbf{\Lambda}) + \mu^2\text{cov}[\mathbf{N}'_k]. \tag{3.55}$$

Since the noise components of $\mathbf{V}'_k$ are mutually uncorrelated, (3.55) is diagonal. It can thus be rewritten as

$$\text{cov}[\mathbf{V}'_k] = (\mathbf{I} - 2\mu\mathbf{\Lambda})^2\text{cov}[\mathbf{V}'_k] + \mu^2(4\xi_{\text{min}}\mathbf{\Lambda}) \tag{3.56}$$

or

$$(\mathbf{I} - \mu\mathbf{\Lambda})\text{cov}[\mathbf{V}'_k] = \mu\xi_{\text{min}}\mathbf{I}. \tag{3.57}$$

When the value of the adaptive constant μ is chosen to be small (as is consistent with a converged solution near the minimum point of the error surface),

$$\mu\mathbf{\Lambda} \ll \mathbf{I}. \tag{3.58}$$

Equation (3.57) thus becomes

$$\text{cov}[\mathbf{V}'_k] = \mu\xi_{\text{min}}\mathbf{I}. \tag{3.59}$$

The covariance of $\mathbf{V}_k$ can now be expressed as follows:

$$\begin{aligned} \text{cov}[\mathbf{V}_k] &= E[\mathbf{V}_k\mathbf{V}_k^T] = E[\mathbf{Q}\mathbf{V}'_k\mathbf{V}'^T_k\mathbf{Q}^{-1}] \\ &= \mathbf{Q}\text{cov}[\mathbf{V}'_k]\mathbf{Q}^{-1} = \mu\xi_{\text{min}}\mathbf{I}. \end{aligned} \tag{3.60}$$

From this we conclude that the components of the weight-vector noise are all of the same variance and are mutually uncorrelated. This derivation of the covariance depends on the assumptions made above and embodied in Eqs. (3.44), (3.46), (3.47), and (3.58). It has been found by experience, however, that (3.60) closely approximates the exact covariance of the weight-vector noise under a considerably wider range of conditions than these assumptions imply. A derivation of bounds on the covariance based on fewer assumptions has been made by Kim and Davisson [33].

3.8 MISADJUSTMENT DUE TO GRADIENT NOISE

Random noise in the weight vector causes an excess MSE. If the weight vector were noise free and converged so that $\mathbf{W}_k = \mathbf{W}^*$, then the MSE would be ξ_{min}. However, this does not occur in actual practice. Because of gradient noise, the weight vector will be noisy and, on the average, will be misadjusted from its optimal setting. The weight vector undergoes Brownian motion about the bottom of the MSE bowl, causing the average MSE to be greater than ξ_{min}.

An expression for MSE in terms of $\mathbf{V}'_k$ is given by (3.40), from which we obtain an expression for excess MSE due to weight vector noise:

$$(\text{excess MSE}) = \mathbf{V}'^T_k\mathbf{\Lambda}\mathbf{V}'_k. \tag{3.61}$$

The average excess MSE is an important quantity. It can be expressed as

$$E[\mathbf{V}'^T_k\mathbf{\Lambda}\mathbf{V}'_k] = \sum_{p=1}^{n}\lambda_p E[(v'_{p_k})^2]. \tag{3.62}$$

where v'_{p_k} is the pth component of $\mathbf{V}'_k$. After adaptive transients die out, $E[\mathbf{V}'_k] = 0$. Therefore, from (3.59) we have

$$E[(v'_{p_k})^2] = \mu\xi_{\min}, \quad \forall p. \tag{3.63}$$

Substitution into (3.62) yields the average excess MSE,

$$E[\mathbf{V}'^T_k \mathbf{\Lambda} \mathbf{V}'_k] = \mu\xi_{\min} \sum_{p=1}^{n} \lambda_p = \mu\xi_{\min} \operatorname{tr} \mathbf{R}. \tag{3.64}$$

We define the *misadjustment* due to gradient noise as the dimensionless ratio of the average excess MSE to the minimum MSE,

$$M \triangleq \frac{\text{average excess MSE}}{\xi_{\min}}. \tag{3.65}$$

For the LMS algorithm, under the conditions assumed above,

$$M = \mu \operatorname{tr} \mathbf{R}. \tag{3.66}$$

This formula works well for small values of misadjustment, 25 percent or less, so that the assumption

$$\mathbf{W}_k \simeq \mathbf{W}^* \tag{3.67}$$

is satisfied. The misadjustment is a useful measure of the cost of adaptability. A value of $M = 10$ percent means that the adaptive system has an MSE only 10 percent greater than $\xi_{\min}$.

It is useful to relate misadjustment to the speed of adaptation and the number of weights being adapted. Since tr $\mathbf{R}$ equals the sum of the eigenvalues,

$$M = \mu \sum_{p=1}^{n} \lambda_p = \mu n \lambda_{\text{ave}} \tag{3.68}$$

where λ_{ave} is the average of the eigenvalues. From (3.43),

$$\lambda_p = \frac{1}{4\mu}\left(\frac{1}{\tau_{p_{\text{mse}}}}\right) \quad \text{or} \quad \lambda_{\text{ave}} = \frac{1}{4\mu}\left(\frac{1}{\tau_{p_{\text{mse}}}}\right)_{\text{ave}}. \tag{3.69}$$

Substituting (3.69) into (3.68) yields

$$M = \frac{n}{4}\left(\frac{1}{\tau_{p_{\text{mse}}}}\right)_{\text{ave}}. \tag{3.70}$$

For the special case where all eigenvalues are equal, the learning curve has only one time constant τ_{mse}, and the misadjustment is given by

$$M = \frac{n}{4\tau_{\text{mse}}} = \mu \operatorname{tr} \mathbf{R}. \tag{3.71}$$

When the eigenvalues are sufficiently similar for the learning curve to be approximately fitted by a single exponential, its time constant may be applied to (3.71) to give an approximate value of M.

Since transients settle in about four time constants, Eq. (3.71) leads to a simple approximate rule of thumb: *The misadjustment approximately equals the number of weights divided by the settling time,* $4\tau_{\text{mse}}$. *The setting time is expressed in number of iteration cycles or number of sampling periods.* Years of experience with adaptive filters convinces one that a 10 percent misadjustment is satisfactory for many engineering designs. Operation with 10 percent misadjustment can generally be achieved with an adaptive settling time equal to 10 times the memory time span of the adaptive transversal filter. Adapting faster will cause more misadjustment. Adapting slower will result in less misadjustment.

3.9 A DESIGN EXAMPLE: CHOOSING NUMBER OF FILTER WEIGHTS FOR AN ADAPTIVE PREDICTOR

Figure 3.4 is a block diagram of an adaptive predictor.[3] Its adaptive filter converts the delayed input $x_{k-\Delta}$ into the undelayed input x_k as best possible in the least squares sense. If the adaptive filter weights are copied into an auxiliary filter having a tapped delay-line structure identical to that of the adaptive filter and the input x_k is applied without delay to this auxiliary filter, the resulting output will be a prediction of the input, a best linear least squares estimate of $x_{k+\Delta}$ (limited by finite filter length and misadjustment).

A computer implementation of the adaptive predictor was made using a simulated input signal x_k obtained by bandpass filtering a white Gaussian signal and adding this to another independent white Gaussian signal. Prediction was one time sample into the future, that is, $\Delta = 1$, using an adaptive filter with five weights, all initially set to zero.

[3]This same predictor was described by Widrow in [4]; it has been used for data compression and speech encoding [34] and for "maximum entropy" spectral estimation [35], [36].

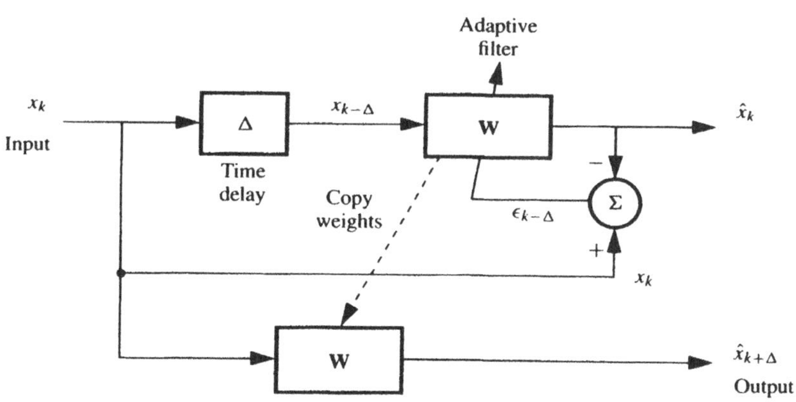

Figure 3.4 An adaptive predictor. From B. Widrow, J.M. McCool, M.G. Larimore, and C.R. Johnson, Jr., "Stationary and nonstationary learning characteristics of the LMS adaptive filter," *Proc. IEEE* ©, Vol. 64, No. 8 (August 1976), pp. 1151–1162.

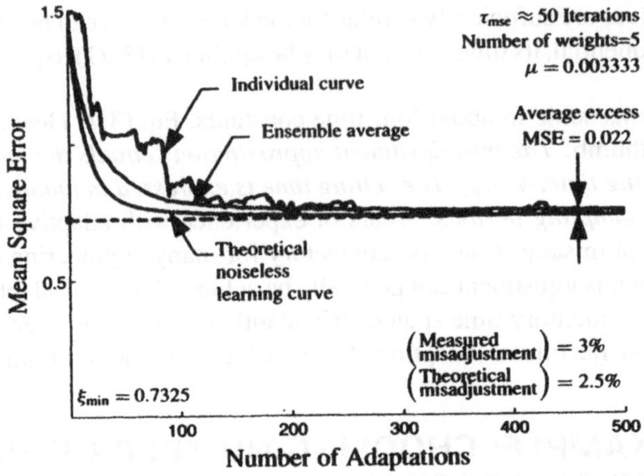

Figure 3.5 Learning curves for adaptive predictor. From B. WIDROW, J.M. MCCOOL, M.G. LARIMORE, and C.R. JOHNSON, Jr., "Stationary and nonstationary learning characteristics of the LMS adaptive filter," *Proc. IEEE* ©, Vol. 64, No. 8 (August 1976), pp. 1151–1162.

Figure 3.5 depicts three learning curves. For each adaptive step, the value of ξ_k corresponding to the current weight vector $\mathbf{W}_k$ was calculated from (3.13) using known values of $\mathbf{R}$ and ξ_{min}, giving the *individual learning curve*. The smooth *ensemble average learning curve* is simply the average of 200 such individual curves, and it approximates the adaptive behavior in the mean. The third curve calculated from (3.41) shows how the process would evolve if perfect knowledge of the gradient were available at each step. It is a noiseless *steepest-descent learning curve*.

Of particular interest is the residual difference, after convergence, between the ensemble average learning curve and the noiseless steepest-descent learning curve. The latter, of course, converges to ξ_{min}. The difference is the excess MSE due to gradient noise, in this case, giving a measured misadjustment of 3 percent. The theoretical misadjustment was $M = 2.5$ percent. The minor discrepancy was due mainly to the fact that the input samples were highly correlated, in violation of the assumption that $E[\mathbf{X}_k\mathbf{X}_{k+l}^T] = 0, \forall\ l \neq 0$, used in the derivation of misadjustment formula (3.71).

The ensemble average learning curve had an effective measured time constant τ_{mse} of about 50 iterations since it fell to within 2 percent of its converged value at around iteration 200.

When all eigenvalues are equal, Eq. (3.43) becomes

$$\tau_{mse} = \frac{1}{4\mu\lambda} = \frac{n}{4\mu \operatorname{tr} \mathbf{R}}. \tag{3.72}$$

Using (3.72) in the present case (although the eigenvalues ranged over a 10 to 1 ratio), we obtained $\tau_{mse} = 50$, which agreed with the experiment. Equation (3.72) gives a formula for an *effective time constant*, useful even when the eigenvalues are disparate.

The performance of the adaptive filter may have improved with an increase in the number of weights. However, for a fixed rate of convergence, larger numbers of weights increase misadjustment. Figure 3.6 shows these conflicting effects. The lowest curve, for

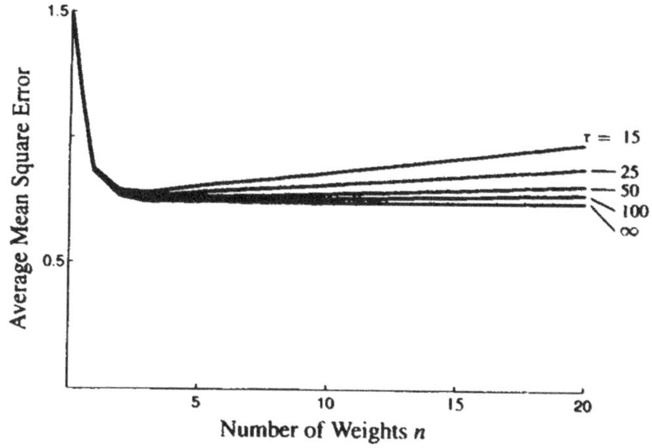

Figure 3.6 Performance versus number of weights and adaptive predictor time constant. From B. WIDROW et. al., "Stationary and nonstationary learning characteristics of the LMS adaptive filter," *Proc. IEEE* ©, Vol. 64, No. 8 (August 1976), pp. 1151–1162

$\tau_{mse} = \infty$, represents idealized noise-free adaptation providing the minimum MSE, $\xi_{min}(n)$, for each value of n. The other curves include average excess MSE due to gradient noise. We define the *average MSE* to be the sum of the minimum MSE and the average excess MSE. Thus

$$(\text{average MSE}) = [1 + M]\xi_{min}(n). \tag{3.73}$$

$\xi_{min}(n)$ goes down with increasing n, but M goes up in proportion to n. Using this formula, theoretical curves have been plotted in Fig. 3.6 for approximate values of τ_{mse} of 100, 50, 25, and 15 iterations. It is apparent from these curves that increasing the number of weights does not always guarantee improved system performance. Experimental points derived by computer simulation have compared very well with theoretical values predicted by (3.73). Typical results are summarized in Table 3.1.

TABLE 3.1 COMPARISON OF THEORETICAL AND EXPERIMENTAL ADAPTIVE PREDICTOR PERFORMANCE

Number of Weights n	Approximate Time Constant τ_{mse}	Average MSE		Misadjustment	
		Theoretical	Experimental	Theoretical	Experimental
5	100	0.742	0.751	1.3%	2.5%
5	50	0.751	0.754	2.5	3.0
5	25	0.769	0.781	5.0	6.6
5	15	0.794	0.824	8.3	12.6
10	100	0.737	0.745	2.5	3.5
10	50	0.755	0.764	5.0	6.2

Appendix A presents alternative derivations of stability bounds for LMS and its misadjustment. These derivations are more rigorous than those presented above. From them, we obtain bounds on μ that ensure stability of the mean and the variance of the weight vector, and formulas that predict misadjustment more accurately than (3.66) when μ is not necessarily small.

3.10 THE EFFICIENCY OF ADAPTIVE ALGORITHMS

We have analyzed the efficiency of the LMS algorithm from the point of view of misadjustment versus rate of adaptation. The question arises: Could another algorithm be devised that would produce less misadjustment for the same rate of adaptation?

Suppose that an adaptive linear combiner is fed N independent input data vectors $\mathbf{X}_1, \mathbf{X}_2, \ldots, \mathbf{X}_k, \ldots, \mathbf{X}_N$ drawn from a stationary ergodic process. Associated with each of these input vectors are their scalar desired responses $d_1, d_2, \cdots, d_N$, respectively, also drawn from a stationary ergodic process. Keeping the weights fixed, a set of N error equations can be written as

$$\epsilon_k = d_k - \mathbf{W}^T\mathbf{X}_k, \quad k = 1, 2, \cdots, N. \tag{3.74}$$

Let the objective be to find a weight vector that minimizes the sum of the squares of the error values based on a sample of N items of data.

Equation (3.74) can be written in matrix form for all of the error values as

$$\mathcal{E} = \mathbf{D} - \mathcal{X}\mathbf{W}, \tag{3.75}$$

where $\mathcal{X}$ is an $N \times n$ rectangular matrix

$$\mathcal{X} \triangleq [\mathbf{X}_1\mathbf{X}_2 \cdots \mathbf{X}_N]^T, \tag{3.76}$$

where $\mathcal{E}$ is an N-element error vector

$$\mathcal{E} \triangleq [\epsilon_1\epsilon_2 \cdots \epsilon_N]^T, \tag{3.77a}$$

and where $\mathbf{D}$ is an N-element desired-response vector

$$\mathbf{D} \triangleq [d_1 d_2 \cdots d_N]^T. \tag{3.77b}$$

A unique solution for the weight vector which brings $\mathcal{E}$ to zero exists only if $\mathcal{X}$ is square and nonsingular. However, the case of greatest interest is that of $N \gg n$. The sum of the squares of the errors is

$$\mathcal{E}^T\mathcal{E} = \mathbf{D}^T\mathbf{D} + \mathbf{W}^T\mathcal{X}^T\mathcal{X}\mathbf{W} - 2\mathbf{D}^T\mathcal{X}\mathbf{W}. \tag{3.78}$$

This sum when multiplied by $1/N$ is an estimate $\hat{\xi}$ of the MSE, ξ. Thus

$$\hat{\xi} = \frac{1}{N}\mathcal{E}^T\mathcal{E} \quad \text{and} \quad \lim_{N\to\infty} \hat{\xi} = \xi. \tag{3.79}$$

Note that $\hat{\xi}$ is a quadratic function of the weights. The parameters of the quadratic form (3.78) are related to properties of the N data samples. $(\mathcal{E}^T\mathcal{E})$ is square and positive definite or positive semidefinite. $\hat{\xi}$ is the small-sample-size MSE function, while ξ is the large-sample-size MSE function. These functions are sketched in Fig. 3.7.

The function $\hat{\xi}$ is minimized by setting its gradient to zero. We will use ∇ here as a symbol for a gradient operator. Accordingly:

$$\nabla\hat{\xi} = 2\mathcal{X}^T\mathcal{X}\mathbf{W} - 2\mathcal{X}^T\mathbf{D}. \tag{3.80}$$

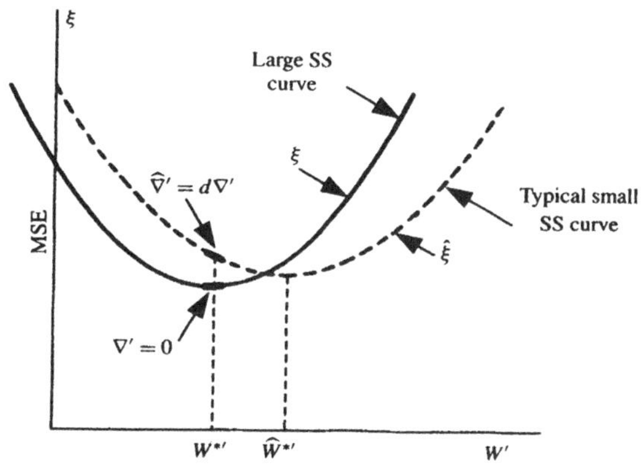

Figure 3.7 Small- and large-sample size MSE curves. From B. WIDROW et. al., "Stationary and nonstationary learning characteristics of the LMS adaptive filter," *Proc. IEEE* ©, Vol. 64, No. 8 (August 1976), pp. 1151–1162

The "optimal" weight vector based only on the N data samples is

$$\widehat{\mathbf{W}}^* = (\mathcal{X}^T\mathcal{X})^{-1}\mathcal{X}^T\mathbf{D}. \tag{3.81}$$

To obtain this result, it is necessary that $(\mathcal{X}^T\mathcal{X})$ be positive definite. Equation (3.81) is a small-sample-size Wiener-Hopf equation. This formula gives the position of the minimum of the small-sample-size bowl. The corresponding formula for the large-sample-size bowl is the Wiener-Hopf equation (3.10).

Making use of the N data samples, we could calculate $\widehat{\mathbf{W}}^*$ by making use of (3.81) by a training process such as LMS, or by some other least squares optimization procedure. Taking the first block of N data samples, we obtain a small-sample-size function $\hat{\xi}_1$ whose minimum is at $\widehat{\mathbf{W}}_1^*$. This could be repeated with a second data sample, giving a function $\hat{\xi}_2$ whose minimum is at $\widehat{\mathbf{W}}_2^*$, and so on. Typically, all the values of $\widehat{\mathbf{W}}^*$ would differ from the true optimum $\mathbf{W}^*$ and would, thereby, be misadjusted.

To analyze the misadjustment, assume that N is large and that the typical small-size curve approximately matches the large-sample-size curve. Therefore,

$$\hat{\xi} \approx \xi \text{ and } (\xi - \hat{\xi}) \triangleq d\xi. \tag{3.82}$$

The true large-sample-size function is

$$\xi = \xi_{\min} + \mathbf{V}'^T\mathbf{\Lambda}\mathbf{V}'. \tag{3.83}$$

The gradient of this function expressed in the primed coordinates is

$$\nabla' = 2\mathbf{\Lambda}\mathbf{V}'. \tag{3.84}$$

A differential deviation in the gradient is

$$(d\nabla') = 2\mathbf{\Lambda}(d\mathbf{V}') + 2(d\mathbf{\Lambda})\mathbf{V}'. \tag{3.85}$$

This deviation could represent the difference in gradients between small- and large-sample-size curves.

Refer to Fig. 3.7. Let $\mathbf{W}' = \mathbf{W}^{*\prime}$, then $\mathbf{V}' = 0$. The gradient of ξ is zero, while the gradient of $\hat{\xi}$ is $\widehat{\nabla}' = d\widehat{\nabla}'$. Using (3.85)

$$(d\nabla') = 2\mathbf{\Lambda}(d\mathbf{V}'). \tag{3.86}$$

From (3.86), the deviation in gradient can be linked to the deviation in position of the small-sample-size curve minimum. Since $(d\mathbf{V}') = (\mathbf{W}^{*\prime} - \widehat{\mathbf{W}}^{*\prime})$, taking averages of (3.86) over an ensemble of small-sample-size curves,

$$\text{cov}[d\nabla'] = 4\mathbf{\Lambda}\text{cov}[d\mathbf{V}']\mathbf{\Lambda}. \tag{3.87}$$

Equation (3.50) indicates that when the gradient is taken from a single sample of the error ϵ_k, the covariance of the gradient noise when $\mathbf{W}' = \mathbf{W}^{*\prime}$ is given by $4\xi_{\min}\mathbf{\Lambda}$. If the gradient were estimated under the same conditions but using N independent error samples, this covariance would be

$$\text{cov}[d\nabla'] = \frac{4}{N}\xi_{\min}\mathbf{\Lambda}. \tag{3.88}$$

Substituting (3.88) into (3.87) yields

$$\text{cov}[d\mathbf{V}'] = \frac{1}{N}\xi_{\min}\mathbf{\Lambda}^{-1}. \tag{3.89}$$

The average excess MSE, an ensemble average, is

$$(\text{average excess MSE}) = E[(d\mathbf{V}')^T\mathbf{\Lambda}(d\mathbf{V}')]. \tag{3.90}$$

Equation (3.89) shows $\text{cov}[d\mathbf{V}']$ to be diagonal, so that

$$(\text{average excess MSE}) = \frac{n}{N}\xi_{\min}. \tag{3.91}$$

Normalizing this with respect to $\xi_{\min}$ gives the misadjustment:

$$M = \frac{n}{N} = \frac{(\text{number of weights})}{(\text{number of independent training samples})}. \tag{3.92}$$

This formula was first presented by Widrow and Hoff [1] in 1960. It has been used for many years in pattern recognition studies. For small values of M (less than 25 percent), it has proven to be very useful. A formula similar to (3.92), based on somewhat different assumptions, was derived by Davisson [37] in 1970.

Although Eq. (3.92) has been derived for training with finite blocks of data, it can be used to assess the efficiency of steady-flow algorithms. Consider an adaptive transversal filter with stationary stochastic inputs, adapted by the LMS algorithm. For simplicity, let all eigenvalues of the input correlation matrix $\mathbf{R}$ be equal. As such, from (3.71),

$$M = \frac{n}{4\tau_{\text{mse}}}. \tag{3.93}$$

The LMS algorithm exponentially weights its input data over time in determining current weight values. If an equivalent uniform averaging window is assumed equal to the adaptive settling time, approximately four time constants, the equivalent data sample taken at any instant by LMS is essentially $N\text{eq} = 4\tau_{\text{mse}}$ samples. Accordingly, for the LMS algorithm,

$$M = \frac{n}{N\text{eq}} = \frac{(\text{number of weights})}{(\text{number of independent training samples})}. \tag{3.94}$$

A comparison of (3.92) and (3.94) shows that when eigenvalues are equal, LMS is about as efficient as a least squares algorithm can be.[4] However, with disparate eigenvalues, the misadjustment is primarily determined by the fastest modes while settling time is limited by the slowest modes. To sustain efficiency with disparate eigenvalues, algorithms similar to LMS have been devised based on Newton's method rather than on steepest descent [39], [40], [41]. Such algorithms premultiply the gradient estimate each iteration cycle by an estimate of the inverse of **R**:

$$\mathbf{W}_{k+1} = \mathbf{W}_k - \mu\widehat{\mathbf{R}^{-1}}\widehat{\nabla}_k, \tag{3.95}$$

or

$$\mathbf{W}_{k+1} = \mathbf{W}_k + 2\mu\widehat{\mathbf{R}^{-1}}\epsilon_k\mathbf{X}_k. \tag{3.96}$$

This algorithm is a steady-state form of *recursive least squares* (RLS). The matrix $\widehat{R^{-1}}$ is recursively updated from the input data [40]. The process causes all adaptive modes to have essentially the same time constant. Algorithms based on this principle are potentially more efficient than LMS but are more difficult to implement. RLS algorithms are described briefly in Appendix E.

3.11 ADAPTIVE NOISE CANCELING: A PRACTICAL APPLICATION FOR ADAPTIVE FILTERING

Many practical applications exist for adaptive FIR filters of the type described in this chapter. Among the more interesting of these applications is adaptive noise canceling. The idea is to subtract noise from a noisy signal to obtain a signal with greatly reduced noise. When done properly with adaptive filtering techniques, substantial improvements in signal-to-noise ratio can be achieved.

Separating a signal from additive noise, even when their respective power spectra overlap, is a common problem in signal processing. Figure 3.8(a) suggests a classical approach to this problem using optimal Wiener or Kalman filtering [42]. The purpose of the optimal filter is to pass the signal s without distortion while stopping the noise n_0. In general, this cannot be done perfectly. Even with the best filter, the signal is distorted, and some noise does go through to the output.

Figure 3.8(b) shows another approach to the problem, using adaptive noise canceling. A *primary input* contains the signal of interest s, and an additive noise n. A *noise reference input* is assumed to be available containing n_1, which is correlated with the original corrupting noise n_0. In Fig. 3.8(b), the adaptive filter receives the reference noise, filters it, and subtracts the result from the primary input. From the point of view of the adaptive filter, the primary input $(s + n_0)$ acts as its desired response and the system output acts as its error. The noise canceler output is obtained by subtracting the filtered reference noise from the primary input. Adaptive noise canceling generally performs better than the classical approach since the noise is subtracted out rather than filtered out. In order to do adaptive noise

[4]Other algorithms based on LMS have been devised to give fast convergence and low misadjustment. They employ a variable μ [38]. Initial values of μ are chosen high for rapid convergence; final values of μ are chosen low for small misadjustment. This works well as long as input statistics are stationary. This procedure and the methods of stochastic approximation on which it is based will not perform well in the nonstationary case however.

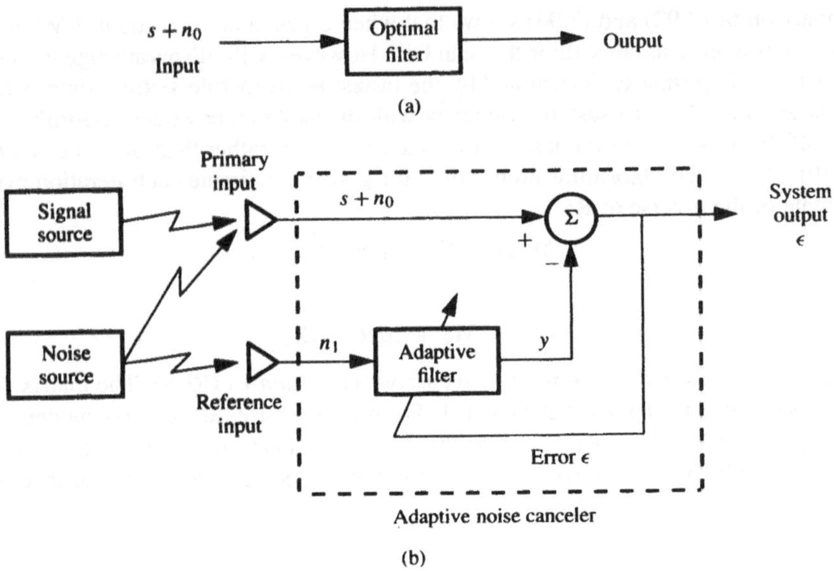

Figure 3.8 Separation of signal and noise: (a) classical approach; (b) adaptive noise-canceling approach. From B. WIDROW et al., "Adaptive noise canceling: Principles and applications," *Proc. IEEE* ©, Vol. 63, No. 12 (December 1975), pp. 1692–1716

canceling, one might expect that some prior knowledge of the signal s or of the noises n_0 and n_1 would be needed by the filter before it could adapt to produce the noise-canceling signal y. A simple argument will show, however, that little or no prior knowledge of s, n_0, or n_1 or of their interrelationships is required, except that n_1 is correlated with n_0 and that s is uncorrelated with both n_0 and n_1.

Assume that s, n_0, n_1, and y are statistically stationary and have zero means. Assume that s is uncorrelated with n_0 and n_1 and suppose that n_1 is correlated with n_0. The output is

$$\epsilon = s + n_0 - y. \tag{3.97}$$

Squaring, one obtains

$$\epsilon^2 = s^2 + (n_0 - y)^2 + 2s(n_0 - y). \tag{3.98}$$

Taking expectations of both sides of Eq. (3.98) and realizing that s is uncorrelated with n_0 and with y, yields

$$\begin{aligned} E[\epsilon^2] &= E[s^2] + E[(n_0 - y)^2] + 2E[s(n_0 - y)] \\ &= E[s^2] + E[(n_0 - y)^2]. \end{aligned} \tag{3.99}$$

Adapting the filter to minimize $E[\epsilon^2]$ will not affect the signal power $E[s^2]$. Accordingly, the minimum output power is

$$\big(E[\epsilon^2]\big)_{\min} = E[s^2] + \big(E[(n_0 - y)^2]\big)_{\min}. \tag{3.100}$$

When the filter is adjusted so that $E[\epsilon^2]$ is minimized, $E[(n_0 - y)^2]$ is therefore also minimized. The filter output y is then a best least squares estimate of the primary noise n_0. Moreover, when $E[(n_0 - y)^2]$ is minimized, $E[(\epsilon - s)^2]$ is also minimized, since, from Eq. (3.97)

$$(\epsilon - s) = (n_0 - y). \tag{3.101}$$

Adjusting or adapting the filter to minimize the total output power is thus tantamount to causing the output ϵ to be a best least squares estimate of the signal s, for the given structure and adjustability of the adaptive filter and for the given reference input. The output ϵ will contain the signal s plus noise. From (3.101) the output noise is given by $(n_0 - y)$. Since minimizing $E[\epsilon^2]$ minimizes $E[(n_0 - y)^2]$, *minimizing the total output power minimizes the output noise power*. Since the signal in the output remains constant, *minimizing the total output power maximizes the output signal-to-noise ratio*.

Adaptive noise canceling has proven to be a very powerful technique for signal processing. The first application was made to the problem of canceling unwanted 60-Hz (or 50-Hz) interference in electrocardiography [43]. The causes of such interference are magnetic induction and displacement currents from the power line, and ground loops. Conventional filtering has been used to combat 60-Hz interference, but the best approach seems to be adaptive noise canceling.

Figure 3.9 shows the application of adaptive noise canceling in electrocardiography. The primary input is taken from the ECG preamplifier; the 60-Hz reference input is taken from a wall outlet. The adaptive filter contains two variable weights, one applied to the reference input directly and the other to a version of it shifted in phase by 90°. The two weighted versions of the reference are summed to form the filter's output, which is subtracted from the primary input. Selected combinations of the values of the weights allow the reference waveform to be changed in magnitude and phase in any way required for optimal cancelation. Two variable weights, or two *degrees of freedom*, are required to cancel the single pure sinusoid.

A typical result of an experiment performed by digital processing is shown in Fig. 3.10. Sample resolution was 10 bits, and sampling rate was 1,000 Hz. Figure 3.10(a) shows the primary input, an electrocardiographic waveform with an excessive amount of 60-Hz interference, and Fig. 3.10(b) shows the reference input from the wall outlet. Figure 3.10(c) is the noise canceler output. Note the absence of interference and the clarity of detail once the adaptive process has converged.

Another useful application of adaptive noise canceling is one that involves canceling interference from the mother's heart when attempting to record clear fetal electrocardiograms [43]. Figure 3.11 shows the location of the fetal and maternal hearts and the placement of the input leads. The abdominal leads provide the primary input (containing fetal ECG and interfering maternal ECG signals), and the chest leads provide multiple reference inputs (containing pure interference, that is, the maternal ECG). Figure 3.12 shows the configuration of the adaptive noise canceler. This is a multiple-reference noise canceler. It works much like a single-channel canceler. Figure 3.13 shows the results. The maternal ECG from the chest leads was adaptively filtered and subtracted from the abdominal signal, leaving the fetal ECG. This was an interesting problem since the fetal and maternal ECG signals had spectral overlap. The two hearts were electrically isolated and worked independently, but the second harmonic frequency of the maternal ECG was close to the fundamen-

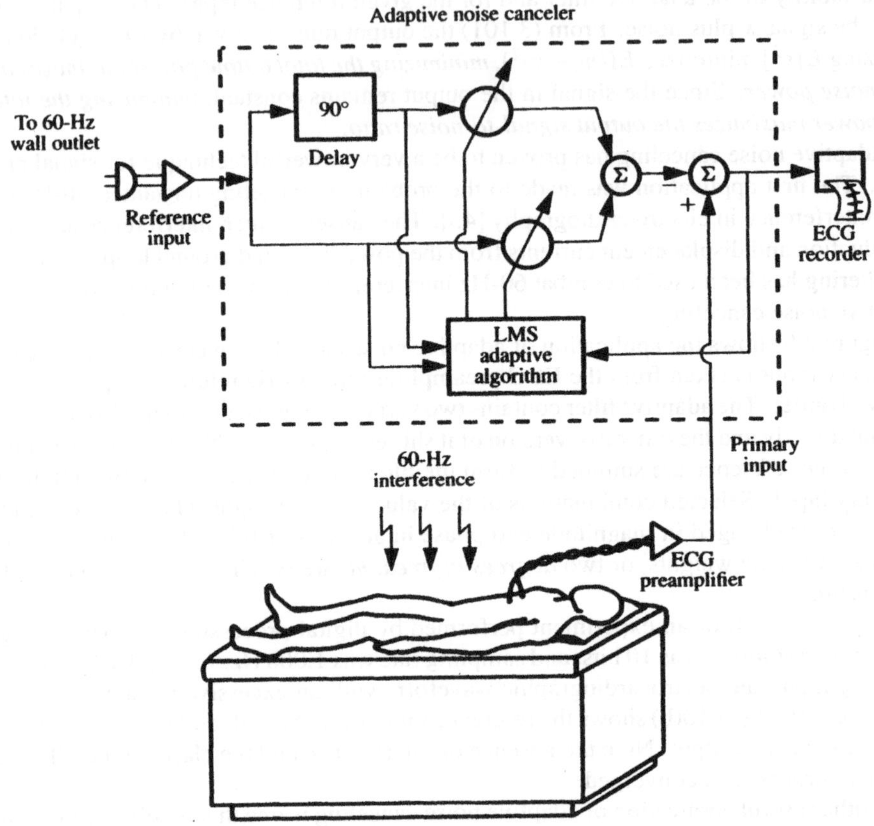

Figure 3.9 Canceling 60-Hz interference in electrocardiography. From B. WIDROW et. al., "Adaptive noise canceling: Principles and applications," *Proc. IEEE* ©, Vol. 63, No. 12 (December 1975), pp. 1692–1716.

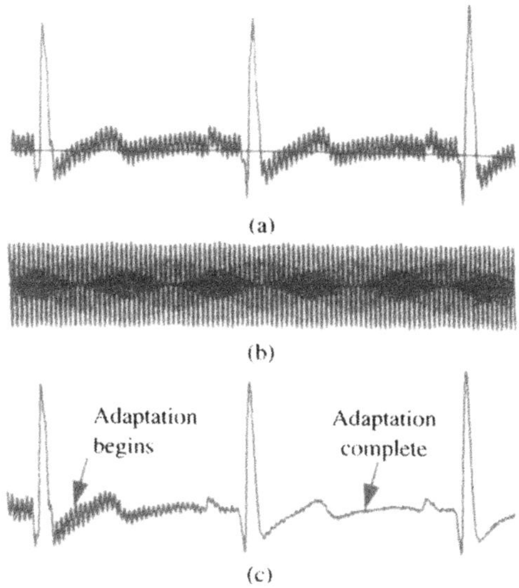

Figure 3.10 Result of electrocardiographic noise-canceling experiment: (a) primary input; (b) reference input; (c) noise canceler output. From B. WIDROW ET. AL., "Adaptive noise canceling: Principles and applications," *Proc. IEEE* ©, Vol. 63, No. 12 (December 1975), pp. 1692–1716.

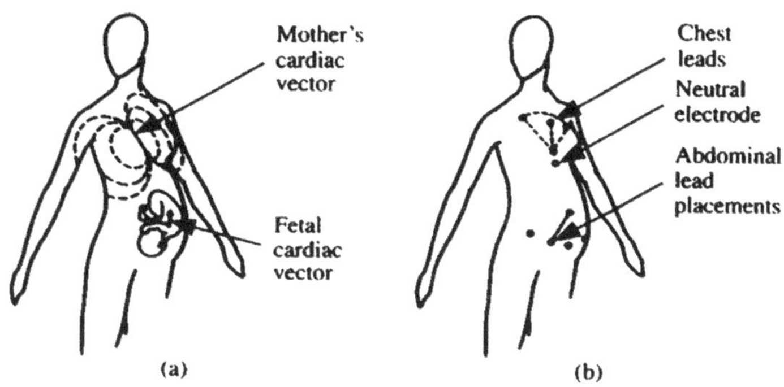

Figure 3.11 Canceling maternal heartbeat in fetal electrocardiography: (a) cardiac electric field vectors of mother and fetus; (b) placement of leads. From B. WIDROW ET. AL., "Adaptive noise canceling: Principles and applications," *Proc. IEEE* ©, Vol. 63, No. 12 (December 1975), pp. 1692–1716.

tal of the fetal ECG. Ordinary filtering techniques would have had great difficulty with this problem.

3.12 SUMMARY

This chapter described the performance characteristics of the LMS adaptive filter, a digital filter composed of a tapped delay line and adjustable weights, whose impulse response is controlled by an adaptive algorithm. For stationary stochastic inputs, the mean square error,

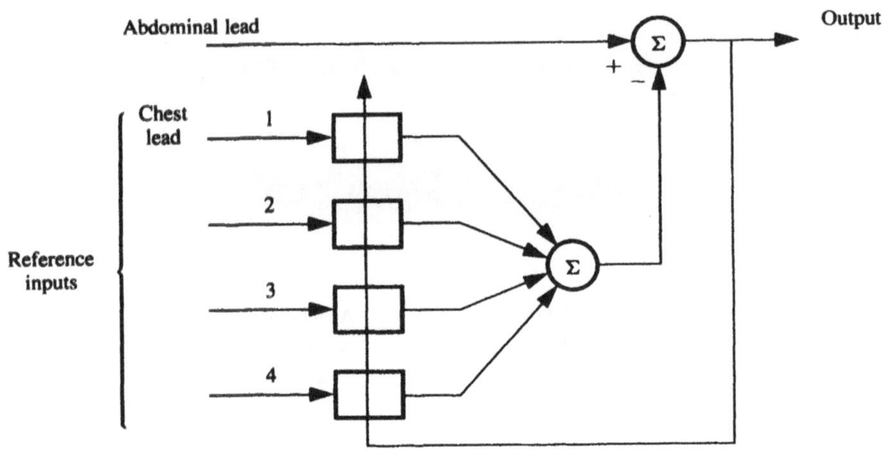

Figure 3.12 Multiple-reference noise canceler used in fetal ECG experiment. From B. WIDROW ET. AL., "Adaptive noise canceling: Principles and applications," *Proc. IEEE* ©, Vol. 63, No. 12 (December 1975), pp. 1692–1716.

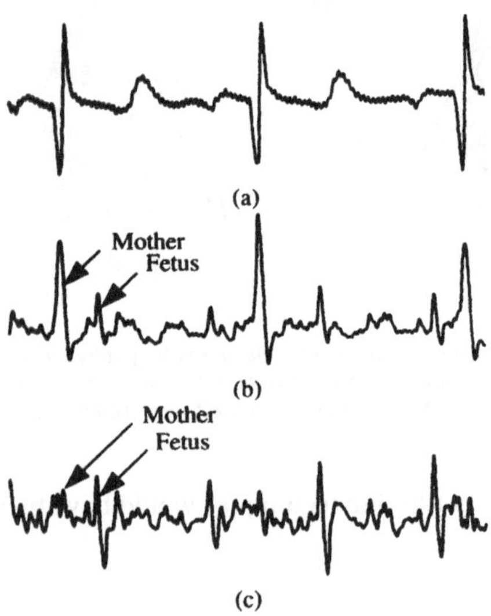

Figure 3.13 Result of fetal ECG experiment (bandwidth, 3–35 Hz; sampling rate, 256 Hz): (a) reference input (chest lead); (b) primary input (abdominal lead); (c) noise canceler output. From B. WIDROW ET. AL., "Adaptive noise canceling: Principles and applications," *Proc. IEEE* ©, Vol. 63, No. 12 (December 1975), pp. 1692–1716.

the difference between the filter output and an externally supplied input called the *desired response*, is a quadratic function of the weights, a paraboloid with a single fixed minimum point that can be sought by gradient techniques. The gradient estimation process is shown to introduce noise into the weight vector that is proportional to the speed of adaptation and number of weights. The effect of this noise is expressed in terms of a dimensionless quantity *misadjustment* that is a measure of the deviation from optimal Wiener performance. It is further shown that for stationary inputs the LMS adaptive algorithm, based on the method of steepest descent, approaches the theoretical limit of efficiency in terms of misadjustment and speed of adaptation when the eigenvalues of the input correlation matrix are equal or close in value. When the eigenvalues are highly disparate ($\lambda_{max}/\lambda_{min} > 10$), an algorithm similar to LMS but based on Newton's method would approach this theoretical limit very closely.

For an adaptive linear combiner or an adaptive FIR filter, the Wiener weight vector is

$$\mathbf{W}^* = \mathbf{R}^{-1}\mathbf{P}. \tag{3.10}$$

The mean square error is

$$\xi = \xi_{\min} + \mathbf{V}^T\mathbf{R}\mathbf{V}. \tag{3.15}$$

The minimum MSE is

$$\xi_{\min} = E[d_k^2] - \mathbf{P}^T\mathbf{W}^*. \tag{3.12}$$

When adapting with the method of steepest descent, the time constant for the pth mode in the weights as they relax toward the Wiener solution is

$$\tau_p = \frac{1}{2\mu\lambda_p}, \quad \text{small } \mu. \tag{3.31}$$

The time constant for the pth mode of the mean square error as it relaxes toward $\xi_{\min}$ is

$$\tau_{p_{mse}} = \frac{1}{4\mu\lambda_p}, \quad \text{small } \mu. \tag{3.43}$$

The LMS algorithm is based on the method of steepest descent and it uses instantaneous gradient estimates based on single samples of data. LMS is given by

$$\mathbf{W}_{k+1} = \mathbf{W}_k + 2\mu\epsilon_k\mathbf{X}_k. \tag{3.35}$$

The condition on the convergence parameter μ for stability of the variance of the weight vector when adapting with LMS is

$$\frac{1}{\operatorname{tr}\mathbf{R}} > \mu > 0. \tag{3.39}$$

When adapting with the LMS algorithm, the misadjustment is

$$M = \mu \operatorname{tr}\mathbf{R}. \tag{3.66}$$

Processes involving adaptation with a finite number N of data samples have a sum of square of errors given by

$$\mathcal{E}^T\mathcal{E} = \mathbf{D}^T\mathbf{D} + \mathbf{W}^T\mathcal{X}^T\mathcal{X}\mathbf{W} - 2\mathbf{D}^T\mathcal{X}\mathbf{W}. \tag{3.78}$$

Mean square error is estimated by

$$\hat{\xi} = \frac{1}{N}\mathcal{E}^T\mathcal{E}, \quad \text{and} \quad \lim_{N\to\infty} \hat{\xi} = \xi. \tag{3.79}$$

A best linear least squares *Wiener* solution based on the finite data sample is

$$\widehat{\mathbf{W}}^* = (\mathcal{X}^T\mathcal{X})^{-1}\mathcal{X}^T\mathbf{D}. \tag{3.81}$$

Misadjustment when adapting with a finite data sample is

$$M = \frac{n}{N} = \frac{(\text{number of weights})}{(\text{number of independent training samples})}. \tag{3.92}$$

This formula applies when the weight vector is chosen to be optimal for the given training data sample. The weight vector goes to the bottom of the *small-sample-size* MSE function. Since MSE is minimized for the finite data sample, the data are used with maximal efficiency. The formula for misadjustment of an FIR filter in steady flow with the LMS adaptation is very similar to (3.92) when the eigenvalues are equal. This suggests that LMS is as efficient in its use of data as an algorithm can be when the eigenvalues are equal or approximately so. With highly disparate eigenvalues, an LMS algorithm based on Newton's method could be the most efficient. Such an algorithm would be

$$\mathbf{W}_{k+1} = \mathbf{W}_k + 2\mu\widehat{\mathbf{R}^{-1}}\epsilon_k\mathbf{X}_k. \tag{3.96}$$

This is a form of recursive least squares, RLS. Good methods for calculating $\widehat{\mathbf{R}^{-1}}$ directly from the input data are given in [41] and are discussed in Appendix E.

Bibliography for Chapter 3

[1] B. Widrow, and M.E. Hoff, "Adaptive switching circuits," in *IRE WESCON Conv. Rec.*, 1960, Pt. 4, pp. 96–104.

[2] B. Widrow, and J.M. McCool, "A comparison of adaptive algorithms based on the methods of steepest descent and random search," *IEEE Trans. Antennas Propag.*, Vol. AP-24, No. 5 (September 1976), pp. 615–637.

[3] B. Widrow, J.M. McCool, M.G. Larimore, and C.R. Johnson, Jr., "Stationary and nonstationary learning characteristics of the LMS adaptive filter," *Proc. IEEE*, Vol. 64, No. 8 (August 1976), pp. 1151–1162.

[4] B. Widrow, "Adaptive filters," in *Aspects of network and system theory*, ed. R.E. Kalman and N. De Claris (New York: Holt, Rinehart and Winston, 1970), pp. 563–587.

[5] B. Widrow, and E. Walach, "On the statistical efficiency of the LMS algorithm with nonstationary inputs," *IEEE Trans. Info. Theory — Special Issue on Adaptive Filtering*, Vol. 30, No. 2, Pt. 1 (March 1984), pp. 211–221.

[6] See references [1–58] of Chapter 1.

[7] See references [1–20] of Chapter 1.

[8] J. KOFORD, and G. GRONER, "The use of an adaptive threshold element to design a linear optimal pattern classifier," *IEEE Trans. Info. Theory*, Vol. IT-12 (January 1966), pp. 42–50.

[9] N. NILSSON, *Learning machines* (New York: McGraw-Hill, 1965).

[10] F.W. SMITH, "Design of quasi-optimal minimum-time controllers," *IEEE Trans. Autom. Control*, Vol. AC-11 (January 1966), pp. 71–77.

[11] B. WIDROW, "Adaptive model control applied to real-time blood-pressure regulation," in *Pattern Recognition and Machine Learning, Proc. Japan-U.S. Seminar on the Learning Process in Control Systems*, ed. K.S. Fu (New York: Plenum Press, 1971), pp. 310–324.

[12] See references [21–30] of Chapter 1.

[13] See references [40–46] of Chapter 1.

[14] P.E. MANTEY, "Convergent automatic-synthesis procedures for sampled-data networks with feedback," TR No. 7663-1, Stanford Electronics Laboratories, Stanford, CA, October 1964.

[15] P.M. LION, "Rapid identification of linear and nonlinear systems," in *Proc. 1966 JACC*, Seattle, WA, August 1966, pp. 605–615, also *AIAA Journal*, Vol. 5 (October 1967), pp. 1835–1842.

[16] R.E. ROSS and G.M. LANCE, "An approximate steepest descent method for parameter identification," in *Proc. 1969 JACC*, Boulder, CO, August 1969, pp. 483–487.

[17] R. HASTINGS-JAMES, and M.W. SAGE, "Recursive generalized-least-squares procedure for online identification of process parameters," *Proc. IEEE*, Vol. 116 (December 1969), pp. 2057–2062.

[18] A.C. SOUDACK, K.L. SURYANARAYANAN, and S.G. RAO, "A unified approach to discrete-time systems identification," *Int'l. J. Control*, Vol. 14, No. 6 (December 1971), pp. 1009–1029.

[19] J.M. MENDEL, *Discrete techniques of parameter estimation: The equation error formulation* (New York: Marcel Dekker, Inc., 1973).

[20] S.J. MERHAV, and E. GABAY, "Convergence properties in linear parameter tracking systems," *Identification and System Parameter Estimation — Part 2, Proc. 3rd IFAC Symp.*, ed. P. Eykhoff (New York: American Elsevier Publishing Co., Inc., 1973), pp. 745–750.

[21] L. LJUNG, and T. SODERSTROM, *Theory and practice of recursive identification* (Cambridge, MA: M.I.T. Press, 1983).

2. Another source of mismatch is due to inadequacies that may exist in the plant input signal, which may not excite all the important plant modes. This can be cured by using a "persistently exciting" dither signal added to the plant input. Dither helps to make the modeling process sure and solid but has the disadvantage of introducing an additional disturbance into the control system.

3. A third source of mismatch is noise in the model weights due to the adaptive process. Finite amounts of data are used by the adaptive process to determine the model weights. Only if the adaptive process were done infinitely slowly, using an infinite amount of real-time data, would there be no weight noise. Fast adaptation results in noise in the weights of $\widehat{\mathbf{P}}_k$.

Neglecting all of these limitations for the moment (they will be addressed below), we proceed to examine idealized plant modeling in the presence of plant noise.

4.1 IDEALIZED MODELING PERFORMANCE

The following analysis of adaptive modeling is idealized in the sense that the adaptive model is assumed to be of infinite length like the plant itself, and it is assumed that all plant modes are excited by the statistically stationary plant input signal. This signal is assumed to be persistently exciting [17] [20]. The adaptation process is assumed to have converged, with no noise present in the adaptive weights.

The adaptive model has an input u_k and it delivers an output $\hat{y}_k$, an estimate of the plant dynamic output response y_k. When converged, the optimized transfer function of the adaptive model will be $\widehat{P}^*(z)$. Its impulse response in vector form will be designated by $\widehat{\mathbf{P}}^*$. A formula for $\widehat{P}^*(z)$ can be obtained from Wiener theory as follows.

Since the impulse response corresponding to $\widehat{P}^*(z)$ is by assumption not restricted to be either causal or finite in length, the unconstrained Wiener solution, from Eq. (2.30), is

$$\widehat{P}^*(z) = \frac{\Phi_{uz}(z)}{\Phi_{uu}(z)}, \tag{4.6}$$

where $\Phi_{uz}(z)$ is the z-transform of the crosscorrelation function $\phi_{uz}(k)$,

$$\begin{aligned} \phi_{uz}(k) &= E[u_j z_{j+k}] \\ \Phi_{uz}(z) &= \sum_{k=-\infty}^{\infty} \phi_{uz}(k) z^{-k}, \end{aligned} \tag{4.7}$$

and where $\Phi_{uu}(z)$ is the z-transform of the autocorrelation function $\phi_{uu}(k)$, as follows:

$$\begin{aligned} \phi_{uu}(k) &= E[u_j u_{j+k}] \\ \Phi_{uu}(z) &= \sum_{k=-\infty}^{\infty} \phi_{uu}(k) z^{-k}. \end{aligned} \tag{4.8}$$

Assume that the plant disturbance n_k is uncorrelated with the plant input u_k and with the plant dynamic output response y_k. As such,

$$\phi_{uz}(k) = E[u_j(y_{j+k} + n_{j+k})] = \phi_{uy}(k). \tag{4.9}$$

Transforming both sides,

$$\Phi_{uz}(z) = \Phi_{uy}(z). \tag{4.10}$$

Substituting into (4.6) yields

$$\widehat{P}^*(z) = \frac{\Phi_{uy}(z)}{\Phi_{uu}(z)}. \tag{4.11}$$

The transform $\Phi_{uy}(z)$ can be expressed in terms of $\Phi_{uu}(z)$ and the plant transfer function $P(z)$. The relation is

$$\Phi_{uy}(z) = \Phi_{uu}(z)P(z). \tag{4.12}$$

Substituting into (4.11) gives

$$\widehat{P}^*(z) = P(z). \tag{4.13}$$

We conclude that in spite of the presence of plant disturbance, the least squares adaptive model will, under the above assumptions, develop a transfer function equal to that of the plant. This result will hold as long as the plant disturbance is uncorrelated with the plant input signal.

The true desired response of the adaptive model is y_k, but this is, of course, unavailable. In Fig. 4.1, the training signal available to the adaptive process as a *desired response* input is the disturbed plant output z_k. Training with z_k gives the same Wiener solution that would have been obtained if it were possible to train with y_k itself.

.2 MISMATCH DUE TO USE OF FIR MODELS

In practice, the linear plant $P(z)$ generally exhibits exponential transient behavior and therefore has an infinite impulse response (IIR). On the other hand the adaptive modeling filter generally used here has a finite impulse response (FIR).[1] Refer to Fig. 4.2. Illustrated is the process of modeling an IIR plant with an FIR model. The result is mismatch, a difference between $\widehat{\mathbf{P}}^*$ and $\mathbf{P}$. Wiener theory can be used to study this mismatch. The IIR plant is shown as an infinite tapped delay-line.

Refer to Fig. 4.2. The Wiener solution (3.10) for the adaptive model can be written in this case as

$$\widehat{\mathbf{P}}^* = [\,\phi_{uu}]^{-1}[\,\phi_{uz}]. \tag{4.14}$$

In more detail, formula (4.14) becomes

$$\begin{bmatrix} \hat{p}_1 \\ \hat{p}_2 \\ \vdots \\ \hat{p}_n \end{bmatrix}^* = \begin{bmatrix} \phi_{uu}(0) & \phi_{uu}(1) & \cdots & \phi_{uu}(n-1) \\ \phi_{uu}(1) & \phi_{uu}(0) & & \phi_{uu}(n-2) \\ \vdots & \vdots & & \vdots \\ \phi_{uu}(n-1) & \phi_{uu}(n-2) & \cdots & \phi_{uu}(0) \end{bmatrix}^{-1} \cdot \begin{bmatrix} \phi_{uz}(0) \\ \phi_{uz}(1) \\ \vdots \\ \phi_{uz}(n-1) \end{bmatrix} \tag{4.15}$$

[1] Adaptive plant modeling need not be restricted to an FIR approach. IIR modeling is certainly possible, giving both advantages and disadvantages. Adapting both poles and zeros is sometimes difficult, particularly when plant disturbance is present. An excellent discussion of IIR modeling is presented by John Shynk in [10].

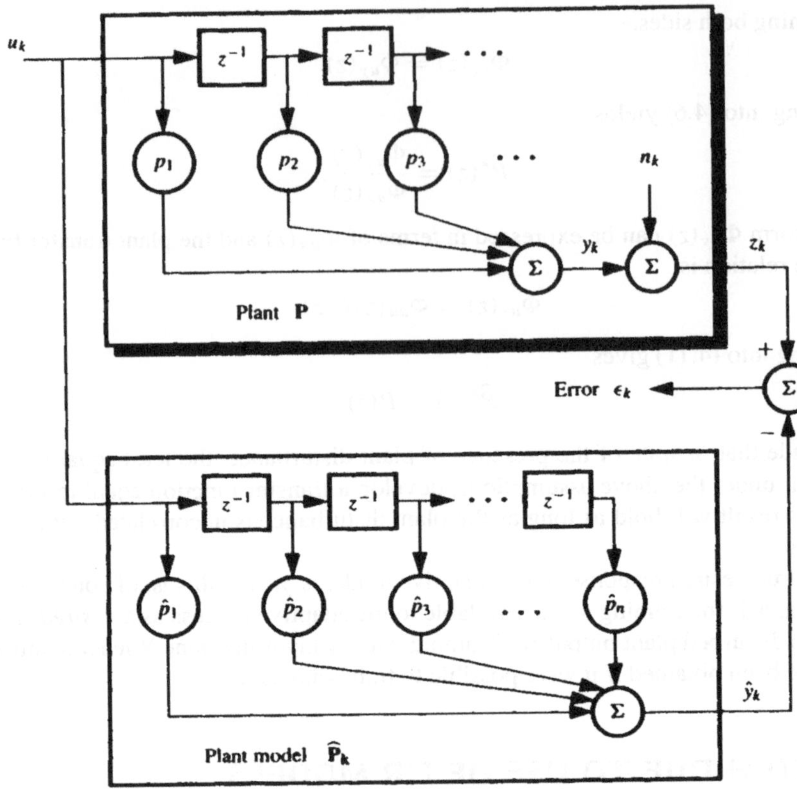

Figure 4.2 An FIR tapped delay-line model for an IIR plant.

Assuming that the plant disturbance n_k is uncorrelated with the plant input u_k, Eq. (4.9) can be used so that the last column of (4.15) can be replaced by a column of ϕ_{uy}'s. Accordingly,

$$\widehat{\mathbf{P}}^* = [\,\phi_{uu}]^{-1}[\,\phi_{uy}]. \tag{4.16}$$

This result is precise but generally hard to interpret on an intuitive basis.

One special case, where the plant input signal u_k is white, is of interest. Here,

$$[\,\phi_{uu}]^{-1} = \begin{bmatrix} \phi_{uu}(0) & & \\ & \ddots & \\ & & \phi_{uu}(0) \end{bmatrix}^{-1} = \frac{1}{\phi_{uu}(0)}\mathbf{I}. \tag{4.17}$$

Making use of (4.14) and using the fact that a white input has an autocorrelation function in the form of a delta function, we may write

$$[\,\phi_{uy}] = \begin{bmatrix} p_1 \\ p_2 \\ \vdots \\ p_n \end{bmatrix} \cdot \phi_{uu}(0). \tag{4.18}$$

Using Eqs. (4.16), (4.17), and (4.18), we conclude that

$$\widehat{\mathbf{P}}^* = \begin{bmatrix} \hat{p}_1 \\ \hat{p}_2 \\ \vdots \\ \hat{p}_n \end{bmatrix}^* = \begin{bmatrix} p_1 \\ p_2 \\ \vdots \\ p_n \end{bmatrix}. \tag{4.19}$$

Therefore, the best FIR model for the IIR plant has an impulse response that exactly matches the plant impulse response over the FIR response duration when the plant input u_k is white. When the plant input is colored but persistently exciting, the matching process will work similarly, but exactly how well can only be determined by making use of Wiener theory.

To obtain good dynamic modeling, it is necessary that the finite duration of the model's impulse response span the plant's impulse response up to the point where the latter decays to a "negligible" level. How long to make the model is always an uneasy question. In a practical situation, one would need to have a rough idea of plant impulse response duration and input statistics in order to choose the number of adaptive weights n to bring the mismatch error essentially to zero.

The parameter n can always be made large enough to bring this error to zero. However, from (3.71) we know that misadjustment is proportional to n. By making n large, the weight vector becomes noisy as a result of adaptation and the only way of reducing this noise is to adapt slowly. Slow convergence of the entire control system would be a result. So, one takes care in the choice of n. An extensive literature exists concerning the determination of the order of an unknown plant [11]–[16].

4.3 MISMATCH DUE TO INADEQUACIES IN THE INPUT SIGNAL STATISTICS; USE OF DITHER SIGNALS

To achieve a close match between the adaptive model and the unknown plant over a specified range of frequencies, the plant input u_k needs to have spectral energy over this range of frequencies. If the plant input has uniform spectral density over the frequencies of interest, then error tolerance will be uniformly tightly held over this frequency range. In many cases, however, the plant input u_k fails to have adequate spectral density at all frequencies where good fit is required. The result is mismatch, the development of a difference between $\widehat{\mathbf{P}}^*$ and $\mathbf{P}$.

Another form of difficulty that often arises in adaptive control systems results from the plant input u_k being nonstationary. Adaptive modeling processes do best with stationary inputs. As an example, suppose that the command input were zero over a long period of time, then stepped up to a constant DC level and held for a long time, then dropped to another constant DC level and held for a long time, and so forth. Level switching might be extremely infrequent and sporadic, but when it takes place, one requires precise response from the adaptive control system. Circumstances of this type present real problems for the modeling process when the plant input is not dynamic and rich in spectral content but is by and large constant.

One way to circumvent these difficulties lies in the use of random dither signal added to the plant input. The dither offers the advantage of providing a known input signal with easily controlled statistical properties, but use of a dither has the disadvantage of introducing more noise into the control process. The optimal choice of dither spectrum and dither power level is an important subject which will be discussed below.

There are several ways that a dither signal can be used in the the modeling process. Three schemes are shown in Fig. 4.3. With the addition of dither, the controller output is no longer the same as the plant input. To distinguish them, we continue to call the plant input u_k, including the dither, and henceforth we call the controller output u'_k.

Scheme A is a very straightforward way to do plant modeling. The dither δ_k is simply added to the controller output to form the plant input u_k. This is an effective scheme when the controller output is a stationary stochastic process and the independent dither is added to achieve a desired spectral character for u_k. Since the dither is an independent random signal, its spectrum adds to that of the controller output to make up the spectrum of u_k. There is an analogy here to the method of flash photography in full daylight, using the flash to fill in shadows.

In contrast to this, schemes B and C of Fig. 4.3 use dither exclusively in effecting the adaptive plant modeling process. The purpose is to assure known stationary statistics for the input modeling signal. When the controller output is nonstationary, one may be better off not including it at all in the modeling process.

Using scheme B with a white dither, the mean square error will be minimized when the impulse response of the adaptive model exactly matches that of the plant over the duration span of the model's impulse response. The minimum mean square error with scheme B is greater than when using either scheme A or the basic ditherless modeling approach of Figs. 4.1 and 4.2. With scheme A, the minimum mean square error will be equal to the plant output disturbance power. With scheme B, the minimum mean square error will be equal to the power of the plant disturbance plus the power of the controller output u'_k after propagating through the plant. The higher value of minimum mean square error will unfortunately cause increased noise in the adaptive model's weights. This is in accord with the theory of misadjustment presented in Chapter 3. Otherwise scheme B is an excellent one.

It is possible to improve on scheme B, and with the cost of a somewhat increased system complexity, one can reduce the weight noise while adapting with the same rate of convergence and with the same dither amplitude as with scheme B. Thus, scheme C has all the good features of scheme B and overcomes the drawback of having an increased minimum mean square error.

Plant modeling with scheme C is block diagrammed in Fig. 4.3(c). The bottom-most filter is the actual adaptive modeling filter. Its transfer function[2] is $\widehat{P}_k(z)$ and its impulse response vector is $\widehat{\mathbf{P}}_k$. Its input is the dither δ_k. Scheme C is similar to scheme B and develops the same converged Wiener solution. The only difference is that scheme C incorporates a technique for eliminating the effects of u'_k propagating through the plant $P(z)$ into the error signal ϵ_k. A filter is used which is an exact digital copy of $\widehat{P}_k(z)$. Its input connects to the controller output u'_k and its output is subtracted from the plant output to provide a desired response training signal for the adaptation of $\widehat{P}_k(z)$ which contains no dynamic components originating with u'_k. Inclusion of the copied $\widehat{P}_k(z)$ filter does not affect the Wiener solution since its output is uncorrelated with the dither δ_k. Inclusion of the copied $\widehat{P}_k(z)$ filter reduces

[2]We define $\widehat{P}_k(z)$ henceforth as the z-transform of the instantaneous values of the weight vector $\widehat{\mathbf{P}}_k$ of the adaptive model at time k.

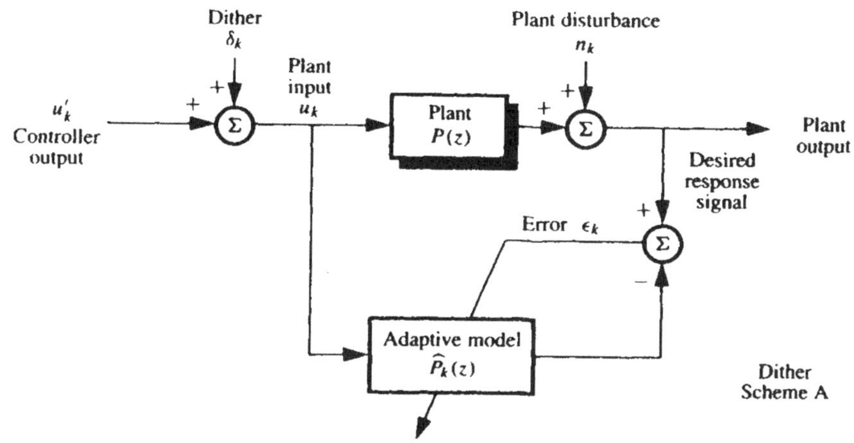

(a)

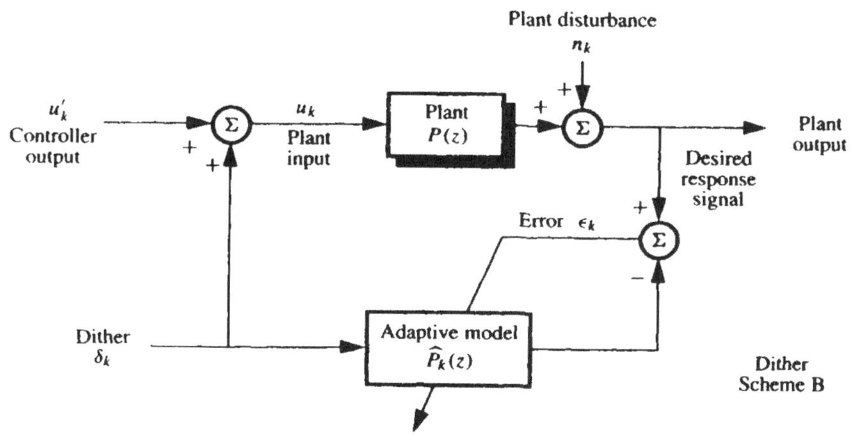

(b)

Figure 4.3 Dithering schemes for plant identification: (a) Dither scheme A; (b) Dither scheme B.

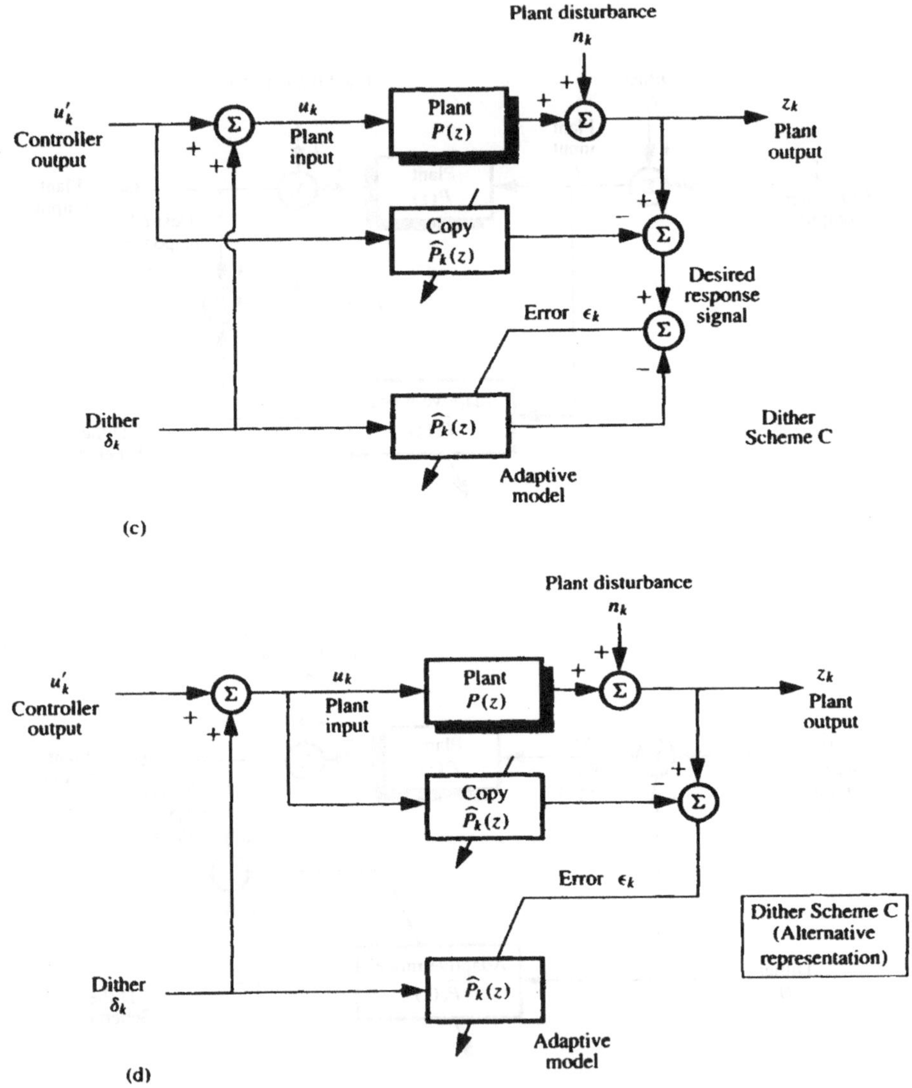

Figure 4.3 (continued). (c) Dither scheme C; (d) Dither scheme C (Alternative representation)

the minimum mean square error, however. Now when $\widehat{P}_k(z)$ converges to be a close match to $P(z)$, the error ϵ_k of the adaptive process is almost exactly equal to the plant output disturbance alone. Scheme C in Fig. 4.3(c) can be redrawn as shown in Fig. 4.3(d). The two configurations yield identical weights for the adaptive model $\widehat{P}_k(z)$. A detailed analysis of schemes A, B, and C is given in Appendix B. Of concern for these schemes are conditions for stability, time constant, and misadjustment.

4.4 ADAPTIVE MODELING SIMULATIONS

To demonstrate the process of adaptive modeling of an unknown plant, simulation experiments have been performed and the results will be described here. For the purposes of experimentation, the plant will be known to us, but "unknown" to the adaptive algorithm.

Chosen for these modeling experiments is the unstable nonminimum-phase plant seen previously in Chapter 1, given by

$$\frac{(s-0.5)}{(s+1)(s-1)}. \tag{4.20}$$

Since we know the plant, we can easily stabilize it. A feedback stabilizer is shown in Fig. 4.4. A root-locus plot is shown in Fig. 4.5. Two values of loop gain k were chosen, $k = 21$ and alternatively, $k = 24$. Both gains yield stabilized conditions. If we represent the transfer function of the stabilized system by $G(s)$, the transfer functions for the two gain values are for $k = 21$,

$$G(s) = \frac{21(s-0.5)}{(s-3.7152)(s-0.1324+0.9601j)(s-0.1424-0.9601j)}, \tag{4.21}$$

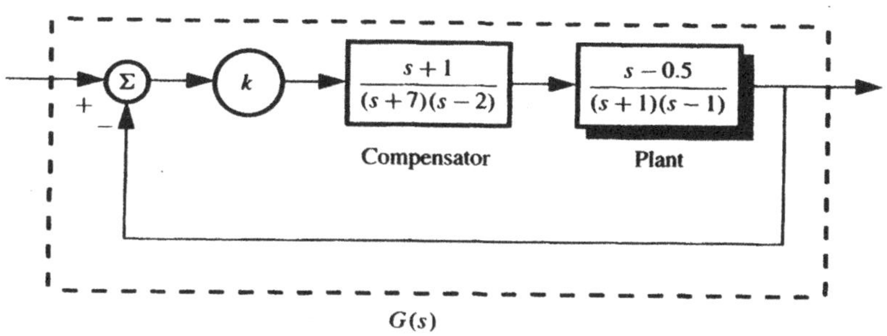

Figure 4.4 Feedback for stabilization of an unstable plant.

and for $k = 24$,

$$G(s) = \frac{24(s-0.5)}{(s+1)^2(s+2)}. \tag{4.22}$$

The poles of these transfer functions are marked on the root-locus plot of Fig. 4.5.

Choosing a compensator and loop gain to achieve stabilization was greatly facilitated by having knowledge of the plant poles and zeros. Without this knowledge, a great deal of experimentation would have been required for successful plant stabilization.

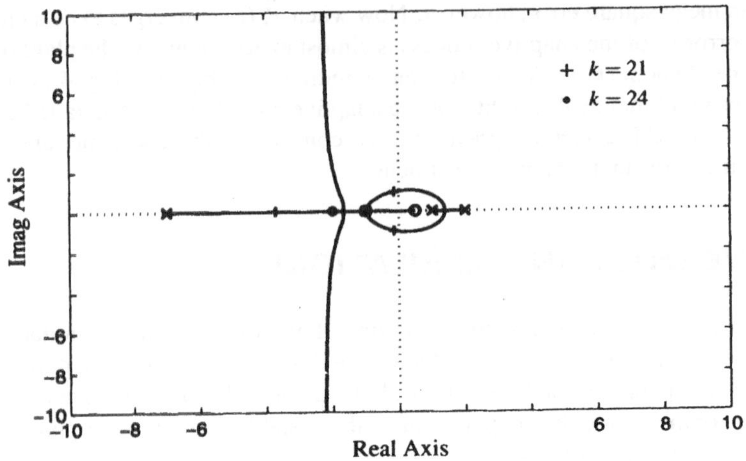

Figure 4.5 Root-locus plot for plant stabilization loop.

In an adaptive inverse control system, the controller will be discrete. A digital-to-analog converter will be needed to take the discrete control signal and use it to drive the plant. For this experiment we choose the DAC to be a zero-order hold. We will model the stabilized plant by including the DAC. The model of the stabilized plant will be discrete, so we will need to sample the plant output for the adaptive modeling process. The basic idea is illustrated by the block diagram of Fig. 4.6. The adaptive model, upon convergence, will closely approximate the discretized $G(s)$, that we will represent by $G(z)$, defined as

$$G(z) \triangleq (1 - z^{-1})\mathcal{Z}\left[\left(\frac{1}{s}\right) G(s)\right]. \tag{4.23}$$

The operator $\mathcal{Z}$ means: take the z-transform. The discretized transfer functions for the two gain values are for $k = 21$,

$$G(z) = \frac{0.906(z - 1.0513)(z + 0.8609)}{(z - 0.9813 + 0.0945j)(z - 0.9813 - 0.0945j)(z - 0.6897)}, \tag{4.24}$$

and for $k = 24$,

$$G(z) = \frac{0.1032(z - 1.0513)(z + 0.8608)}{(z - 0.9048)^2(z - 0.8187)}. \tag{4.25}$$

The impulse response corresponding to $G(z)$ for $k = 21$ is plotted in Fig. 4.7. The impulse response of the adaptive model, obtained in accord with the scheme of Fig. 4.6, is shown in Fig. 4.8. The input to the DAC was a random, zero-mean, first-order Markov process. Learning is demonstrated by the plot of squared error over time, shown in Fig. 4.9. The error became very small, and the resulting adaptive model closely fits the ideal response of Fig. 4.7.

A strong disturbance was added to the plant output (random, zero-mean, also first-order Markov) to test the ability of the modeling process of Fig. 4.6. Allowing enough time for convergence, the adaptive process was momentarily stopped, and a snapshot of the impulse response of the adaptive model was taken and this is plotted in Fig. 4.10. Some dis-

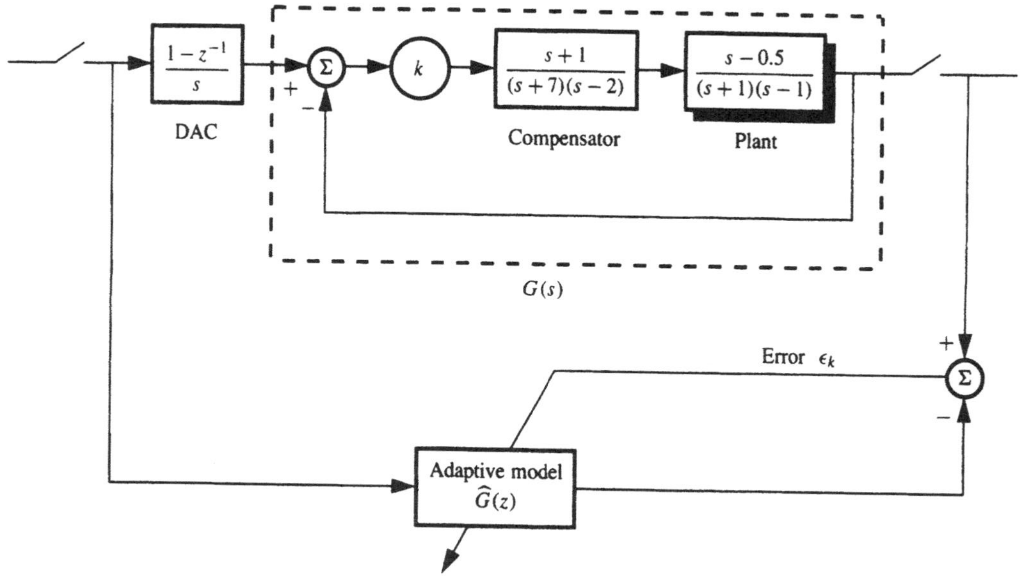

Figure 4.6 Adaptive modeling of the stabilized plant.

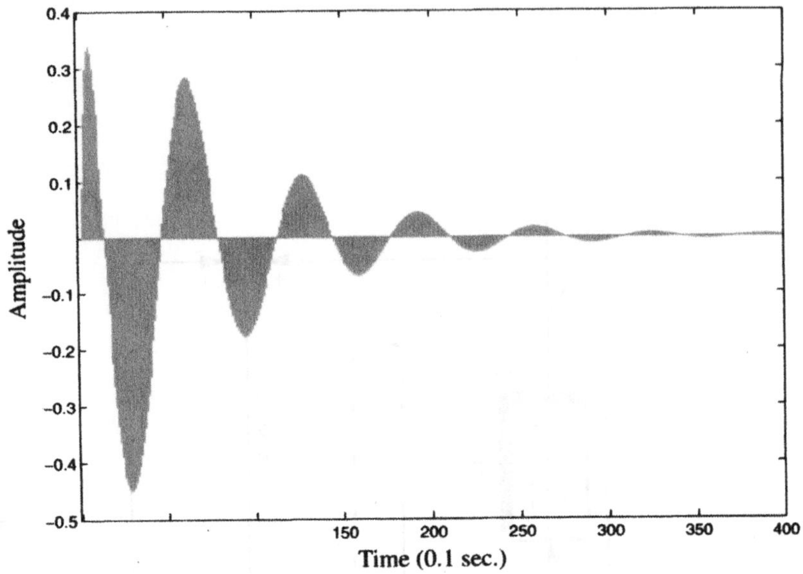

Figure 4.7 Impulse response of discretized stabilized nonminimum-phase plant with $k = 21$.

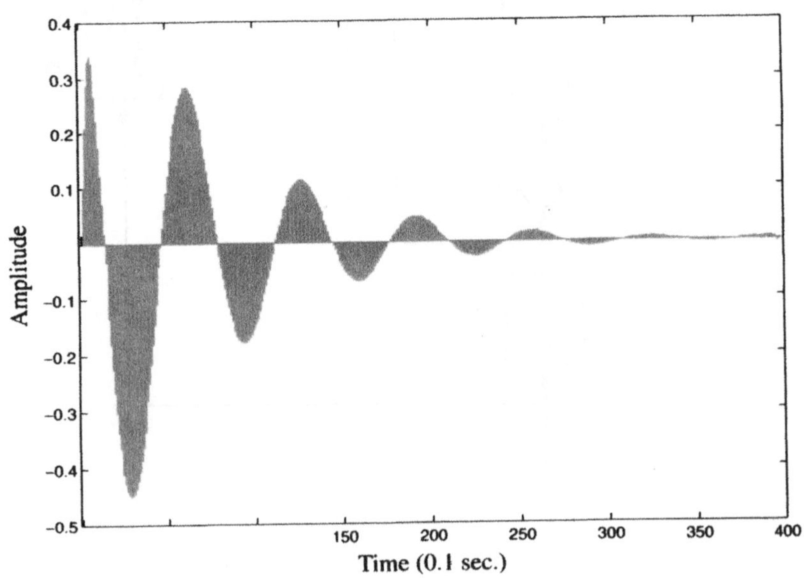

Figure 4.8 Adaptive model of discretized stabilized nonminimum-phase plant with $k = 21$.

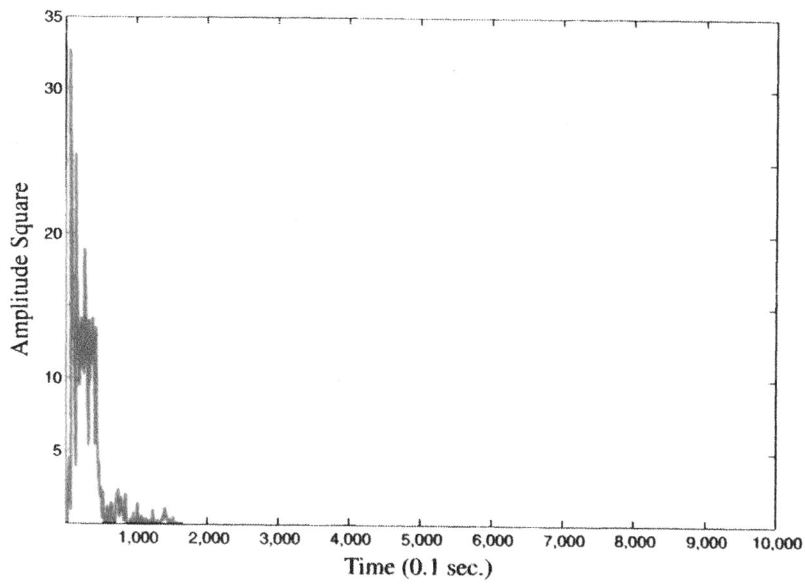

Figure 4.9 Square of the error versus iteration cycle number for the adaptive modeling process of Fig 4.6 with $k = 21$.

tortion in the impulse response is evident, but the similarity to the ideal impulse response is clear. The distortion disappears when μ of the LMS learning algorithm is made smaller. Reducing μ 10 times reduces the distortion to an almost invisible level.

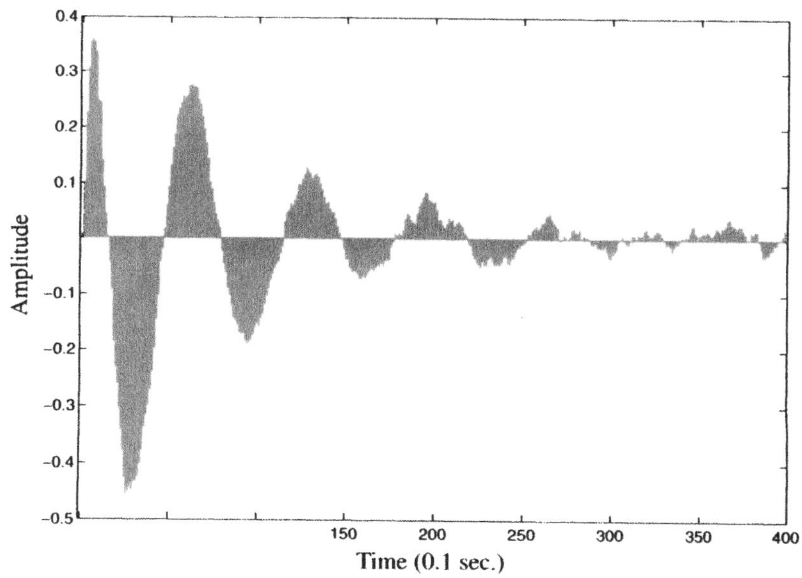

Figure 4.10 Adaptive model of discretized stabilized nonminimum-phase plant subject to random disturbance, with $k = 21$.

The modeling scheme of Fig. 4.6 relies on the natural plant input signal to excite the plant. An experiment was made to demonstrate learning not with the plant input signal, but with a dither signal, utilizing scheme C. A block diagram of the experimental learning process is shown in Fig. 4.11. The learned impulse response is shown in Fig. 4.12. It matches very closely to the ideal impulse response of Fig. 4.7. Scheme C works very well. Adding a strong first-order Markov disturbance to the plant output results in some distortion in the learned impulse response, and this can be seen in the plot of Fig. 4.13. Making μ small causes the distortion to become small, as before.

All of these experiments were run again for a different case, this time with $k = 24$. The discretized transfer function $G(z)$ is given by (4.25). The impulse response is plotted in Fig. 4.14. Modeling the stabilized plant in accord with the scheme of Fig. 4.6, the adaptive impulse response that results is plotted in Fig. 4.15. The learning process is illustrated by the plot of Fig. 4.16 which shows the square of the error versus the number of adaptation cycles.

Disturbance was added to the plant output (zero-mean, first-order Markov) and the modeling process (with $k = 24$) was repeated. Allowing time for convergence, a snapshot of the adaptive model was taken and plotted in Fig. 4.17. Distortion in the adaptive impulse response due to plant disturbance is visible, but this goes away when μ is made smaller.

Using a dither signal to train the adaptive model in accord with scheme C, the result shown in Fig. 4.18 was obtained (with $k = 24$). The block diagram of the training process is shown in Fig. 4.11. Comparing Fig. 4.18 with the ideal response of Fig. 4.14 demonstrates how well the learning process works. Adding a random, zero-mean, first-order Markov disturbance to the plant output, allowing enough time for scheme C to converge, and taking a snapshot of the adaptive model's impulse response, some distortion appears in this impulse response, as shown in Fig. 4.19. This distortion diminishes as μ is reduced in size and disappears when μ is made very small.

These experiments show that stabilized plants can be modeled with FIR adaptive filters. When the natural signal driving the plant is adequate for modeling, the scheme of Fig. 4.6 can be used. When dither is necessary, either scheme A, B, or C could be used. Scheme C has been used in these experiments and it is diagrammed in Fig. 4.11. Plant disturbance does not bias the modeling solution but could cause the model's weights to be noisy. To reduce the effects of the noise, one needs to adapt more slowly by reducing the value of μ.

4.5 SUMMARY

In this chapter, several modeling methods were introduced which were based on adaptation with available control signals, with dither signals, and with a combination of dither and available signals. All of these methods have their advantages and their limitations, and they all attempt to cause the adaptive model $\widehat{P}_k(z)$ to be a "good match" to the plant $P(z)$. The difference between the model and the plant is the mismatch which may be defined both in terms of transfer functions and weight vectors as

$$\Delta P_k(z) \triangleq \widehat{P}_k(z) - P(z); \quad \Delta \mathbf{P}_k \triangleq \widehat{\mathbf{P}}_k - \mathbf{P}. \tag{4.26}$$

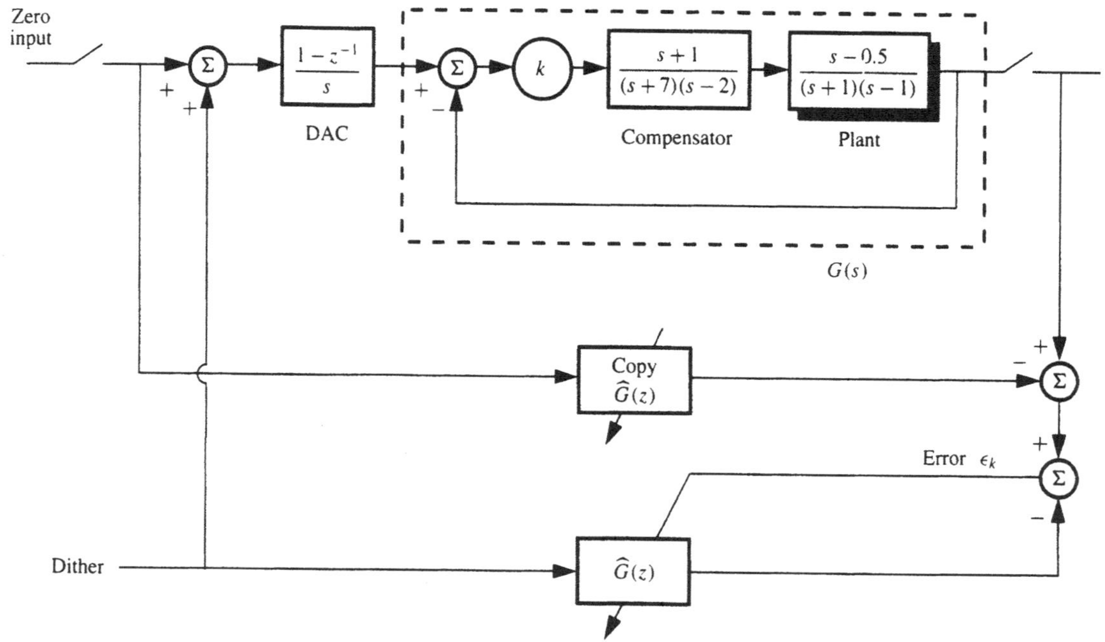

Figure 4.11 Modeling the stabilized plant using dither scheme C.

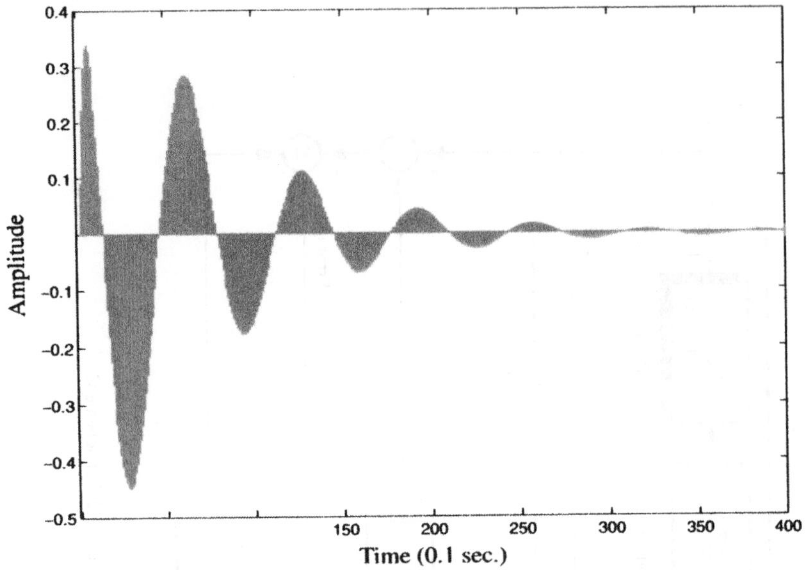

Figure 4.12 Adaptive model of discretized stabilized nonminimum-phase plant with $k = 21$, obtained with dither scheme C.

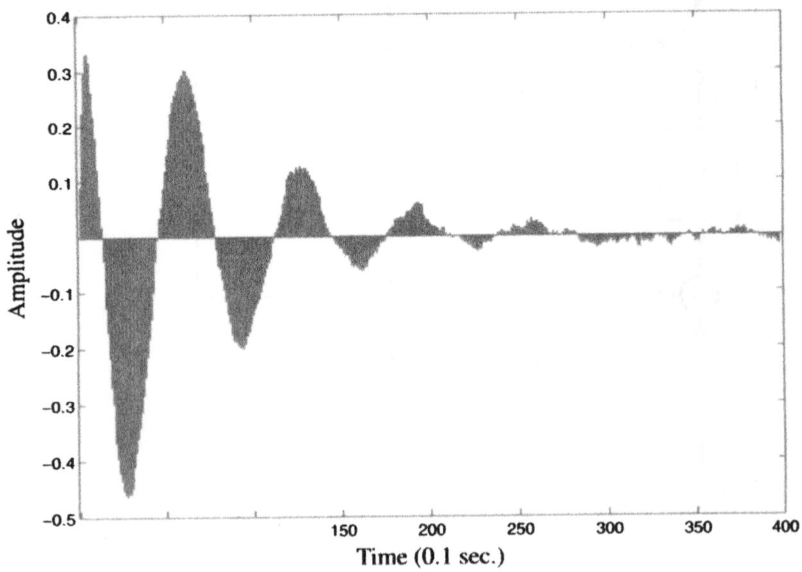

Figure 4.13 Adaptive model of randomly disturbed discretized stabilized nonminimum-phase plant with $k = 21$, obtained with dither scheme C.

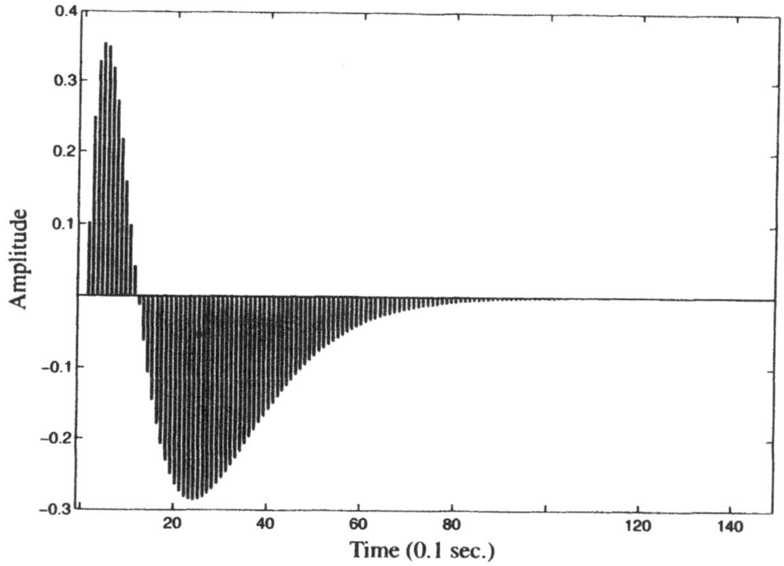

Figure 4.14 Impulse response of discretized stabilized nonminimum-phase plant with $k = 24$.

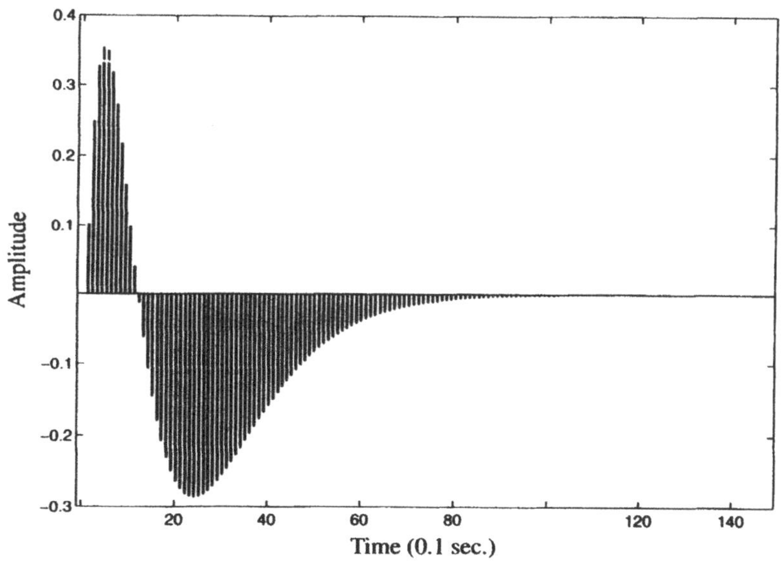

Figure 4.15 Adaptive model of discretized stabilized nonminimum-phase plant with $k = 24$.

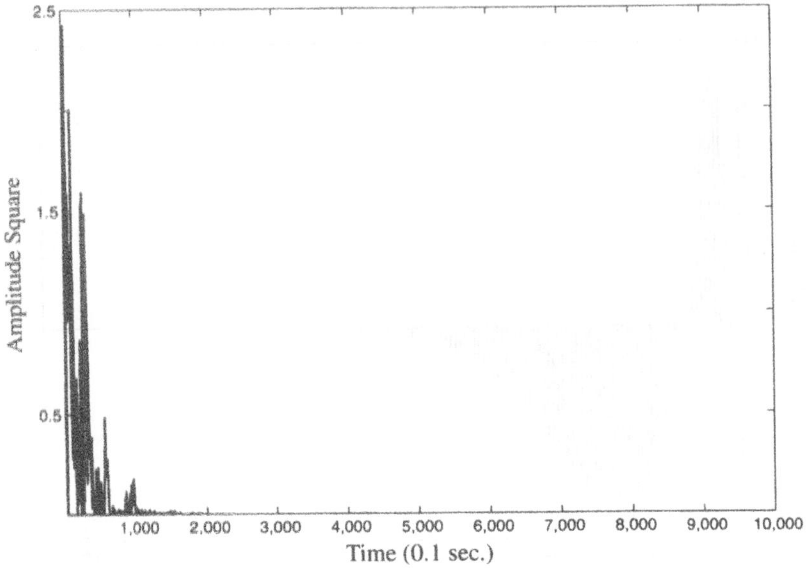

Figure 4.16 Square of the error versus iteration cycle number for the adaptive modeling process of Fig 4.6 with $k = 24$.

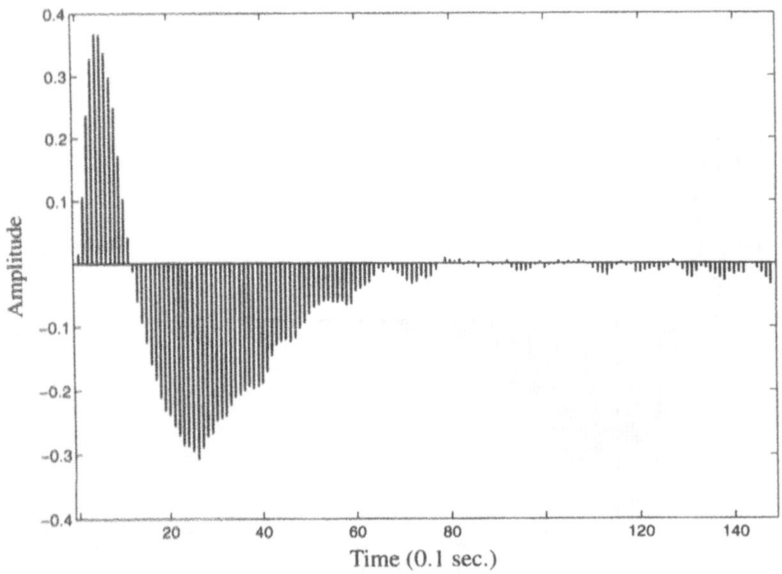

Figure 4.17 Adaptive model of discretized stabilized nonminimum-phase plant subject to random disturbance with $k = 24$.

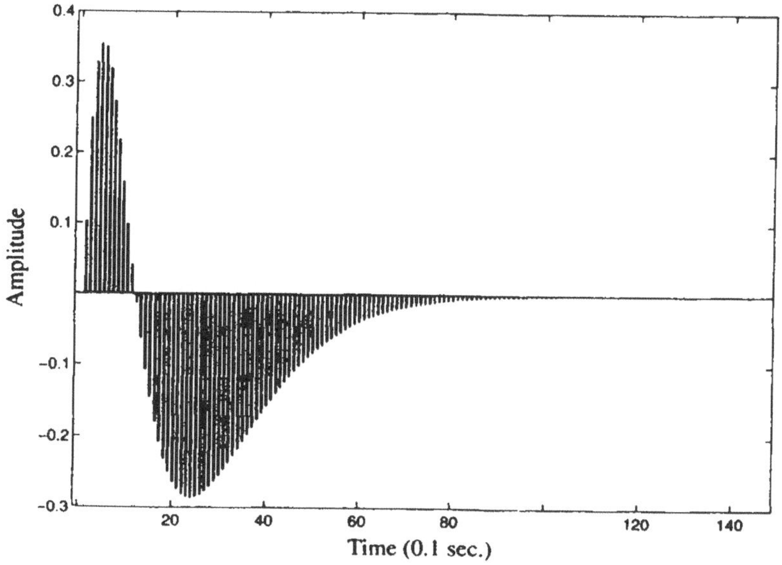

Figure 4.18 Adaptive model of discretized stabilized nonminimum-phase plant with $k = 24$, obtained with dither scheme C.

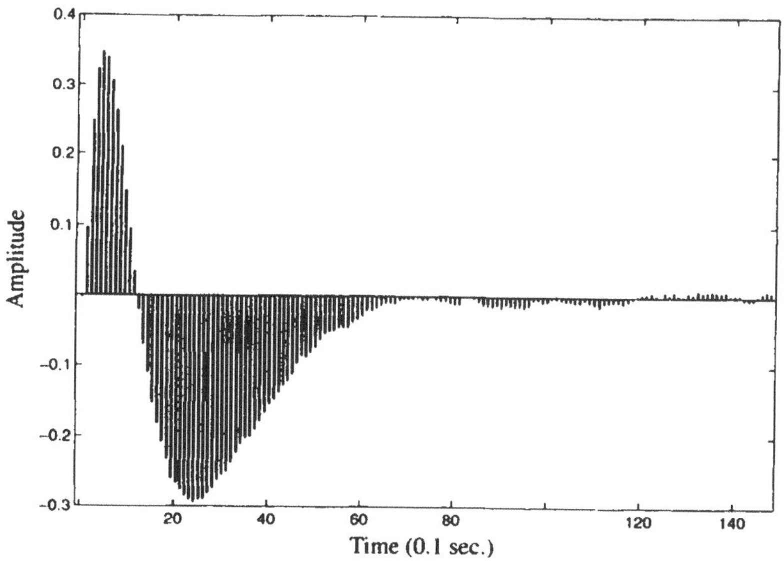

Figure 4.19 Adaptive model of randomly disturbed, discretized, stabilized nonminimum-phase plant with $k = 24$, obtained with dither scheme C.

We assume henceforth that the adaptive model will have a large enough number of weights, so that with appropriate weight settings, the difference in the dynamic responses of plant and plant model will be negligible. When the model weights are determined by an adaptive least squares process, the weights converge in the mean to a Wiener solution. With a large number of model weights, the Wiener solution is essentially equal to the plant impulse response and therefore when the adaptive process converges, the mean of the mismatch is

$$E[\Delta\mathbf{P}_k] = 0. \tag{4.27}$$

Weight noise is always present during adaptation, during initializing transients, and in steady state as the model weight vector randomly hovers about the Wiener solution. Thus, when the adaptive process converges, the covariance of the modeling error is

$$\text{cov}[\Delta\mathbf{P}_k] = \begin{bmatrix} \text{weight noise} \\ \text{covariance matrix} \end{bmatrix}. \tag{4.28}$$

This covariance matrix is obtained for scheme C, with white dither, from Eq. (4.31).

Knowledge of the mean and the covariance of the mismatch will be useful in determining the mean and covariance of the inverse modeling error, which will subsequently be related to the overall error of the adaptive inverse control system.

This chapter introduces and explains dither schemes A, B, and C. They represent distinctly different approaches to the use of dither signals for plant modeling. Mathematical analyses of schemes A, B, C are presented in Appendix B. Further explanation of their means of behavior is given there. A comparative analysis, citing advantages and disadvantages of these schemes, is discussed. Performance characteristics of these schemes are given in Table B.3 of Appendix B. Scheme C is often the method of choice when using a dither signal as an aid to the modeling process.

For scheme C, it is shown in Appendix B that

Stable range for μ

$$\frac{1}{n(E[u_k'^2] + E[\delta_k^2])} > \mu > 0. \tag{4.29}$$

Unique time constant

$$\tau = \frac{1}{2\mu E[\delta_k^2]}. \tag{4.30}$$

Weight noise covariance matrix

$$\begin{bmatrix} \text{weight noise} \\ \text{covariance matrix} \end{bmatrix} = \frac{\mu E[n_k^2]}{1 - \mu n(E[u_k'^2] + E[\delta_k^2])}\mathbf{I}. \tag{4.31}$$

Bibliography for Chapter 4

[1] B. Widrow, and M.E. Hoff, "Adaptive switching circuits," in *IRE WESCON Conv. Rec.*, Pt. 4, pp. 96–104.

[2] B. WIDROW, and J.M. MCCOOL, "A comparison of adaptive algorithms based on the methods of steepest descent and random search," *IEEE Trans. Antennas Propag.*, Vol. AP-24, No. 5 (September 1976), pp. 615–637.

[3] B. WIDROW, J.M. MCCOOL, M.G. LARIMORE, and C.R. JOHNSON, JR., "Stationary and nonstationary learning characteristics of the LMS adaptive filter," *Proc. IEEE*, Vol. 64, No. 8 (August 1976), pp. 1151–1162.

[4] B. WIDROW, "Adaptive filters," in *Aspects of network and system theory*, ed. R.E. Kalman and N. De Claris (New York: Holt, Rinehart and Winston, 1970), pp. 563–587.

[5] B. WIDROW, and E. WALACH, "On the statistical efficiency of the LMS algorithm with nonstationary inputs," *IEEE Trans. Info. Theory — Special Issue on Adaptive Filtering*, Vol. 30, No. 2, Pt. 1 (March 1984), pp. 211–221.

[6] M. MORF, T. KAILATH, and L. LJUNG, "Fast algorithms for recursive identification," in *Proceedings 1976 Conf. on Decision and Control*, Clearwater Beach, FL, December 1976, pp. 916–921.

[7] L.J. GRIFFITHS, "A continuously adaptive filter implemented as a lattice structure," in *Proc. IEEE Int'l. Conf. on Acoust., Speech, Signal Processing*, Hartford, CT, May 1977, pp. 683–686.

[8] D.T.L. LEE, M. MORF, and B. FRIEDLANDER, "Recursive least squares ladder estimation algorithms," *IEEE Trans. on Circuits and Systems*, Vol. CAS-28, No. 6 (June 1981), pp. 467–481.

[9] B. FRIEDLANDER, "Lattice filters for adaptive processing," *Proc. IEEE*, Vol. 70, No. 8 (August 1982), pp. 829–867.

[10] J.J. SHYNK, "Adaptive IIR filtering," *IEEE ASSP Magazine*, Vol. 6. No. 2 (April 1989), pp. 4–21.

[11] H. AKAIKE, "Maximum likelihood identification of Gaussian autoregressive moving average models," *Biometrika*, Vol. 60 (1973), pp. 255–265.

[12] H. AKAIKE, "A new look at the statistical model identification," *IEEE Trans. Auto. Control*, Vol. AC-19 (1974), pp. 716–723.

[13] J. RISSANEN, "Modeling by shortest data description," *Automatica*, Vol. 14 (1978), pp. 465–471.

[14] G. SCHWARTZ, "Estimating the dimension of a model," *Ann. Stats.*, Vol. 6 (1978), pp. 461–464.

[15] M.B. PRIESTLEY, *Spectral analysis and time series*, Vols. 1 and 2 (New York: Academic Press, 1981).

[16] S. HAYKIN, *Adaptive filter theory* (Englewood Cliffs, NJ: Prentice Hall, 1986).

[17] L. LJUNG, *System identification* (Englewood Cliffs, NJ: Prentice Hall, 1987).

[18] S. SASTRY and M. BODSON, *Adaptive control* (Englewood Cliffs, NJ: Prentice Hall, 1989).

[19] K.J. ÅSTRÖM and B. WITTENMARK, *Adaptive Control* (Reading, MA: Addison-Wesley, 1989).

[20] G.F. FRANKLIN, J.D. POWELL, and M.L. WORKMAN, *Digital control of dynamic systems*, 2nd ed. (Reading, MA: Addison-Wesley, 1990). Useful comments on the subject of persistent excitation are given in chap. 8, Section 8.4.

Chapter 5

Inverse Plant Modeling

.0 INTRODUCTION

Control concepts taught in Chapter 1 involve the use of adaptive plant inverses as controllers in feedforward control configurations. Pursuing these ideas, our next step is the development of general techniques for finding inverses of plants that need to be controlled. We shall restrict our development to apply only to stable plants. If the plant of interest is unstable, conventional feedback should be applied to stabilize it. Then the combination of the plant and its feedback stabilizer can be regarded as an equivalent stable plant. The subject is discussed in detail in Appendix D. Only linear, single-input single-output (SISO) systems will be treated here. Nonlinear and MIMO systems will be discussed subsequently in Chapters 10 and 11.

The plant generally has poles and zeros. The inverse of the plant therefore should have zeros and poles. If the plant is minimum-phase, that is, has all of its zeros inside the unit circle in the z-plane, then the inverse will be stable with all of its poles inside the unit circle. If the plant is nonminimum-phase, then some of the poles of the inverse will be outside the unit circle and the inverse will be unstable. In general, not knowing whether the plant is minimum-phase or not, there is uncertainty about the feasibility of making a plant inverse. This uncertainty can for the most part be overcome and excellent inverses can be made in practice by using the appropriate adaptive inverse modeling techniques.

We shall begin with a discussion of inverse modeling of minimum-phase plants, then discuss inverse modeling of nonminimum-phase plants. Model-reference inverse modeling will be described next. The effects of plant disturbance will then be considered, and online and offline inverse modeling will be illustrated. Finally, the effects of gradient noise upon the plant inverse model will be analyzed, leading to expressions for noise in the weights of the inverse model and for the variance of the dynamic system error of the entire control system.

1 INVERSES OF MINIMUM-PHASE PLANTS

Refer to Fig. 5.1. The plant is represented by $P(z)$. The plant inverse, to be used ultimately as the controller, shall be designated by $C(z)$ if it is ideal, or by $\widehat{C}(z)$ if it is obtained in a practical way and is not quite perfect. Assume that plant $P(z)$ is minimum-phase, having all of its zeros inside the unit circle in the z-plane. As such, a perfect inverse $C(z) = 1/P(z)$ would exist. The reciprocal of $P(z)$ would be both stable and causal. An adaptive algorithm, block diagrammed in Fig. 5.1, would provide an inverse $\widehat{C}(z)$ which closely approximates $C(z)$, given that the adaptive filter has sufficient flexibility, that is, sufficient number of degrees of freedom. Adjusting $\widehat{C}(z)$ to minimize the mean square error would bring the error close to zero and would cause $\widehat{C}(z)$ to be almost equal to $C(z)$.

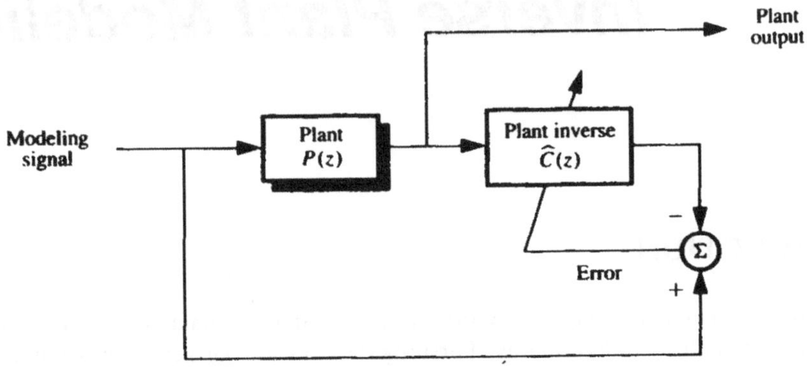

Figure 5.1 Forming a plant inverse.

Suppose for example that the plant has the transfer function

$$P(z) = \frac{1 + \frac{1}{2}z^{-1}}{1 - z^{-1} + \frac{3}{4}z^{-2}}. \tag{5.1}$$

This plant is causal and it is stable since both of its poles are inside the unit circle in the z-plane. It is minimum-phase since its zero is also inside the unit circle. The reciprocal of this plant is a perfect inverse,

$$C(z) = \frac{1}{P(z)} = \frac{1 - z^{-1} + \frac{3}{4}z^{-2}}{1 + \frac{1}{2}z^{-1}}. \tag{5.2}$$

The transfer function $C(z)$ can be expanded by long division, according to

$$C(z) = 1 - \frac{3}{2}z^{-1} + \frac{3}{2}z^{-2} - \frac{3}{4}z^{-3} + \frac{3}{8}z^{-4} - \frac{3}{16}z^{-5} + \cdots. \tag{5.3}$$

It is clear from (5.2) that $C(z)$ is stable since its pole is inside the unit circle. $C(z)$ is causal, as evidenced by expansion (5.3). Referring now to Fig. 5.1, if the adaptive inverse filter $\widehat{C}(z)$ had an infinitely long impulse response, it could perfectly realize $C(z)$. If it had a finite impulse response but a very long one, the difference between $\widehat{C}(z)$ and $C(z)$ would be negligible.

We will pause for a moment to demonstrate that the unconstrained Wiener solution for the adaptive filter transfer function $\widehat{C}(z)$ will be equal to the inverse of $P(z)$. Let the modeling signal in Fig. 5.1 be white, with unit power. The z-transform of this signal's autocorrelation function is therefore unity. From Eq. (2.22), the z-transform of the plant output autocorrelation function is

$$P(z)P(z^{-1}). \tag{5.4}$$

This is also the transform of the autocorrelation of the input to the adaptive filter $\widehat{C}(z)$.

Referring to Fig. 5.1, the desired response for $\widehat{C}(z)$ can be seen to be the modeling signal itself. The z-transform of the crosscorrelation between the input to $\widehat{C}(z)$ and the desired response can be obtained by using (2.15) and (2.12). This is

$$P(z^{-1}). \tag{5.5}$$

The unconstrained Wiener solution is given by (2.30). For this case, it is

$$\widehat{C}(z) = \frac{P(z^{-1})}{P(z)P(z^{-1})} = \frac{1}{P(z)} = C(z). \tag{5.6}$$

Minimizing mean square error therefore provides the correct inverse.

INVERSES OF NONMINIMUM-PHASE PLANTS

Next we consider the example of a plant whose transfer function is

$$P(z) = \frac{1 + 2z^{-1}}{1 - z^{-1} + \frac{3}{4}z^{-2}}. \tag{5.7}$$

This well-behaved plant is causal and stable, but nonminimum-phase because its zero lies outside the unit circle. The inverse of this plant is

$$C(z) = \frac{1 - z^{-1} + \frac{3}{4}z^{-2}}{1 + 2z^{-1}}. \tag{5.8}$$

It is evident that $C(z)$ is unstable, since its pole lies outside the unit circle. Using such a $C(z)$ as an open-loop controller would be disastrous. But there is a way to alleviate the situation which is based on the theory of two-sided Laplace transforms.

$C(z)$ can be expanded in two ways:

$$C(z) = \frac{1 - z^{-1} + \frac{3}{4}z^{-2}}{1 + 2z^{-1}} = 1 - 3z^{-1} + \frac{27}{4}z^{-2} - \frac{27}{2}z^{-3} + 27z^{-4} + \cdots \tag{5.9}$$

$$C(z) = \frac{1 - z^{-1} + \frac{3}{4}z^{-2}}{1 + 2z^{-1}} = \frac{3}{8}z^{-1} - \frac{11}{16} + \frac{27}{32}z - \frac{27}{64}z^{2} + \frac{27}{128}z^{3} + \cdots \tag{5.10}$$

The first expansion corresponds to a causal but unstable inverse. Not good. The second expansion corresponds to a noncausal inverse, but at least it is stable. $C(z)$ in either of these forms would not result from Wiener optimization. The unstable form (5.9) would cause the mean square error to be infinite. Minimization of mean square error would never give this solution. The second expansion (5.10), the stable one, is noncausal and therefore not realizable by the causal filter $\widehat{C}(z)$. The first two terms in Eq. (5.10) are causal, however, and would be realizable. If the remainder of the terms were relatively small, then the series could be approximated by the first two terms and a causal filter would be able to realize a useful approximate inverse. This is not the case with the present example however, but the idea is suggestive. Before pursuing this idea, let us see what the causal Wiener solution would be for this particular example. We will use the Shannon-Bode approach. Figure 5.2 illustrates how this is done. The modeling signal is assumed to be white, of zero-mean, and of unit variance. The plant is specified by (5.7). The causal plant inverse is, in accord with Shannon-Bode, comprised of a *whitening filter* and an optimized causal filter. This latter filter is easily obtained since, having a white input, it can be optimized without regard to causality and then the noncausal part of its impulse response can be simply deleted. The whitening filter is chosen to be stable, causal, and minimum-phase so that its inclusion in the system before the optimized filter does nothing irreversible to the signal. The inverse of the whitening filter is stable and causal.

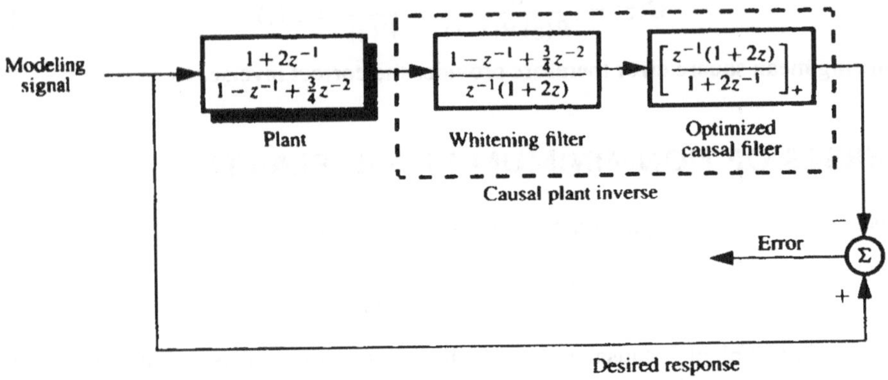

Figure 5.2 An example of Shannon-Bode design of a causal Wiener plant inverse.

For the present example, using a white modeling signal, the z-transform of the autocorrelation of the plant output is

$$\frac{(1+2z^{-1})(1+2z)}{(1-z^{-1}+\frac{3}{4}z^{-2})(1-z+\frac{3}{4}z^{2})}. \tag{5.11}$$

After multiplying and factoring, the transfer function of the proper whitening filter is obtained as

$$\frac{(1-z^{-1}+\frac{3}{4}z^{-2})}{z^{-1}(1+2z)} = \frac{(\frac{1}{2}-\frac{1}{2}z^{-1}+\frac{3}{8}z^{-2})}{(1+\frac{1}{2}z^{-1})}. \tag{5.12}$$

Without regard to causality, the transfer function of the optimized filter may be obtained by inspection of Fig. 5.2 as the inverse of the product of transfer functions of the plant and the whitening filter. This is

$$\frac{z^{-1}(1+2z)}{(1+2z^{-1})}. \tag{5.13}$$

The next step is to find a stable impulse response corresponding to (5.13) and then delete the noncausal portion. Making an expansion of (5.13),

$$\frac{z^{-1}(1+2z)}{(1+2z^{-1})} = \frac{1}{2} + \frac{3}{4}z - \frac{3}{8}z^{2} + \frac{3}{16}z^{3} - \cdots. \tag{5.14}$$

The expansion (5.14) is stable, but every term corresponds to a noncausal impulse response sample except the first one. Deleting the noncausal part yields

$$\left[\frac{z^{-1}(1+2z)}{1+2z^{-1}}\right]_+ = \frac{1}{2}. \tag{5.15}$$

The causal Wiener solution therefore has the transfer function

$$\frac{1 - z^{-1} + \frac{3}{4}z^{-2}}{4(1+\frac{1}{2}z^{-1})}. \tag{5.16}$$

The minimum mean square error can be shown to be $\frac{3}{4}$. This is a very large mean square error compared to unity, the mean square value of the modeling signal. The causal Wiener filter will not make a very good controller. The difficulty comes from attempting to force the nonminimum-phase plant to respond instantaneously to the command input. In such cases, delayed responses can generally be accomplished more precisely, with much lower error.

Plants having transport delay and other characteristics that fall under the general description of nonminimum-phase cannot be made to respond instantly to sudden changes in the command input. The best thing to do is to adapt the controller to provide delayed plant responses to the modeling signal. The idea is illustrated in Fig. 5.3. The delayed plant inverse $\widehat{C}_\Delta(z)$ is adaptively chosen so that when its transfer function is multiplied by the plant transfer function $P(z)$, the product will be a best fit to $z^{-\Delta}$, the transfer function of a Δ-units of time delay.

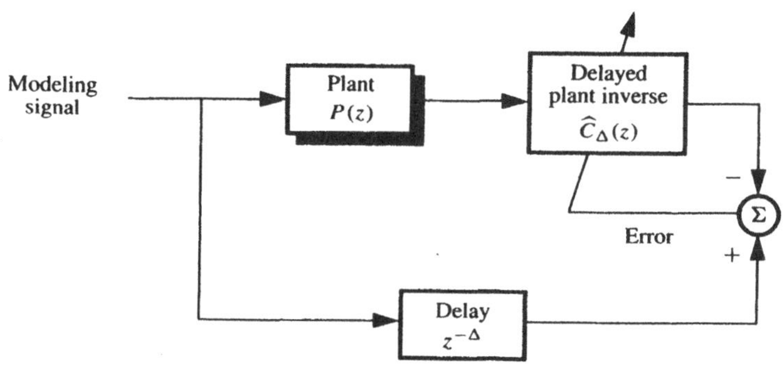

Figure 5.3 Forming a delayed plant inverse.

We will rework the above example. Refer to Fig. 5.2 but incorporate a delay $z^{-\Delta}$ in the desired response signal path, as is done in Fig. 5.3. The plant stays the same, the plant input and output stay the same, and so the whitening filter specified by (5.12) remains the same. Ignoring the requirement of causality for the moment, the transfer function of the optimized filter is obtained as the inverse of the product of the plant and whitening filter transfer functions multiplied by $z^{-\Delta}$. The result is

$$z^{-\Delta}\frac{z^{-1}(1+2z)}{(1+2z^{-1})} = z^{-\Delta}\left(\frac{1}{2} + \frac{3}{4}z - \frac{3}{8}z^2 + \frac{3}{16}z^3 - \cdots\right). \tag{5.17}$$

Now let us find a causal solution. The number of causal terms will be equal to $\Delta + 1$. For example, let $\Delta = 4$. It follows that

$$\left[\frac{z^{-5}(1+2z)}{(1+2z^{-1})}\right]_+ = \frac{1}{2}z^{-4} + \frac{3}{4}z^{-3} - \frac{3}{8}z^{-2} + \frac{3}{16}z^{-1} - \frac{3}{32}. \tag{5.18}$$

The transfer function for the new causal Wiener filter, a delayed plant inverse, is

$$\left[\frac{1 - z^{-1} + \frac{3}{4}z^{-2}}{z^{-1}(1+2z)}\right]\left[\frac{z^{-(\Delta+1)}(1+2z)}{(1+2z^{-1})}\right]_+. \tag{5.19}$$

The minimum mean square error for $\Delta = 4$ can be shown to be approximately 0.003. This represents very low error. The new causal Wiener filter would thus make a very good controller. If a much larger value of Δ were chosen however, many more terms in the series would be included and an even more perfect delayed inverse would be the result. But the delay in the overall control system response would of course be greater if the inverse filter were used as a controller. With infinite delay, the delayed inverse would be perfect but useless from a practical point of view. It is clear that increasing Δ reduces the minimum mean square error. For any nonminimum-phase plant, this would generally be the case. For any minimum-phase plant, $\Delta = 0$ would suffice, except when the plant has more poles than zeros, then $\Delta = 1$ would suffice.[1]

The above analysis is based on the assumption that the adaptive inverse filter $\widehat{C}_\Delta(z)$ is causal and has infinite impulse response (IIR). If $\widehat{C}_\Delta(z)$ is causal and of finite impulse response (FIR), then increasing Δ beyond the point of any reasonable need could be harmful. One would in effect cause the impulse response to be "pushed out of the time window" of $\widehat{C}_\Delta(z)$. Figure 5.4 shows the impulse response for the plant represented by Eq. (5.7). Figure 5.5(a)–(e) shows delayed inverse impulse responses for various values of Δ. These plots have been developed for $\widehat{C}_\Delta(z)$ being an FIR tapped delay line with 11 weights. Over a broad middle range of Δ, from about 5 through 9, the optimized impulse response has approximately the same shape except for delay. Outside that range, the choice of Δ affects the shape of the impulse response. Figure 5.6 shows a plot of minimum mean square error versus Δ for the above example.

Experience with the formation of inverses for a variety of plants of widely varying complexity has shown that the shape of the curve of Fig. 5.6 is typical, with a wide almost flat minimum. Choice of Δ is generally not critical within the flat minimum. With no other information, a good rule of thumb would make Δ about equal to half the chosen length of $\widehat{C}_\Delta(z)$. From there on one could experiment with Δ, generally keeping this parameter as small as possible, consistent with low adaptive inverse modeling error.

The smallest value of Δ that minimized the minimum MSE plotted in Fig. 5.6, that is, $\Delta = 6$, gave an excellent delayed inverse with extremely low error. Convolving the inverse impulse response with the plant impulse response provided the impulse response of their cascade, and this is plotted in Fig. 5.7(a). If the inverse were perfect, this plot would show zero response at all times except for a unit response at the chosen value of Δ, that is, $\Delta = 6$. Imperfections in the inverse would be evidenced by small *side lobes* existing about the main

[1] When the analog plant is discretized, more poles than zeros causes the discrete impulse response to begin after a delay of one sample period.

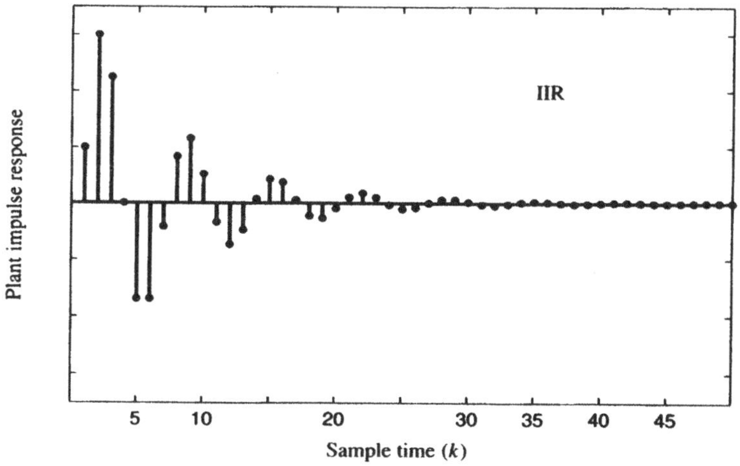

Figure 5.4 Plant impulse response.

spike at $k = 6$. Such imperfections are almost nonexistent in this plot. The step response of the cascaded plant and delayed inverse is shown in Fig. 5.7(b). The result is an almost perfect step, delayed by 6 units of time. This is the step response of the corresponding adaptive inverse control system.

From the simple analytical example studied above, one can see the possibilities for finding excellent delayed inverses for both minimum-phase and nonminimum-phase plants. We have used Shannon-Bode theory to determine optimal causal IIR transfer functions for $\widehat{C}_\Delta(z)$ and to determine associated values of mean square error. This theory can only be used when there exists a precise representation of the plant $P(z)$ in pole-zero form. In the real world, $P(z)$ would generally not be known. Instead of using Shannon-Bode theory, a simple adaptive least squares algorithm would be used to determine the best $\widehat{C}_\Delta(z)$. The adaptive solution would closely approximate the true Shannon-Bode result. Although Shannon-Bode works with IIR filters, the adaptive inverse filter would generally be an FIR filter. The FIR filter architecture allows adaptation with the LMS algorithm, and there being no poles, there would be no stability problem.

5.3 MODEL-REFERENCE INVERSES

An adaptive process is illustrated in Fig. 5.8 for finding a model-reference plant inverse $\widehat{C}_M(z)$. The goal of this process is to obtain a controller $\widehat{C}_M(z)$ that, when used to drive the plant, would result in a control system whose overall transfer function would closely match the transfer function $M(z)$ of a given reference model. The delayed inverse modeling scheme of Fig. 5.3 is a special case, where the model response is a simple delay with a transfer function of $z^{-\Delta}$.

To demonstrate how the model-reference inverse works, the following experiment was done. Figure 5.9 shows the step response of an exemplary plant $P(z)$. It is stable, it has transport delay, and it has a damped oscillatory response. Figure 5.10 shows the step

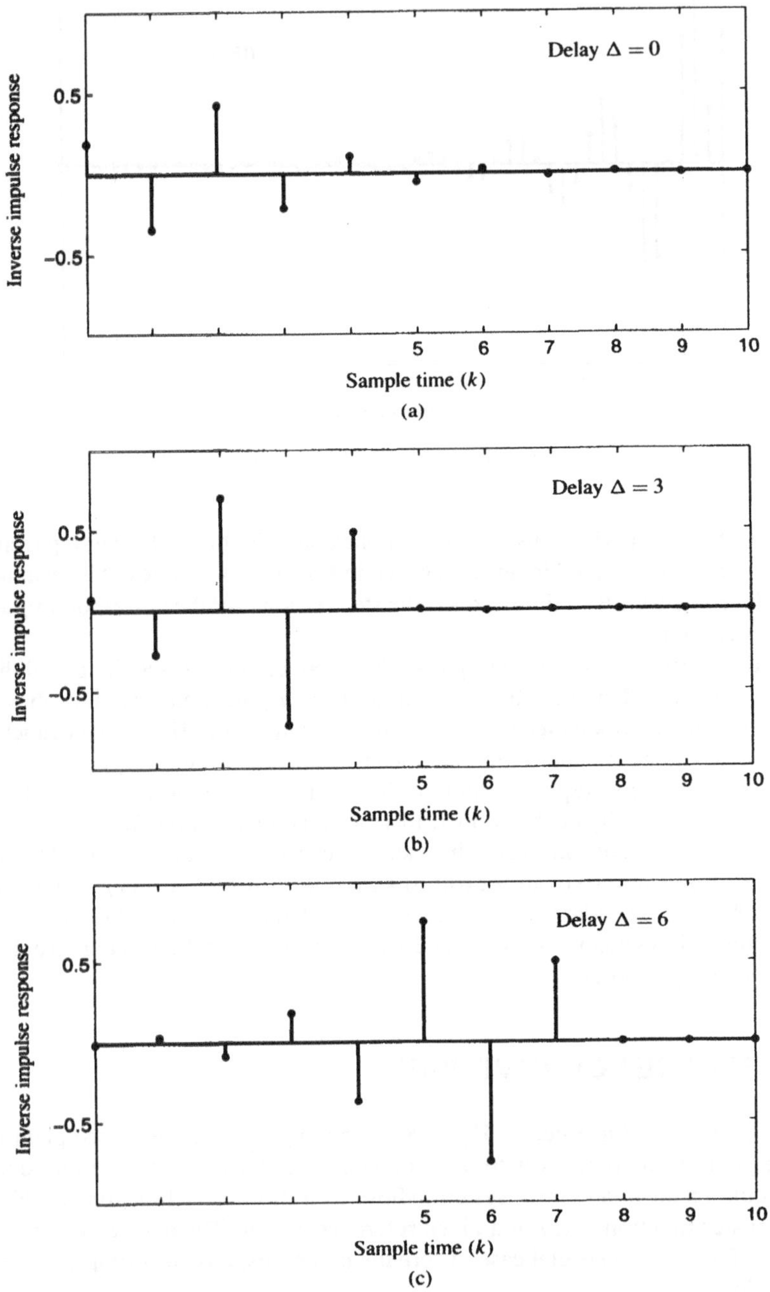

Figure 5.5 Inverse impulse response for various values of delay Δ. Inverse filter is FIR with $m = 11$ weights: (a) $\Delta = 0$ samples; (b) $\Delta = 3$ samples; (c) $\Delta = 6$ samples;

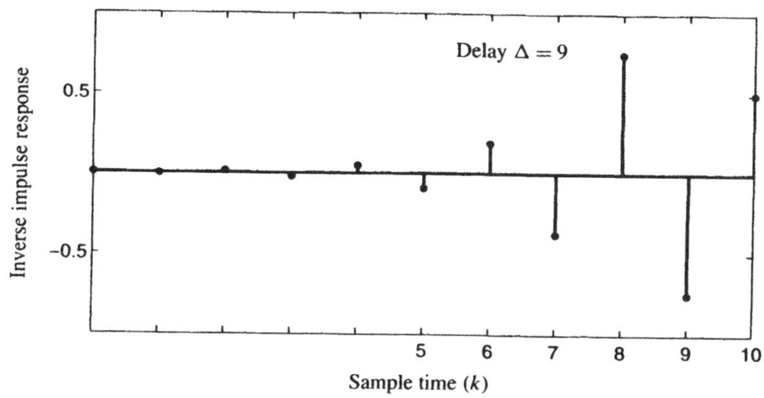

(d)

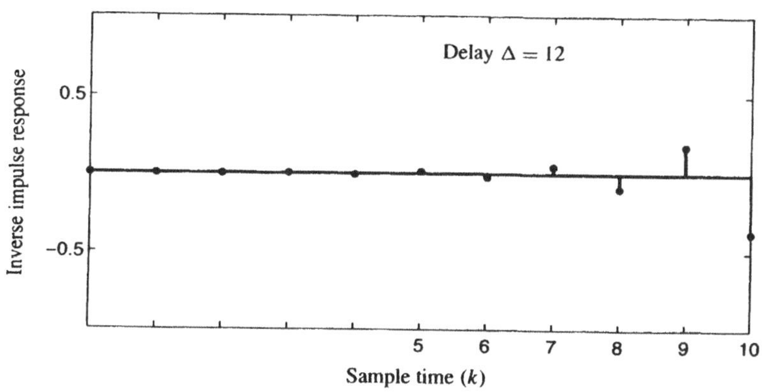

(e)

Figure 5.5 (continued) (d) $\Delta = 9$ samples; (e) $\Delta = 12$ samples.

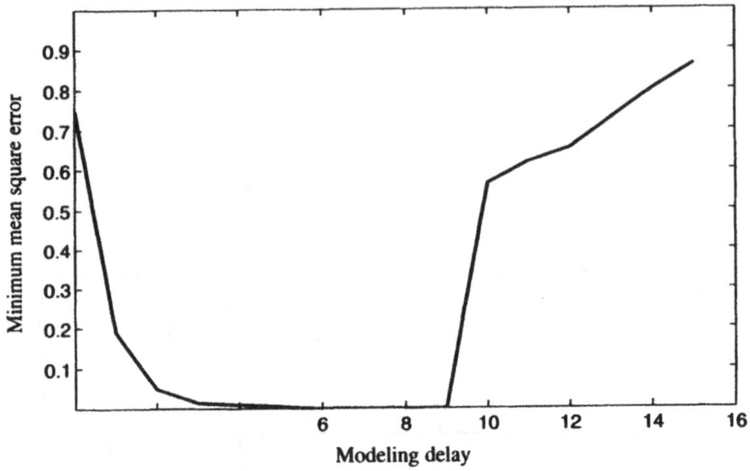

Figure 5.6 Minimum mean square error versus inverse modeling delay Δ.

response of a reference model $M(z)$, designed to have the same transport delay and have an overdamped response. Figure 5.10 also shows the step response of the cascade of the model-reference inverse $\widehat{C}_M(z)$ and the plant $P(z)$. It can be seen that the cascaded step response matches very closely to the model step response, thus illustrating the effectiveness of the model-reference inverse process. For this experiment, the plant transfer function was

$$P(z) = \frac{2.4z^{-1}(1 - 0.8z^{-1})}{(1 + 0.6z^{-1})(1 - 0.7z^{-1})}, \tag{5.20}$$

and the transfer function of the reference model was

$$M(z) = \frac{0.25z^{-1}}{(1 - 0.5z^{-1})^2}. \tag{5.21}$$

For the inverse to have been accurate, it was necessary that the reference model have a delay at least as great as that of the plant, or to have a slow step-response rise having sufficient equivalent delay. As this condition was met, the step response of the controller and plant closely matched that of the reference model.

.4 INVERSES OF PLANTS WITH DISTURBANCES

The inverse modeling process becomes somewhat more complicated when the plant is subject to disturbance. The inverse modeling scheme shown in Fig. 5.11, very much like that of Fig. 5.8, is the simplest approach. Unfortunately, this approach does not work very well with a disturbed plant. The reason is that the plant disturbance, going directly into the input of the adaptive inverse filter $\widehat{C}(z)$, biases the converged Wiener solution and prevents the formation of the proper inverse.

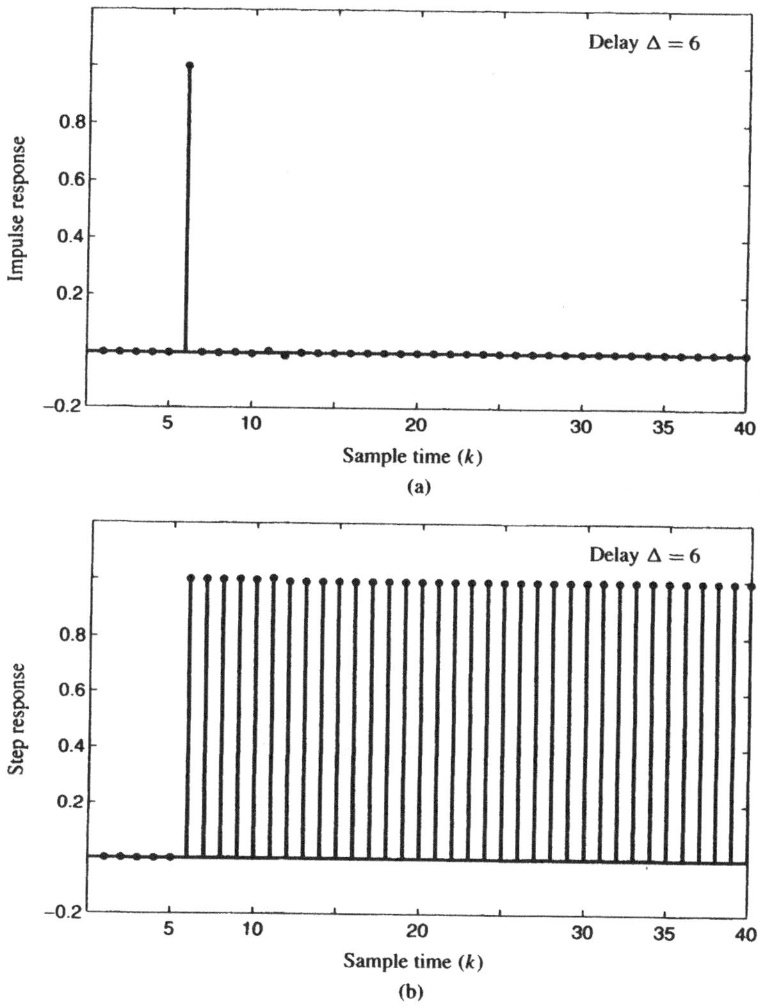

Figure 5.7 Response of cascade of plant and delayed inverse: (a) impulse response; (b) step response.

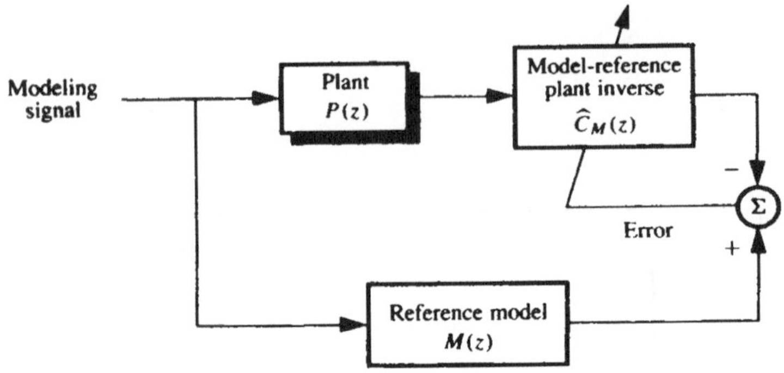

Figure 5.8 Finding model-reference plant inverse.

Without plant disturbance, the Wiener solution would be

$$\hat{C}(z) = \frac{\Phi_{yd}(z)}{\Phi_{yy}(z)} = \frac{M(z)}{P(z)}. \tag{5.22}$$

This would be the result that we seek. Equation (5.22) is in accord with Eqs. (2.30) and (5.6), generalized for the model-reference case. Note that y_k is the plant output without disturbance (as in Fig. 5.8) and d_k is the reference model output.

The situation is different in the presence of plant disturbance. In Fig. 5.11, the disturbed plant output is z_k. The Wiener solution can be written as

$$\hat{C}(z) = \frac{\Phi_{zd}(z)}{\Phi_{zz}(z)}. \tag{5.23}$$

Since the plant disturbance n_k is uncorrelated with d_k and z_k, Eq. (5.23) can be rewritten as

$$\hat{C}(z) = \frac{\Phi_{yd}(z)}{\Phi_{yy}(z) + \Phi_{nn}(z)}. \tag{5.24}$$

The second term in the denominator causes a bias so that, for the disturbed plant,

$$\hat{C}(z) \neq \frac{M(z)}{P(z)}. \tag{5.25}$$

The inverse modeling scheme shown in Fig. 5.12 overcomes the bias problem by making use of an adaptive model of the plant. Instead of finding a model-reference inverse of the plant $P(z)$, the model-reference inverse is taken from $\hat{P}(z)$. The idea is that $\hat{P}(z)$ has essentially the same dynamic response as $P(z)$ but is free of disturbance. A close approximation to the desired model-reference inverse, unbiased by plant disturbance, is thereby obtained.

It should be recalled from a discussion in Chapter 4 that plant disturbance does not affect the Wiener solution when the plant is directly modeled to obtain $\hat{P}(z)$. Plant disturbance does affect the Wiener solution, however, when doing inverse modeling by the method of Fig. 5.11.

The inverse modeling scheme of Fig. 5.12 has been tried and tested with disturbed plants and found to work very well. It should be noted, however, that the existence of high levels of plant disturbance forces one to adapt $\hat{P}(z)$ slowly to keep noise in the weights

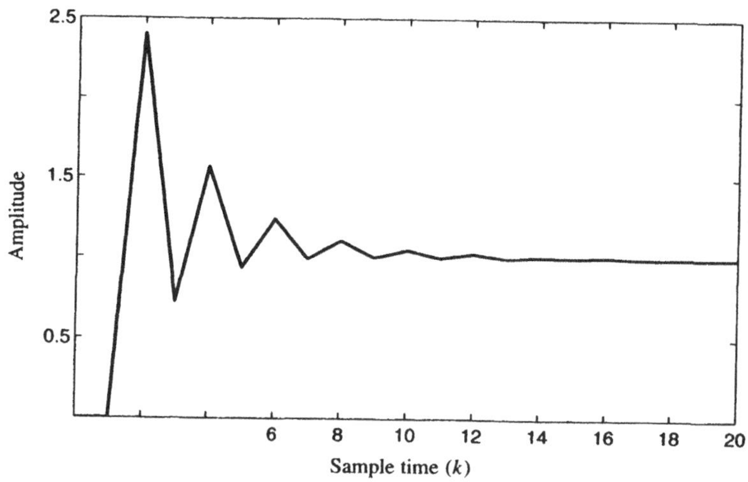

Figure 5.9 Step response of the plant used in the example of model-reference control.

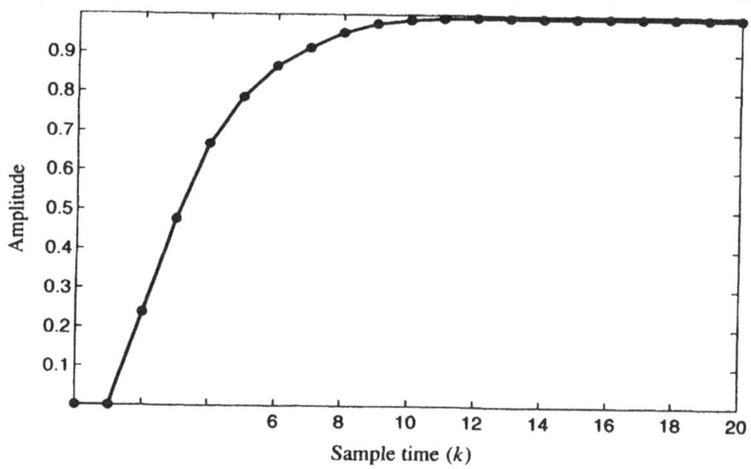

Figure 5.10 Step response of the controlled plant (•) superimposed on the desired step response of the reference model, illustrating successful adaptation.

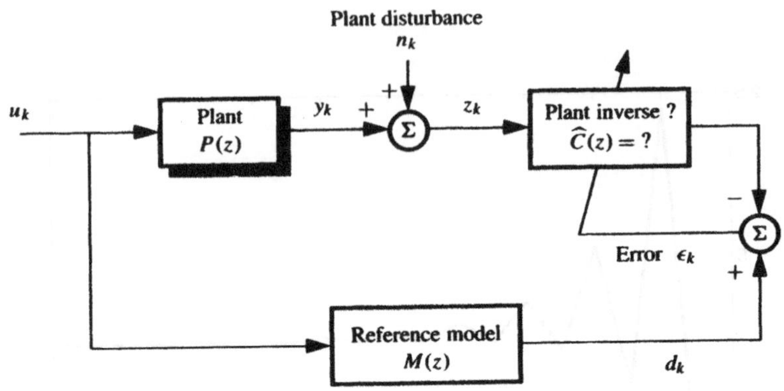

Figure 5.11 An incorrect method for inverse modeling of a plant with disturbance.

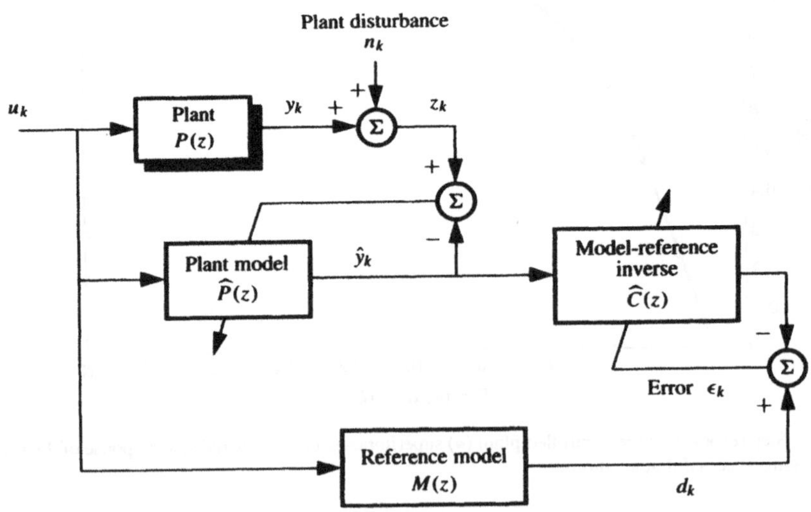

Figure 5.12 A proper method for online inverse modeling of a plant with disturbance.

small. It is important to have low noise in the weights of $\widehat{P}(z)$ so that the weights of $\widehat{C}(z)$ will also have low noise.

The system of Fig. 5.12 adapts the weights of $\widehat{C}(z)$ *online*. The input to $\widehat{P}(z)$ is u_k, the actual plant input. The output of $\widehat{P}(z)$ drives the input of $\widehat{C}(z)$. The adaptive process for obtaining $\widehat{C}(z)$ will always be lagging behind that for obtaining $\widehat{P}(z)$. The two adaptive processes work in cascade. To allow the adaptive process for $\widehat{C}(z)$ to keep up with the changes in $\widehat{P}(z)$ without lag, the *offline* scheme of Fig. 5.13 is proposed.

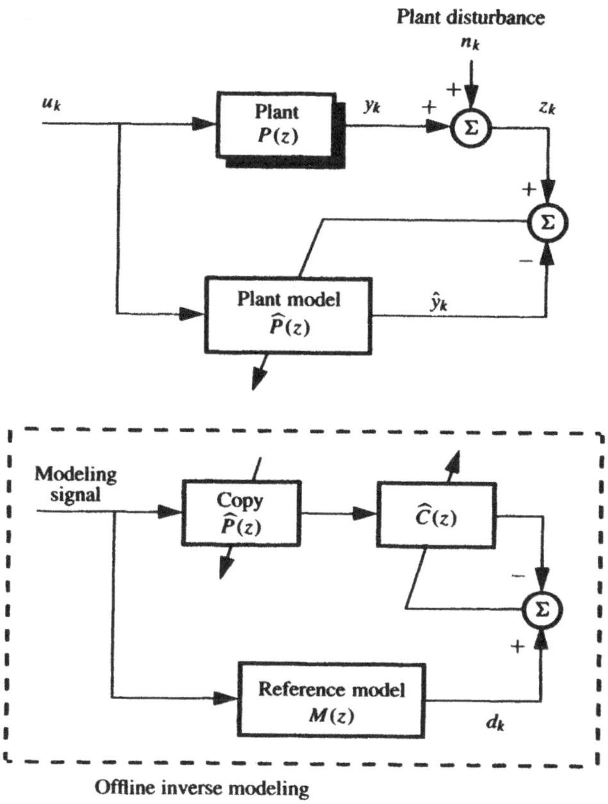

Figure 5.13 An offline process for inverse modeling of a plant with disturbance.

This scheme uses one of the methods of Chapter 4 to obtain $\widehat{P}(z)$. An exact digital copy of $\widehat{P}(z)$ is then used in an offline process to obtain $\widehat{C}(z)$. Computer-generated noise drives the copy of $\widehat{P}(z)$, and its output drives $\widehat{C}(z)$. The same noise drives the reference model $M(z)$, and its output is compared with the output of $\widehat{C}(z)$ to obtain an error signal. $\widehat{C}(z)$ is adapted to minimize the mean square of this error.

The offline inverse modeling process can be run much faster than real time, so fast, in principle, that as instantaneous values of $\widehat{P}(z)$ are obtained from the direct modeling process, the offline inverse modeling process provides corresponding values of $\widehat{C}(z)$, essentially instantaneously.

The noise used to drive the offline inverse modeling process could be white, or better yet, it could be spectrally shaped and designed to provide best results. This noise is called the *modeling signal.*

An alternative to the scheme of Fig. 5.13 would be a method for directly calculating $\widehat{C}(z)$ from knowledge of $\widehat{P}(z)$ and $M(z)$. Wiener theory could be used for the calculation. In some cases, this would be a good idea. In other cases, computation of $\widehat{C}(z)$ with Wiener theory would be computationally intensive.

5.5 EFFECTS OF MODELING SIGNAL CHARACTERISTICS ON THE INVERSE SOLUTION

Assume that $\widehat{C}(z)$ has an impulse response that is not constrained, that it is allowed to have an infinite impulse response in both directions in time. Then $\widehat{C}(z)$ can be adjusted to be equal to $C(z)$, assuming the availability of a perfect $\widehat{P}(z) = P(z)$. As such, any modeling signal that has at least some energy at all frequencies will cause $\widehat{C}(z)$ to be equal to $C(z)$. On the other hand, when $\widehat{C}(z)$ is restricted to be causal and FIR, not enough weights, or degrees of freedom in $\widehat{C}(z)$ are available to perfectly match $C(z)$. Under these conditions, the spectral shape of the modeling signal could have a considerable influence on the frequency response of $\widehat{C}(z)$.

In general, the simplest choice of modeling signal would be white noise. The effect of this choice can be visualized in the following way.

Refer to Fig. 5.13. There will in general be a difference between the frequency response curve of $M(z)$ and that of $\widehat{P}(z) \cdot \widehat{C}(z)$. Optimizing $\widehat{C}(z)$ with a white modeling signal will cause transfer function differences to be weighted equally at all frequencies causing the area of the difference of the two frequency response curves to be minimized. Using a non-white modeling signal causes frequency response differences to be weighted more heavily at frequencies where the modeling signal has higher power density.

The modeling signal characteristics should be chosen to be like those of the command input signal of the entire control system. The controller should be trained to produce the best overall control response. If no a priori knowledge of the command input signal is available, one could simply assume a default position by choosing a white, zero-mean modeling signal.

5.6 INVERSE MODELING ERROR

Thus far we have discussed various forms of inverse, the plant inverse itself, the delayed plant inverse, and the model-reference plant inverse all for unknown plants with and without plant disturbance. In the disturbed plant case, we have assumed that an exact model of the plant $P(z)$ is available to the inversion process. In reality, the plant model $\widehat{P}_k(z)$ would be available instead.

The question is, what effect does mismatch in $\widehat{P}_k(z)$, caused by noise in the adaptive weights, have upon $\widehat{C}_k(z)$? The mismatch $\Delta P_k(z)$ is related by

$$\begin{aligned} \widehat{P}_k(z) &= P(z) + \Delta P_k(z), \quad \text{or} \\ \widehat{\mathbf{P}}_k &= \mathbf{P} + \Delta \mathbf{P}_k. \end{aligned} \tag{5.26}$$

The mismatch in $\widehat{P}_k(z)$ when using the LMS algorithm to find $\widehat{\mathbf{P}}$ has been characterized in Eq. (3.60) by

$$E[\Delta \mathbf{P}_k] = 0, \quad \text{and} \tag{5.27}$$

$$\text{cov}[\Delta \mathbf{P}_k] = \begin{bmatrix} \text{weight} \\ \text{noise} \\ \text{covariance} \\ \text{matrix} \end{bmatrix} = \beta \mathbf{I} = \mu \xi_{\min} \mathbf{I}, \tag{5.28}$$

where β is defined as the individual weight noise variance. The weight noise covariance matrix is a scalar matrix. The weight noise components have equal power and are uncorrelated from weight to weight.

Various modeling schemes were suggested to obtain $\widehat{P}(z)$, with and without dither. Without dither, the weight noise covariance matrix is obtained from Eq. (3.60). With dither using schemes A, B, or C, expressions for the weight noise covariance matrices have been derived and can be obtained from Appendix B. These expressions are all closely approximated by (5.28) for small μ.

The errors in $\widehat{P}(z)$ at each time k cause errors in $\widehat{C}_k(z)$ at each time k. The offline inverse modeling process of Fig. 5.13 causes $\widehat{C}_k(z)$ to instantaneously track and follow $\widehat{P}_k(z)$. Our present objective is to find the relationship between these errors. The errors in $\widehat{C}_k(z)$ have been characterized by

$$\widehat{C}_k(z) = C(z) + \Delta C_k(z), \quad \text{or} \tag{5.29}$$

$$\widehat{\mathbf{C}}_k = \mathbf{C} + \Delta \mathbf{C}_k. \tag{5.30}$$

When $\widehat{C}_k(z)$ is obtained in actual practice by inversion of $\widehat{P}_k(z)$, $\Delta \mathbf{C}_k$ contains truncation effects which give it a bias, plus noise resulting from noise in $\widehat{\mathbf{P}}_k$. Assume that these errors are small and additive. The covariance of $\Delta \mathbf{C}_k$ will be

$$\text{cov}[\Delta \mathbf{C}_k] = \begin{bmatrix} \text{a function of} \\ \text{cov}[\Delta \mathbf{P}_k] \end{bmatrix}. \tag{5.31}$$

The mean of $\Delta \mathbf{C}_k$ will be assumed to be

$$E[\Delta \widehat{\mathbf{C}}_k] = (\text{truncation effects}) \approx 0. \tag{5.32}$$

The reason for Eq. (5.32) is that, in practice, $M(z)$ includes enough smoothing and delay and the impulse response of $\widehat{C}_k(z)$ is chosen to be long enough to keep truncation in $\Delta C_k(z)$ small.

It is useful to relate $\Delta C_k(z)$ to $\Delta P_k(z)$. Refer to Fig. 5.13 and to its offline inverse modeling process. Recall that the ideal inverse would be

$$C(z) = \frac{M(z)}{P(z)}. \tag{5.33}$$

The actual model-reference inverse in Fig. 5.13 is $\widehat{C}_k(z)$ not $C(z)$. Since truncation effects are negligible, $\widehat{C}_k(z)$ will be causal and of the form

$$\widehat{C}_k(z) = \frac{M(z)}{\widehat{P}_k(z)}. \tag{5.34}$$

Causality and small truncation effects require proper choice of $M(z)$, that is, proper choice of its internal delay, smoothness of response, and so forth.

We shall now relate $\Delta C_k(z)$ to $\Delta P_k(z)$. The relation is

$$\begin{aligned}\Delta C_k(z) &= \widehat{C}_k(z) - C(z)\\ &= \frac{M(z)}{P(z)+\Delta P_k(z)} - \frac{M(z)}{P(z)}\\ &\approx -\frac{M(z)\cdot \Delta P_k(z)}{P^2(z)}.\end{aligned} \tag{5.35}$$

Since $\widehat{C}_k(z)$ will be used as the controller of the plant $P(z)$ in a feedforward cascade configuration (illustrated in Fig. 1.3), it is possible to use the above expression for $\Delta\widehat{C}_k(z)$ to determine the overall error of the plant control process. This determination is our next objective.

5.7 CONTROL SYSTEM ERROR DUE TO INVERSE MODELING ERROR

Successful control will give the cascade of $\widehat{C}_k(z)$ and $P(z)$ a dynamic response that closely matches the dynamic response of the reference model $M(z)$. Figure 5.14 shows this cascade and $M(z)$, both driven by the command input $I(z)$ of the control system. The difference in the outputs of $P(z)$ and $M(z)$ may be called the *dynamic system error* $\underline{\epsilon}_k(z)$. This is a dynamic error in the sense that, all else being fixed, it is proportional to the command input $I(z)$. If $\Delta C_k(z)$ were zero, the dynamic system error would also be zero. In general, the dynamic system error is

$$\begin{pmatrix}\text{dynamic}\\ \text{system}\\ \text{error}\end{pmatrix} = \underline{\epsilon}_k(z) = -I(z)\cdot \Delta C_k(z)\cdot P(z). \tag{5.36}$$

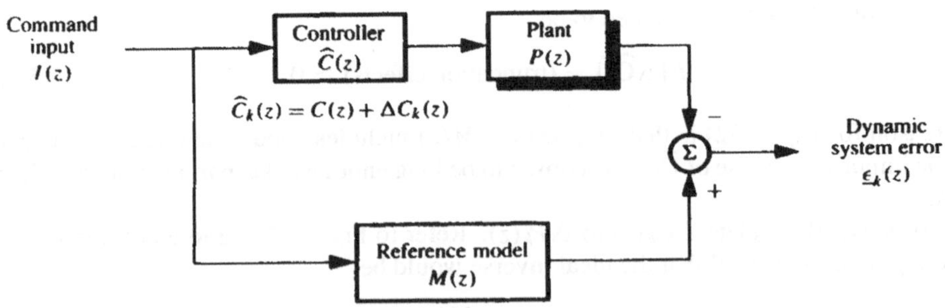

Figure 5.14 Diagram for the definition of dynamic system error.

We are using transforms because we assume as before that changes in $\Delta C_k(z)$ caused by changes in $\Delta P_k(z)$ take place slowly, especially with μ small.

Our next objective is to find the variance of $\underline{\epsilon}_k(z)$. We can replace $\Delta C_k(z)$ in (5.36) by substituting (5.35). The result is

$$\underline{\epsilon}_k(z) = \frac{I(z) \cdot M(z) \cdot \Delta P_k(z)}{P(z)}. \tag{5.37}$$

This is the transform of the dynamic system error. It can be thought of as the output of the slowly time variable system which is pictured in Fig. 5.15. The command input $I(z)$ is applied at point A in this figure, and a filtered version appears at point B. This in turn is applied to the filter $\Delta P_k(z)$ whose weights slowly fluctuate over time k, uncorrelatedly with each other, each weight having zero-mean and variance β. Of interest is the variance at point C. This is equal to the variance at point B multiplied by β multiplied by the number of weights n of $\Delta P_k(z)$ (equal to the number of weights of the plant model $\hat{P}(z)$). The transform of the autocorrelation function at point B is

$$\begin{aligned}\Phi_{BB}(z) &= I(z)I(z^{-1})\frac{M(z)M(z^{-1})}{P(z)P(z^{-1})} \\ &= \cdots + z\phi_{BB}(-1) + \phi_{BB}(0) + z^{-1}\phi_{BB}(1) + \cdots.\end{aligned} \tag{5.38}$$

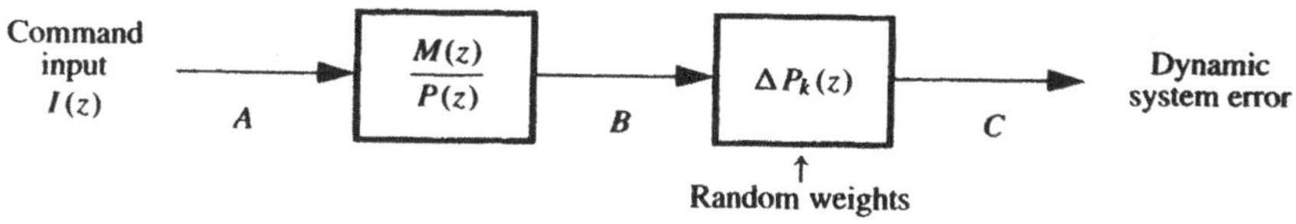

Figure 5.15 Construction of dynamic system error due to fluctuation in $\hat{P}(z)$.

The variance at point B is $\phi_{BB}(0)$. The variance at point C is therefore

$$\begin{pmatrix}\text{variance of} \\ \text{dynamic} \\ \text{system} \\ \text{error}\end{pmatrix} = [\phi_{BB}(0)]n\beta. \tag{5.39}$$

The system designer can predict this error variance given some notion of the command input $I(z)$ and of the plant characteristics $P(z)$. The reference model-response $M(z)$ is known. The adaptive weight noise variance β of $\hat{P}(z)$ is the product of μ and $\xi_{\min}$ of the plant modeling process, in accord with Eqs. (A.39) and (B.54).

In the above discussion, we have assumed that both $\widehat{P}_k(z)$ and $\widehat{C}_k(z)$ were chosen long enough so that there would be no truncation effects. Making $\widehat{P}_k(z)$ short would be advantageous, as long as there would be no truncation effects. This is clear from Eq. (5.39), which shows the dynamic system error variance to be proportional to the number of weights of $\widehat{P}_k(z)$. In this derivation, it has been assumed that $\widehat{C}_k(z)$ is very long, essentially infinite.

A generalization of this theory, too lengthy and detailed to include here, shows that as long as there are no truncation effects, making $\widehat{C}(z)$ short could also be advantageous. The general formula for dynamic error variance obtained from this theory is

$$\begin{pmatrix} \text{variance of} \\ \text{dynamic} \\ \text{system} \\ \text{error} \end{pmatrix} = [\phi_{BB}(0)] \cdot \min(m, n) \cdot \beta, \tag{5.40}$$

where $\min(m, n)$ means to use the smaller of the number of weights, m of $\widehat{C}_k(z)$, or n of $\widehat{P}_k(z)$. This result has been checked by simulation studies and it works very well. An example will be presented in the next section.

Equation (5.40) is simple and easy to use. The variance β is the product of μ and $\xi_{\min}$, and $\xi_{\min}$ is equal to the variance of the plant output noise when modeling without dither or when modeling with dither schemes A or C. Scheme B would require a special calculation. By knowing the number of weights in $\widehat{P}(z)$ and $\widehat{C}(z)$, one can determine $\min(m, n)$. Finally, $\phi_{BB}(0)$ is the variance resulting from the command input going through

$$\frac{M(z)}{P(z)} = C(z). \tag{5.41}$$

In the actual control system, this is essentially equal to the variance resulting from $I(z)$ going through $\widehat{C}(z)$. This is in fact the variance of the control signal applied to the plant.

5.8 A COMPUTER SIMULATION

In order to verify Eq. (5.40), we have performed a computer simulation of the system of Fig. 5.14. The plant was chosen to be

$$P(z) = \frac{1 - \frac{1}{3}z^{-1}}{1 - \frac{1}{2}z^{-1}}. \tag{5.42}$$

The reference model was chosen to be

$$M(z) = 1. \tag{5.43}$$

The plant controller $\widehat{C}_k(z)$ was computed as follows:

a. The plant model $\widehat{P}_k(z)$ was obtained by an adaptive process using dithering scheme C of Fig. 4.3(c).

b. The optimal controller $\widehat{C}_k(z)$ was computed algebraically at each time k from $\widehat{P}_k(z)$ using Eq. (5.34).

For each time k, the weights of $\widehat{C}_k(z)$ were used in order to compute the dynamic system error. Various simulation parameters were chosen as follows:

a. The command input was a zero-mean white signal of unit power, $E[i_k^2] = 1$.

b. The dither was a zero-mean white noise of unit power, $E[\delta_k^2] = 1$.

c. The adaptation constant of the process for finding $\widehat{P}_k(z)$ was set at $\mu = 0.001$.

d. The plant disturbance was chosen to be zero-mean white noise.

e. The number of weights of $\widehat{P}(z)$, n, was a variable parameter. The number of weights of the controller $\widehat{C}(z)$, m, was a variable parameter. The plant disturbance power $E[n_k^2]$ was a variable parameter.

The system was allowed to run for 2,000 samples (about four time constants) to achieve steady-state performance before taking data. Then the average variance of the dynamic system error was measured. Experimental values were compared with theoretical ones obtained from Eq. (5.40).

In order to use (5.40), we needed values for β and for $[\phi_{BB}(0)]$. The value of β is the product of μ and the minimum mean square error in modeling for $\widehat{P}(z)$. In our case,

$$\xi_{\min} = E[n_k^2],$$

which was a variable parameter, and

$$\beta = \mu\xi_{\min} = (0.001)E[n_k^2]. \tag{5.44}$$

The value of $[\phi_{BB}(0)]$ was approximated as the output power of the ideal controller $C(z)$ fed with the command input. For this case, the ideal controller was

$$C(z) = \frac{M(z)}{P(z)} = \frac{1 - \frac{1}{2}z^{-1}}{1 - \frac{1}{3}z^{-1}}. \tag{5.45}$$

The impulse response of $C(z)$ was

$$c_k = \begin{cases} 0, & k < 0 \\ 1, & k = 0 \\ -\frac{1}{6}\left(\frac{1}{3}\right)^k, & k \geq 1 \end{cases}. \tag{5.46}$$

Since the command input signal was white,

$$\begin{aligned} \phi_{BB}(0) &= E[i_k^2]\begin{pmatrix} \text{sum of squares of} \\ \text{impulses of impulse} \\ \text{response of } C(z) \end{pmatrix} \\ &= 1 \cdot \left(\frac{33}{32}\right) \\ &\approx 1. \end{aligned} \tag{5.47}$$

Using (5.44) and (5.47), the variance of the dynamic system error was estimated for each choice of the design parameters. The results are summarized in Table 5.1. For a wide range of choices of β, m, and n, the simulation results matched the theoretical predictions to within about 25 percent.

EXAMPLES OF OFFLINE INVERSE MODELING OF NONMINIMUM-PHASE PLANTS

The unstable nonminimum-phase plant described in Section 4.4 was stabilized with feedback. The block diagram of the stabilization loop is shown in Fig. 4.4. Two cases were

TABLE 5.1 VARIANCE OF DYNAMIC SYSTEM ERROR AS A FUNCTION OF n (NUMBER OF WEIGHTS OF PLANT MODEL), m (NUMBER OF WEIGHTS OF THE CONTROLLER), AND β (VARIANCE OF NOISE IN A SINGLE PLANT MODEL WEIGHT)

			Variance of Dynamic System Error	
β	n	m	Estimated	Theoretical
10^{-5}	10	10	0.000100	0.000137
10^{-5}	50	10	0.000100	0.000063
10^{-5}	10	50	0.000100	0.000074
10^{-5}	50	50	0.000500	0.000605
10^{-4}	10	10	0.00100	0.00094
10^{-4}	50	10	0.00100	0.00104
10^{-4}	10	50	0.00100	0.00076
10^{-4}	50	50	0.00500	0.00542

studied in Chapter 4, with $k = 21$ and $k = 24$. The discretized impulse response of the stabilized plant is damped and oscillatory for $k = 21$ (see Fig. 4.7). This impulse response is much less oscillatory for $k = 24$ (see Fig. 4.14). For both of these cases, inverse modeling was done using the offline method illustrated in Fig. 5.13. Results are presented in this section.

For $k = 21$, the discretized transfer function $G(z)$ of the stabilized plant is given by Eq. (4.24). A 400-weight adaptive model of $G(z)$ was made, and its impulse response is shown in Fig. 4.8. A delayed inverse of this impulse response has been obtained. Using a delay of 50 sampling periods, a 100-weight inverse has been taken, and the resulting impulse response is shown in Fig. 5.16. To demonstrate that the inverse is really an inverse, its impulse response in Fig. 5.16 was convolved with the true discretized impulse response of the stabilized plant, shown in Fig. 4.7. The result is very good. The first 100 samples of the convolution are shown in Fig. 5.17.

A result almost as good could have been obtained with substantially less delay, say 10 sampling periods rather than 50. The reason for this can be determined from inspection of Fig. 5.16. The inverse impulse response is almost zero until about 40 sampling periods, 10 sampling periods before 50.

Noise was injected into the plant to create the effect of plant disturbance. A 400-weight adaptive model was made whose impulse response is shown in Fig. 4.10. Distortion in the model was caused by the disturbance. This distortion got smaller as the rate of adaptation was made smaller. Using the distorted model of Fig. 4.10, a 100-weight inverse with a delay of 50 sampling periods was taken. This inverse impulse response is shown in Fig. 5.18. Convolving it with the true discretized impulse response of the stabilized plant yields the result shown in Fig. 5.19. This is not a perfect delayed unit impulse. Although the inverse is far from perfect, it is still highly effective as a controller. This subject is addressed in Chapter 6.

Changing the gain in the plant stabilization loop to $k = 24$, the discretized transfer function $G(z)$ of the stabilized plant is given by Eq. (4.25). A 150-weight adaptive model of $G(z)$ was made, and its impulse response is shown in Fig. 4.15. With a delay of 50 sampling periods, a 100-weight inverse of the 150-weight model of $G(z)$ has been taken, and

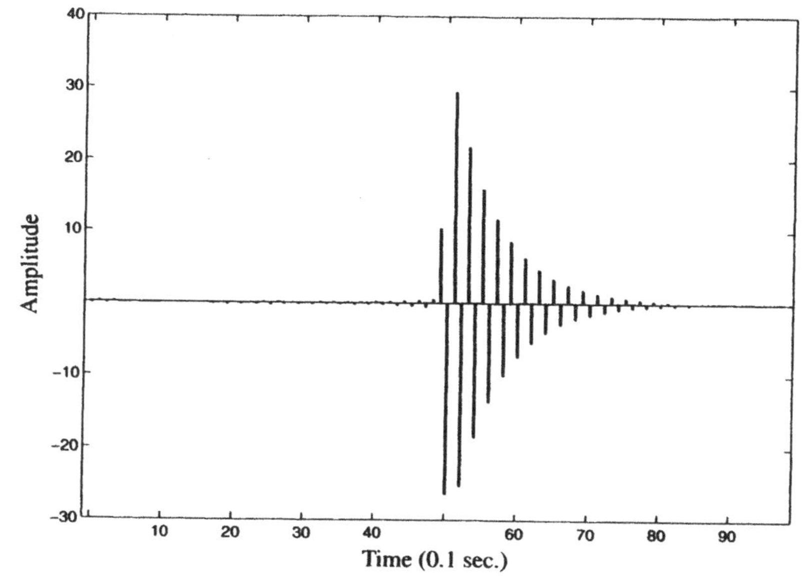

Figure 5.16 Delayed inverse of discretized stabilized ($k = 21$) nonminimum-phase plant model (Fig. 4.8).

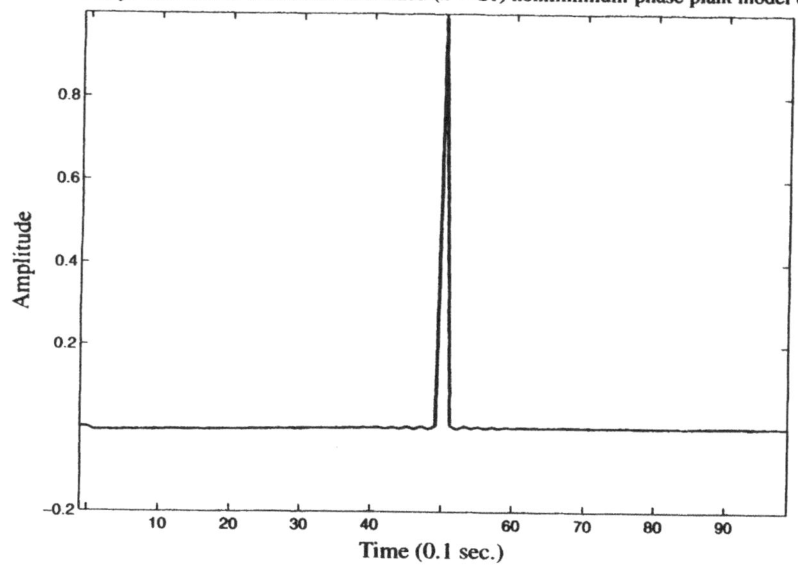

Figure 5.17 Convolution of true $G(z)$ for $k = 21$ with delayed inverse of its 400-weight model (Fig. 4.8).

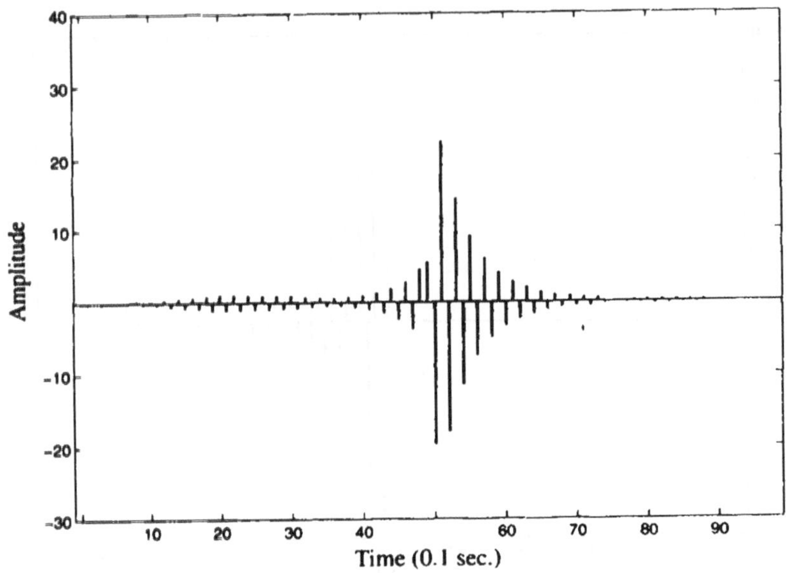

Figure 5.18 Delayed inverse of discretized stabilized ($k = 21$) nonminimum-phase plant model subject to disturbance (Fig. 4.10).

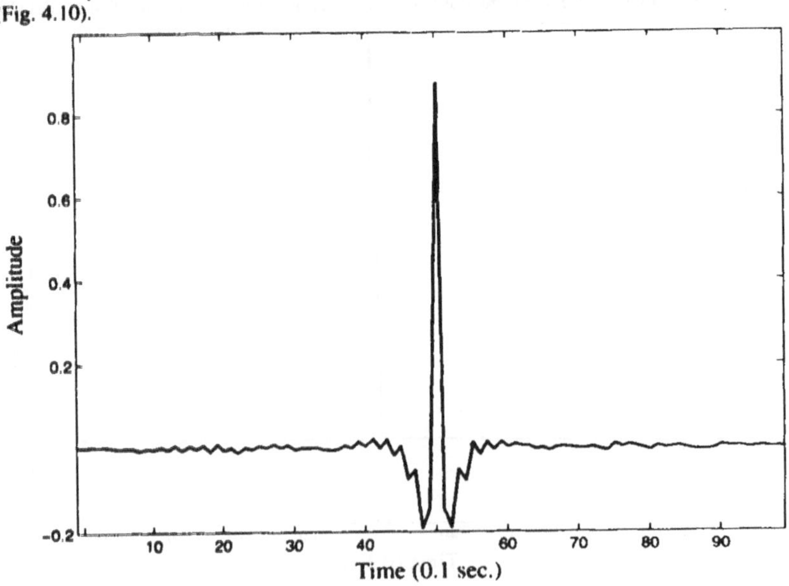

Figure 5.19 Convolution of true $G(z)$ for $k = 21$ with delayed inverse of its 400-weight model affected by plant disturbance (Fig. 4.10).

the impulse response of this inverse is shown in Fig. 5.20. Convolving this with the impulse response of the true $G(z)$ inverse with the impulse response of the true $G(z)$, the result of Fig. 5.21 is obtained. It is clear from the convolution that the inverse filter would make a superb controller.

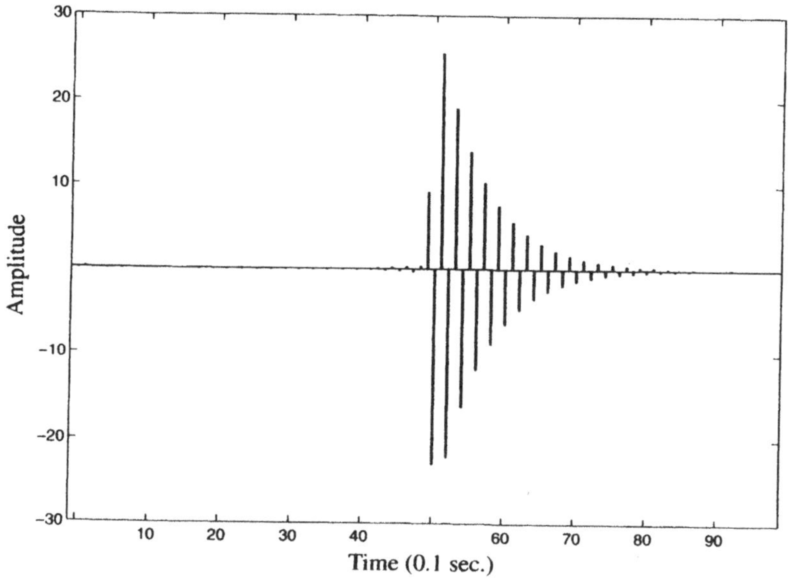

Figure 5.20 Delayed inverse of discretized stabilized ($k = 24$) nonminimum-phase plant model (Fig. 4.15).

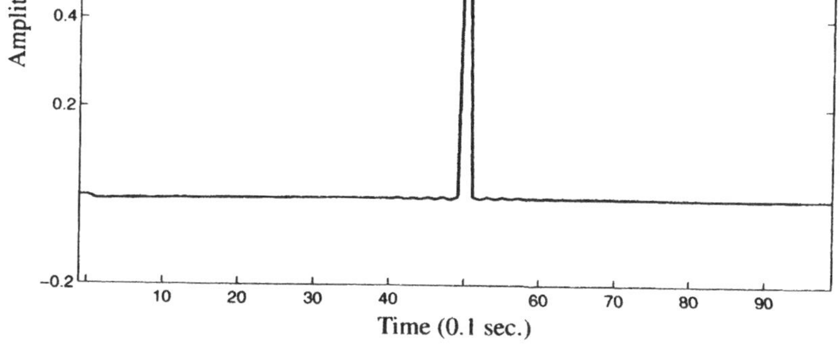

Figure 5.21 Convolution of true $G(z)$ for $k = 24$ with delayed inverse of its 150-weight model (Fig. 4.15).

Adding noise to the plant output to simulate the effects of plant disturbance, a noisy 150-weight model of $G(z)$ is obtained as in Fig. 4.17. The impulse response of the 100-weight delayed inverse is shown in Fig. 5.22. Convolving these gives the result of Fig. 5.23. The convolution is not a perfect impulse. In spite of this, the inverse filter is capable of excellent performance as a controller, as will be demonstrated in Chapter 6. The model of $G(z)$, $\widehat{G}(z)$, would have been less noisy and this would have given a much more precise inverse if the adaptation of $\widehat{G}(z)$ were done more slowly with a smaller value of μ.

5.10 SUMMARY

An adaptive least squares algorithm can be used to set the weights of a filter to approximate the inverse of an unknown plant. If the plant is minimum-phase, an accurate inverse can be readily made. If the plant is nonminimum-phase, an accurate inverse can be made providing a delayed response. The technique can be easily extended to form model-reference inverses, so that the plant and inverse when cascaded have a transfer function that closely matches that of a selected reference model. The inverse filter can be used as a controller to drive the plant.

Generally, physical plants have disturbance. Adapting to form a plant inverse or a model-reference plant inverse in the most straightforward way results in a biased inverse when the plant is disturbed. Techniques to overcome this difficulty were presented in this chapter. The inverse was based on a model of the plant which was not biased by the plant disturbance.

Gradient noise in the weights of $\widehat{P}_k(z)$ causes noise in the weights of $\widehat{C}_k(z)$ since the latter is obtained from the former. Noise in the weights of $\widehat{C}_k(z)$ causes error in the dynamic response of the overall control system. Analysis of this error has resulted in the formula

$$\begin{pmatrix} \text{variance of dynamic} \\ \text{system error} \end{pmatrix} = [\phi_{BB}(0)] \cdot \min(m, n) \cdot \beta. \tag{5.40}$$

The system designer may not have all of the factors of (5.40) exactly, but in most cases would have some idea about the sizes of these elements and would therefore be able to make a good estimate of the variance of the dynamic system error. This is important in the determination of overall system performance. Equation (5.40) was checked with regard to a simulated adaptive inverse control system, and experiment and theory agreed in a wide variety of cases to within 25 percent.

In the next chapter, we shall examine various ways of using the inverse plant model.

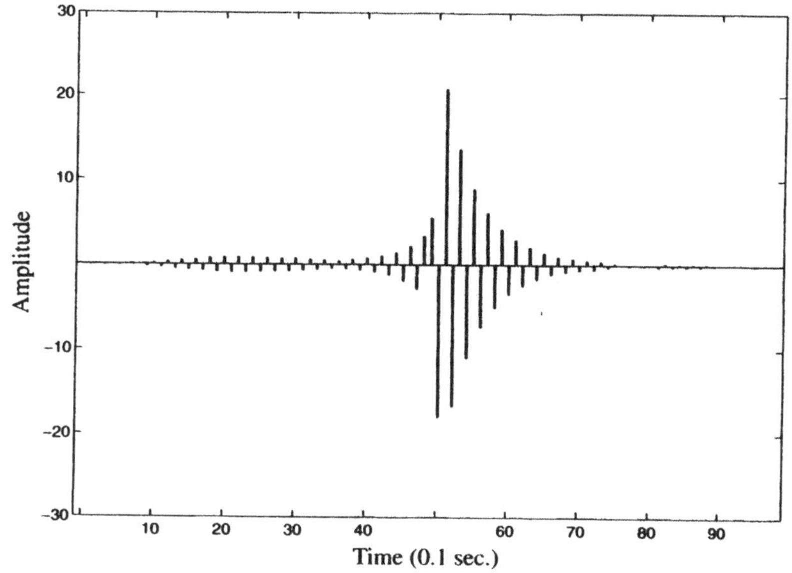

Figure 5.22 Delayed inverse of discretized stabilized ($k = 24$) nonminimum-phase plant model subject to disturbance (Fig. 4.17).

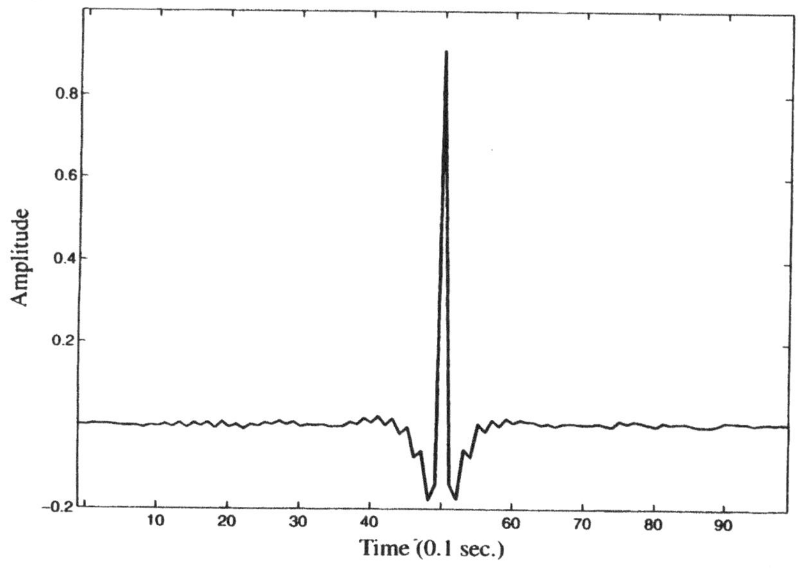

Figure 5.23 Convolution of true $G(z)$ for $k = 24$ with delayed inverse of its 150-weight model affected by plant disturbance (Fig. 4.17).

Chapter 6

Adaptive Inverse Control

6.0 INTRODUCTION

An adaptive inverse control system is diagrammed in Fig. 6.1. If the controller were ideal, its transfer function would be

$$C(z) = \frac{M(z)}{P(z)}. \tag{6.1}$$

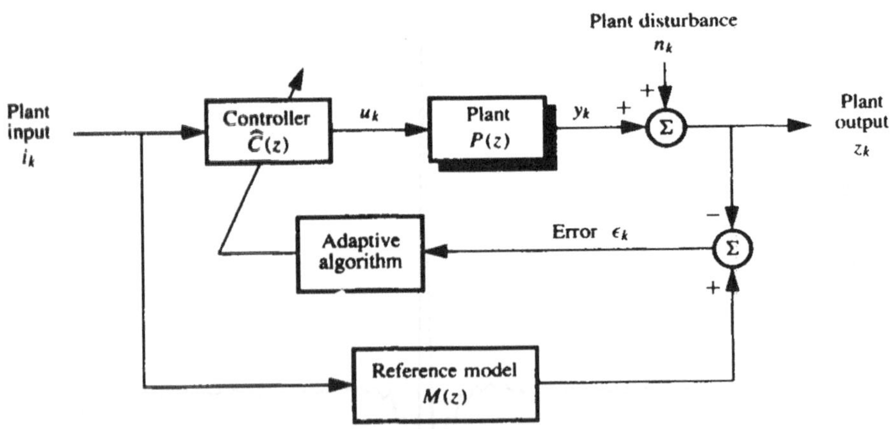

Figure 6.1 An adaptive inverse control system that works well but adapts slowly.

The adaptive controller will generally not be ideal; its transfer function can therefore be designated as

$$\hat{C}(z) = C(z) + \Delta C(z). \tag{6.2}$$

The controller weight vector can be expressed accordingly as

$$\hat{\mathbf{C}} = \mathbf{C} + \Delta\mathbf{C}. \tag{6.3}$$

The LMS algorithm cannot be used to adapt the controller of Fig. 6.1. Many other adaptive algorithms can be used, however, to automatically adjust the weights of $\widehat{C}(z)$ of Fig. 6.1. Examples are the differential steepest-descent (DSD) algorithm and the linear random search algorithm (LRS) of reference [1]. When using these algorithms, changes in the controller weights are made to minimize the measured mean square error. Each time the controller weights are changed, time must be allowed for statistical equilibrium to develop in the plant before measuring MSE.

The DSD algorithm is based on the method of steepest descent. It uses a gradient vector which is obtained one component at a time, and each gradient component is obtained by measuring MSE with the corresponding weight increased and held for some time, then decreased and held for some time. The LRS algorithm tries random changes in the weight vector. After each trial change, the MSE is measured and compared with the measured MSE before the trial change. The actual weight vector change is made equal to the trial change multiplied by the MSE difference (before and after the trial change). If the trial change causes an improvement in performance, a lowering of MSE, the actual weight change will be in the trial direction and proportion to the improvement. If the trial change causes a reduction in performance, then the actual change will be opposite in direction to the trial direction and proportional to the reduction in performance. The DSD and LRS algorithms perform similarly, except that LRS converges twice as slowly as DSD when both adapt with the same level of misadjustment. Both of these algorithms converge extremely slowly compared to LMS when they are all set to adapt with the same level of misadjustment.

It would be desirable to use the LMS algorithm because it is much faster than LRS and DSD. It cannot be used directly because the available error ϵ_k of Fig. 6.1 is an error referred to the plant output.[1] LMS really needs an error referred to the plant input, that is, to the adaptive controller output. To get an appropriate error for LMS implementation, one would need to apply ϵ_k to the inverse of the plant $P(z)$, thus requiring the solution in order to get the solution. The system of Fig. 6.1 is not our system of choice.

In order to be able to make use of the LMS algorithm and other high-speed adaptive processes, the inverse modeling configuration of Fig. 6.2 has been devised. The plant and its inverse model are commuted, so that the error ϵ_k is directly available for the adaptation of $\widehat{C}(z)$. Once $\widehat{C}(z)$ is obtained, an exact digital copy can be used as a controller for the plant. This adaptive control system concept was first proposed in reference [2].

The system of Fig. 6.2 works very well as long as there is no plant disturbance. If plant disturbance is present, its effect is to bias the Wiener solution so that $\widehat{C}(z)$ will not be a proper controller. The disturbance that appears at the plant output adds a component to the covariance of the input signal of the adaptive inverse model, directly affecting the Wiener solution for $\widehat{C}(z)$. So what should one do? There are a number of choices, and the approach indicated in Fig. 6.3 offers the possibility of rapid adaptation and proper control even in the presence of plant disturbance.

The control system of Fig. 6.3 is based on the inverse modeling scheme of Fig. 5.13. It works in the following way. A model $\widehat{P}_k(z)$ of the plant $P(z)$ is formed, using in this case dither scheme A. An offline process can be used to obtain controller $\widehat{C}_k(z)$ from a digital copy of $\widehat{P}_k(z)$, and the reference model $M(z)$. The offline process, illustrated in

[1] Nor can any of the exact least squares lattice algorithms be used in this application for the same reasons.

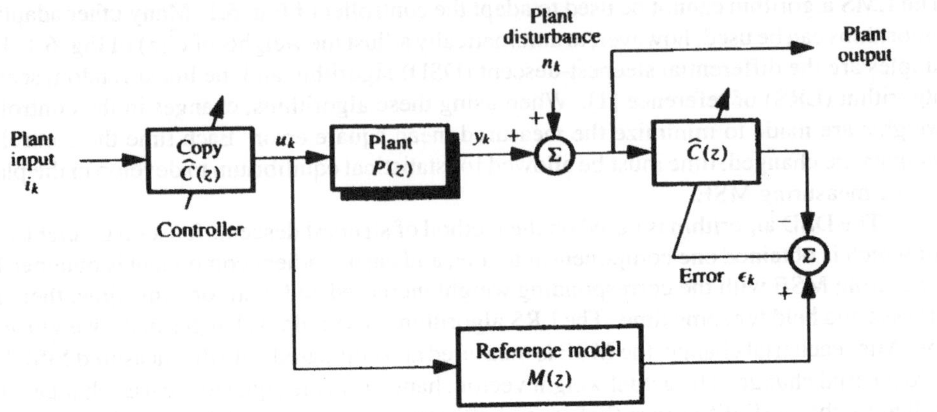

Figure 6.2 An adaptive inverse control system with commuted plant and inverse model. It works well only when plant disturbance has low level.

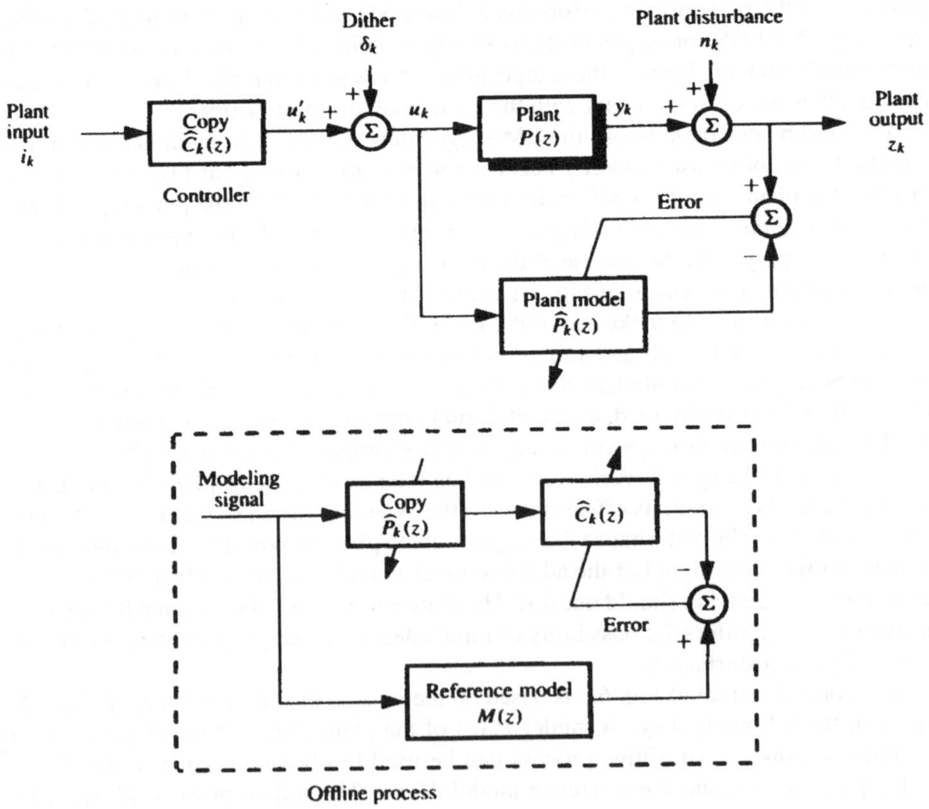

Figure 6.3 An adaptive inverse control system with offline inverse modeling adapts rapidly and works well even with plant disturbance.

Fig. 6.3, adapts $\widehat{C}_k(z)$ so that the output of the cascade of $\widehat{P}_k(z)$ and $\widehat{C}_k(z)$ becomes a best least squares match to the output of the reference model $M(z)$.[2] Both the cascade and the reference model are driven simultaneously by the same *modeling signal*. This signal is synthesized to have an appropriate spectral character, like that anticipated for the command input.

The process for finding $\widehat{C}(z)$ could also be nonadaptive. Fundamentally, $\widehat{C}_k(z)$ is deterministically related to $\widehat{P}_k(z)$ and $M(z)$ for any specified modeling signal spectrum.[3] Now given $\widehat{C}_k(z)$, an exact digital copy of it can be used as a controller, as shown in Fig. 6.3. The result is a controller and plant having an overall dynamic response which closely approximates the optimal dynamic response of the reference model $M(z)$.

The offline process of Fig. 6.3 forms a model-reference inverse of the plant model $\widehat{P}_k(z)$. We have used the model $\widehat{P}_k(z)$ rather than the plant $P(z)$ because the output of the real $P(z)$ is generally corrupted by plant disturbance. Since $\widehat{P}_k(z)$ does not perfectly match $P(z)$ at all times, use of $\widehat{P}(z)$ in determination of $\widehat{C}(z)$ causes errors in $\widehat{C}(z)$. Also, even if $\widehat{P}(z)$ were perfect, there would generally be limitations preventing the perfect realization of $C(z) = M(z)/P(z)$. These limitations will be explored next. We will assume at first a perfect $\widehat{P}_k(z) = P(z)$ and proceed to find best inverses of $\widehat{P}(z)$. Errors in the inversion process will be analyzed. Then the additional effects of errors in $\widehat{P}_k(z)$ will be considered.

6.1 ANALYSIS

Figure 6.4 is a block diagram of an adaptive inverse control system that we will consider for purposes of analysis. The method of plant modeling used here is dither scheme C. Other block diagrams could be drawn incorporating other modeling methods, with or without dither, but we have selected scheme C for presentation and analysis because we believe that this scheme is likely to be a very useful one in the future.

An offline process is shown in Fig. 6.4 for the calculation of controller $\widehat{C}_k(z)$ from a copy of the plant model $\widehat{P}_k(z)$. As discussed above, this offline process could be based on an adaptive algorithm using a modeling signal (which could be white or colored noise), or equivalently $\widehat{C}_k(z)$ could be found by an analytic process such as Levinson's algorithm [3]. A copy of $\widehat{C}_k(z)$ is shown serving as the plant's controller.

The reference model $M(z)$ is used in Fig. 6.4 in the offline inverse modeling process. Another copy of this reference model could be used symbolically for system error monitoring. The dotted portion of the system block diagram suggests this possibility. The overall system error is $E(z)$, and this consists of a sum of four components:

i. Plant disturbance

ii. Dither noise filtered through the plant

[2]When given $\widehat{P}_k(z)$ and $M(z)$, the offline process finds $\widehat{C}_k(z)$. Using an artificial modeling signal, the offline process, if it is an adaptive one, can work at much faster than real-time rates to determine $\widehat{C}_k(z)$. Its convergence time is so short that it does not affect the convergence time of the entire control system.

[3]If the modeling signal is white, $\widehat{C}_k(z) = M(z)/\widehat{P}_k(z)$. $\widehat{C}_k(z)$ could be obtained algebraically, without adaptation. However, when the modeling signal is not white or when $P(z)$ is nonminimum-phase or when $\widehat{C}_k(z)$ is to be realized as an FIR filter, a Wiener solution is needed. An adaptive solution is easy to get. An algebraic solution could also be easily obtained if a computer were available to calculate "R" and "P" and to implement Eq. (3.10).

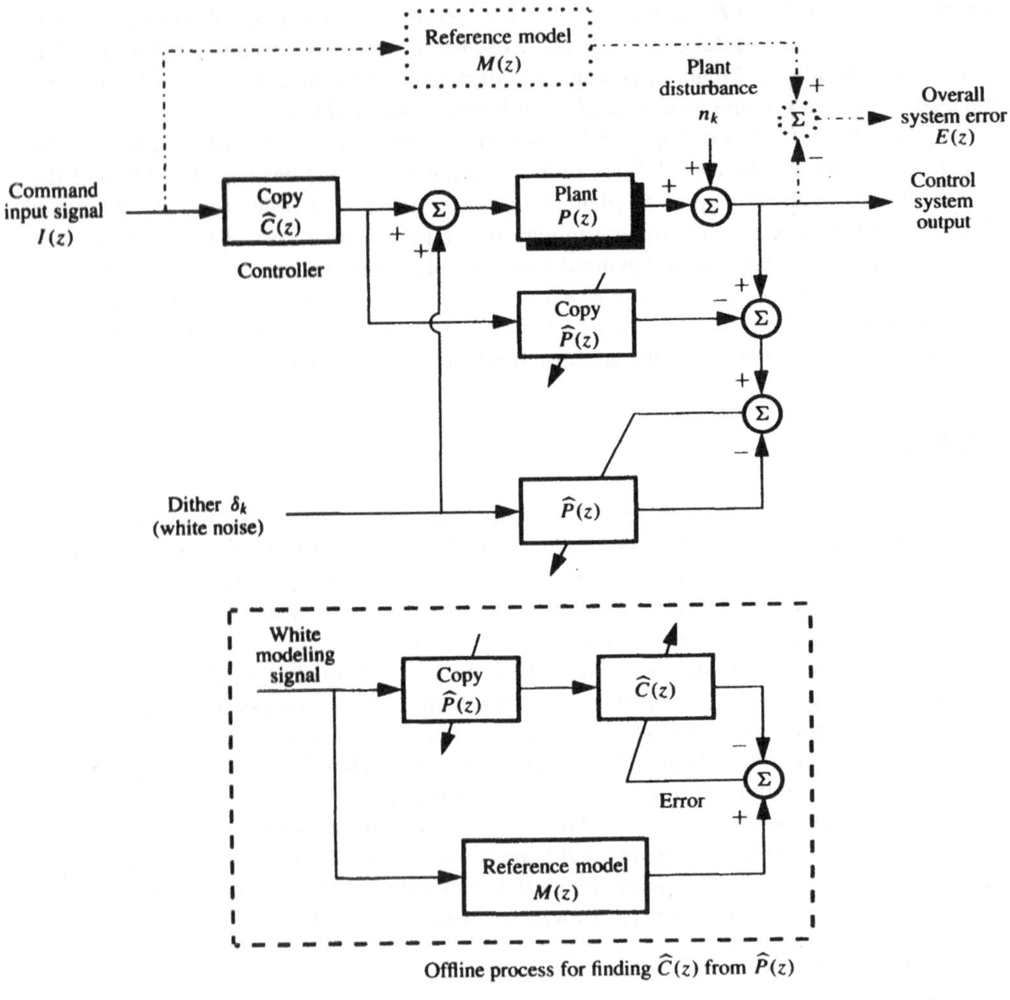

Figure 6.4 A model-reference adaptive inverse control system.

iii. System error due to truncation of $\widehat{P}$ and/or $\widehat{C}$

iv. Dynamic system error

Component (i) is plant output disturbance whose mean square value is

$$\begin{pmatrix} \text{plant} \\ \text{output} \\ \text{disturbance} \\ \text{power} \end{pmatrix} = E[n_k^2]. \tag{6.4}$$

Component (ii) is noise at the plant output due to dither which, if white, would be given by

$$\begin{aligned} \begin{pmatrix} \text{dither noise} \\ \text{at plant output} \end{pmatrix} &= E[\delta_k^2] \cdot \begin{bmatrix} \text{sum of squares of impulses} \\ \text{of impulse response of } P(z) \end{bmatrix} \\ &= E[\delta_k^2] \cdot \left[\sum_{i=0}^{\infty} p_i^2 \right]. \end{aligned} \tag{6.5}$$

Component (iii) is the system error at the plant output due to truncation effects, and *it is assumed to be small or negligible* as a result of an appropriate choice of number of weights n of $\widehat{P}_k(z)$, choice of number of weights m of $\widehat{C}_k(z)$, and an appropriate choice of the reference model $M(z)$.

Component (iv) is the dynamic system error at the plant output, and its power is given by Eq. (5.40).

The overall system error power is the sum of the above described components. It is

$$\begin{pmatrix} \text{overall} \\ \text{system} \\ \text{error} \\ \text{power} \end{pmatrix} = E[n_k^2] + E[\delta_k^2] \cdot \left[\sum_{i=0}^{\infty} p_i^2 \right] + \min(m, n) \cdot \beta \cdot [\phi_{BB}(0)]. \tag{6.6}$$

The first term in (6.6) is the plant disturbance power $E[n_k^2]$, which is independent of the design parameters of the control system. Of special interest are the last two terms in (6.6), which together comprise the excess error power due to imperfection of the control system:

$$\begin{pmatrix} \text{excess} \\ \text{error} \\ \text{power} \end{pmatrix} \triangleq E[\delta_k^2] \cdot \left[\sum_{i=0}^{\infty} p_i^2 \right] + \min(m, n) \cdot \beta \cdot [\phi_{BB}(0)]. \tag{6.7}$$

Minimization of this excess error power is the next objective.

The control system of Fig. 6.4 has only a small number of adjustable parameters once the sampling rate is fixed and the numbers of weights m and n are chosen. The remaining adjustables are the value of μ in the LMS algorithm used in obtaining $\widehat{P}_k(z)$ and the power of the dither signal $E[\delta_k^2]$. The time constant of the adaptive modeling process for $\widehat{P}_k(z)$ is actually the learning time constant for the entire system. This time constant should be chosen to be as long as possible, but not so long that the adaptive process would not be able to keep up with the natural variations in plant characteristics. Some knowledge of the rate of change of plant characteristics would be helpful in fixing the value of the adaptive time constant τ for $\widehat{P}(z)$.

With τ fixed, choosing a specific value of dither power is a simple optimization problem. The objective is to minimize (6.7). Combining Eqs. (B.34) and (B.53) with (6.7) yields, for large values of τ (or small values of μ),

$$\begin{pmatrix}\text{excess}\\\text{error}\\\text{power}\end{pmatrix} = E[\delta_k^2]\cdot\left[\sum_{i=0}^{\infty} p_i^2\right] + \min(m,n)[\phi_{BB}(0)]\frac{E[n_k^2]}{2\tau E[\delta_k^2]}. \tag{6.8}$$

To find the optimal value of $E[\delta_k^2]$, Eq. (6.8) is differentiated with respect to $E[\delta_k^2]$ and the derivative is set to zero. The best choice of dither power is[4]

$$E[\delta_k^2]_{\text{opt}} = \sqrt{\frac{E[n_k^2][\phi_{BB}(0)]\cdot\min(m,n)}{2\tau\sum_{i=0}^{\infty}p_i^2}}. \tag{6.9}$$

The two terms of (6.7) are equal under these conditions:

$$E[\delta_k^2]_{\text{opt}}\cdot\sum_{i=0}^{\infty}p_i^2 = \min(m,n)[\phi_{BB}(0)]\frac{E[n_k^2]}{2\tau E[\delta_k^2]_{\text{opt}}}. \tag{6.10}$$

The two sides of this equation are identified with the plant output power due to dither noise and the plant output power due to dynamic system error. Accordingly, with dither power optimized,

$$\begin{pmatrix}\text{dither noise}\\\text{power at}\\\text{plant output}\end{pmatrix} = \begin{pmatrix}\text{dynamic system error}\\\text{at the plant output}\end{pmatrix}. \tag{6.11}$$

This is the criterion for optimality for the dither power when using scheme C. The minimum value of (6.8) is obtained by using the optimal dither power. The result is

$$\begin{pmatrix}\text{excess}\\\text{error}\\\text{power}\end{pmatrix}_{\min} = \sqrt{\frac{2}{\tau}E[n_k^2]\cdot\phi_{BB}(0)\left[\sum_{i=0}^{\infty}p_i^2\right]\min(m,n)}. \tag{6.12}$$

Figure 6.5 is a generic plot of overall system error power versus dither power based on Eq. (6.8).

6.2 COMPUTER SIMULATION OF AN ADAPTIVE INVERSE CONTROL SYSTEM

In order to demonstrate the workability of an adaptive inverse control system, do plant modeling with scheme C, gain experience with the optimization of dither power, and verify Eqs. (6.8)–(6.12), a computer simulation of the system of Fig. 6.4 was performed. The

[4] An exception to this arises when the plant disturbance has low power, requiring low dither power, in accord with (6.9). The use of low dither power in turn forces a large value of μ, in accord with (B.34). Making μ large could cause instability. [See (B.50)]. The maximum μ in such cases would be limited by stability considerations. Having thus chosen μ and τ, the dither power would be determined by (B.34) rather than (6.9). In any event, the system output error power is obtained from (6.6).

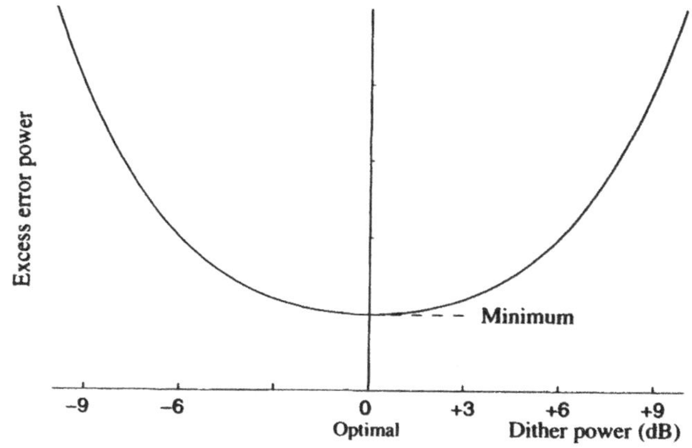

Figure 6.5 Excess error power versus dither power.

system worked very well and proved to be self-contained and quite useful. The formulas worked well also.

The following is a description of the experiment performed. A minimum-phase plant was selected for this test,

$$P(z) = \frac{1}{2} \cdot \frac{1 - \frac{1}{3}z^{-1}}{1 - \frac{1}{2}z^{-1}}. \tag{6.13}$$

The plant disturbance was white, with zero-mean and power of 10, $E[n_k^2] = 10$. The reference model was chosen to be $M(z) = 1$. The command input was a zero-mean white signal of unit power. Scheme C was used for modeling the plant, and instead of using the offline process to obtain $\widehat{C}(z)$ from $\widehat{P}(z)$ and $M(z)$, an algebraic method based on equation (5.22) was used. The simulation was run with different values of adaptation constant μ and with different values of dither power $E[\delta_k^2]$.

The time constant of the adaptive process for $\widehat{P}_k(z)$ depends upon the parameters μ and $E[\delta_k^2]$. These parameters were chosen so that in all cases, the adaptive time constant was 5,000 samples. Accordingly,

$$\tau = \frac{1}{2\mu E[\delta_k^2]} = 5,000. \tag{6.14}$$

There were 10 weights for $\widehat{P}(z)$, and 20 weights for $\widehat{C}(z)$. Each simulation was let run for at least three time constants before excess error was measured.

Theoretically, the optimal dither power was determined by (6.9). To have used this equation, we needed to know $E[n_k^2]$, $[\phi_{BB}(0)]$, $\min(m, n)$, τ, and $\sum_{i=0}^{\infty} p_i^2$. We knew most of this, since $E[n_k^2] = 10$, and $\min(m, n) = 10$ because $m = 20$ and $n = 10$, and τ was set to have the value of 5,000. We needed only to find $[\phi_{BB}(0)]$ and $\sum_{i=0}^{\infty} p_i^2$ in order to have used the formula.

Since the command input was white with unit power, $[\phi_{BB}(0)]$ equaled unity multiplied by the sum of squares of the impulses of the impulse response of the ideal controller. The ideal controller was

$$C(z) = \frac{1}{P(z)} = 2\frac{1 - \frac{1}{2}z^{-1}}{1 - \frac{1}{3}z^{-1}}. \tag{6.15}$$

The sum of squares turned out to be 2.12. Accordingly, $[\phi_{BB}(0)] = 2.12$.

The transfer function of the plant was given by (6.13). The sum of squares of the impulses of its impulse response had been calculated, and this was $\sum_{i=0}^{\infty} p_i^2 = 0.26$. We then had what was needed to apply (6.9). Accordingly,

$$E[\delta_k^2]_{\text{opt}} = \sqrt{\frac{(10)(2.12)(10)}{(2)(5000)(0.26)}} = 0.4. \tag{6.16}$$

From (6.14), we obtained, the optimal adaptation constant as

$$\mu = \frac{1}{2\tau E[\delta_k^2]} = \frac{1}{2(5000)(0.4)} = 2.5(10)^{-4}. \tag{6.17}$$

From (6.12), the minimum excess error power was

$$\begin{aligned}\begin{pmatrix}\text{excess}\\ \text{error}\\ \text{power}\end{pmatrix} &= \sqrt{\frac{2}{\tau}E[n_k^2][\phi_{BB}(0)]\left[\sum_{i=0}^{\infty} p_i^2\right]\min(m,n)}\\ &= \sqrt{\left(\frac{2}{5000}\right)(10)(4.12)(0.26)(10)}\\ &= 0.207.\end{aligned} \tag{6.18}$$

In order to verify (6.16), (6.17), and (6.18), the control system was operated with optimal dither power in accord with (6.16). Then dither power was increased by a factor of 2, and then by a factor of 4. Returning to optimal, the dither power was then reduced by a factor of 2 and further by a factor of 4. The excess error power was measured in each case. The results are given in Table 6.1.

TABLE 6.1 EXCESS ERROR POWER AS A FUNCTION OF DITHER POWER

Dither	Excess Error Power (from Eq. (6.7))	
Power [dB]	Theoretical [dB]	Experimental [dB]
-6	2.12	1.9
-3	1.25	1.44
0_{optimal}	1	0.92
3	1.25	1.34
6	2.12	2.13

Referring to Table 6.1, it may be noted that the dither power is given in dB, with the theoretically optimum dither power of 0.4 represented by 0 dB. The other dither powers of the experiment differ by 3-dB increments. Theoretical and experimental values of excess error power are given in the table. The values are normalized, so that the theoretically minimal value of excess error power given by (6.18) had a normalized value of 1. Equation (6.8) did not turn out to be perfect, but it did give us a good "ballpark" prediction of what was to happen.

6.3 SIMULATED INVERSE CONTROL EXAMPLES

The stabilized nonminimum-phase plant that was described in the previous chapter has been adaptively modeled, and inverse controllers have been obtained for values of the stabilization loop gain set at $k = 21$ and $k = 24$. The plant modeling was done without dither, only using the command input signal. In this section, we will use simulation techniques to study several adaptive inverse control systems involving the same plant. We will study tracking ability and control effort.

Figure 6.6 shows the convolution of the adaptive inverse impulse response with the true discretized impulse response of $G(z)$, the stabilized plant, with $k = 24$. The plant had no disturbance. The inverse filter had 100 weights and a delay of 50 sampling periods. Although the plant was nonminimum-phase, the inverse filter was an excellent one as is evidenced by the convolution plot, an almost perfect unit impulse delayed by 50 sampling periods. A comparison of the plant output with the command input delayed by 50 sampling periods is shown in Fig. 6.7. The command input was a zero-mean first-order Markov process. Excellent tracking can be seen. The control effort, the input signal to the plant, is shown in Fig. 6.8.

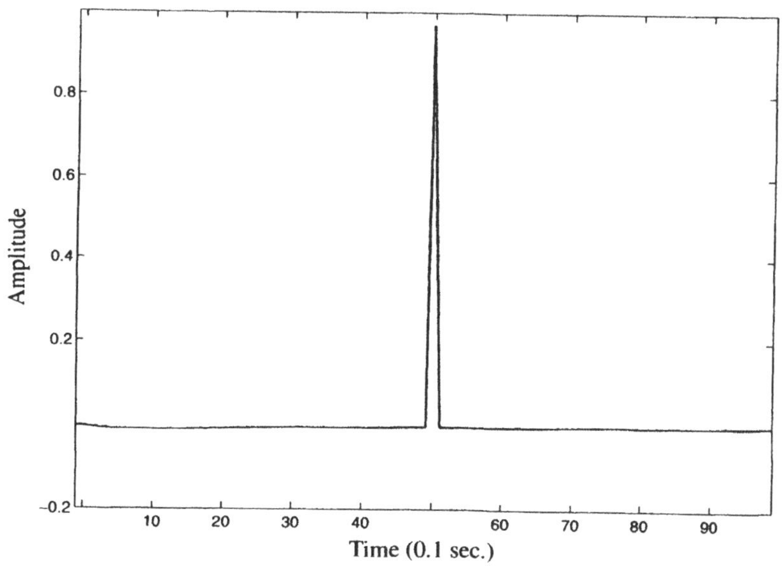

Figure 6.6 Convolution of impulse response of true $G(z)$ for $k = 24$ with adaptive delayed inverse impulse response. The plant had no disturbance.

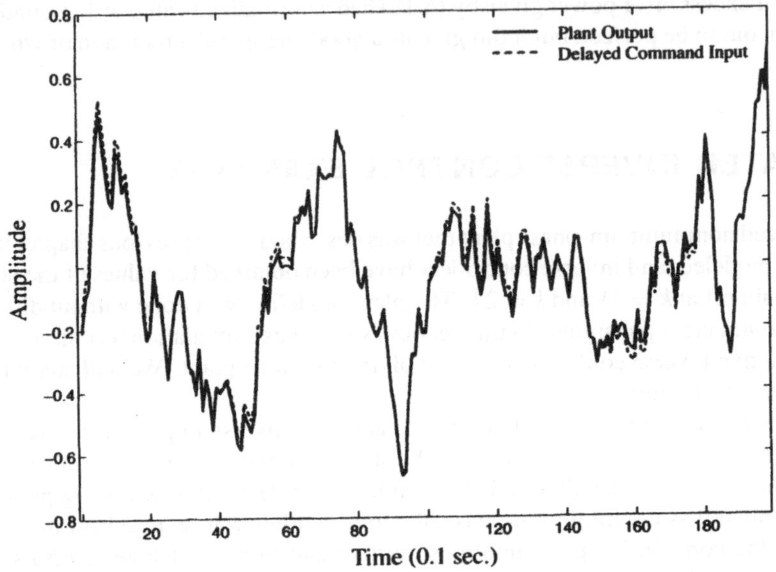

Figure 6.7 Tracking performance: A comparison of the plant output with the command input signal delayed by 50 sampling periods.

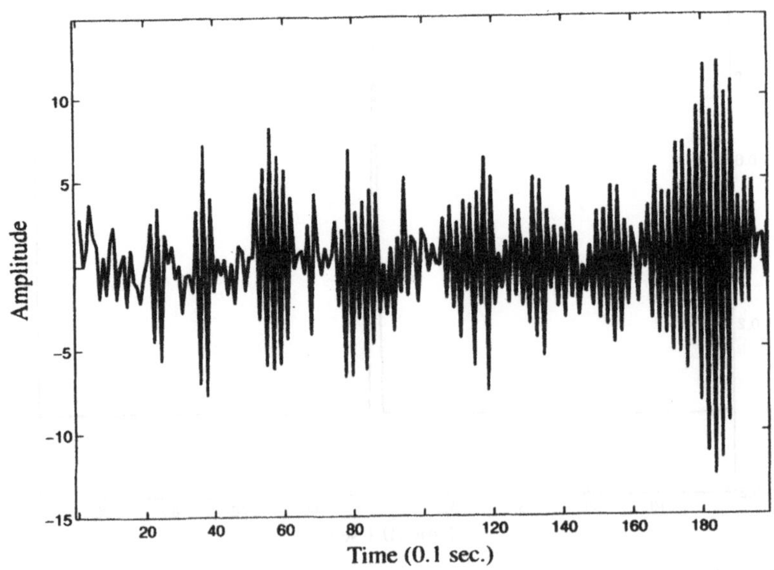

Figure 6.8 The control effort: The plant input signal.

Figure 6.9 shows the convolution of the adaptive inverse impulse response with the true discretized impulse response of $G(z)$, with $k = 24$. In this case, the plant was subject to disturbance. This caused significant noise to exist in the weights of the adaptive plant model $\widehat{P}$ and in the inverse model $\widehat{C}$. The noise in $\widehat{C}$ caused some imperfection in the convolution, which can be seen in the plot of Fig. 6.9. By slowing the convergence of $\widehat{P}$, the weight noises could have been made as small as one wished, but the penalty would have been slow adaptation. In any event, the noisy $\widehat{C}$ was used as a controller with good results. The tracking performance is shown in Fig. 6.10, and it is almost as good as that of Fig. 6.7 for the plant without disturbance. To make this comparison, the plant disturbance has been subtracted from the plant output. Control of the plant disturbance itself is the subject of Chapter 8.

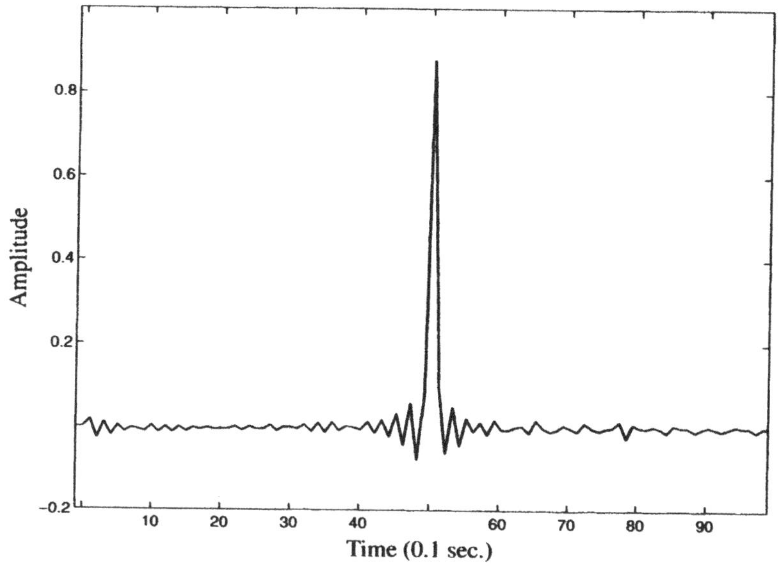

Figure 6.9 Convolution of impulse response of true $G(z)$ for $k = 24$ with adaptive delayed inverse impulse response. The plant was subject to disturbance.

Good dynamic control of the plant is therefore feasible even when $\widehat{C}$ is not perfect. Figure 6.11 shows a plot of the control effort. The amplitude of this effort was almost the same as that for the case of the plant without disturbance. This is not so surprising, because dynamic control of the plant is done about as well with the imperfect $\widehat{C}$ as with the almost perfect $\widehat{C}$.

One reason for using a smooth reference model $M(z)$ is limitation of the amplitude of the control effort. When $M(z)$ is a simple delay, the system step response is a step. To achieve such a sharp response, a strong control effort may be required. This could result in nonlinear overall response and could sometimes even cause physical damage.

To explore the effects of smoothing in $M(z)$ upon control effort, further simulations were done. $M(z)$ was made low-pass with a delay by cascading a one-pole filter with a delay of 50 sampling periods. Figure 6.12 shows the impulse response of $M(z)$ and the impulse response of $\widehat{C}(z)$ convolved with that of $G(z)$. The fit is not perfect, but very close. With

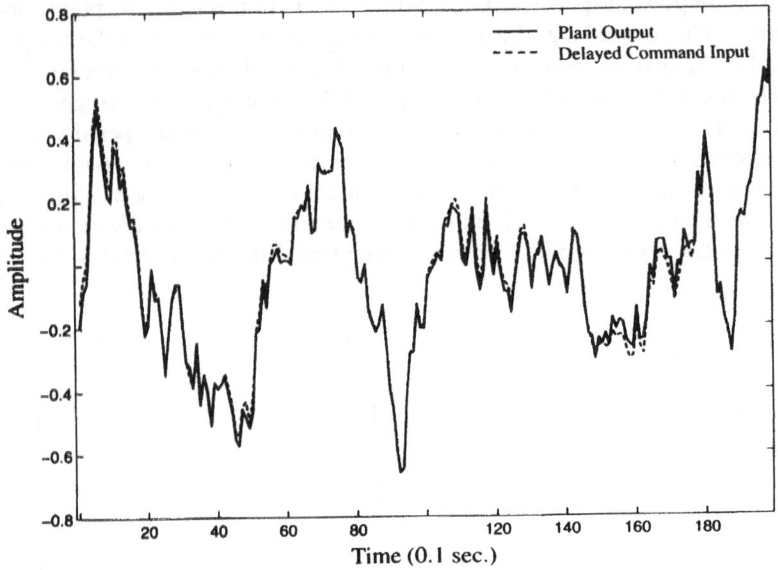

Figure 6.10 Tracking performance: A comparison of the plant output (with plant disturbance subtracted) with the command input signal delayed by 50 sampling periods.

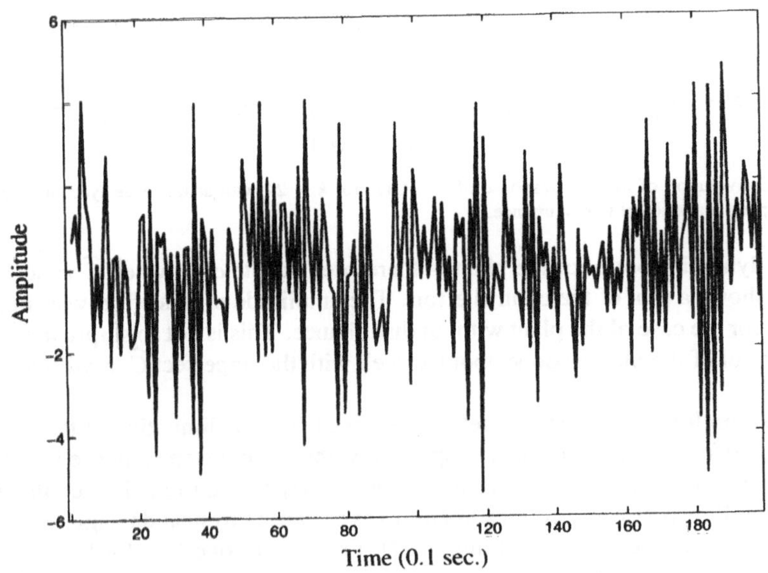

Figure 6.11 The control effort: The plant input signal.

a first-order Markov command input like the one used in the previous examples, very good tracking resulted. Figure 6.13 shows how closely the output of the plant matched the output of $M(z)$.

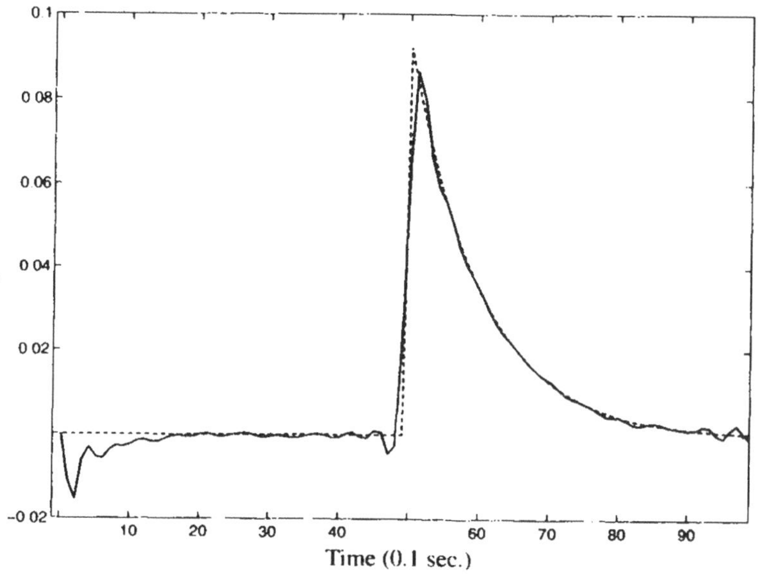

Figure 6.12 Convolution of impulse response of $G(z)$ for $k = 24$ with impulse response of model-reference inverse $\hat{C}(z)$. The plant had no disturbance. The dotted curve is the impulse response of $M(z)$.

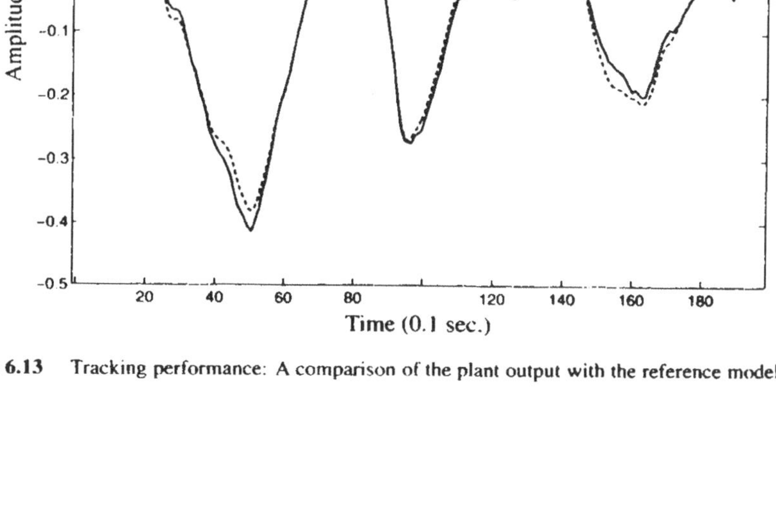

Figure 6.13 Tracking performance: A comparison of the plant output with the reference model output.

The signal driving the plant, the control effort, is shown in Fig. 6.14. It is interesting to compare this with the control effort signal shown previously in Fig. 6.8, where the reference model $M(z)$ was a simple delay. The low-pass smoothing in $M(z)$ gave it a much slower reacting output, causing the plant output to be much slower reacting, and the same for the control effort. It is interesting to note that the low-pass filtering in $M(z)$ reduced the amplitude of the control effort by about an order of magnitude.

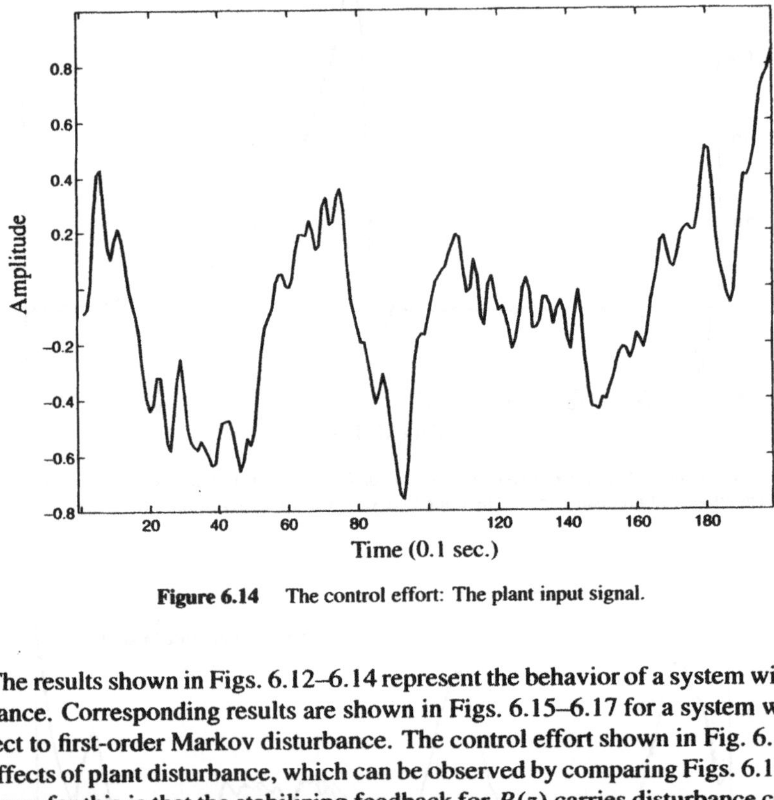

Figure 6.14 The control effort: The plant input signal.

The results shown in Figs. 6.12–6.14 represent the behavior of a system without plant disturbance. Corresponding results are shown in Figs. 6.15–6.17 for a system whose plant is subject to first-order Markov disturbance. The control effort shown in Fig. 6.17 exhibits some effects of plant disturbance, which can be observed by comparing Figs. 6.17 and 6.14. The reason for this is that the stabilizing feedback for $P(z)$ carries disturbance components at the plant output back into the plant input. A plant without this feedback would have the same control effort with disturbance or without disturbance. The system being implemented is like that of Fig. 6.3 or 6.4.

For all of the above cases, the command input has been a zero-mean first-order Markov process. It is interesting to challenge the adaptive inverse control system by adding a variable mean to the command input. This was tried with the nonminimum-phase plant, with $k = 24$. The reference model $M(z)$ was a simple delay of 50 sampling periods. The inverse controller $\widehat{C}$ had 100 weights. The convolution of the impulse response of $\widehat{C}$ with that of $G(z)$ is shown in Fig. 6.6, and it looks like a perfect impulse delayed by 50 sampling periods. Excellent tracking of the command input is shown in Fig. 6.7, and this is no surprise. However, when a step is added to the unbiased command input, the tracking results are poor.

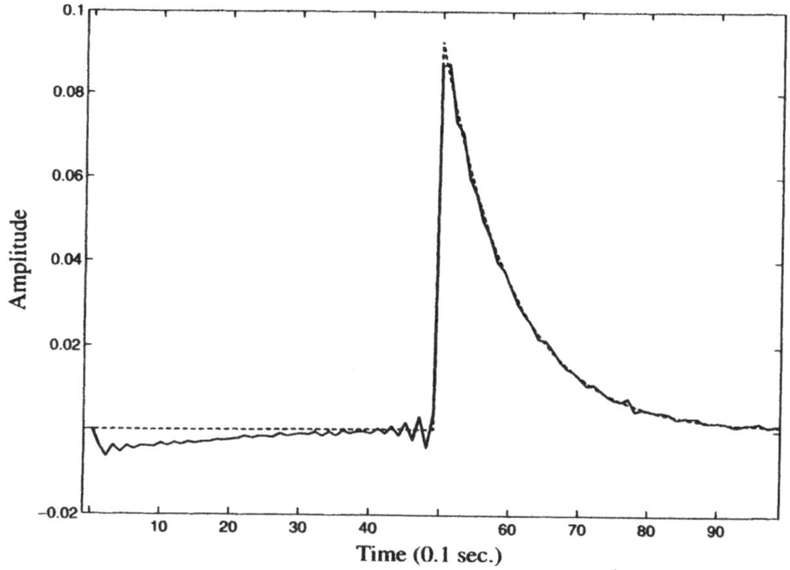

Figure 6.15 Convolution of impulse response $G(z)$ for $k = 24$ with impulse response of model-reference inverse. The plant was subject to disturbance. The dotted curve is the impulse response of $M(z)$.

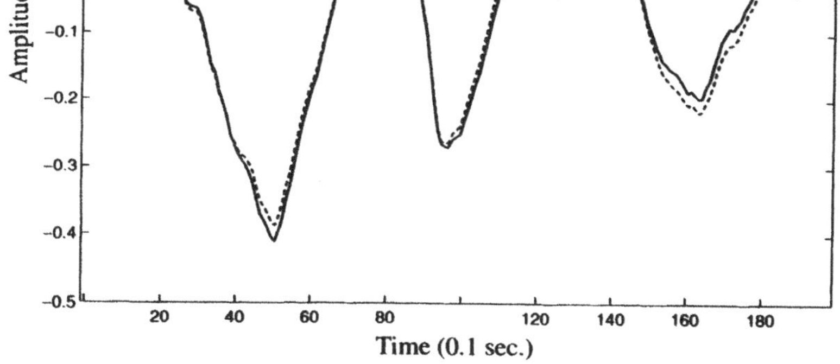

Figure 6.16 Tracking performance: A comparison of the plant output (with plant disturbance subtracted) with the reference model output.

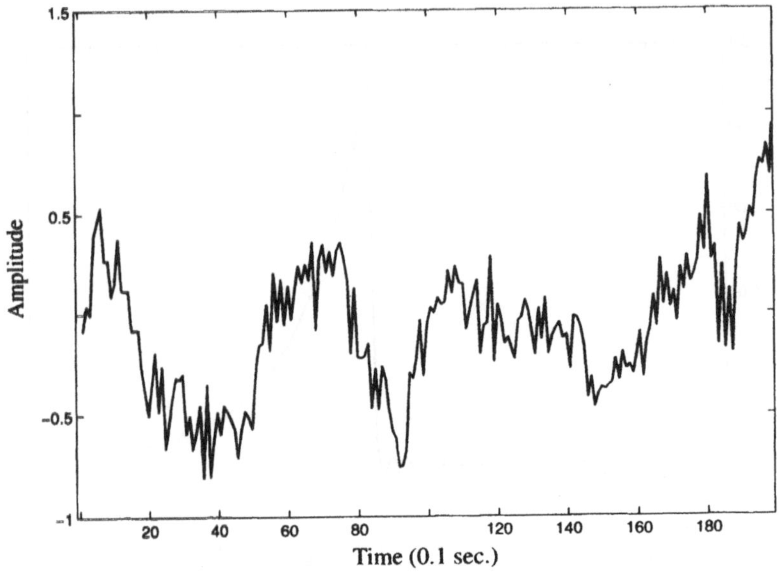

Figure 6.17 The control effort: The plant input signal.

In order to get a perfect step response, the DC gain from command input to plant output would need to be unity. The DC gain can be found from the convolution of Fig. 6.6, taking the sum of its constituent impulse values. Even though the convolution looks almost perfect, the sum of the impulses turns out to be 0.68. Tracking is shown in Fig. 6.18. The system has not been trained to deal with this kind of command input. A much better DC response in $\widehat{C}$ is needed.

Since the plant is nonminimum-phase, a better controller response at zero frequency can be obtained by increasing the number of weights in $\widehat{C}$, and by increasing the modeling delay. The number of weights was increased to 200, and the delay was increased to 100 sampling periods. Checking the sum of the impulses of the new convolution shows this to be 0.98. The corresponding tracking result is excellent, and this can be seen in Fig. 6.19. This result is an especially good one, since the system still has not been specifically trained to deal with a command input signal containing step functions. It has been trained with zero-mean first-order Markov inputs only.

6.4 APPLICATION TO REAL-TIME BLOOD PRESSURE CONTROL

A problem was presented by Dr. Mark Yelderman and Dr. William New,[5] formerly with the department of anesthesia at the Stanford University Medical Center, concerning blood pressure regulation during postsurgical recovery. Appropriate drug infusion is often done to prevent the patient's blood pressure from exceeding a selected limit while at the same

[5] Dr. New is a founder of Nellcor Inc. of Hayward, CA. Nellcor manufactures a blood oximeter instrument that gives real-time readings of the percentage of oxygen saturation in the blood.

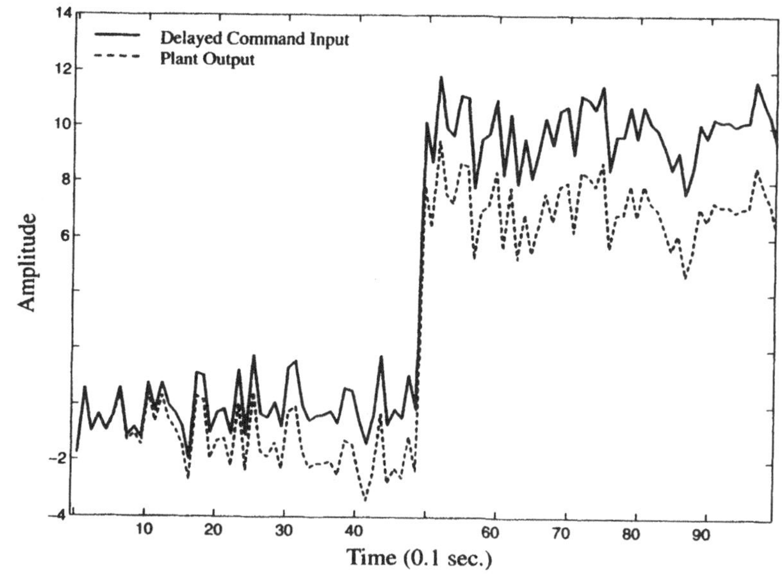

Figure 6.18 Tracking performance: A comparison of the plant output with the command input signal delayed by 50 sampling periods. A step was added at time zero to the unbiased command input. $\hat{C}$ had 100 weights.

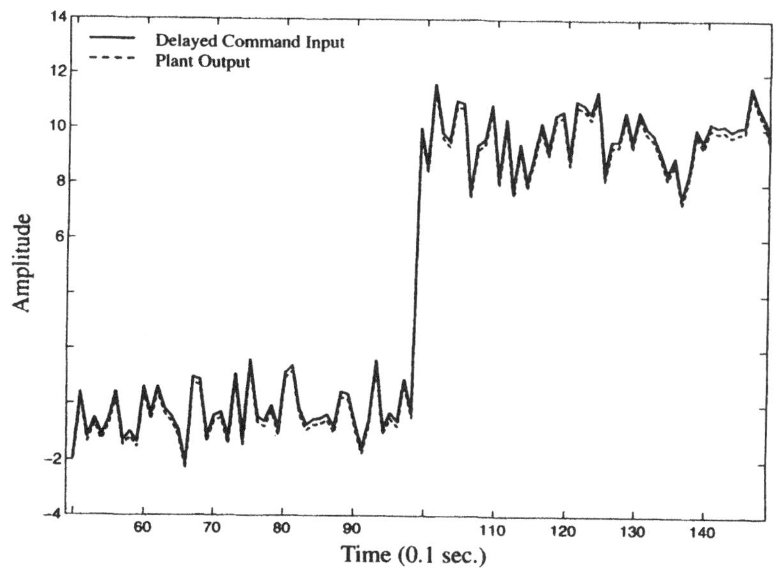

Figure 6.19 Tracking performance: A comparison of the plant output with the command input signal delayed by 100 sampling periods. A step was added at time zero to the unbiased command input. $\hat{C}$ had 200 weights.

time ensuring that the blood pressure does not fall to a value too low to ensure adequate circulation through critical organs. The problem is to determine in real time the proper drug flow rate.

Accordingly, we constructed a microprocessor-based automatic inverse controller to regulate blood pressure. The system used an intravenous (IV) drip controller which was fed a digital setpoint. With very little lag, the setpoint determined the number of drops per minute of drug solution flowing in the IV. An arterial catheter connected to a pressure transducer provided a signal proportional to instantaneous blood pressure. These devices are shown symbolically in Fig. 6.20. The transfer function and impulse response of experimental animals have been observed with this apparatus. A typical impulse response of a 22-kg dog to a 2-mg dose of sodium nitroprusside, a pressure reducing vasodilator drug, is shown in Fig. 6.21. Notice that the impulse response was negative going, corresponding to reduced pressure. The dog behaved like a "black box," with drug flow as its input and with arterial blood pressure as its output.

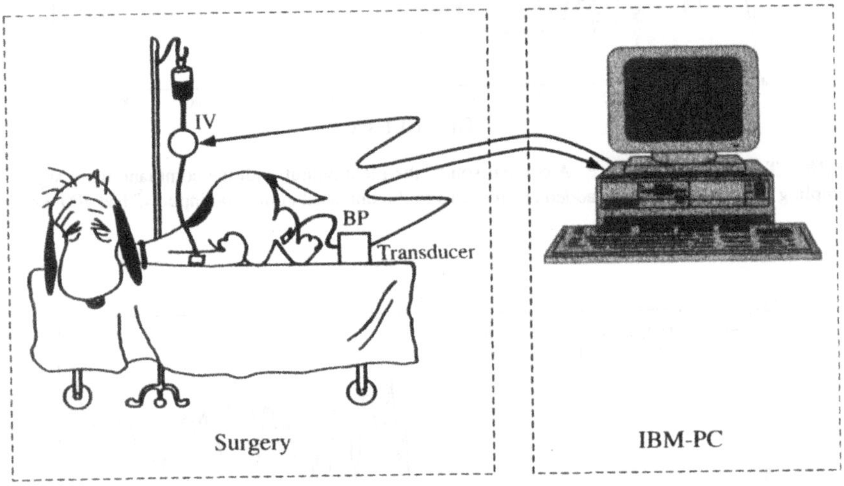

Figure 6.20 Pictorial representation of the experimental system. From B. WIDROW and S.D. Stearns, *Adaptive signal processing*, (Englewood Cliffs, NJ: Prentice Hall, 1985).

An adaptive control system like the one of Fig. 6.4 was simulated on a laboratory computer to control the dog's blood pressure. A recorded version of the dog's impulse response, $\widehat{P}_k(z)$, obtained by real-time adaptation, was used to find $\widehat{C}_k(z)$. A reference model $M(z)$ was chosen to have the step response shown in Fig. 6.22. After convergence of $\widehat{P}_k(z)$ and computation of $\widehat{C}_k(z)$, the step response of the entire system was as shown in Fig. 6.23. The difference between this step response and the step response of the reference model of Fig. 6.22 was almost imperceptible.

When controlling the recorded impulse response, the simulated drug flow was allowed to go positive or negative to obtain the result of Fig. 6.23. Control of this type can be physically achieved by using a combination of drugs. Notwithstanding this, unidirectional control with a single drug is possible, although not as precise. Using the same adaptive control system of Fig. 6.4 except that the simulated drug flow was always positive and was zero

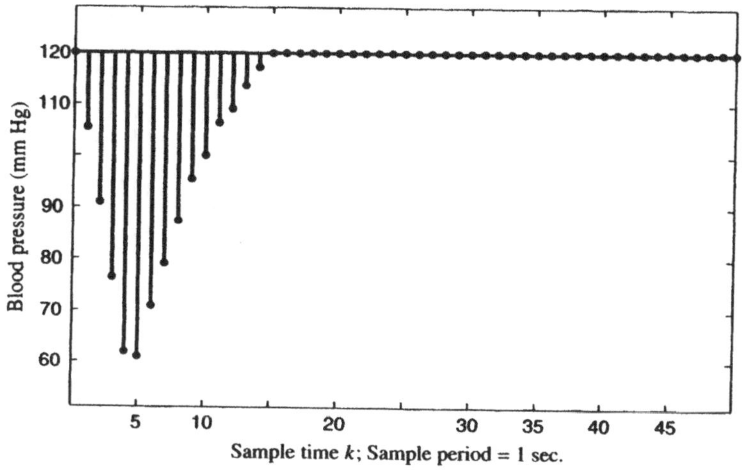

Figure 6.21 Impulse response of dog; 22-kg dog, 2-mg dose of sodium nitroprusside.

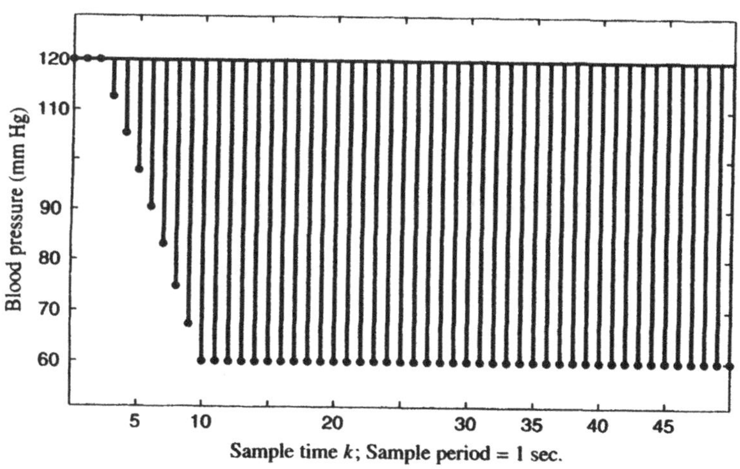

Figure 6.22 Step response of reference model.

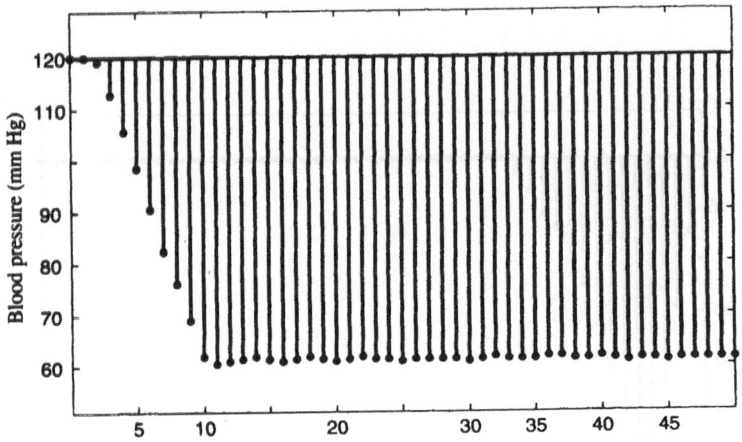

Figure 6.23 Step response of controller and dog. Bidirectional driving function.

when the system called for negative flow, the step response shown in Fig. 6.24 was obtained. The difference between this response and the model response is visible but slight.

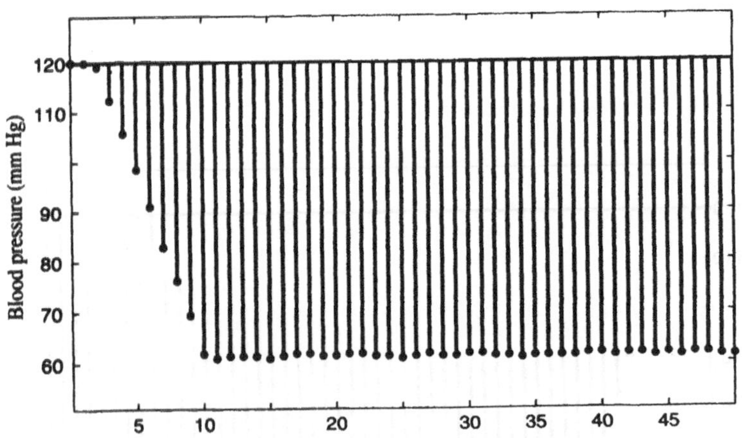

Figure 6.24 Step response of controller and dog. Unidirectional driving function.

The adaptive control scheme of Fig. 6.4 appears to be robust and relatively trouble free. Much more work needs to be done, however, to understand its behavior fully and in particular to understand how it works in a nonlinear control context. The dynamic response of the mammalian blood pressure system is certainly not linear, yet ignoring this led to reasonable results. On the experimental side, a fully automatic blood pressure regulating system is under development as sketched in Fig. 6.20. Using an IBM personal computer close to the dog during the surgery, several teams of Stanford students working with Professors

Yelderman, New, and Widrow have successfully controlled blood pressure in real time. All of the animals have survived the experiments.

After the typical experiment, the dog begins to wake up and must be quickly released from constraints on the operating table. Animals are frightened by confinement. The dog is placed on the floor and is initially wobbly when trying to walk. The effects of anesthesia wear off very quickly with dogs. The students doing the experiment are anxious to clean up the mess, unplug the computer, get their notebooks in order, and so forth. But the dog wants to play, with tail wagging in full force. The students oblige, of course.

.5 SUMMARY

This chapter shows how inverse plant models and model-reference inverse plant models can be used as controllers in adaptive control systems. An analysis of error is presented. Formulas are derived for optimal power levels for dither noise used in the plant modeling process and for error power at the plant output. These formulas were checked by experiment, and comparisons between experiment and theory show close agreement.

Some of the key formulas are the following:

The power of the difference between the plant output and the reference model output is given by

$$\begin{pmatrix} \text{overall} \\ \text{system} \\ \text{error} \\ \text{power} \end{pmatrix} = E[n_k^2] + E[\delta_k^2] \cdot \left[\sum_{i=0}^{\infty} p_i^2 \right] + \min(m, n) \cdot \beta \cdot [\phi_{BB}(0)]. \tag{6.6}$$

The optimal dither is determined by

$$E[\delta_k^2]_{\text{opt}} = \sqrt{\frac{E[n_k^2][\phi_{BB}(0)] \cdot \min(m, n)}{2\tau \sum_{i=0}^{\infty} p_i^2}}. \tag{6.9}$$

The next chapter presents other means for achieving adaptive inverse control, usable when the plant is noisy and usable when there is error in $\widehat{P}(z)$.

Bibliography for Chapter 6

[1] B. WIDROW, and J.M. MCCOOL, "A comparison of adaptive algorithms based on the methods of steepest descent and random search," *IEEE Trans. on Antennas and Propag.*, Vol. AP-24, No. 5 (September 1976), pp. 615–637.

[2] B. WIDROW, J. MCCOOL, and B. MEDOFF, "Adaptive Control by Inverse Modeling," in *Conf. Rec. of 12th Asilomar Conf. on Circuits, Systems and Computers* Santa Clara, CA (November 1978) pp. 90–94.

[3] N. Levinson, "The Wiener RMS (root-mean-square) error criterion in filter design and prediction," *J. Math. and Phys.*, Vol. 25 (1947) pp. 261–278.

Chapter 7

Other Configurations for Adaptive Inverse Control

7.0 INTRODUCTION

A very effective adaptive inverse control system is the one diagrammed in Fig. 6.4. Assuming that $\widehat{P}$ and $\widehat{C}$ have adequate numbers of weights, this system develops, by adaptation, a controller that when cascaded with the plant provides a very close dynamic match to the reference model. However, if truncation effects were significant, the system of Fig. 6.4 would develop a controller whose impulse response and transfer function would be biased relative to that of the ideal controller, the one that minimizes the mean square of the *overall system error*. It is the purpose of this chapter to introduce new forms of adaptive inverse control giving inverse controllers that are much less sensitive to truncation and other types of error in $\widehat{P}$.

Figure 7.1 is similar to Fig. 6.4, except that the process for finding $\widehat{C}$ is online rather than offline. There is not much advantage to this, except perhaps that the modeling signal for finding $\widehat{C}$ is a natural one. The real reason for introducing the system of Fig. 7.1 here is to provide contrast for the system of Fig. 7.2. The latter is an adaptive inverse control system that is significantly different from the system of Fig. 6.4. Figures 7.1 and 7.2 are almost alike, only differing in the source of error signal for the adaptation of $\widehat{C}(z)$. But the system of Fig. 7.1 is highly sensitive to errors in $\widehat{P}$, whereas the system of Fig. 7.2 is highly insensitive to errors in $\widehat{P}$.

In this chapter, we shall analyze two systems named *filtered-X* and *filtered-*ϵ. Both are capable of adaptive inverse control with low sensitivity to plant model errors.

7.1 THE FILTERED-X LMS ALGORITHM

The system of Fig. 7.2 does online adaptation of $\widehat{C}(z)$ and uses the overall system error in the adaptation process. A copy of $\widehat{C}(z)$ is the system controller. This algorithm for adapting $\widehat{C}(z)$ is called the filtered-X LMS algorithm. It is derived in Widrow and Stearns ([1], pp. 288–300), and a number of applications are given there. This algorithm gets its name from the fact that the input to the adaptive filter providing the X-vector is filtered. Early references to this algorithm in its various forms are [2,3,4].

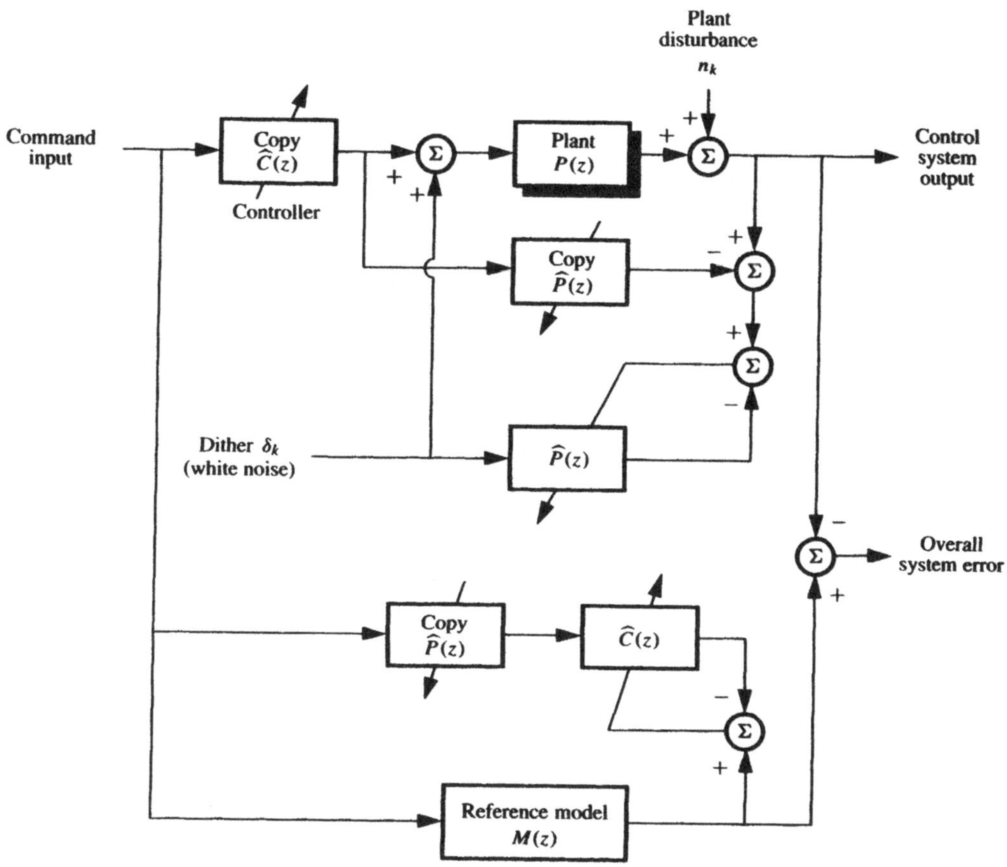

Figure 7.1 An adaptive inverse control system with online adaptation of $\widehat{C}(z)$.

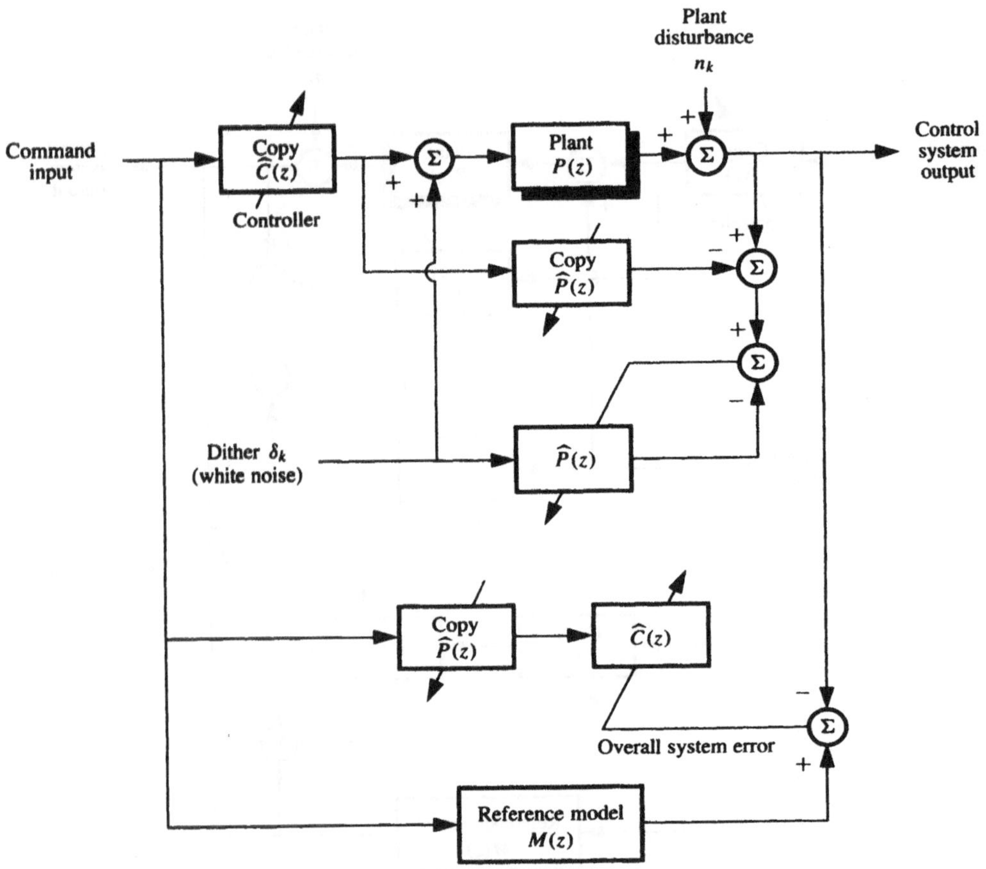

Figure 7.2 An adaptive inverse control system with online adaptation of $\widehat{C}(z)$ using overall system error (filtered-X LMS algorithm).

Even if $\widehat{P}(z)$ is truncated and is not a perfect match to $P(z)$, the adaptive algorithm of Fig. 7.2 will tend to minimize the overall system error by optimizing the choice of $\widehat{C}(z)$. We will show that the converged value of $\widehat{C}(z)$ is relatively insensitive to error in $\widehat{P}(z)$ as long as $\widehat{P}(z)$ is more or less similar to $P(z)$. This is not true for the systems of Figs. 6.4 and 7.1.

In order to demonstrate the insensitivity of the converged $\widehat{C}(z)$ to errors in $\widehat{P}(z)$, we simplify Fig. 7.2 by interchanging the plant with its controller as in Fig. 7.3. The plant disturbance keeps the same intensity and point of injection in the system. The *dither noise* is δ_k propagated through the plant $P(z)$. Figure 7.3 will be used for the analysis of the system of Fig. 7.2.

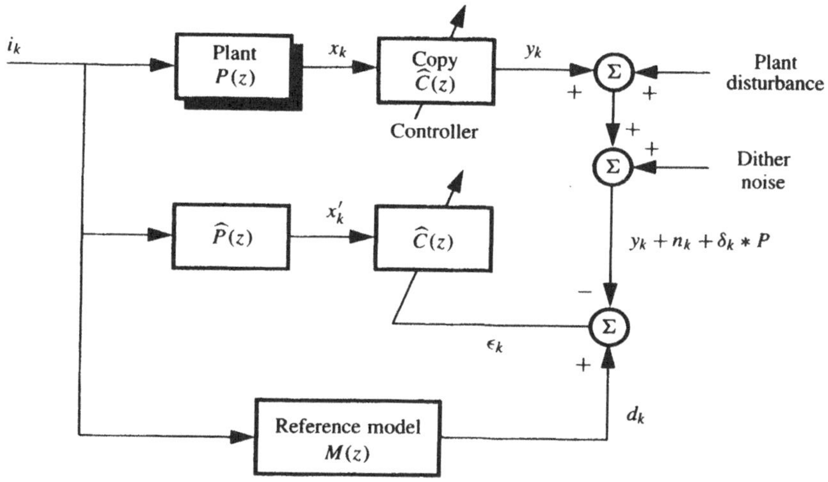

Figure 7.3 A diagram which is useful for the analysis of the system of Fig. 7.2 (filtered-*X* LMS algorithm).

Using the same notation as in Fig. 7.3, we will analyze the algorithm for adapting $\widehat{C}(z)$. We will show that this algorithm causes the weights of $\widehat{C}(z)$ to form the best linear least squares model-reference inverse of the plant $P(z)$. The filter $\widehat{C}(z)$ converges to an unusual Wiener solution. The mathematical techniques of Chapter 2 will be used to derive this solution.

Suppose that the impulse response of $\widehat{C}$ is not constrained. The Wiener solution is obtained as follows. Refer to Fig. 7.3. Recall that the crosscorrelation function between the error and the input signal is zero for a Wiener filter. Also recall that the instantaneous gradient used by the LMS algorithm when adapting $\widehat{C}$ is proportional to the product of the error ϵ_k and the input signal vector defined as $\mathbf{X}'_k$. For the system of Fig. 7.3, repeated and continual use of the LMS algorithm to adapt $\widehat{C}(z)$ causes the product of ϵ_k and $\mathbf{X}'_k$ to become small and eventually the expected value of this product goes to zero. Strange as it may seem, adapting $\widehat{C}(z)$ with a least squares algorithm does not cause the mean square of the error ϵ_k to be minimized if $\widehat{P}(z)$ differs from $P(z)$. The reason is that the output of $\widehat{C}(z)$ goes nowhere and is not a part of the error ϵ_k. Instead, least squares adaptation simply causes the correlation between ϵ_k and $\mathbf{X}'_k$ to go to zero. Accordingly,

$$0 = E[x'_k \epsilon_{k+m}]$$

$$
\begin{aligned}
&= E[x'_k(d_{k+m} - y_{k+m} - n_{k+m} - \delta_{k+m} * P)] \\
&= E[x'_k(d_{k+m} - y_{k+m})] \qquad (7.1) \\
&= E[x'_k d_{k+m} - \sum_{l=-\infty}^{\infty} \hat{c}_l x_{k-l+m} x'_k] \\
&= \phi_{x'd}(m) - \sum_{l=-\infty}^{\infty} \hat{c}_l \phi_{x'x}(m-l).
\end{aligned}
$$

This is the Wiener-Hopf equation for the filter $\widehat{C}(z)$. In getting this result, we assumed that both plant disturbance and dither noise at the plant output are uncorrelated with x'_k.

Equation (7.1) involves a simple convolution and may be easily z-transformed:

$$\Phi_{x'd}(z) = \widehat{C}(z)\Phi_{x'x}(z). \qquad (7.2)$$

From this we obtain

$$\widehat{C}(z) = \frac{\Phi_{x'd}(z)}{\Phi_{x'x}(z)}. \qquad (7.3)$$

It is useful to compare this result with the basic Wiener solution of Eq. (2.30).

In order to find $\widehat{C}(z)$, it is necessary to obtain $\Phi_{x'd}(z)$ and $\Phi_{x'x}(z)$. Referring once again to Fig. 7.3,

$$
\begin{aligned}
\Phi_{x'd}(z) &= \Phi_{ii}(z)\widehat{P}(z)\widehat{P}(z^{-1})\frac{M(z)}{\widehat{P}(z)} \qquad (7.4) \\
&= \Phi_{ii}(z)\widehat{P}(z^{-1})M(z).
\end{aligned}
$$

Also, from Fig. 7.3,

$$
\begin{aligned}
\Phi_{x'x}(z) &= \Phi_{ii}(z)\widehat{P}(z)\widehat{P}(z^{-1})\frac{P(z)}{\widehat{P}(z)} \qquad (7.5) \\
&= \Phi_{ii}(z)\widehat{P}(z^{-1})P(z).
\end{aligned}
$$

Therefore, the Wiener solution is

$$\widehat{C}(z) = \frac{\Phi_{x'd}(z)}{\Phi_{x'x}(z)} = \frac{M(z)}{P(z)}. \qquad (7.6)$$

This is the perfect solution! Adapting $\widehat{C}(z)$ (as in both Figs. 7.2 and 7.3) leads to the perfect solution. Note that the value of $\widehat{P}(z)$ has no effect on the Wiener solution.

In practice, the perfect solution would generally not be obtained. One reason is that $\widehat{C}(z)$ would not be unconstrained but would be causal and FIR. Another reason why the perfect solution might not be achieved is that the adaptive algorithm might not converge. Obtaining convergence for $\widehat{C}(z)$ with the system of Fig. 7.1 is straightforward. Convergence for $\widehat{C}(z)$ with the system of Fig. 7.2 is only straightforward if $\widehat{P}(z) = P(z)$. Otherwise, the convergence process involves some unknowns. The only reason for wanting $\widehat{P}$ to be similar to P itself is to ensure convergence of $\widehat{C}(z)$.

To gain a better understanding of the convergence process, let us apply the LMS algorithm to $\widehat{C}(z)$ of Fig. 7.3. The effects of plant disturbance and dither noise at the plant output will drop out, so they are omitted at the outset. We obtain

$$\widehat{\mathbf{C}}_{k+1} = \widehat{\mathbf{C}}_k + 2\mu(d_k - \mathbf{X}_k^T\widehat{\mathbf{C}}_k)\mathbf{X}'_k$$

$$\begin{aligned} &= \widehat{\mathbf{C}}_k + 2\mu d_k \mathbf{X}'_k - 2\mu \mathbf{X}'_k \mathbf{X}_k^T \widehat{\mathbf{C}}_k \\ &= (\mathbf{I} - 2\mu \mathbf{X}'_k \mathbf{X}_k^T)\widehat{\mathbf{C}}_k + 2\mu d_k \mathbf{X}'_k. \end{aligned} \tag{7.7}$$

Taking expectation of both sides yields

$$E[\widehat{\mathbf{C}}_{k+1}] = (\mathbf{I} - 2\mu \mathbf{R}')E[\widehat{\mathbf{C}}_k] + 2\mu \mathbf{P}', \tag{7.8}$$

$$\text{where } \mathbf{R}' \triangleq E[\mathbf{X}'_k \mathbf{X}_k^T], \text{ and} \tag{7.9}$$

$$\mathbf{P}' \triangleq E[d_k \mathbf{X}'_k]. \tag{7.10}$$

To obtain this result, we have assumed that $\mathbf{X}_k$ and $\mathbf{X}'_k$ are uncorrelated with $\widehat{\mathbf{C}}_k$.

If (7.8) converges, then

$$\lim_{k\to\infty} E[\widehat{\mathbf{C}}_{k+1}] = \lim_{k\to\infty} E[\widehat{\mathbf{C}}_k],$$

and we have

$$\lim_{k\to\infty} E[\widehat{\mathbf{C}}_k] = (\mathbf{R}')^{-1}\mathbf{P}'.$$

It is evident from (7.8) that convergence of $\widehat{\mathbf{C}}$ in the mean corresponds to

$$\lim_{k\to\infty} (\mathbf{I} - 2\mu \mathbf{R}')^k = 0. \tag{7.11}$$

It should be noted that the matrix $\mathbf{R}'$ is generally not symmetric. It is possible that $\widehat{P}(z)$ could be chosen with sufficient error that the resulting $\mathbf{R}'$-matrix, when entered into (7.11), would cause instability regardless of the choice of μ. In most cases, however, this would not be a problem. Particularly when $\widehat{P}(z)$ is close to $P(z)$. Then $\mathbf{R}'$ becomes close to being symmetric and achieving stability would be easy.

When using the filtered-X LMS algorithm, it is not necessary to wait for full convergence of $\widehat{P}(z)$ before beginning to adapt $\widehat{C}(z)$. The reason is that a perfect $\widehat{P}(z)$ is not necessary for the proper adaptation of $\widehat{C}(z)$, as long as the adaptive process is stable. Convergence of the entire system of Fig. 7.2 is generally not much slower than convergence of $\widehat{P}(z)$ alone.

Calculation of the noise in the weights of $\widehat{C}(z)$ is a complicated issue. This weight noise causes modulation of the command input and in turn causes noise at the output of $\widehat{C}(z)$. This propagates through the plant and causes noise at the plant output. A calculation of this noise has been made and the results will be discussed below. Stability conditions for the adaptation of $\widehat{C}(z)$ of Fig. 7.2 have been obtained and will also be discussed below.

THE FILTERED-ϵ LMS ALGORITHM

The system of Fig. 7.4 represents another approach to finding $\widehat{C}(z)$ that does not critically rely on the accuracy of $\widehat{P}$. The algorithm for finding $\widehat{C}(z)$ according to this approach is called the filtered-ϵ LMS algorithm. To the best of our knowledge, there is no prior literature for this algorithm.

An ideal controller is pictured in Fig. 7.4. If this were the actual controller instead of $\widehat{C}$ copy, the mean square of the overall system error ϵ_k would be minimized. The objective

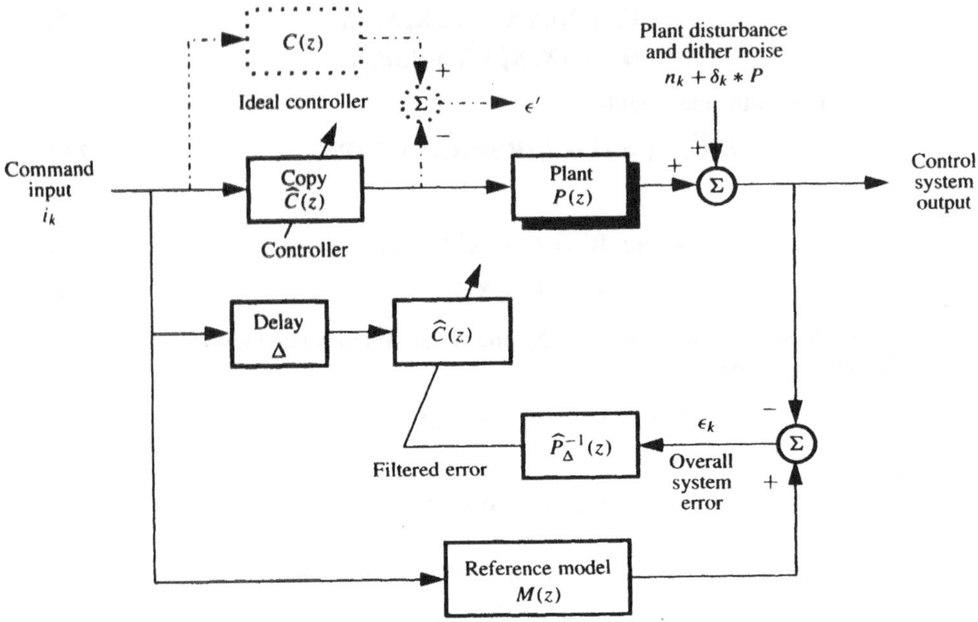

Figure 7.4 The filtered-ϵ LMS algorithm.

is to make $\widehat{C}$ be as close as possible to the ideal $C(z)$. The difference between the outputs of $\widehat{C}(z)$ copy and $C(z)$, both driven by the command input, is therefore an error signal. We can call this error ϵ'. If ϵ' were available for the adaptation of $\widehat{C}(z)$, the result would be a $\widehat{C}(z)$ that would adapt rapidly and directly toward $C(z)$. The error ϵ' is an ideal error signal for $\widehat{C}(z)$. The only problem is that it is unavailable.

Refer again to Fig. 7.4. A *filtered error*, obtained by filtering the overall system error ϵ_k, is used for adaptation in place of ϵ'. The filter is a delayed inverse of $P(z)$. If this inverse were perfect and if the plant were free of disturbance, then the filtered error would be exactly equivalent to ϵ' for adaptation purposes. In spite of its limitations, the filtered error does a fine job for adaptation of $\widehat{C}(z)$. The reasons for this are (a) the plant disturbance is uncorrelated with the plant input and the command input, and (b) errors in $\widehat{P}_\Delta^{-1}(z)$ are not critical to the correct convergence of $\widehat{C}(z)$, as will be shown below.

Since a delayed inverse of $\widehat{P}(z)$, represented by $\widehat{P}_\Delta^{-1}(z)$, is used to filter the overall system error ϵ, the input to $\widehat{C}(z)$ must be correspondingly delayed so that the input and error of $\widehat{C}(z)$ are time synchronized. As $\widehat{C}(z)$ adapts by LMS, its weights are copied into the controller. The output of $\widehat{C}(z)$ goes nowhere. Only its weights are used by the control system.

Figure 7.5 shows more detail regarding the implementation of the filtered-ϵ LMS algorithm. The plant model $\widehat{P}(z)$ is obtained in this case by using dither scheme C. (Other plant modeling schemes could have been used, with or without dither.) An offline process utilizes a copy of $\widehat{P}(z)$ to obtain its inverse. Generally, a modeling delay will be advantageous to use, and a delayed inverse $\widehat{P}_\Delta^{-1}(z)$ results. There is no performance penalty from

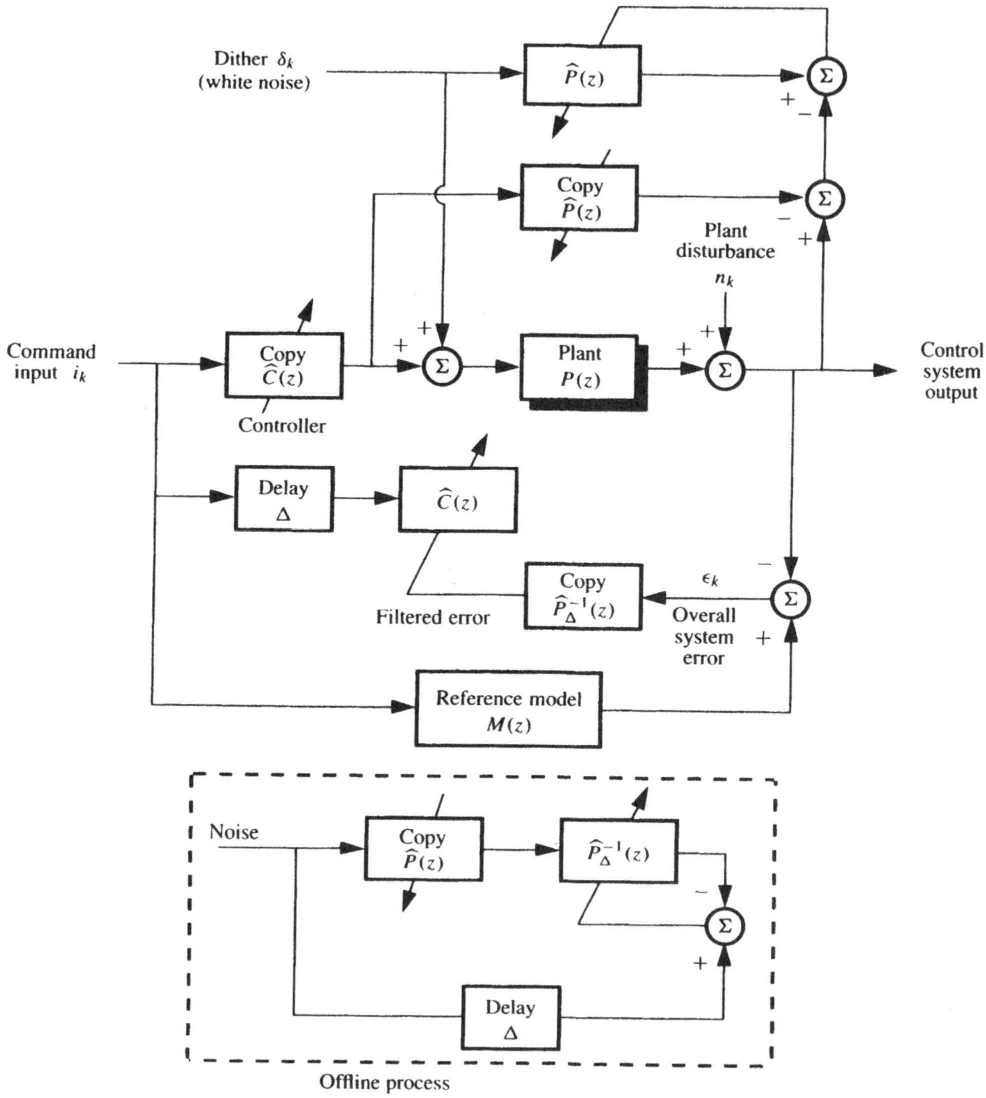

Figure 7.5 The filtered-ϵ LMS algorithm with details.

the delay Δ as long as the input to $\widehat{C}(z)$ undergoes the same delay. We assume that the command input i_k is stationary.

It should be noted that if the plant $P(z)$ is minimum-phase, an excellent inverse is obtainable without delay Δ. As such, the filtered error of Fig. 7.5 would not be delayed and this error could be used to adapt the controller directly. The system could be simplified. There would be no need for the separate filters, $\widehat{C}(z)$ and its copy, only the controller would be needed.

Next, we analyze the filtered-ϵ LMS algorithm and show that when convergent, this algorithm causes the weights of $\widehat{C}(z)$ to form the best linear least squares model-reference inverse of the plant $P(z)$.

Let the impulse response of $\widehat{C}(z)$ be unconstrained, and let us find its Wiener solution. To do this, we refer to Fig. 7.4. Assume that the adaptive process has converged and that the weights of $\widehat{P}(z)$, $\widehat{P}_{\Delta}^{-1}(z)$, and $\widehat{C}(z)$ are essentially static. As such, the block diagram of Fig. 7.4 can be modified in stages to be equivalent to the block diagrams of Figs. 7.6 and 7.7, leaving unchanged the input signal and the error signal of the filter $\widehat{C}(z)$. The modifications have been made by commuting the various filtering operations, permissible with linear time-invariant systems.

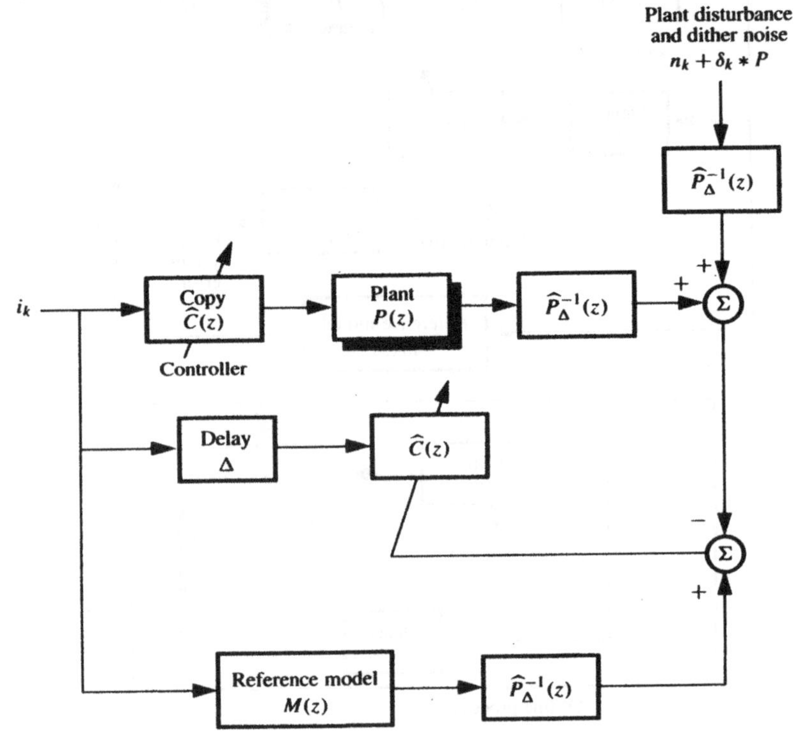

Figure 7.6 Block diagram equivalent to that of Fig. 7.4.

Comparison of Fig. 7.7 with Fig. 7.3 reveals great similarities between the filtered-X LMS and filtered-ϵ LMS algorithms. But there are differences. The input signal i_k is filtered in Fig. 7.7 and the noise (the sum of the disturbance at the plant output and the dither at the plant output) is also filtered. This is not the case in Fig. 7.3. However, the structure of the filters surrounding $\widehat{C}(z)$, comprised of the cascade of $P(z)$ with $\widehat{C}(z)$, the cascade of $\widehat{P}(z)$ with $\widehat{C}(z)$, and $M(z)$, and their interconnections, is identical when comparing Figs. 7.3 and 7.7.

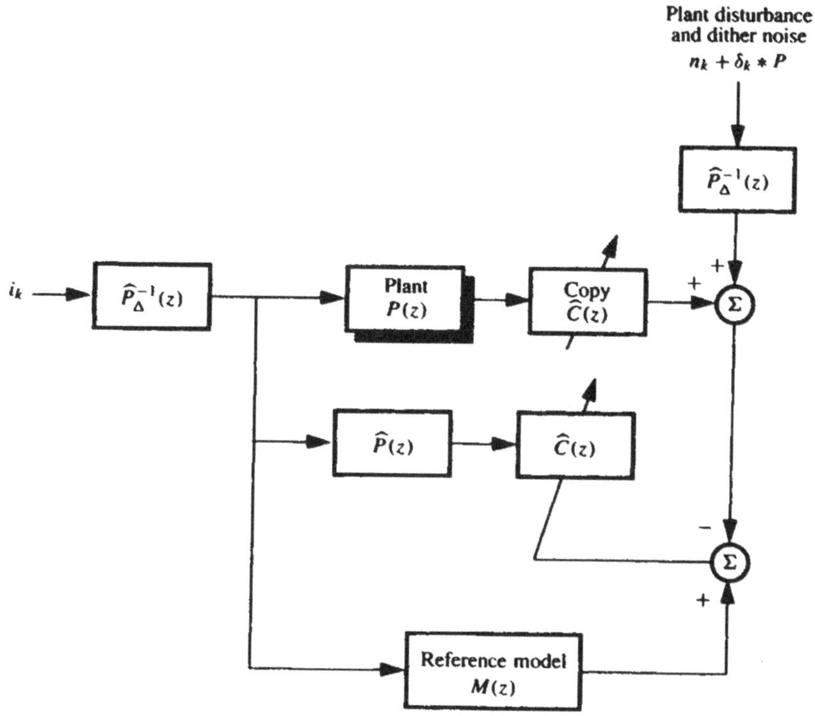

Figure 7.7 Block diagram equivalent to that of Fig. 7.6.

With the filtered-X LMS algorithm (Fig. 7.3), the unconstrained Wiener solution for $\widehat{C}(z)$ was given by (7.6) as

$$\widehat{C}(z) = \frac{M(z)}{P(z)}. \tag{7.12}$$

This is an ideal solution, and is unaffected by lack of precision in $\widehat{P}(z)$. The derivation of (7.6) made no assumptions about the input i_k and about the plant and dither noise except that the noises are independent of input i_k. Although the input and noise are filtered by $\widehat{P}_\Delta^{-1}(z)$ in Fig. 7.7, the same assumptions are valid, and once again the Wiener solution is given by (7.12). We can therefore conclude that the filtered-ϵ LMS algorithm yields a Wiener solution that is unaffected by choice of $\widehat{P}(z)$ and by choice of $\widehat{P}_\Delta^{-1}(z)$. Errors in $\widehat{P}(z)$ and in $\widehat{P}_\Delta^{-1}(z)$ will affect stability and convergence of the adaptive algorithm but will not affect the Wiener solution when convergence takes place. This is a pleasing result.

7.3 ANALYSIS OF STABILITY, RATE OF CONVERGENCE, AND NOISE IN THE WEIGHTS FOR THE FILTERED-ϵ LMS ALGORITHM

The filtered-ϵ LMS algorithm can be analyzed and its adaptive behavior can be approximately determined by establishing certain analogies between its behavior and the behavior of scheme C, a dithering algorithm for finding $\hat{P}(z)$. Since these are two completely different systems having totally different uses, the discovery of similarity in their behaviors is a surprising result. Scheme C is described in Chapter 4, diagrammed in Fig. 4.3(c), and is analyzed in some detail in Appendix B.

Once again, we shall make block diagrams of the filtered-ϵ LMS algorithm and commute operators in the block diagrams. Adaptive filters have time-variant impulses responses and are normally not commutable with time-invariant linear filters. If we assume, however, that the impulse responses of the adaptive filters are close to their Wiener solutions and are therefore almost constant, an approximate analysis can be made by allowing commutation of adaptive filtering (linear with fixed weight vector) with time-invariant linear filtering.

Accordingly, we begin with the block diagram for filtered-ϵ LMS in Fig. 7.4, and go immediately to the equivalent diagram of Fig. 7.7. From there, by a series of small steps, we go from Fig. 7.7 to Figs. 7.8, 7.9, 7.10, 7.11, 7.12, and finally to 7.13. The reader can verify the equivalence between the block diagram of Fig. 7.13 and that of Fig. 7.4 by tracing the steps.

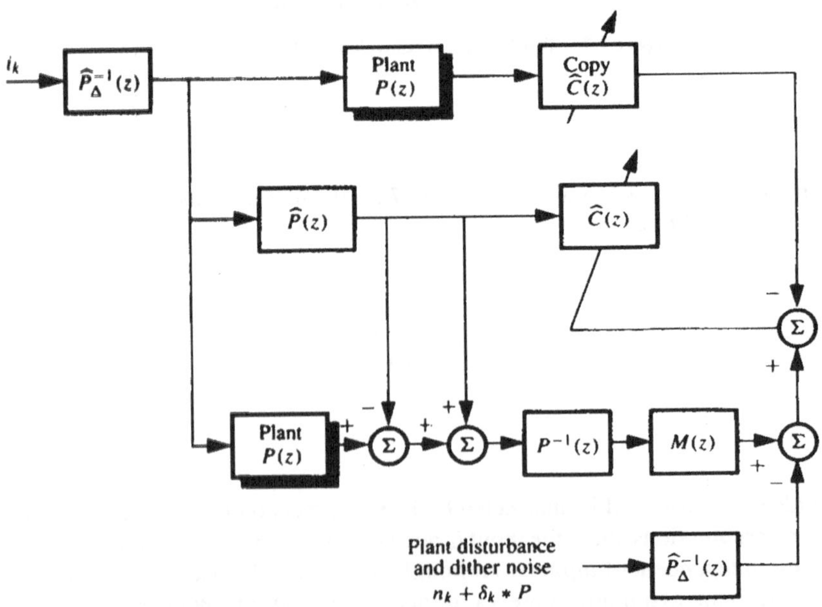

Figure 7.8 Block diagram equivalent to that of Fig. 7.7.

Although scheme C and the filtered-ϵ LMS algorithm are completely different and have different purposes and applications, there is a strong resemblance between the block

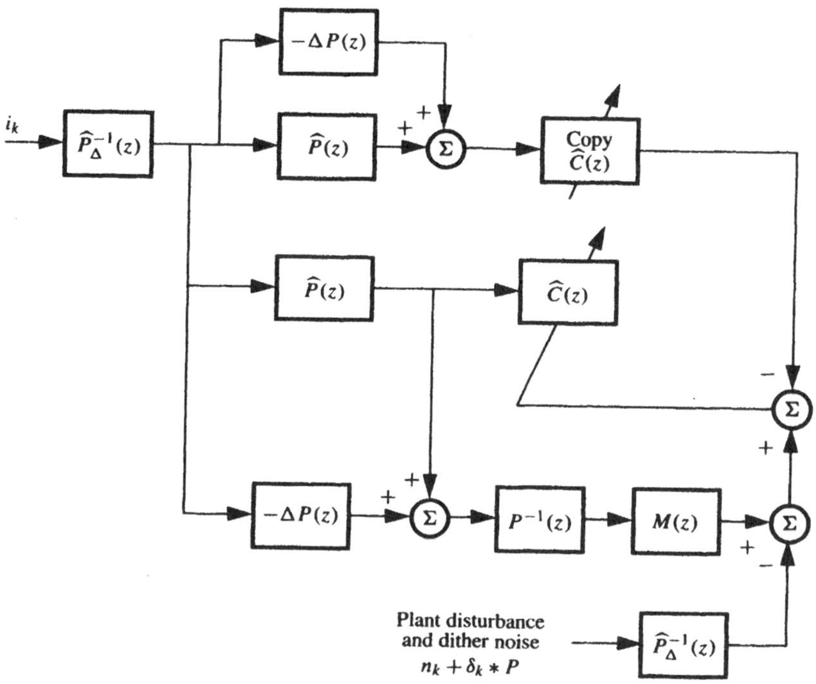

Figure 7.9 Block diagram equivalent to that of Fig. 7.8.

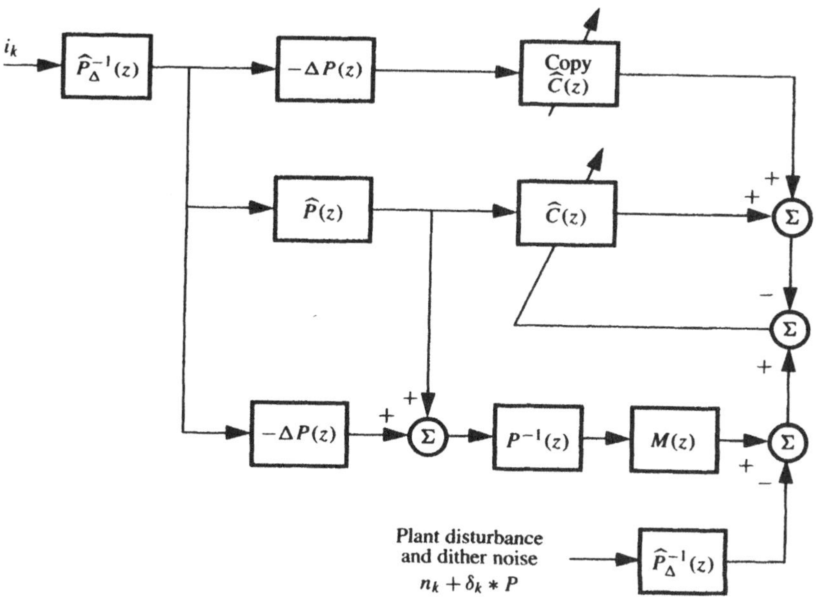

Figure 7.10 Block diagram equivalent to that of Fig. 7.9.

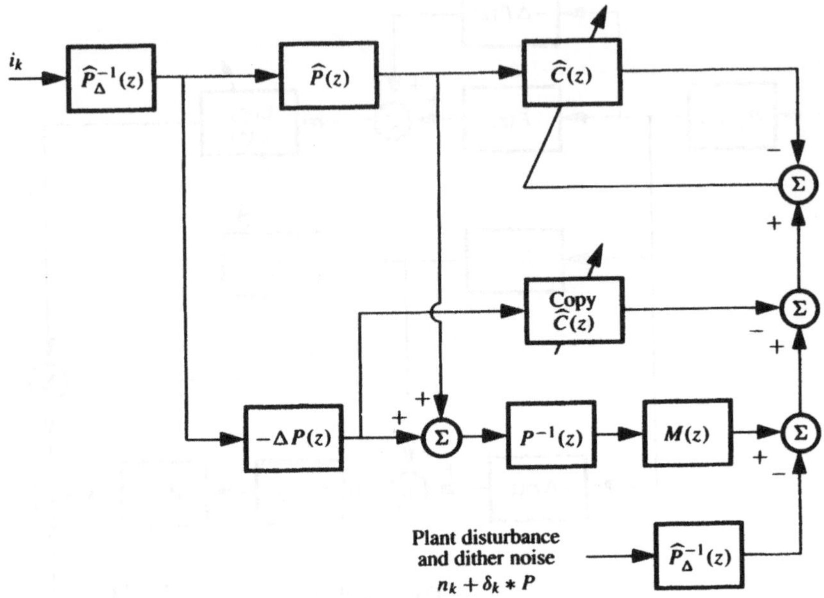

Figure 7.11 Block diagram equivalent to that of Fig. 7.10.

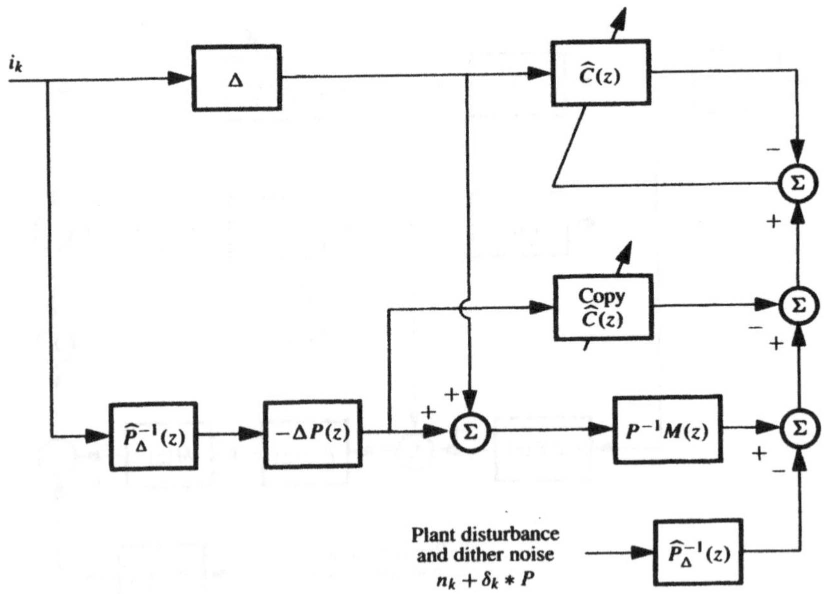

Figure 7.12 Block diagram equivalent to that of Fig. 7.11.

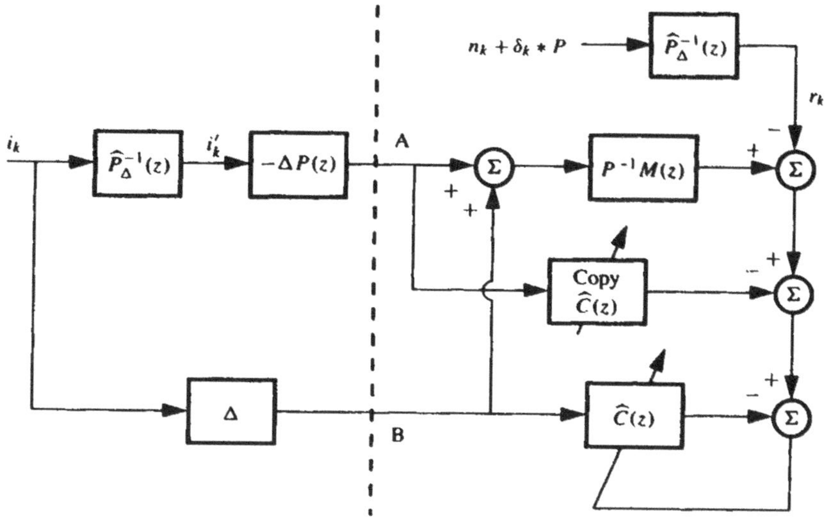

Figure 7.13 Block diagram equivalent to that of Fig. 7.12.

diagram of Fig. 7.13 and that of scheme C in Fig. 4.3(c). To the right of the dotted line in Fig. 7.13 is a block diagram that is structurally identical to the block diagram of Fig. 4.3(c).

In Fig. 7.13, the signal on line A is uncorrelated with that on line B. The reason for this is that the weight noise $\Delta P(z)$ in the vector $\hat{P}(z)$ is random and uncorrelated from component to component and uncorrelated with the command input i_k. The signal on line A is noise, while the signal on line B is a delayed version of the command input. The situation is somewhat different in Fig. 4.3(c), with the signal on the line analogous to A being the controller output, and with the signal on the line analogous to B being independent dither noise. The key agreement is that the signals on lines A and B are uncorrelated with each other in both cases. Furthermore, the noise in scheme C is plant disturbance, and the analogous noise in Fig. 7.13 is a combination of plant disturbance and noise at the plant output. In both cases, these noises are uncorrelated with the signals on lines A and B.

Since the input signals and noises have analogous properties and since the structure of scheme C is analogous to that of filtered-ϵ LMS, we obtain from Appendix B the means to analyze the filtered-ϵ LMS algorithm.

The time constants for filtered-ϵ LMS can be obtained by making use of Eq. (B.32). [The input signal on line B in Fig. 7.13 is not white, so Eq. (B.34) cannot be used.] An equation of the same form as (B.32) is Eq. (A.7) whose solution yielded an expression for time constants (A.13a). We use this solution as follows:

$$\tau_p \cong \frac{1}{2\mu_2\lambda_p}, \tag{7.13}$$

where μ_2 is the LMS adaptation constant for $\hat{C}(z)$ with filtered-ϵ LMS, and λ_p is the pth eigenvalue of the autocorrelation matrix of the input to $\hat{C}(z)$ in Fig. 7.13 or Fig. 7.4. This input is the command input i_k.

Our next concern is convergence of the variance of $\widehat{C}$. The conditions for convergence of the variance can be obtained directly from (B.50) of Appendix B. In (B.50), an expression for the total power into the plant, $E[u_k^2]$, appears. Comparing Figs. 4.3(c) and 7.13, the analogous power for the filtered-ϵ algorithm is the sum of the powers on lines A and B. The power on line B is $E[i_k^2]$. The power on line A is $E[(i_k')^2]\mu_1 n E[n_k^2]$, where the first factor is the power of the output of $\widehat{P_\Delta^{-1}}(z)$ driven by i_k, μ_1 is the adaptation constant of the plant modeling process that obtains $\widehat{P}(z)$, n is the number of weights of $\widehat{P}(z)$, and the last factor is the plant disturbance power. The noise variance in each component of $\Delta P(z)$ is given by (B.53) which, for small μ_1, reduces to $\mu_1 E[n_k^2]$. The total power on lines A and B is therefore

$$E[i_k^2] + E[(i_k')^2]\mu_1 n E[n_k^2]. \tag{7.14}$$

From (B.50), the condition for convergence of the variance of the filtered-ϵ LMS algorithm is

$$0 < \mu_2 < \frac{1}{m\left[E[i_k^2] + E[(i_k')^2]\mu_1 n E[n_k^2]\right]}, \tag{7.15}$$

where m is the number of weights in the controller $\widehat{C}(z)$.

The weight noise covariance matrix can be gotten from (B.54). For the filtered-ϵ LMS algorithm,

$$\begin{bmatrix} \text{weight noise} \\ \text{covariance matrix} \end{bmatrix} = \frac{\mu_2 \xi_{\text{min}}}{1 - \mu_2 m\left[E[i_k^2] + E[(i_k')^2]\mu_1 n E[n_k^2]\right]}\mathbf{I}. \tag{7.16}$$

For small μ_2, this becomes

$$\begin{bmatrix} \text{weight noise} \\ \text{covariance matrix} \end{bmatrix} = \mu_2 \xi_{\text{min}} \mathbf{I}. \tag{7.17}$$

To get ξ_{min}, refer to Fig. 7.13. This is the power of the noise r_k coming from the filter $\widehat{P_\Delta^{-1}}(z)$. This noise can be written as

$$\begin{aligned} r_k &= (\widehat{P_\Delta^{-1}}) * (n_k + \delta_k * P) \\ &\approx \widehat{P_\Delta^{-1}} * n_k + \delta_{k-\Delta}. \end{aligned} \tag{7.18}$$

The symbols P and $\widehat{P_\Delta^{-1}}$ represent impulse responses of the corresponding filters. The minimum mean square error is therefore

$$\xi_{\text{min}} = E[r_k^2] = E[\delta_k^2] + E[(n_k * \widehat{P_\Delta^{-1}})^2]. \tag{7.19}$$

The dither power is known. The plant disturbance power is generally known, but knowledge of the plant inverse $\widehat{P_\Delta^{-1}}(z)$ would be required to evaluate or estimate the second term of (7.19). In any event,

$$\begin{bmatrix} \text{weight noise} \\ \text{covariance matrix} \end{bmatrix} = \mu_2 \left(E[\delta_k^2] + E[(n_k * \widehat{P_\Delta^{-1}})^2]\right) \mathbf{I}. \tag{7.20}$$

With knowledge of the noise in the weights of $\widehat{C}(z)$, we can calculate the noise that this causes at the plant output when the filtered-ϵ LMS algorithm is practiced. The dynamic system error is caused by the command input i_k going through $\widehat{C}(z)$, then driving the plant $P(z)$. The noise in $\widehat{C}(z)$ randomly modulates i_k, thereby creating a noise at the output of $\widehat{C}(z)$. This noise propagates through $P(z)$ giving a plant output noise component, the dynamic system error.

Figures 7.14(a) and (b) are helpful in evaluating the dynamic system error. The actual mechanism generating this error is represented by Fig. 7.14(a). Assuming that μ_2 is small and that ΔC_k (z) changes slowly, commutation is possible and Fig. 7.14(b) also represents the mechanism generating the dynamic system error. Using Fig. 7.14(b), the variance of the dynamic system error is obtained as

$$\begin{pmatrix} \text{variance of} \\ \text{the dynamic} \\ \text{system error} \end{pmatrix} = E[(i_k'')^2]\mu_2 m \xi_{\min}. \tag{7.21}$$

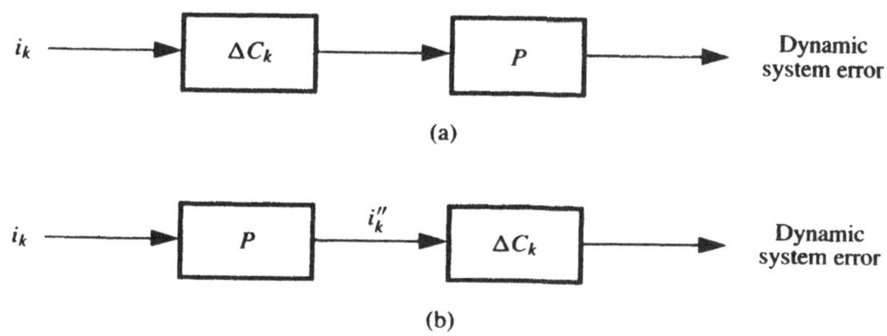

Figure 7.14 Evaluation of dynamic system error with filtered-ϵ LMS.

$\xi_{\min}$ is given by (7.19). $E[(i_k'')^2]$ is the power of the output of the plant driven by the command input i_k. Some knowledge of plant dynamics is needed to calculate or estimate this. Equation (7.21) can be rewritten as

$$\begin{pmatrix} \text{variance of} \\ \text{the dynamic} \\ \text{system error} \end{pmatrix} = E[(i_k * P)^2]\mu_2 m \left(E[\delta_k^2] + E[(n_k * \widehat{P}_\Delta^{-1})^2]\right). \tag{7.22}$$

This completes our analysis of the filtered-ϵ LMS algorithm and its application to adaptive inverse control.

7.4 SIMULATION OF AN ADAPTIVE INVERSE CONTROL SYSTEM BASED ON THE FILTERED-ϵ LMS ALGORITHM

In order to verify the effectiveness of the filtered-ϵ LMS and to verify Eqs. (7.13), (7.16), and (7.22), which describe key behavioral aspects of the algorithm, a simulation of the adaptive inverse control system of Fig. 7.5 was undertaken. The specific plant chosen for the experiment was

$$P(z) = \frac{1}{1 - \frac{1}{2}z^{-1}}. \tag{7.23}$$

The reference model was chosen to be

$$M(z) = 1. \tag{7.24}$$

Various simulation parameters were chosen as follows:

a. The command input was a zero-mean white signal of unit power, that is, $E[i_k^2] = 1$.

b. Dither (used in scheme C) was a zero-mean white noise of unit power, that is, $E[\delta_k^2] = 1$.

c. The plant noise was zero-mean and white. Its power was a variable parameter for the simulation.

d. The number of weights for $\widehat{P}(z)$ was $n = 10$, and the number of weights for $\widehat{C}(z)$ was $m = 10$.

These parameters were constant throughout the series of filtered-ϵ experiments. The variable simulation parameters were the adaptation constant μ_1, for adapting $\widehat{P}(z)$, μ_2 for adapting $\widehat{C}(z)$, and the plant noise power $E[n_k^2]$. For each variation of these parameters, the simulation was run for a few thousand data samples, about four time constants, to achieve steady flow before making measurements.

The initial condition for the controller transfer function was $\widehat{C}(z) = 2$, deliberately chosen to be far from the optimal transfer function. The convergence of $\widehat{C}(z)$ onto the ideal $C(z) = 1/P(z)$ was noted. The time constant of the learning process was obtained by observing the changes of the weight vector $\widehat{\mathbf{C}}_k$ as it adapted toward the ideal weight vector $\mathbf{C}$. The variance of the weights in steady state was measured, as was the variance of the dynamic system error. Each simulation was run 10 times and the results were averaged. These results are presented in Table 7.1, along with theoretical predictions from (7.13), (7.16), and (7.22).

In this case, Eqs. (7.13), (7.16), and (7.22) were able to be utilized quite easily. Indeed, in order to utilize (7.13), we needed to know only μ_2 (chosen by the designer) and λ_p-eigenvalues of the command input's autocorrelation matrix. Since the command input was a white zero-mean unit-power signal, all of its eigenvalues were equal to 1. Hence, in this case,

$$\tau = \frac{1}{2\mu_2}. \tag{7.25}$$

In order to compute $\xi_{\min}$, we needed to know the dither power (which was set to unity). We also needed to know $E[(n_k * \widehat{P}_\Delta^{-1})^2]$, which is the output power of the plant inverse if it were fed with the plant disturbance as an input signal. This can be approximated as follows:

$$\widehat{P}_\Delta^{-1}(z) \approx P_\Delta^{-1}(z) = \left(1 - \frac{1}{2}z^{-1}\right)z^{-\Delta}. \tag{7.26}$$

TABLE 7.1 EXPERIMENTAL VERIFICATION OF THE FILTERED-ϵ LMS ALGORITHM

Simulation Number	Variable Parameters			Time Constant of $\widehat{C}$		Weight Noise Variance for $\widehat{C}$		Dynamic System Error Variance	
	μ_1	μ_2	$E[n_k^2]$	Estimated	Measured	Estimated	Measured	Estimated	Measured
1	0	0.001	1	500	727	0.0227	0.0234	0.03	0.0351
2	0	0.0003	1	1667	2015	0.00667	0.0053	0.009	0.0043
3	0.001	0.001	1	500	655	0.0227	0.0225	0.03	0.028
4	0.001	0.001	10	500	604	0.137	0.177	0.18	0.147
5	0.001	0.001	100	500	unstable	1.289	—	1.68	—
6	0.01	0.001	1	500	519	0.0228	0.0331	0.03	0.0307
7	0.02	0.001	1	500	unstable	0.0228	—	0.03	—
8	0.001	0.01	1	50	92	0.25	0.253	0.3	0.265
9	0.001	0.06	1	8	—	3.44	3.78	1.8	4.43
10	0.001	0.1	1	unstable	unstable	—	—	—	—

The plant disturbance was assumed to be white. Accordingly,

$$E[(n_k * \widehat{P}_\Delta^{-1})^2] \approx E[n_k^2]\begin{pmatrix}\text{sum of squares}\\ \text{of impulses of}\\ \text{impulse responses}\\ \text{of } \widehat{P}_\Delta^{-1}(z)\end{pmatrix} \tag{7.27}$$

$$= E[n_k^2]\left(\frac{5}{4}\right).$$

Substituting into (7.19),

$$\xi_{\min} = E[\delta_k^2] + E[(n_k * \widehat{P}_\Delta^{-1})^2] \tag{7.28}$$

$$= 1 + \frac{5}{4}E[n_k^2].$$

Knowing $\xi_{\min}$, we were able to compute the noise variance of the weights of the controller. From (7.16),

$$\begin{pmatrix}\text{variance of the}\\ \text{noise in the}\\ \text{controller weights}\end{pmatrix} = \frac{\mu_2 m \xi_{\min}}{1 - \mu_2 m\left[E[i_k^2] + E[(i_k')^2]\mu_1 n E[n_k^2]\right]}. \tag{7.29}$$

All the parameters were known in (7.29) except for $E[(i_k')^2]$, which was the output power of $\widehat{P}_\Delta^{-1}(z)$ driven by the command input i_k. Since the command input was white, zero-mean, and of unit power,

$$E[(i_k')^2] = E[i_k^2]\begin{pmatrix}\text{sum of squares of}\\ \text{impulses of impulse}\\ \text{response of } \widehat{P}_\Delta^{-1}(z)\end{pmatrix} \tag{7.30}$$

$$= 1 \cdot \left(\frac{5}{4}\right).$$

Now (7.29) was able to be evaluated:

$$\begin{pmatrix}\text{variance of the}\\ \text{noise in the}\\ \text{controller weights}\end{pmatrix} = \frac{\mu_2 \cdot 10\left(1 + \frac{5}{4}E[n_k^2]\right)}{1 - \mu_2 \cdot 10\left(1 + \frac{5}{4}\mu_1(10)E[n_k^2]\right)}. \tag{7.31}$$

We wanted to estimate the variance of the dynamic system error, using (7.22). Everything was known except for $E[(i_k * P)^2]$. This was obtained in the following way. The command input i_k was a white, zero-mean, unit power signal. The vector **P** is of course the plant impulse response vector. Accordingly,

$$E[(i_k * P)^2] = 1 \cdot \begin{pmatrix}\text{sum of squares of}\\ \text{impulses of impulse}\\ \text{response of plant } \mathbf{P}\end{pmatrix} \tag{7.32}$$

$$= \frac{4}{3}.$$

Equation (7.22) was able to be evaluated in accord with

$$\begin{aligned}\begin{pmatrix}\text{variance of the}\\ \text{dynamic system error}\end{pmatrix} &= E[(i_k * P)^2]\mu_2 m\left(E[\delta_k^2] + E[(n_k * \widehat{P}_\Delta^{-1})^2]\right)\\ &= \frac{4}{3}\mu_2(10)\left(1+\frac{5}{4}E[n_k^2]\right)\\ &= \frac{40}{3}\mu_2\left(1+\frac{5}{4}E[n_k^2]\right).\end{aligned} \tag{7.33}$$

Equations (7.25), (7.31), and (7.33) allowed us to predict the time constant for the controller weights, the variance of the controller weight noise, and the variance of the dynamic system error for each choice of μ_1, μ_2, and $E[n_k^2]$. The theoretical predictions and the experimental measurements were summarized and compared in Table 7.1. The following commentary will help in making an interpretation of the results of Table 7.1.

In simulations 1 and 2, the adaptation constant μ_1 for the plant model $\widehat{P}(z)$ was kept constant at the value $\mu_1 = 0$. The idea was to demonstrate that the filtered-ϵ algorithm was indeed insensitive to errors in $\widehat{P}(z)$. Fixed errors were deliberately introduced into $\widehat{P}(z)$ in accord with

$$\widehat{P}(z) = P(z) + 0.4\left(1 + z^{-1} + \cdots + z^{-9}\right). \tag{7.34}$$

If the controller $\widehat{C}(z)$ was computed based on the erroneous plant model of (7.34), then the dynamic system error variance (due to error in $\widehat{P}(z)$) would have been quite large, that is, 0.119. However, application of filtered-ϵ adaptation reduced this variance to much lower levels. The effect of the fixed error in $\widehat{P}(z)$ was rendered inconsequential, and only the noise in the weights of $\widehat{C}(z)$ due to adaptation contributed to the dynamic system error variance. Instead of this variance being 0.119 for experiment 2, it turned out to be only 0.0043 (the theoretical prediction was 0.009).

Next, we examined the impact of plant disturbance on system performance. In simulations 3,4, and 5, both μ_1 and μ_2 were kept constant and the value of $E[n_k^2]$ was varied. The theoretical time constant for $\widehat{C}_k(z)$ remained unchanged. Experimental time constants for experiments 3 and 4 were close to theoretical. Weight noise variance and dynamic system error variance checked well between theory and experiment. But simulation 5 was unstable. What happened was that the plant model $\widehat{P}(z)$ was so far away from $P(z)$ because of such large plant disturbance ($E[n_k^2] = 100$), instability resulted. The only way that this could have been prevented would have been to adapt $\widehat{P}(z)$ with much smaller μ_1 and to not attempt to adapt $\widehat{C}(z)$ until $\widehat{P}(z)$ converged closer to $P(z)$.

The parameters μ_2 and $E[n_k^2]$ were kept constant for simulations 3, 6, 7, and the parameter μ_1 was varied. As expected from theory, there was no significant change in either time constant or weight noise variance or dynamic system error variance. Results of theory and experiment agreed well. But simulation 7 went unstable. In this case, instability was predicted from Eq. (7.15). The value of μ_1 was made large enough for instability to occur.

Finally, in simulations 3, 8, 9, and 10, we maintained μ_1 and $E[n_k^2]$ unchanged while μ_2 was increased from 0.001 to 0.1. Theory predicts that the time constant of $\widehat{C}_k(z)$ would be reduced in proportion to increase in μ_2. At the same time, the weight noise variance of $\widehat{C}_k(z)$ and the variance of the dynamic system error would increase with increase in μ_2, as predicted by Eqs. (7.16) and (7.22). Once again, the theory was confirmed by the experimental results.

For $\mu_2 = 0.1$, the system became unstable. This was predictable from (7.15), which gave the stable range for μ_2 as

$$0 < \mu_2 < 0.0988. \tag{7.35}$$

In general, as can be seen from the results of Table 7.1, Eqs. (7.13), (7.16), and (7.22) provided excellent prediction of the behavior of the filtered-ϵ control system. It is also clear that the stability criterion of (7.15) is quite accurate.

The exceptional situation is the case of instability caused by very large plant model error, due to large plant disturbance or due to a large value μ_1. Then the condition of (7.15) is not tight enough to guarantee stability. The reason is that for large errors in $\widehat{P}(z)$, filtered error ceases to approximate the correct error for adapting the controller. Hence, controller adaptation can no longer be regarded as an approximation of the conventional LMS algorithm, and the entire process becomes unstable. While designing a filtered-ϵ system, one must keep the plant model error from becoming too large. However, once this goal is achieved, the overall system error becomes insensitive to the error in the plant model. Then the entire system is robust and behaves well, as predicted by theory.

.5 EVALUATION AND SIMULATION OF THE FILTERED-X LMS ALGORITHM

A complete mathematical analysis of the adaptive inverse control system based on the filtered-X LMS algorithm of Fig. 7.2 is not presented in this book. Although the method of analysis for the filtered-X is similar to that for the filtered-ϵ algorithm, it is significantly more complicated. Without going into the details, some results are given next.

For the inverse control system based on filtered-X and using scheme C for plant modeling, the range of μ_1 that allows convergence of $\widehat{P}(z)$ is given by (B.50), with $E[u_k^2]$ equal to the plant input signal power and n equal to the number of weights of $\widehat{P}(z)$.

Convergence of the variance of $\widehat{C}(z)$ is obtained if μ_2 is chosen from within the range

$$0 < \mu_2 < \frac{1}{m[E[(i_k * P)^2] + E[i_k^2]\mu_1 n E[n_k^2]]}, \tag{7.36}$$

where m equals number of weights of $\widehat{C}(z)$, n equals number of weights of $\widehat{P}(z)$, μ_2 is the adaptation constant of $\widehat{C}(z)$, μ_1 is the adaptation constant of $\widehat{P}(z)$, $E[n_k^2]$ is the plant output disturbance power, $E[i_k^2]$ is the power of the command input signal, and $E[(i_k * P)^2]$ is the output power of the plant if it were driven by the command input.

The time constant for the weights of $\widehat{P}(z)$ is obtained from (B.34). $E[\delta_k^2]$ is the dither power. The learning time constants for $\widehat{C}(z)$ are given by (7.37),

$$\tau_p = \frac{1}{2\mu_2\lambda_p}, \tag{7.37}$$

where μ_2 is the adaptation constant of $\widehat{C}(z)$, and λ_p is the pth eigenvalue of the autocorrelation matrix of the output signal from the plant $P(z)$ when driven by the command input i_k.

The variance of the dynamic system error is given by (7.38):

$$\begin{pmatrix}\text{variance of}\\ \text{the dynamic}\\ \text{system error}\end{pmatrix} = E[(i_k * P)^2]\mu_2 m[E[\delta_k^2]\sum_{i=1}^{n} p_i^2 + E[n_k^2]]\left(1 + n\mu_1 E[n_k^2]\right), \quad (7.38)$$

where μ_1 and μ_2 are adaptation constants of $\widehat{P}(z)$ and $\widehat{C}(z)$, n and m are the numbers of the weights of $\widehat{P}(z)$ and $\widehat{C}(z)$, $E[n_k^2]$ is the plant disturbance power, $E[\delta_k^2]$ is the dither power, $\sum_{i=1}^{n} p_i^2$ is the sum of the squares of the weights of $P(z)$ or approximately the sum of the squares of the weights of $\widehat{P}(z)$, and $E[(i_k * P)^2]$ is the power of the output of the plant if it were driven by the command input signal.

In order to verify Eqs. (7.36)–(7.38), we have constructed in computer simulation the system of Fig. 7.2 (filtered-X LMS algorithm). We have chosen to experiment with exactly the same plant, same plant disturbance, same dither, same control system command input, and same number of weights for $\widehat{P}(z)$, and same number of weights for $\widehat{C}(z)$ as was described in Section 7.4 for the filtered-ϵ system. The plant and reference model are specified by (7.23) and (7.24) respectively, and the experimental parameters still hold for the present simulation.

As before, we chose various design parameters to vary: the plant model $\widehat{P}(z)$ adaptation constant μ_1, the controller $\widehat{C}(z)$ adaptation constant μ_2, and the plant disturbance power $E[n_k^2]$. For each set of parameter values, the system of Fig. 7.2 was run starting with the initial value for $\widehat{C}(z) = 2$. Time constants for the convergence of $\widehat{C}(z)$ to $C(z)$ were obtained. After allowing the system to run for at least four time constants to allow the system to settle, the variance of weight noise in $\widehat{C}(z)$ and the variance of the dynamic system error were measured. For each set of parameter values, 10 different simulations were run and averaged. Results are presented in Table 7.2.

TABLE 7.2 EXPERIMENTAL VERIFICATION OF THE FILTERED-X LMS ALGORITHM

Simulation Number	Variable Parameters			Average Time Constant		Dynamic System Error	
	μ_1	μ_2	$E[n_k^2]$	Estimated	Measured	Estimated	Measured
1	0	0.001	1	375	494	0.0310	0.0253
2	0	0.003	1	1250	2118	0.0093	0.0075
3	0.001	0.001	1	375	368	0.0314	0.0398
4	0.001	0.001	10	375	403	0.1662	0.1781
5	0.001	0.001	100	375	350	2.702	2.604
6	0.001	0.001	1000		unstable		
7	0.01	0.001	1	375	410	0.0342	0.0405
8	0.02	0.001	1		unstable		
9	0.001	0.01	1	38	25	0.9142	0.6311
10	0.001	0.03	1		unstable		

Theoretical predictions of system behavior were obtained by making use of Eqs. (7.36)–(7.38), and they were compared with the experiment. The theoretical expressions were once again easy to use, and they gave predictions that agreed very well with the experimental results. In utilizing Eq. (7.36), we needed to know $E[(i_k * P)^2]$. This was already calculated to have a value of 4/3, given by (7.32). Substituting (7.32) into (7.36) yields the stable range of μ_2,

$$0 < \mu_2 < \frac{1}{10\left(\frac{4}{3} + \mu_1 \cdot 10 \cdot E[n_k^2]\right)}. \tag{7.39}$$

Similarly, substituting (7.32) into (7.38) and recalling that the dither power was $E[\delta_k^2] = 1$, we have

$$\begin{pmatrix} \text{variance of} \\ \text{dynamic} \\ \text{system error} \end{pmatrix} = \mu_2 \cdot \frac{40}{3} \cdot \left(\frac{4}{3} + E[n_k^2]\right)\left(1 + 10\mu_1 E[n_k^2]\right). \tag{7.40}$$

In applying (7.37), we needed to know the eigenvalues of the autocorrelation matrix of the output signal of the plant if it were driven by the command input signal i_k. The matrix is

$$\mathbf{R}_1 = \frac{4}{3}\begin{bmatrix} 1 & \frac{1}{2} & \frac{1}{4} & \cdots & \frac{1}{2^9} \\ \frac{1}{2} & 1 & \frac{1}{2} & \cdots & \frac{1}{2^8} \\ \vdots & \vdots & \vdots & & \vdots \\ \frac{1}{2^9} & \frac{1}{2^8} & \frac{1}{2^7} & \cdots & 1 \end{bmatrix}. \tag{7.41}$$

The smallest eigenvalue is 0.483, and the largest eigenvalue is 3.577. Hence eigenvalue spread is relatively small. Accordingly, we approximated all the time constants by a single value,

$$\tau \approx \frac{1}{2\mu_2\lambda_{\text{ave}}} = \frac{m}{2\mu_2 \text{tr}\,(\mathbf{R}_1)}. \tag{7.42}$$

Hence,

$$\tau \approx \frac{10}{2\mu_2 \cdot 10 \cdot \frac{4}{3}} = \frac{3}{8\mu_2}. \tag{7.43}$$

From (7.39), (7.40), and (7.43), for each set of values of the parameters μ_1, μ_2, and $E[n_k^2]$, we computed the variance of the dynamic system error and the time constant of $\widehat{C}(z)$, and we determined the stable range of μ_2. Results are in Table 7.2.

For simulations 1 and 2, we set $\mu_1 = 0$ and set $\widehat{P}(z)$ in accordance with (7.34). The system was not allowed to adapt and correct the erroneous plant model. The idea was to test and demonstrate the insensitivity of the filtered-X algorithm to errors in the plant model. Indeed, without utilization of the filtered-X approach (or alternatively the filtered-ϵ approach), if the controller had been computed based on the erroneous plant model, the dynamic system error variance would have been 0.119. However, by using filtered-X, the dynamic system error variance was much smaller and was independent of the error in $\widehat{P}(z)$. The table shows close correlation between experimental values of the variance and the theoretical values predicted by (7.40).

Next, we examined the impact of variation in the power of the plant disturbance on system performance. In simulations 3, 4, 5, and 6, μ_1 and μ_2 were kept constant and $E[n_k^2]$ was varied. The time constants remained constant. The dynamic system error variance increased with increase in $E[n_k^2]$, as predicted by Eq. (7.38). For very large plant disturbance power (in this case, for $E[n_k^2] = 1000$), the system became unstable. We concluded that for filtered-X LMS, as is the case with filtered-ϵ LMS, plant disturbance power must be considered in the determination of the stable ranges of μ_1 and μ_2. This fact is in accordance with (7.36).

In simulations 3, 7, and 8, μ_2 and $E[n_k^2]$ were kept constant and μ_1 was increased from 0.001 to 0.02. As expected, there was no significant change in time constant of $\widehat{C}(z)$ or in dynamic system error variance. However, for large values of μ_2 (in this case $\mu_2 = 0.02$), the system became unstable, as predicted from (7.36).

Finally, in simulations 3, 9, and 10, μ_1 and $E[n_k^2]$ were kept constant while μ_2 was increased from 0.001 to 0.03. The time constant of $\widehat{C}(z)$ decreased and the variance of the dynamic system error increased with increase of μ_2, in accord with Eqs. (7.37) and (7.38). For $\mu_2 = 0.03$, the system became unstable. Hence, the stable range for μ_2 was

$$0 < \mu_2 < 0.03. \tag{7.44}$$

The theoretical range was, from (7.39),

$$0 < \mu_2 < 0.075. \tag{7.45}$$

The agreement was not perfect, but not bad.

For a wide range of μ_1 and μ_2, Eqs. (7.36)–(7.38) generally provide excellent prediction of the behavior of the filtered-X LMS algorithm. In cases where the adaptation constants μ_1 and μ_2 are close to the ends of their respective stable ranges (such as in simulation 9), the dynamic system error variance turns out to be somewhat larger than predicted.

Equation (7.36) provides a good estimate (within a factor of 2) of the stable range of μ_2. However, similarly to what we saw with the filtered-ϵ LMS algorithm, when the error in the plant model becomes very large (either due to large plant disturbance power or to large plant model $\widehat{P}(z)$ adaptation constant μ_1), the filtered-X system ceases to approximate the LMS convergence process. As a result, the entire system becomes unstable (even if μ_2 was chosen well within the range determined by (7.36)). However, for small enough errors in $\widehat{P}(z)$, the system is stable, robust, and completely insensitive to errors in the plant model.

By comparing Tables 7.1 and 7.2, we conclude that the filtered-X and filtered-ϵ algorithms performed similarly to one another. For a given level of dynamic system error variance, the filtered-X algorithm provided somewhat smaller time constants on average. The filtered-ϵ algorithm had only a single time constant. Regarding the experiments reported in the tables, the time constant of the filtered-ϵ algorithm was half that of the slowest mode of the filtered-X algorithm, but it was larger than that of the fastest mode of filtered-X. Because physical plants tend to be low-pass, there generally will be substantial eigenvalue spread for the matrix $\mathbf{R}_1$. Under worst case conditions, filtered-X can be slower than filtered-ϵ. However, on average, filtered-X will be somewhat faster than filtered-ϵ.

These conclusions are intuitive and of a rule of thumb nature. In each specific case, both algorithms should be considered. They should be analyzed both theoretically and by computer simulation before the best configuration is chosen.

7.7 A PRACTICAL EXAMPLE: ADAPTIVE INVERSE CONTROL FOR NOISE-CANCELING EARPHONES

It is difficult to communicate with a person working in a high-noise environment. The situation is helped if the person is wearing earphones and communication is done by feeding the audio signal to the earphones. But even that form of help is often not enough. Many of us have experienced the difficulty of trying to understand the sound track of a video shown on an airplane flight. The strong ambient noise propagates through the earphones and competes with the audio signal.

Further improvements can be achieved by utilizing various schemes for noise canceling. A simple nonadaptive scheme is shown in Fig. 7.15. A microphone attached to the outer face of the earphone receives the ambient noise. The microphone signal is applied to a fixed filter and the filtered signal is subtracted from the audio signal driving the earphone. If the fixed filter is designed properly to have the right gain and phase shift as a function of frequency, a significant noise reduction is possible. Commercial versions of this scheme exist, with the fixed filter being an amplifier with a small gain and a small delay.

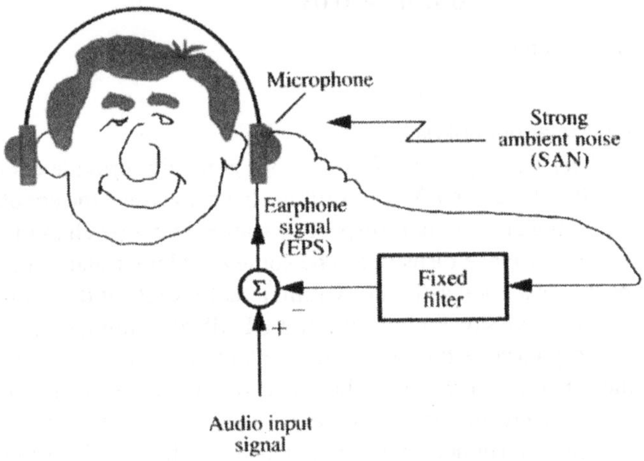

Figure 7.15 A nonadaptive noise-canceling earphone.

The limitations of the scheme of Fig. 7.15 result from variations in people's head and ear shapes and variations in how the earphones are worn. There are also manufacturing imperfections that result in variations in the microphone and earphone characteristics. The design of the fixed filter involves many compromises.

Better results can be obtained with adaptive techniques. The basic idea is illustrated in Fig. 7.16. Two microphones are used in the pictured system, one on the outer face of the earphone and the other on the inner face, close to the ear canal. The signals obtained from these microphones are labeled M_1 and M_2, respectively. Microphone number one receives the strong ambient noise (SAN). Microphone number two receives the signal entering the ear. Both microphone signals are input to an adaptive noise canceling system. The audio input signal containing the useful audio information goes through this system to form the earphone signal (EPS), but ambient noise components are adaptively filtered and subtracted from it. The goal is to reduce and minimize (in the mean square sense) ambient noise components present at the entrance of the ear canal and at the same time to allow the audio input signal to exist there.

Figure 7.17 describes the physical situation. Strong ambient noise is received by microphone number one, a transducer whose transfer function is represented by $[M_1/\text{SAN}]$

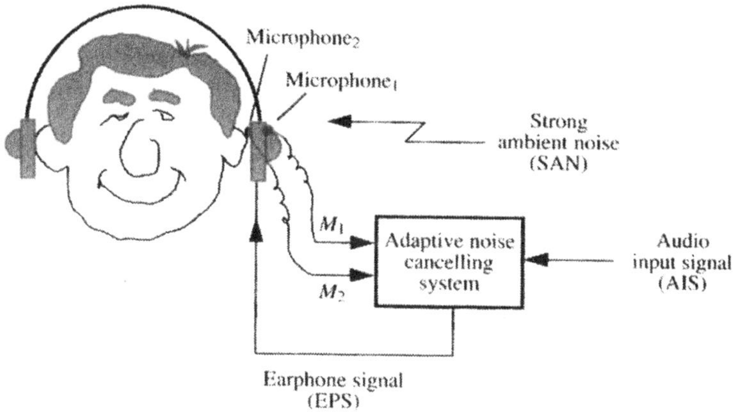

Figure 7.16 An adaptive noise-canceling earphone.

and whose output is represented by the signal M_1. The ambient noise propagates through the structure of the earphone and resonates in the cavity between the earphone and the ear. The resulting acoustic signal is transduced by microphone number two. The transfer function between the ambient noise and the signal M_2 is represented by $[M_2/\text{SAN}]$. One other signal component is sensed by microphone number two, and that is caused by the earphone signal (EPS) electrically actuating the loudspeaker in the earphone.[1] The transfer function between the earphone signal and the output of microphone number two is designated by $[M_2/\text{EPS}]$. In Fig. 7.17, the formation of signal M_2 is therefore represented as a sum of the ambient noise components and the earphone signal components. The earphone signal itself equals the audio input signal minus the ambient noise signal after adaptive filtering. The corresponding adaptive filter is designated as $\widehat{C}$, analogous to the controller of an adaptive inverse control system.

The weights of $\widehat{C}$ are to be adjusted to minimize the mean square of M_2, which is taken by the adaptive process to be the overall system error. The adaptive process reduces the power of only the ambient noise components in M_2. The audio input components are uncorrelated with the noise, and they add to the power of M_2. The adaptive process leaves these components unaffected.

An entire noise canceling system, based on the filtered-ϵ LMS algorithm, is shown in Fig. 7.18. The error M_2 could not be used directly as the error signal for adaptation of the adaptive noise canceling filter $\widehat{C}$. Filtering the error is necessary because the transfer function from the output of $\widehat{C}$ back to M_2 is not a simple gain of minus one.[2] Instead, the gain is $-[M_2/\text{EPS}]$. Therefore, it is necessary to model this gain and do a filtered-ϵ algorithm.

[1]Because of physical isolation, we assume that the level of earphone signal (EPS) received by microphone number one is negligible.

[2]If this gain were -1, then the adaptive system would be a simple adaptive noise canceler as described by Widrow, and colleagues [5], and by Widrow and Stearns [1].

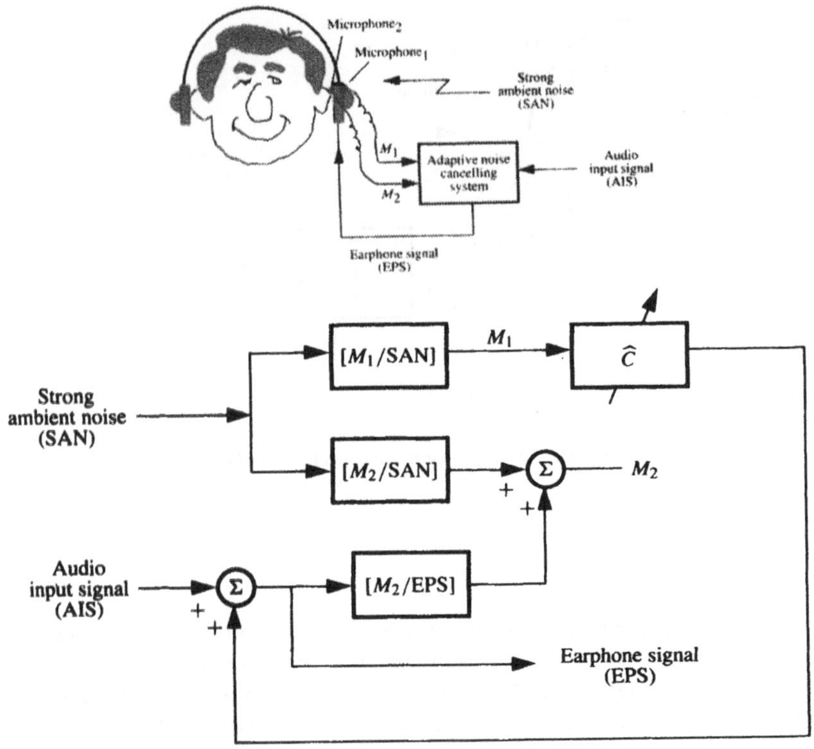

Figure 7.17 Some details of adaptive noise-canceling earphone.

In Fig. 7.18, the audio input signal is used in modeling $[M_2/\text{EPS}]$. We can represent the results of the modeling as

$$\widehat{P} \triangleq \widehat{[M_2/\text{EPS}]}.$$

The filtered $\widehat{P}$ has the audio input signal as its input. The microphone signal M_2 is its desired response. The delayed inverse of $\widehat{P}$ is obtained offline. This inverse is used as the error filter to adapt $\widehat{C}$. A copy of $\widehat{C}$ does the actual ambient noise filtering for the canceling operation.

Using the methodology of Fig. 7.18, it should be possible to devise a more comfortable earphone that would not need to fit as tightly as present earphones do and could incorporate an air "breathing space" between the ear and the earphone. This would be the best airplane earphone ever, but it would be expensive.

7.7 AN EXAMPLE OF FILTERED-X INVERSE CONTROL OF A MINIMUM-PHASE PLANT

The stabilized minimum-phase plant that was studied previously in Chapter 1 has been controlled by means of the filtered-X algorithm. Figure 7.19 is a block diagram showing the plant and its stabilization loop driven by a zero-order-hold DAC. The stabilized plant and

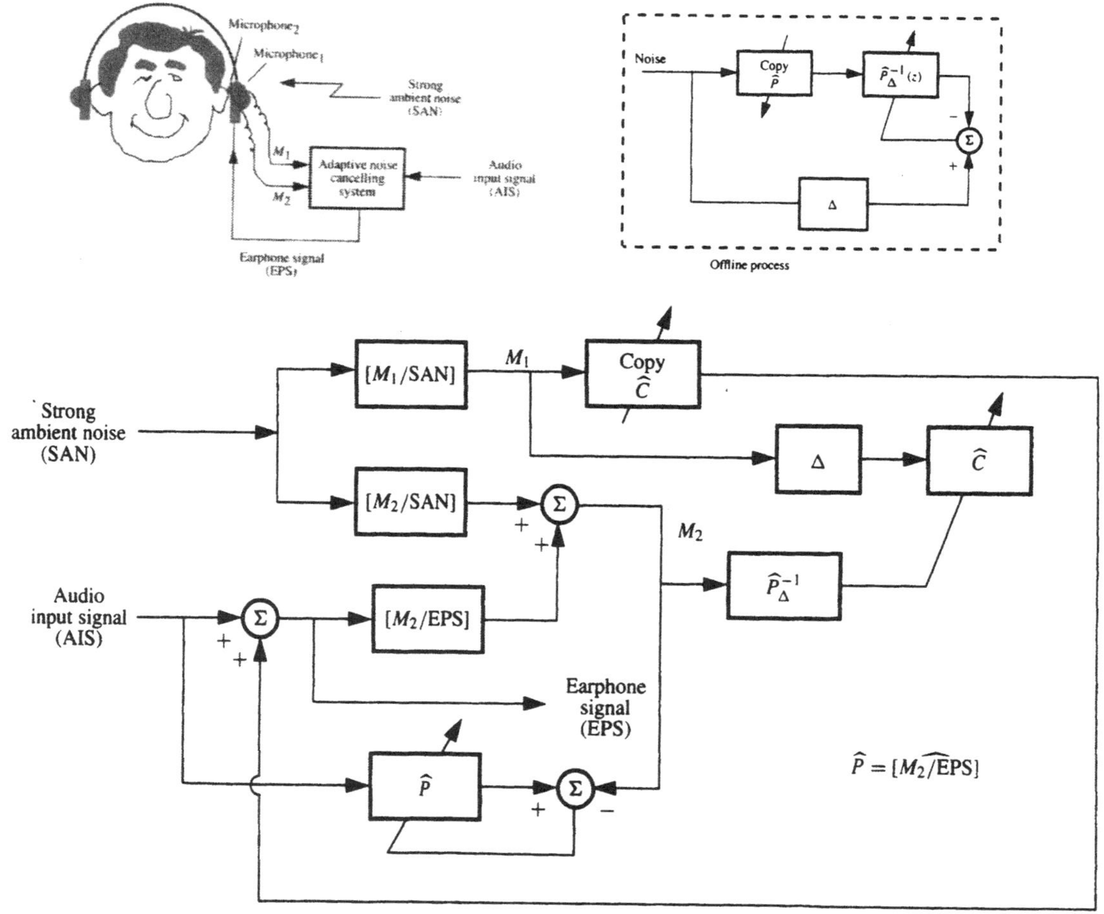

Figure 7.18 Complete details of adaptive noise-canceling earphone.

DAC are modeled by the adaptive filter $\widehat{G}(z)$. A copy of $\widehat{G}(z)$ is used with the filtered-X algorithm to generate $\widehat{C}(z)$. In simulation, the command input signal was random, zero-mean, slowly varying, and first-order Markov. The plant disturbance was first-order Markov noise. The reference model $M(z)$ was chosen for this experiment to be a unit delay.

Figure 7.20 shows the theoretical input response $G(z)$ of the stabilized plant and the DAC. The theoretical impulse response of the inverse with a unit delay is shown in Fig. 7.21. The convolution of the two is shown in Fig. 7.22. The convolution yields a unit impulse, with a unit delay.

The stabilized plant and the DAC were adaptively modeled, as represented in Fig. 7.19. At a given moment of time, the weights of $\widehat{G}(z)$ were recorded and then plotted. The result is shown in Fig. 7.23. This is a very noisy version of the true theoretical impulse response of Fig. 7.20. The noise in the weights of $\widehat{G}(z)$ is caused by rapid adaptation, and is increased by the plant disturbance. Using this noisy $\widehat{G}(z)$ with the filtered-X algorithm to obtain $\widehat{C}(z)$, the result is plotted in Fig. 7.24. The convolution of $\widehat{C}(z)$ with the true impulse response $G(z)$ is shown in Fig. 7.25. This is a remarkably good result, almost a perfect unit impulse with a unit delay, and it compares very well with the convolution shown in Fig. 7.22. The controller $\widehat{C}(z)$ is almost unaffected by noise in $\widehat{G}(z)$. That is one of the beautiful aspects of the filtered-X algorithm.

Using $\widehat{C}(z)$ as a controller, as in Fig. 7.19, excellent tracking of the command input signal took place. Figure 7.26 shows superposed plots of the command input signal, delayed by one sample time, and the plant output, with the plant disturbance subtracted. There is almost no difference. Figure 7.27 shows superposed plots of the command input signal, delayed by one sample time, and the actual plant output, including the plant disturbance. The difference is almost completely due to the plant disturbance. Canceling this disturbance will be the subject of the next chapter.

8 SOME PROBLEMS IN DOING INVERSE CONTROL WITH THE FILTERED-X LMS ALGORITHM

The filtered-X LMS algorithm requires that the input signal to $\widehat{C}(z)$ be filtered by a model of the plant. Above, we have studied the use of an unstable plant being stabilized by a feedback loop, shown in Fig. 7.19. The input to $\widehat{C}(z)$ is filtered by $\widehat{G}(z)$ which is low-pass. Filtering by $\widehat{G}(z)$ causes the spectrum of the input to $\widehat{C}(z)$ to be far from white. The covariance matrix of the inputs to the weights of $\widehat{C}(z)$ has a wide eigenvalue spread. In some cases of this type, the eigenvalue spread could be so great that convergence with the steepest descent LMS algorithm would be impossible to obtain in a practical length of time.

The magnitude of the frequency response of the minimum-phase $\widehat{G}(z)$ of Fig. 7.19 is shown in Fig. 7.28. Although this passes all frequencies, it yields an input signal to $\widehat{C}(z)$ having an eigenvalue spread of 30,000 to 1.[3] Convergence was slow but sure in this case. On the other hand, when doing the filtered-X LMS algorithm with the nonminimum-phase plant (stabilized as in Fig. 1.17 with $k = 24$, and with $M(z)$ chosen to be a 50 time unit delay),

[3]The spectrum of the input to a transversal filter determines the eigenvalue spread of the **R**-matrix. The ratio of the largest to the smallest eigenvalue equals the ratio of the peak value of the spectral density to the lowest value of the spectral density [6,7].

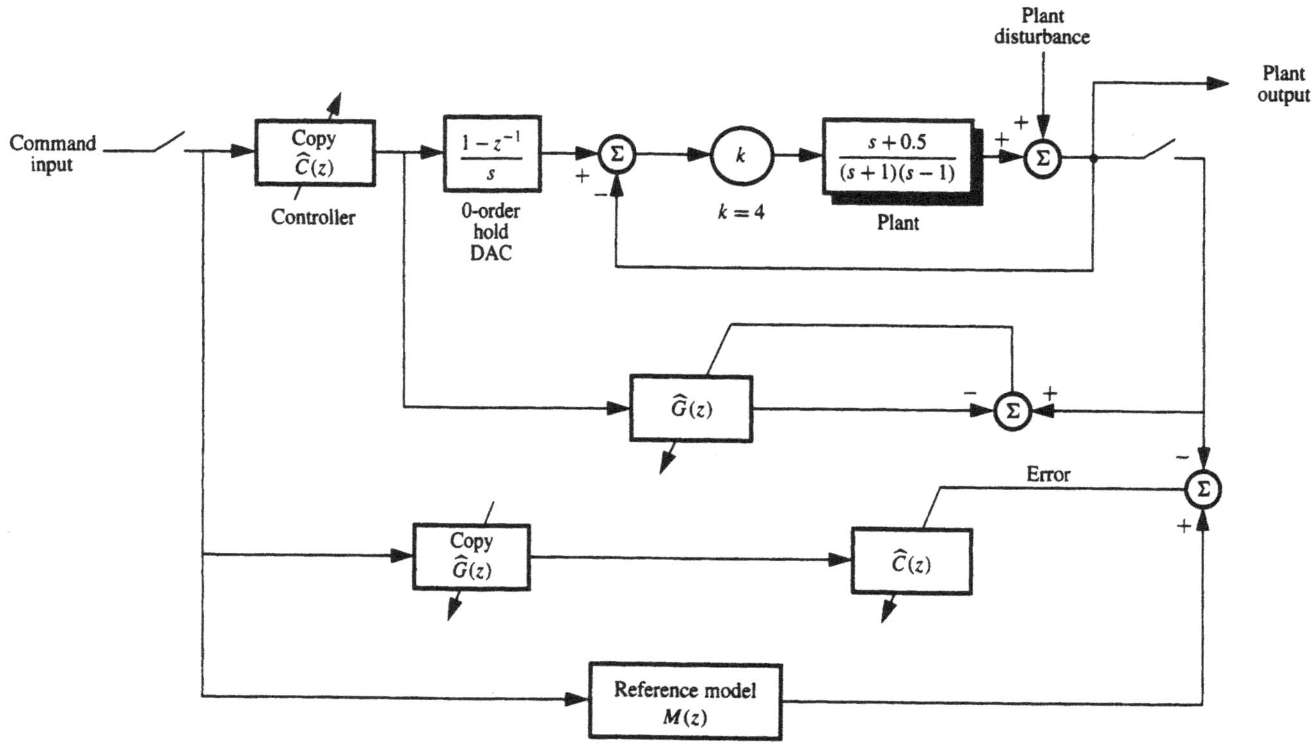

Figure 7.19 Filtered-X control of stabilized minimum-phase plant.

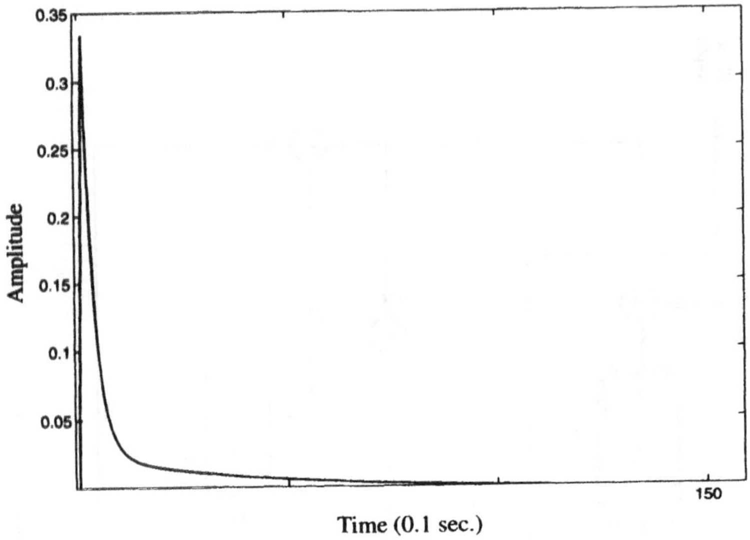

Figure 7.20 Theoretical impulse response of $G(z)$, minimum-phase plant of Fig. 7.19.

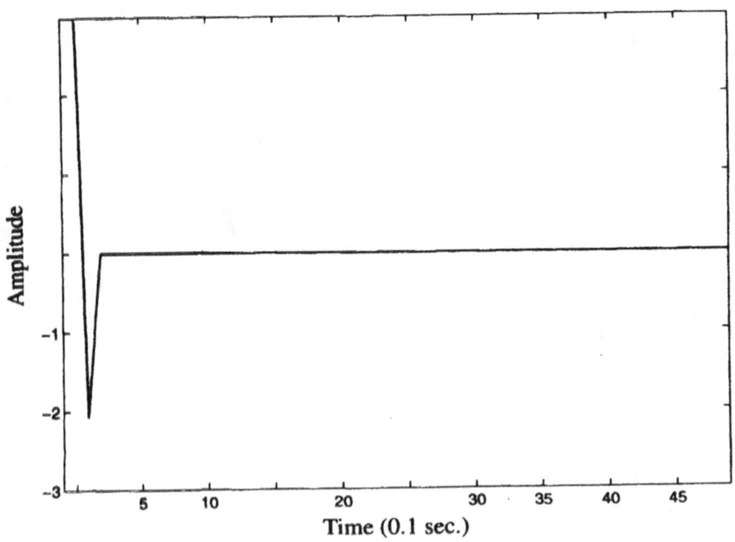

Figure 7.21 Theoretical unit-delay inverse of $G(z)$.

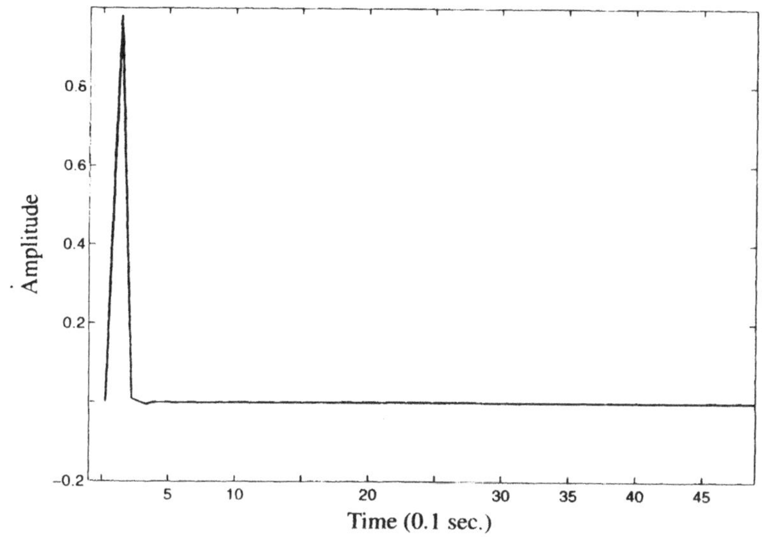

Figure 7.22 Convolution of theoretical $G(z)$ with theoretical unit-delay inverse.

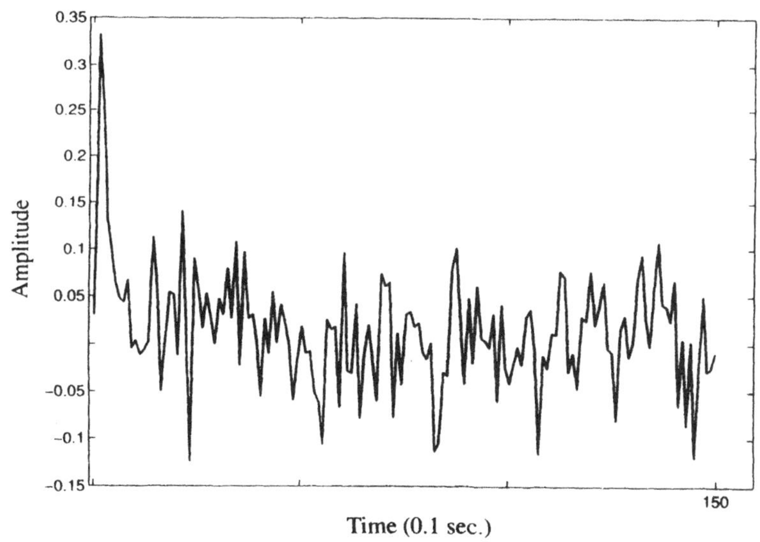

Figure 7.23 A noisy $\widehat{G}(z)$.

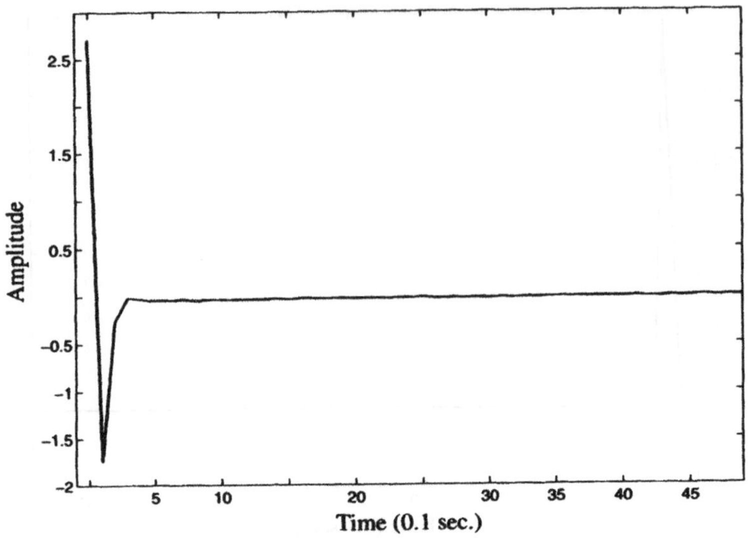

Figure 7.24 $\widehat{C}(z)$ obtained with noisy $\widehat{G}(z)$.

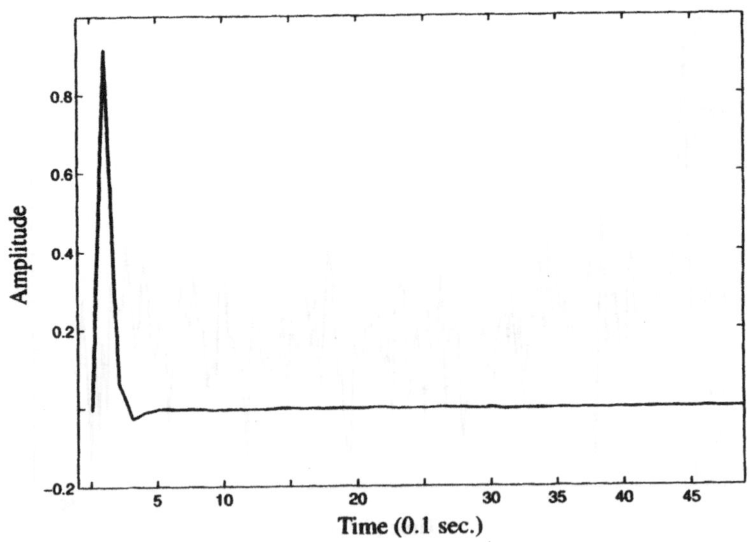

Figure 7.25 Convolution of $\widehat{C}(z)$ with the true $G(z)$.

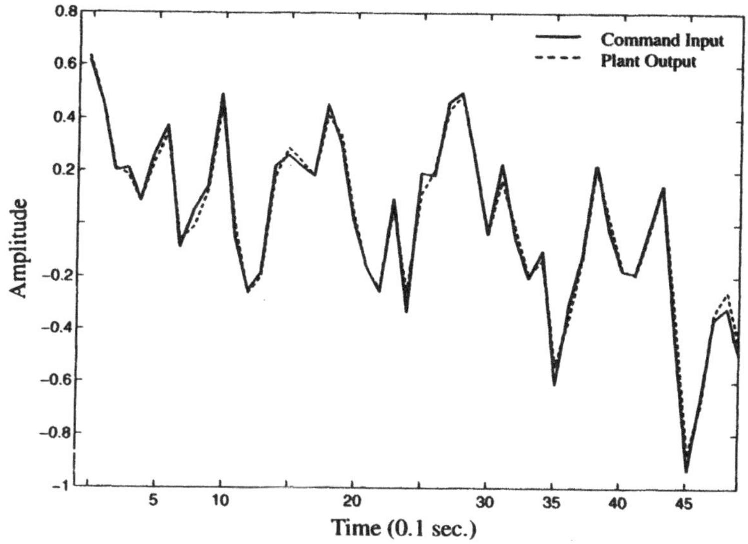

Figure 7.26 Tracking performance of filtered-X inverse control of stabilized minimum-phase plant.

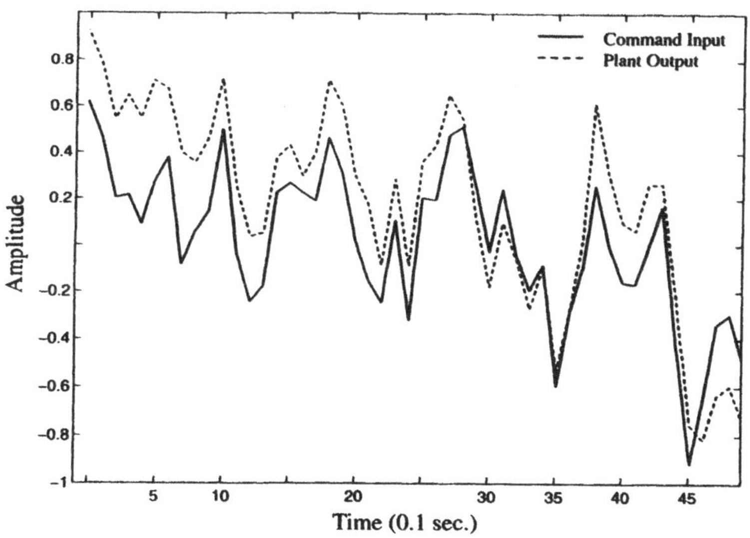

Figure 7.27 Command input signal, delayed by one sample period and plant output including plant disturbance.

$\widehat{G}(z)$ did not pass a significant amount of midrange and high-frequency energy. Its transfer function magnitude is shown in Fig. 7.29. The input to $\widehat{C}(z)$ had an eigenvalue spread of 10^8 to 1. Although there was no trouble in adaptively modeling $G(z)$ to get $\widehat{G}(z)$, shown in Fig. 7.30, using it with the filtered-X algorithm to find $\widehat{C}(z)$ did not work well. After 10^6 adaptations of the filtered-X LMS algorithm, the resulting $\widehat{C}(z)$ is shown in Fig. 7.31. Convolution of this $\widehat{C}(z)$ with the true $G(z)$ is shown in Fig. 7.32. It is clear from this result that things have been going in the right direction. The convolution looks like an impulse. But instead of being a unit impulse, it is less than that with a way to go. For this case, the filtered-X LMS algorithm would not be an appropriate choice. It takes too long to converge.

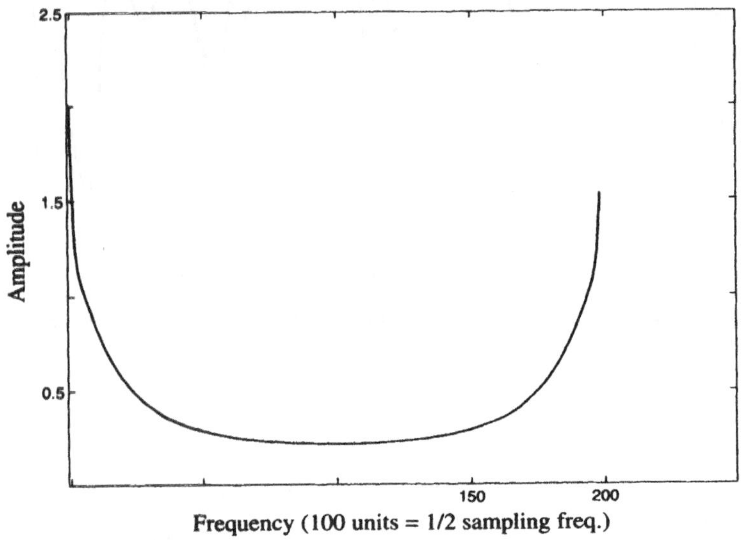

Figure 7.28 Magnitude of transfer function of $\widehat{G}(z)$ of stabilized minimum-phase plant.

7.9 INVERSE CONTROL WITH THE FILTERED-X ALGORITHM BASED ON DCT/LMS

To get improved convergence when the **R**-matrix of the input to the adaptive filter has a very large eigenvalue spread, algorithms other than LMS are often used. Please refer to Appendix E for a description of the RLS and the DCT/LMS algorithms which are designed for this purpose.

An attempt was made to use the RLS algorithm to do filtered-X, and this was not successful. We do not know yet why it did not work. However, the DCT/LMS algorithm was tried and this was very successful.

There was a dramatic improvement in the speed of convergence of $\widehat{C}(z)$ for the minimum-phase plant described in Section 7.8 above. When using the LMS algorithm, the eigenvalue spread of the inputs to the weights was 30,000 to 1. When using the DCT/LMS algorithm, however, the eigenvalue spread of the inputs to the weights was only 3 to 1.

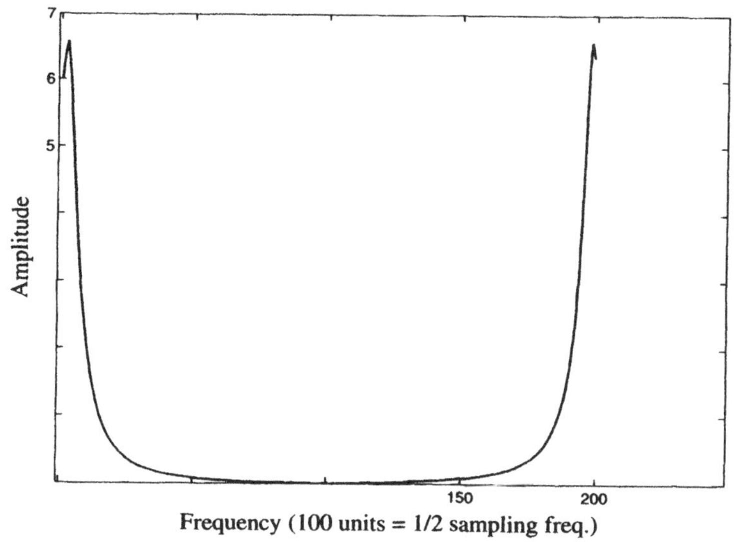

Figure 7.29 Magnitude of transfer function of $\widehat{G}(z)$ of stabilized nonminimum-phase plant.

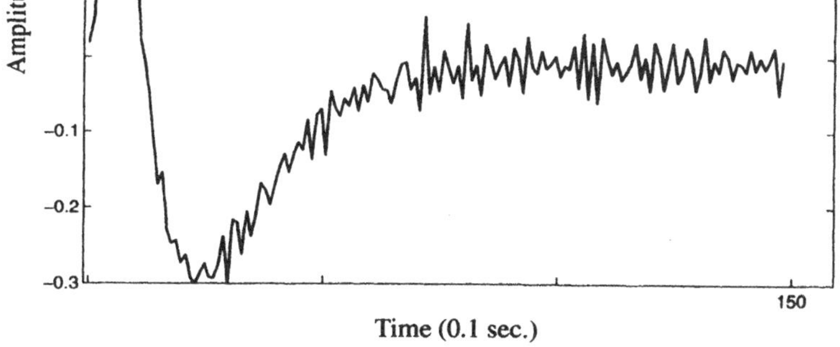

Figure 7.30 $\widehat{G}(z)$ for the stabilized nonminimum-phase plant with $k = 24$.

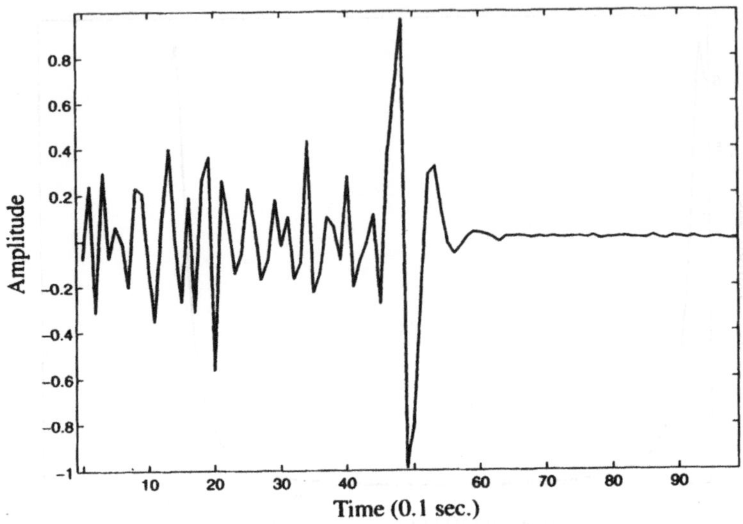

Figure 7.31 $\widehat{C}(z)$ for the stabilized nonminimum-phase plant, $k = 24$, after 10^6 iterations of the filtered-X algorithm.

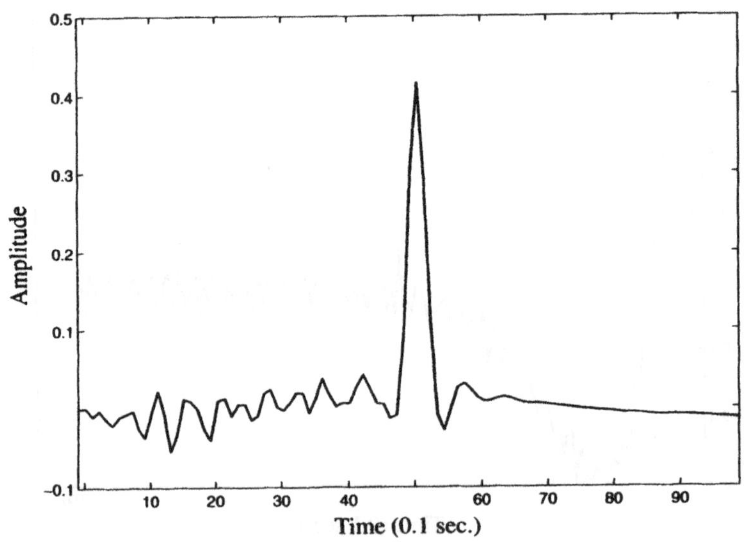

Figure 7.32 Convolution of $\widehat{C}(z)$ with the true $G(z)$.

There was an even more dramatic improvement in convergence of $\widehat{C}(z)$ for the nonminimum-phase plant described in Section 7.8. When using the filtered-X LMS algorithm to find $\widehat{C}(z)$, the eigenvalue spread of the inputs to the weights was 10^8 to 1. When using the filtered-X DCT/LMS algorithm, the eigenvalue spread was reduced to roughly 10^2 to 1. Convergence became possible, where before it required an impractically large number of adaptations. Figure 7.33 is a block diagram showing the filtered-X algorithm using DCT/LMS to control the stabilized nonminimum-phase plant.

Typical results using this technique are shown in Figs. 7.34–7.38. Figure 7.34 shows the impulse response of $\widehat{G}(z)$ that was obtained by DCT/LMS adaptation. This $\widehat{G}(z)$ was used with filtered-X to obtain $\widehat{C}(z)$, shown in Fig. 7.35.

Adaptation of $\widehat{C}(z)$ was done by DCT/LMS. Convolving the impulse response of $\widehat{C}(z)$ with that of the true $G(z)$ yields the result shown in Fig. 7.36. This looks like a good impulse, delayed by 50 sampling periods, matching $M(z)$ which was set to be a pure delay. A closer study of the convolution shows that the impulse has side lobes, and that its amplitude is about 0.8. The reason for the imperfections is that $\widehat{C}(z)$ was not adapted long enough.

To get an almost perfect match to $M(z)$ over a wide range of frequencies would require more adaptation to make the convolution more like a unit impulse. But the command input signal was first-order Markov and low-pass. So, although the convolution was not a perfect impulse, excellent tracking of the command input still took place. The system learned to make the error small for the given command input signal.

A sample of plant output including the plant disturbance is compared in Fig. 7.37 with the same plant output displayed with the disturbance removed. The plots indicate that the plant disturbance was substantial. Figure 7.38 shows the command input, delayed in accord with $M(z)$, and the plant output plotted with the disturbance removed. It can be seen that the tracking is very good indeed, almost perfect.

7.10 INVERSE CONTROL WITH THE FILTERED-ϵ ALGORITHM BASED ON DCT/LMS

The filtered-ϵ algorithm was used to control the stabilized nonminimum-phase plant that was studied previously, in Chapter 1. Figure 7.39 is a block diagram of the system. The input to $\widehat{C}(z)$, as shown in the block diagram, is the command input itself, not filtered as with the filtered-X algorithm, except for a time delay of Δ. The command input was first-order Markov. The eigenvalue spread was much less severe than with the filtered-X approach, "only" 340 to 1. To speed up the convergence process, LMS was first replaced with RLS and this did not work. The DCT/LMS algorithm worked very well and was used in the filtered-ϵ experiments for adapting $\widehat{C}(z)$. The eigenvalue spread of the inputs to the weights was 2 to 1.

Figure 7.40 is the impulse response of $\widehat{G}(z)$ obtained by LMS. Figure 7.41 is the impulse response of $\widehat{G}_{\Delta}^{-1}(z)$, the delayed inverse of $\widehat{G}(z)$, which could have been obtained with LMS by the offline process of Fig. 7.39 but was instead obtained offline with DCT/LMS. Figure 7.42 shows the impulse response of $\widehat{C}(z)$, obtained by filtered-ϵ. $M(z)$ was chosen to be a simple delay. Figure 7.43 shows the convolution of the impulse response of $\widehat{C}(z)$ with that of the true $G(z)$. The convolution is not a perfect delayed impulse, but it is close enough to allow excellent input-output tracking. Figure 7.44 shows the output of the plant

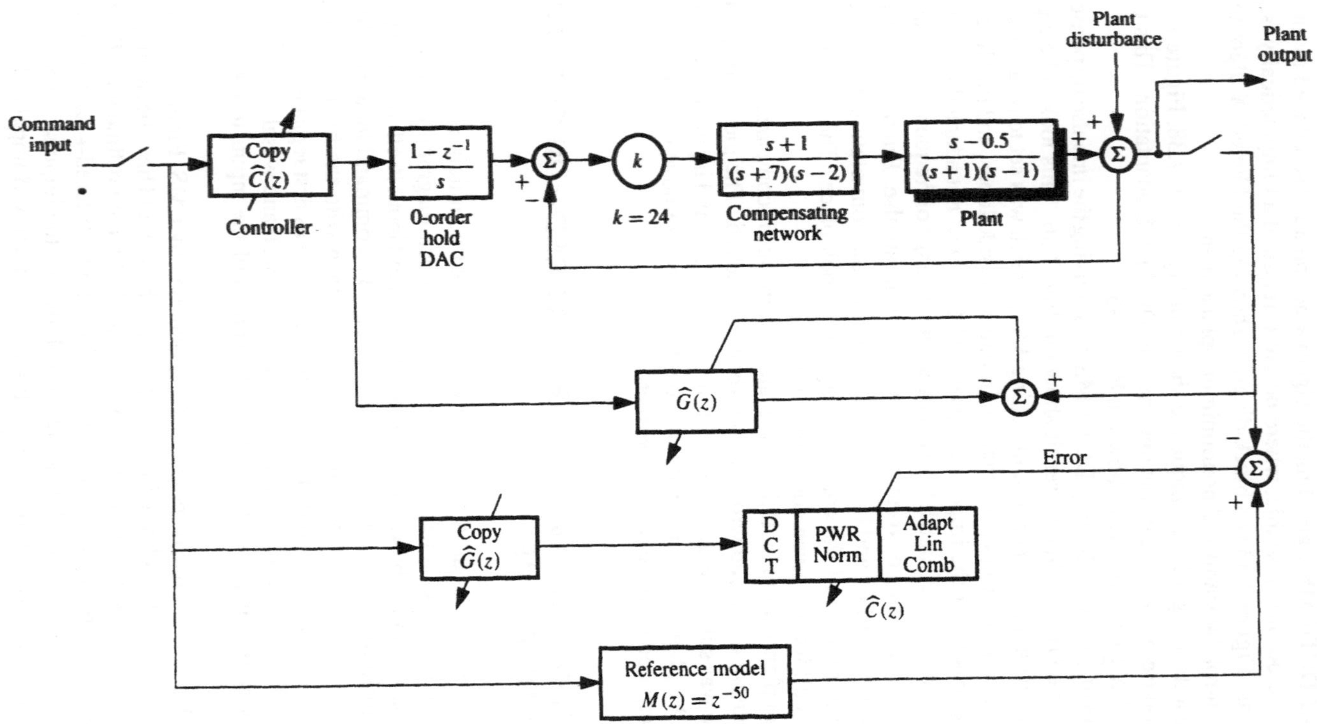

Figure 7.33 Filtered-X control of stabilized nonminimum-phase plant using the DCT/LMS algorithm.

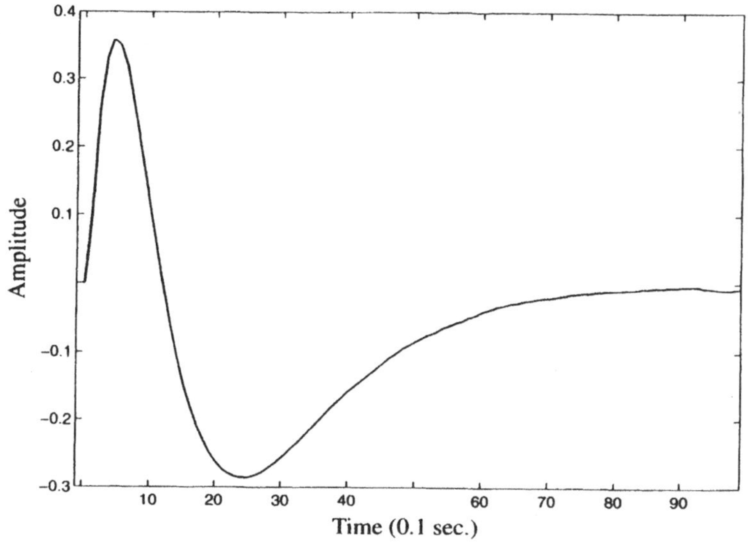

Figure 7.34 Impulse response of $\widehat{G}(z)$ for the stabilized nonminimum-phase plant.

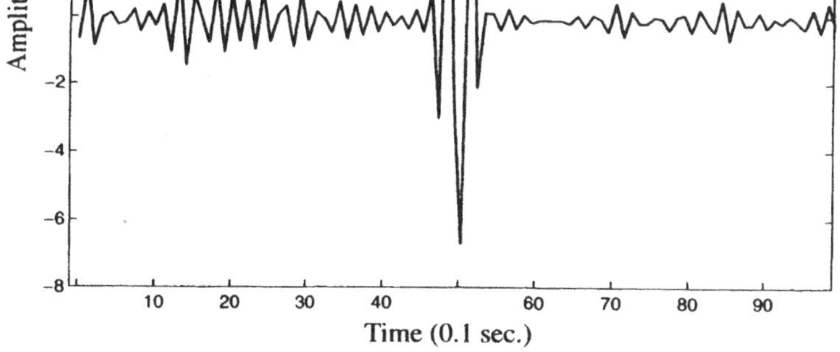

Figure 7.35 Impulse response of $\widehat{C}(z)$ for the stabilized nonminimum-phase plant, obtained by filtered-X with DCT/LMS adaptation.

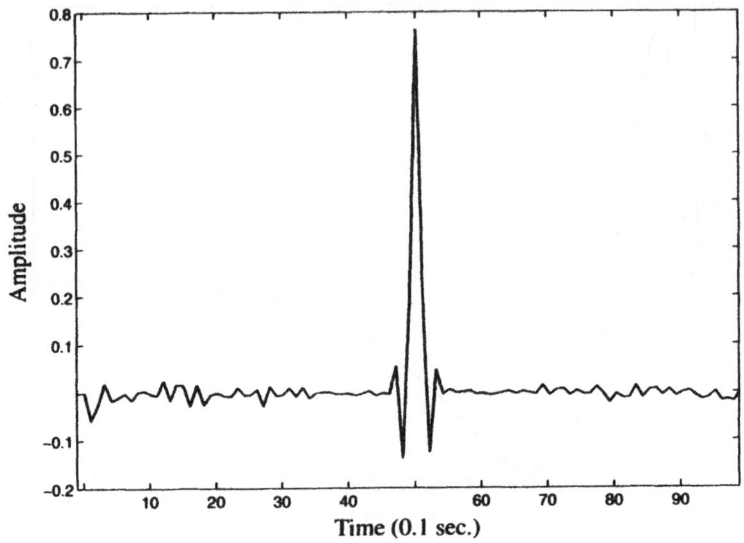

Figure 7.36 Convolution of impulse response of $\widehat{C}(z)$ with that of $G(z)$.

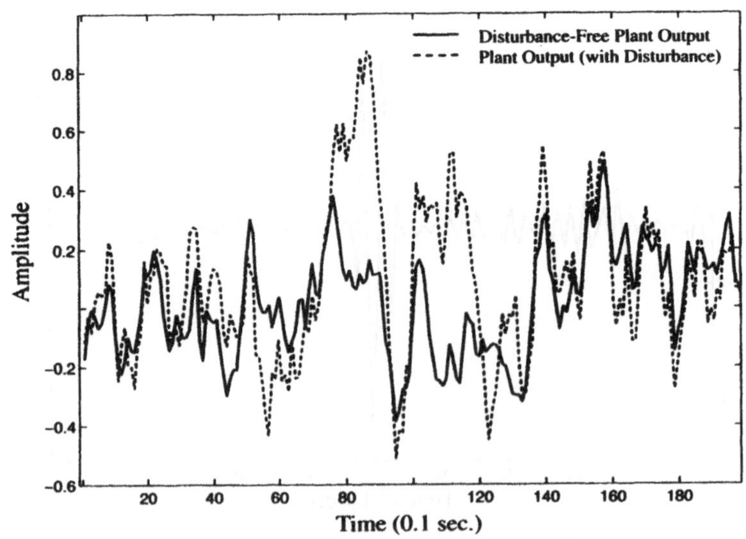

Figure 7.37 Plant output, and plant output with plant disturbance removed.

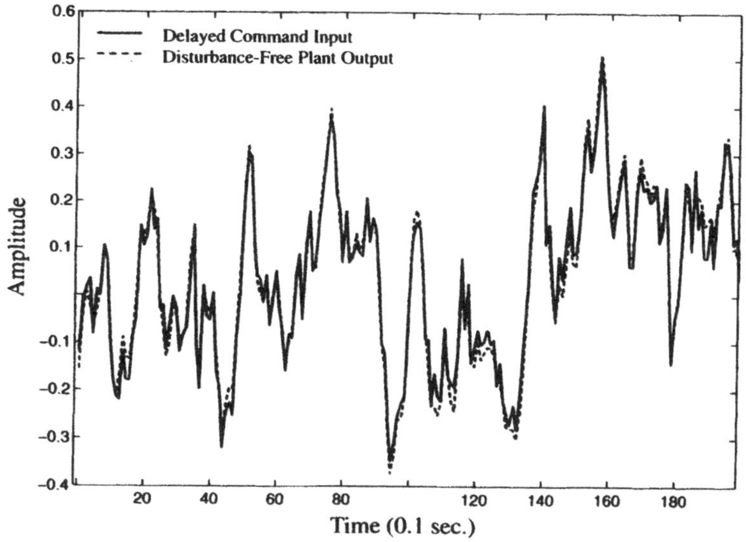

Figure 7.38 Tracking performance: Command input signal and plant output with disturbance removed.

including the plant disturbance, and the same output of the plant plotted with the disturbance removed. From this one can see that the disturbance was substantial. Figure 7.45 shows the command input, delayed to correspond to $M(z)$, and the plant output plotted with the plant disturbance removed. It can be seen that not perfect but very good tracking is taking place.

.11 SUMMARY

We conclude this chapter by comparing the performance characteristics of adaptive inverse control systems based on the filtered-X and filtered-ϵ LMS algorithms with the system of Fig. 6.4.

The system of Fig. 6.4 is constructed so that $\widehat{P}(z)$ has enough weights, that is, is long enough to be able to model P with a negligible truncation error. Error in $\widehat{P}(z)$ will bias the choice of $\widehat{C}(z)$. The filtered-ϵ and filtered-X algorithms, on the other hand, will not give biased choices of $\widehat{C}(z)$ even if $\widehat{P}(z)$ has some truncation error. They both cause $\widehat{C}(z)$ to converge in the mean to the ideal controller, the one that minimizes the mean square of the overall system error in Fig. 6.4. But error in $\widehat{P}(z)$ could cause instability in the adaptive process that yields $\widehat{C}(z)$. This does not happen with the system of Fig. 6.4.

When designing an adaptive inverse control system, one could follow the approach of Fig. 6.4 and use a long $\widehat{P}(z)$. Controlling stability would be easy, but to reduce the effects of weight noise, adaptation of $\widehat{P}(z)$ would need to be slow. Adaptation of $\widehat{C}(z)$, done offline at high speed, would not slow down the overall adaptive process. Therefore, $\widehat{C}(z)$ would converge as fast as $\widehat{P}(z)$.

Another possibility would be to use the filtered-ϵ algorithm. Now $\widehat{P}(z)$ would not be so long, and its speed of convergence could be allowed to go higher. $\widehat{P}(z)$ would need to be long enough and close enough to $P(z)$ to enable the algorithm computing $\widehat{C}(z)$ to be

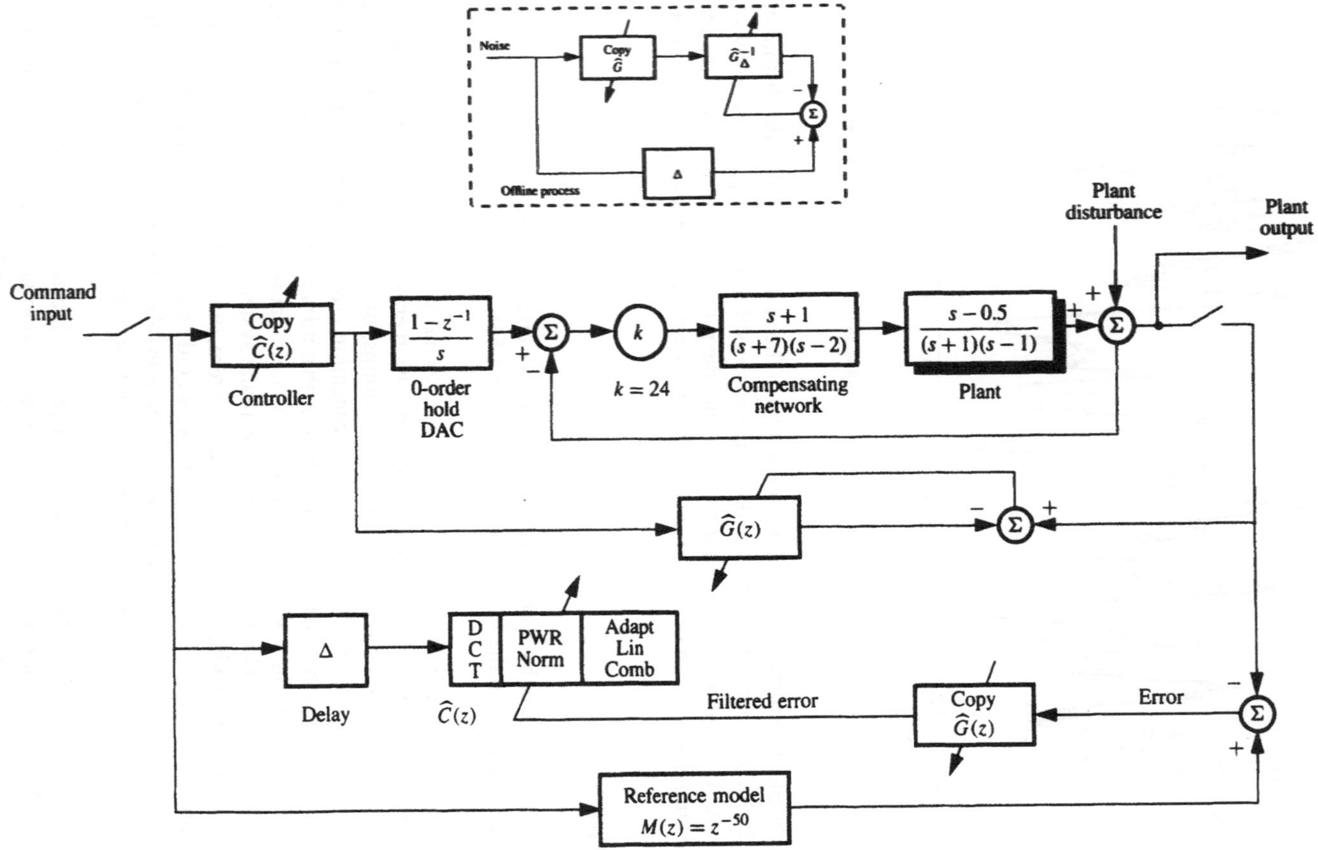

Figure 7.39 Filtered-ϵ control of stabilized nonminimum-phase plant using the DCT/LMS algorithm.

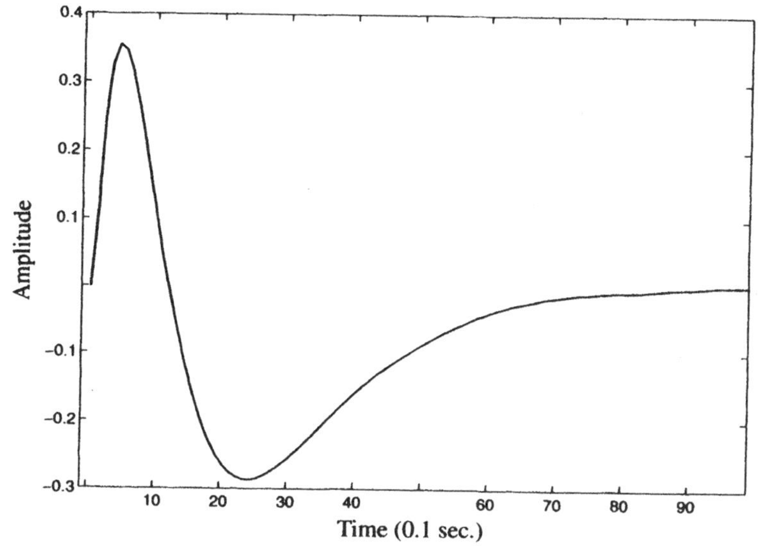

Figure 7.40 Impulse response of $\widehat{G}(z)$, nonminimum-phase plant with feedback stabilization.

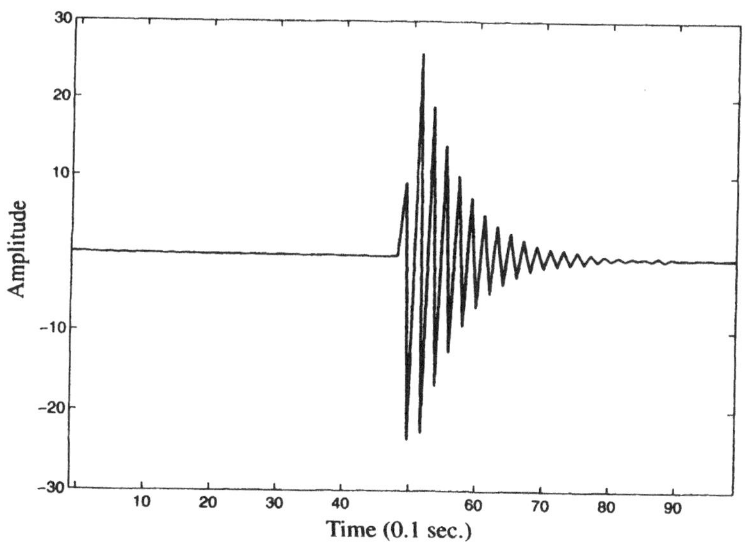

Figure 7.41 Impulse response of $\widehat{G}(z)_{\Delta}^{-1}$, the delayed inverse of $\widehat{G}(z)$.

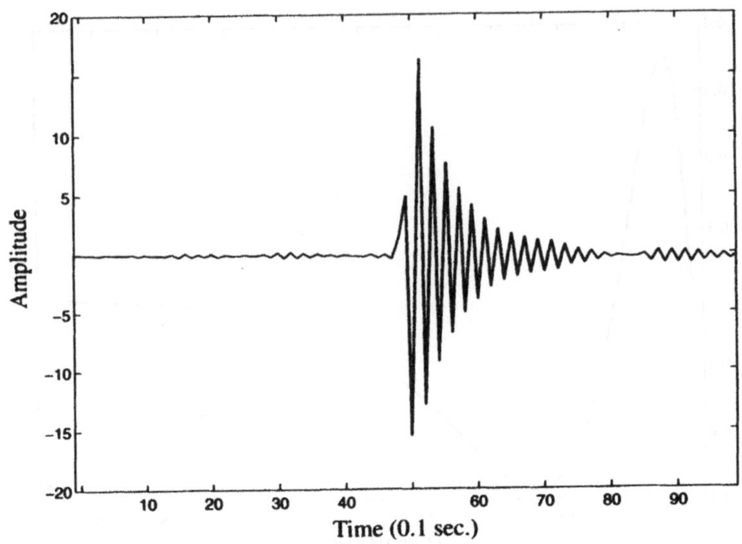

Figure 7.42 Impulse response of $\widehat{C}(z)$ for nonminimum-phase plant obtained by DCT/LMS. $M(z)$ was a simple delay.

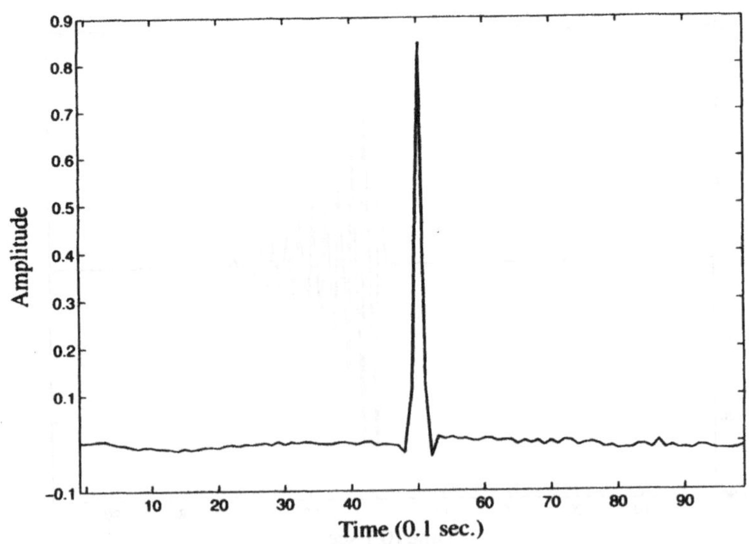

Figure 7.43 Convolution of the impulse response of $\widehat{C}(z)$ with the impulse response of the true $G(z)$ of the stabilized nonminimum-phase plant.

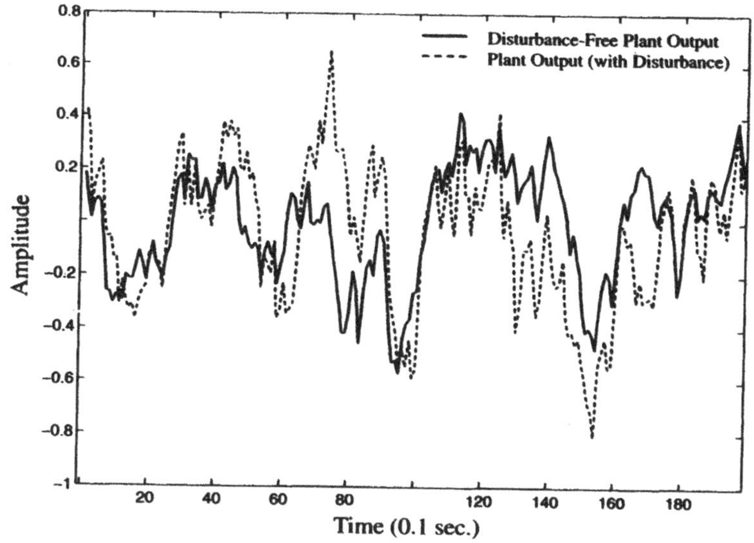

Figure 7.44 The plant output including its disturbance, and the same plant output plotted with the disturbance removed.

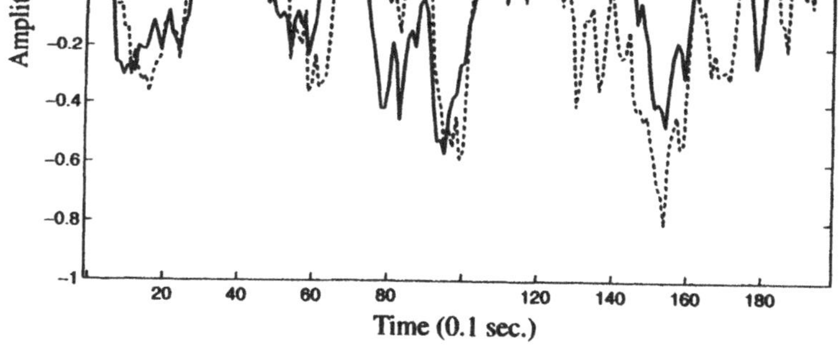

Figure 7.45 Tracking performance: Comparing delayed command input signal with plant output plotted with plant disturbance removed.

stable. The same is true for the filtered-X algorithm. All of these algorithms work well, and trade-offs need to be made when choosing.

For the system of Fig. 6.4, the range of μ for stability of the variance of $\widehat{P}(z)$ is given by Eq. (B.50):

$$0 < \mu < \frac{1}{n(E[\delta_k^2] + E[(u'_k)^2])} = \frac{1}{nE[u_k^2]}, \tag{B.50}$$

where $E[u_k^2]$ is the power of the input signal to the plant and n is the number of weights of $\widehat{P}$ (z). The learning time constant of the weights of $\widehat{P}(z)$ are given by (B.34) as

$$\tau = \frac{1}{2\mu E[\delta_k^2]}, \tag{B.34}$$

where $E[\delta_k^2]$ is the dither power. The variance of the dynamic system error at the plant output due to the noise of the weights of $\widehat{C}(z)$ which, in turn, is due to the noise of the weights in $\widehat{P}(z)$ is given by

$$\begin{pmatrix} \text{variance of} \\ \text{dynamic system error} \end{pmatrix} = \phi_{BB}(0) \cdot \min(m, n) \cdot \beta, \tag{5.40}$$

where $\phi_{BB}(0)$ is the power output of the controller $\widehat{C}(z)$, m is the number of weights in $\widehat{C}(z)$ and n is the number of weights of $\widehat{P}(z)$, $\min(m, n)$ is the smaller number of the two, and where β equals the product of μ and $E[n_k^2]$, the plant output disturbance power.

For the adaptive inverse control system based on the filtered-ϵ LMS algorithm, shown in Fig. 7.5, the range of μ_1 that allows convergence of the variance of $\widehat{P}(z)$ is given by (B.50). Once again, $E[u_k^2]$ is the power of the plant input signal and n is the number of weights of $\widehat{P}(z)$. Convergence of the variance of $\widehat{C}(z)$ is assured if μ_2 is chosen within the range

$$0 < \mu_2 < \frac{1}{n\left[E[i_k^2] + E[(i'_k)^2]\mu_1 n E[n_k^2]\right]} \quad , \tag{7.15}$$

where m is the number of weights of $\widehat{C}(z)$, n is the number of weights of $\widehat{P}(z)$, $E[n_k^2]$ is the plant output disturbance power, $E[i_k^2]$ is the power of the system command input, and $E[i'_k]$ is the output power of a filter $\widehat{P_\Delta^{-1}}(z)$ driven by the command input i_k. The filter $\widehat{P_\Delta^{-1}}(z)$ is the delayed plant inverse which is obtained by an offline adaptive process. Stability of this simple adaptive process is easy to obtain by making use of (A.36). The time constant for the weights of $\widehat{P}(z)$ is obtained from (B.34), and once again $E[\delta_k^2]$ is the dither power. The time constants for $\widehat{C}(z)$ are given by (7.13).

$$\tau_p \cong \frac{1}{2\mu_2\lambda_p} \tag{7.13}$$

where μ_2 is the adaptation constant for $\widehat{C}(z)$ and λ_p is the pth eigenvalue of the autocorrelation matrix of the input to $\widehat{C}(z)$, that is, the command input i_k. The variance of the dynamic system error is given by (7.22) as

$$\begin{pmatrix} \text{variance of} \\ \text{the dynamic} \\ \text{system error} \end{pmatrix} = E[(i_k * P)^2]\mu_2 m[E[\delta_k^2] + E[(n_k * \widehat{P_\Delta^{-1}})^2]], \tag{7.22}$$

where $E[(i_k * P)^2]$ is the power of the plant output if the plant were driven by the command input i_k, μ_2 is the adaptation constant of $\widehat{C}(z)$, m is the number of weights of $\widehat{C}(z)$, $E[\delta_k^2]$ is the dither power, and $E[(n_k * \widehat{P}_\Delta^{-1})^2]$ is the power output of the delayed plant inverse if it were driven by the plant output noise. This is equivalent to the plant disturbance referred to the plant input instead of to the plant output.

Analysis of the adaptive inverse control system based on the filtered-X LMS algorithm of Fig. 7.2 is not shown in this book. Although the methods of analysis for this system are similar to those for the filtered-ϵ algorithm, they are significantly more complicated. Without going into the details, some results are presented in Section 7.5 and summarized here.

For the adaptive inverse control system based on filtered-X, the range of μ that allows convergence of $\widehat{P}(z)$ is given by (B.50), with $E[u_k^2]$ equal to the plant input signal power and n equal to the number of weights of $\widehat{P}(z)$. Convergence of the variance of $\widehat{C}(z)$ is obtained if μ_2 is chosen from within the range

$$0 < \mu_2 < \frac{1}{m[E[(i_k * P)^2] + E[i_k^2]\mu_1 n E[n_k^2]]}, \tag{7.36}$$

where m equals number of weights of $\widehat{C}(z)$, n equals number of weights of $\widehat{P}(z)$, μ_2 is the adaptation constant of $\widehat{C}(z)$, μ_1 is the adaptation constant of $\widehat{P}(z)$, $E[n_k^2]$ is the power of the plant disturbance, $E[i_k^2]$ is the power of the command input signal, and $E[(i_k * P)^2]$ is the output power of the plant if it were driven by the command input. The time constant for the weights of $\widehat{P}(z)$ is obtained from (B.34), and $E[\delta_k^2]$ is the dither power. The time constants for $\widehat{C}$ are given by

$$\tau_p = \frac{1}{2\mu_2\lambda_p}, \tag{7.37}$$

where μ_2 is the adaptation constant of $\widehat{C}(z)$, and λ_p is the pth eigenvalue of the autocorrelation matrix of the output signal from the plant $P(z)$ if it were driven by the command input i_k. The variance of the dynamic system error is given by

$$\begin{pmatrix}\text{variance of}\\ \text{the dynamic}\\ \text{system error}\end{pmatrix} = E[(i_k * P)^2]\mu_2 m[E[\delta_k^2]\sum_{i=1}^{n} p_i^2 + E[n_k^2]]\left(1 + n\mu_1 E[n_k^2]\right), \tag{7.38}$$

where μ_1 and μ_2 are adaptation constants of $\widehat{P}(z)$ and $\widehat{C}$ (z), n and m are the numbers of weights of $\widehat{P}(z)$ and $\widehat{C}$ (z), $E[n_k^2]$ is the power of the plant disturbance at the plant output, $E[\delta_k^2]$ is the dither power, $\sum_{i=1}^{n} p_i^2$ is the sum of the weights of $P(z)$ or approximately the sum of the weights of $\widehat{P}(z)$, and $E[(i_k * P)^2]$ is the power of the output of the plant if it were driven by the command input signal.

Stability, time constants, and dynamic system error can be determined for adaptive inverse control systems by making use of the equations of this summary. Most of the parameters of the formulas are readily available to the system designer. Some need to be estimated. Exact values will not always be available. The best way to choose from these algorithms would be to analyze all three, with the formulas and by computer simulation.

When using either the filtered-ϵ or the filtered-X algorithm for adaptive inverse control, errors in the plant model $\widehat{P}(z)$, designated by $\Delta P(z)$, have no effect upon the Wiener solution for the controller $\widehat{C}(z)$. As long as the adaptive process that generates $\widehat{C}(z)$ is stable, errors in the plant model do not disturb the optimality of $\widehat{C}(z)$. Adaptive feedback minimizing the overall system error gives rise to these surprising results.

This chapter and the previous chapter have presented methods for controlling the dynamic response of plants with and without disturbance. The effects of plant disturbance have been calculated, but nothing has been done about it per se, that is, to mitigate its effects at the plant output. The next chapter addresses the issue of reduction of plant disturbance by adaptive cancelation.

Bibliography for Chapter 7

[1] B. WIDROW and S.D. STEARNS, *Adaptive signal processing* (Englewood Cliffs, NJ: Prentice Hall, 1985).

[2] J.C. BURGESS, "Active adaptive sound control in a duct: A computer simulation," *J. Acoust. Soc. Am.*, Vol. 70, No. 3 (September 1981), pp. 715–726.

[3] B. WIDROW, D. SCHUR, and S. SHAFFER, "On adaptive inverse control," in *Conf. Rec. of 15th Asilomar Conf. on Circuits, Systems and Computers*, Santa Clara, CA (November 1981), pp. 185–189.

[4] B. WIDROW, "Adaptive inverse control," in *Proc. of the 2nd IFAC Workshop on Adaptive Systems in Control and Signal Processing* (Lund, Sweden, July 1986), plenary paper, Pergamon Press pp. 1–5.

[5] B. WIDROW, J.M. MCCOOL, J.R. GLOVER, JR., J. KAUNITZ, C. WILLIAMS, R.H. HEARN, J.R. ZEIDLER, E. DONG, JR., and R.C. GOODLIN, "Adaptive noise cancelling: Principles and applications," *Proc. IEEE*, Vol. 63, No. 12 (December 1975), pp. 1692–1716.

[6] U. GRENANDER and G. SZEGO, *Toeplitz forms and their applications* (Berkeley, CA: University of California Press, 1958).

[7] R.M. GRAY, "Toeplitz and circulant matrices: II," Information Systems Laboratory Technical Report No. 6504-1, Stanford University, April 1977.

[8] Dozens of applications of the Filtered-X LMS algorithm are given in the *Proc. of the Conf. on Recent Advances in Active Control of Sound and Vibration*, Virginia Polytechnic Institute, April 1991 (Lancaster, PA: Technomics Publishing Co.), Fax: (717) 295-4538.

Chapter 8

Plant Disturbance Canceling

8.0 INTRODUCTION

Methods for controlling plant dynamics have been described in the previous chapters. These methods have no effect on plant disturbance, which would simply appear unimpeded at the plant output. A feedback scheme for plant disturbance canceling which does not alter plant dynamics is suggested in Figs. 1.4 and 1.5. The purpose of this chapter is to develop plant disturbance canceling techniques based on this scheme. The goal is to minimize plant output disturbance power without changing plant dynamics.

In control theory, it is most common to control plant response and plant disturbance in one process. With adaptive inverse control, however, it is convenient to treat these problems independently. In this way, the dynamic control process is not compromised by the need to reduce plant disturbance. Furthermore, the plant disturbance reduction process is not compromised by the needs of dynamic control.

A plant disturbance canceling system is diagrammed in Fig. 8.1. Disturbance cancelation is accomplished by this system in the following manner. A copy of $\widehat{P}_k(z)$, a very close, disturbance-free match[1] to $P(z)$, is fed the same input as the plant $P(z)$. The difference between the disturbed output of the plant and the disturbance-free output of $\widehat{P}_k(z)$ is a very close approximation to the plant output disturbance n_k. The approximate n_k is then input to the filter $z^{-1}Q_k(z)$ which is a best least squares inverse of $\widehat{P}_k(z)$. The output of $z^{-1}Q_k(z)$ is subtracted from the plant input to effect cancelation of the plant disturbance. Unit delays z^{-1} were placed in front of $Q_k(z)$'s in recognition of the fact that digital feedback links must have at least one unit of delay around each loop.[2] Thus, the current value of the plant disturbance n_k can be used only for the cancelation of future values of plant disturbance and cannot be used for instantaneous self-cancelation. The effects of these unit delays are small when the system is operated with a high sampling rate, however.

[1] $\widehat{P}_k(z)$ is "disturbance free" in the sense that plant disturbance n_k is absent. Noise in the weights of $\widehat{P}_k(z)$ due to adaptation will of course exist. However, noise in the weights can be made small by making μ small (i.e., making adaptation slow).

[2] In physical systems where the plant is analog, if the plant has more poles in the s-plane than zeros, the discretized form of the plant, $P(z)$, would have at least a unit delay in its response to an impulse input. As such, the unit delay in line with $Q_k(z)$ would be unnecessary and should be left out. Including it would cause a loss in performance.

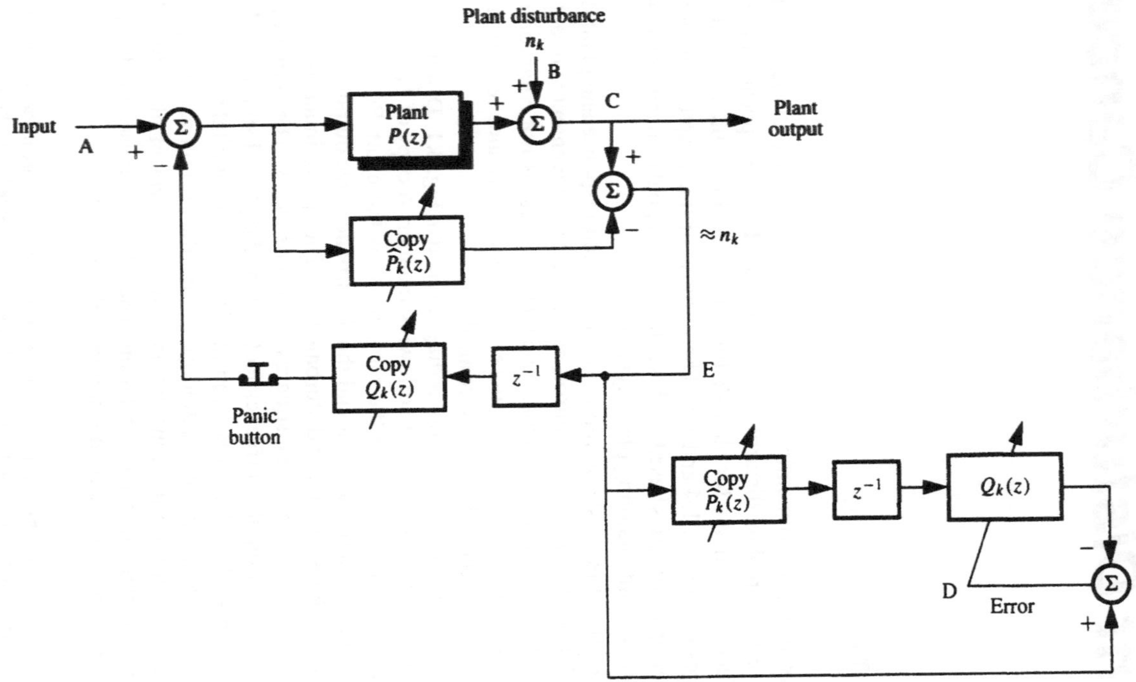

Figure 8.1 An adaptive system for canceling plant disturbance.

The system of Fig. 8.1 is comprised of two parts. One part does the actual disturbance canceling. The other part performs the inverse modeling of $\hat{P}_k(z)$ to obtain $z^{-1}Q_k(z)$. There really is a third part to this system, not shown in the figure, that models the plant to obtain $\hat{P}_k(z)$. This portion of the system will be incorporated subsequently.

The adaptive plant disturbance canceler of Fig. 8.1 differs markedly from the conventional adaptive noise canceler described in Chapter 3 and illustrated in Fig. 3.8. The conventional noise canceler gets its noise reference signal externally and uses it for canceling by *feedforward filtering* and subtracting. The plant disturbance canceler gets its disturbance reference from the plant output and uses it for canceling by *feedback filtering* and subtracting from the plant input. It tries to cancel out its own disturbance reference signal. Nothing like this happens with the conventional adaptive noise canceler. The adaptive plant disturbance canceler represents a wholly new concept in noise canceling for a completely different type of noise (disturbance) control problem.

1 THE FUNCTIONING OF THE ADAPTIVE PLANT DISTURBANCE CANCELER

Exactly how the system of Fig. 8.1 cancels plant disturbance can be explained algebraically. Suppose that the system has converged and that $\hat{P}_k(z)$ and $Q_k(z)$ are now fixed. Referring to Fig. 8.1, it is useful to note that the transfer function from the input at point A to the plant output at point C is

$$H_{AC}(z) = \frac{P(z)}{1 + z^{-1}P(z)\cdot Q_k(z) - z^{-1}\hat{P}_k(z)\cdot Q_k(z)}. \tag{8.1}$$

From this transfer function, one can see that when $\hat{P}_k(z) = P(z)$, the transfer function from point A to point C is equal to the transfer function $P(z)$ of the plant itself. The feedback circuit of the plant disturbance canceler therefore does not change the plant dynamics when $\hat{P}_k(z)$ matches $P(z)$, regardless of the value of $Q_k(z)$.

Furthermore, when $\hat{P}_k(z)$ matches $P(z)$, the stability of Eq. (8.1) is assured. (Note that $P(z)$ itself has been assumed to be stable.) On the other hand, if $\hat{P}_k(z)$ does not match $P(z)$, then values of $Q_k(z)$ can easily be found to make the disturbance canceler become unstable.

It is most important to have $\hat{P}_k(z)$ converged and sustained close to $P(z)$ before turning on the disturbance canceling feedback. While running, if there is a sudden catastrophic change in $P(z)$, it may be necessary to shut off the feedback until $\hat{P}_k(z)$ can have an opportunity to lock back onto $P(z)$. A *panic button* switch should be included in the feedback path in case of instability. It would be better to suffer the effects of plant disturbance for a while rather than go unstable.

Our next interest is with the transfer function from the point of injection of plant disturbance n_k, point B, to the plant output, point C. It can easily be shown to be

$$H_{BC}(z) = \frac{1 - z^{-1}\hat{P}_k(z)\cdot Q_k(z)}{1 + z^{-1}P(z)\cdot Q_k(z) - z^{-1}\hat{P}_k(z)\cdot Q_k(z)}. \tag{8.2}$$

Furthermore, the transfer function from point B to point D can easily be shown to be

$$H_{BD}(z) = \frac{1 - z^{-1}\widehat{P}_k(z) \cdot Q_k(z)}{1 + z^{-1}P(z) \cdot Q_k(z) - z^{-1}\widehat{P}_k(z) \cdot Q_k(z)}. \tag{8.3}$$

These turn out to be identical transfer functions. We therefore deduce that the plant disturbance component at point D is equal to the plant disturbance component at point C. Inspection of Fig. 8.1 allows one to conclude that when $\widehat{P}_k(z)$ exactly matches $P(z)$, only plant disturbance will be present at point E and only filtered plant disturbance will be present at point D. Minimizing the power at point D, as is done by the adaptive process for finding $Q_k(z)$, thereby causes the power of the plant disturbance component at point D to be minimized. This corresponds to the power of the plant disturbance at the plant output point C also being minimized. Therefore, the adaptive process finds the best value of $Q_k(z)$ to be copied and used in the disturbance canceling feedback process for the purpose of minimizing plant output disturbance. The only question that remains is whether or not the system configuration shown in Fig. 8.1 is optimal. Could some other system arrangement allow another adaptive process to do an even better job of disturbance cancelation?

8.2 PROOF OF OPTIMALITY FOR THE ADAPTIVE PLANT DISTURBANCE CANCELER

Fig. 8.2(a) shows the plant with additive output disturbance n_k. The plant output disturbance is assumed to have been generated by an original disturbance l_k, which is white and of unit power, acting as a driving function input to a linear filter whose transfer function is $\Gamma(z)$. This disturbance model is illustrated in Fig. 8.2(b). The filter $\Gamma(z)$ must be stable in order to have finite plant output disturbance. Without loss of generality, this filter will be assumed to be stable, minimum-phase, and causal. A plant output disturbance n_k having any desired power spectrum can be generated by the application of a suitably chosen white disturbance source l_k to a stable, causal, minimum-phase filter $\Gamma(z)$. It is useful to note that the reciprocal of $\Gamma(z)$ is also a causal, stable filter, and it is a whitening filter for n_k.

For control purposes, the only access that one has to the disturbed plant is its input port. In order to reduce the plant output disturbance, one can only add disturbance to the plant input. This input disturbance must be properly related to the source l_k of the plant output disturbance in order to effect output disturbance reduction. If the disturbance canceling system is to be a linear one, a theoretically optimal plant disturbance canceler will be configured as shown in Fig. 8.2(c), with $F^*(z)$ optimized to minimize plant disturbance power at point F. Subsequently, we shall demonstrate that the behavior of this optimal disturbance canceler is equivalent to that of the disturbance canceler of Fig. 8.1.

One cannot generally implement the optimal scheme of Fig. 8.2(c). The reason is that one would not usually have access to the source noise l_k. An equivalent scheme is shown in Fig. 8.2(d) which uses the actual plant output disturbance, not the source noise.

Assuming for the moment that one did have access to the source noise as in Fig. 8.2(c), the source noise could be filtered[3] and applied to the plant input to do the best possible job

[3] In Figs. 8.2(c) and (d), we actually assume access to l_{k-1} and n_{k-1}, respectively. Delays of z^{-1} are included in the noise flow paths because the current value of the plant noise can only be used for the cancelation of future

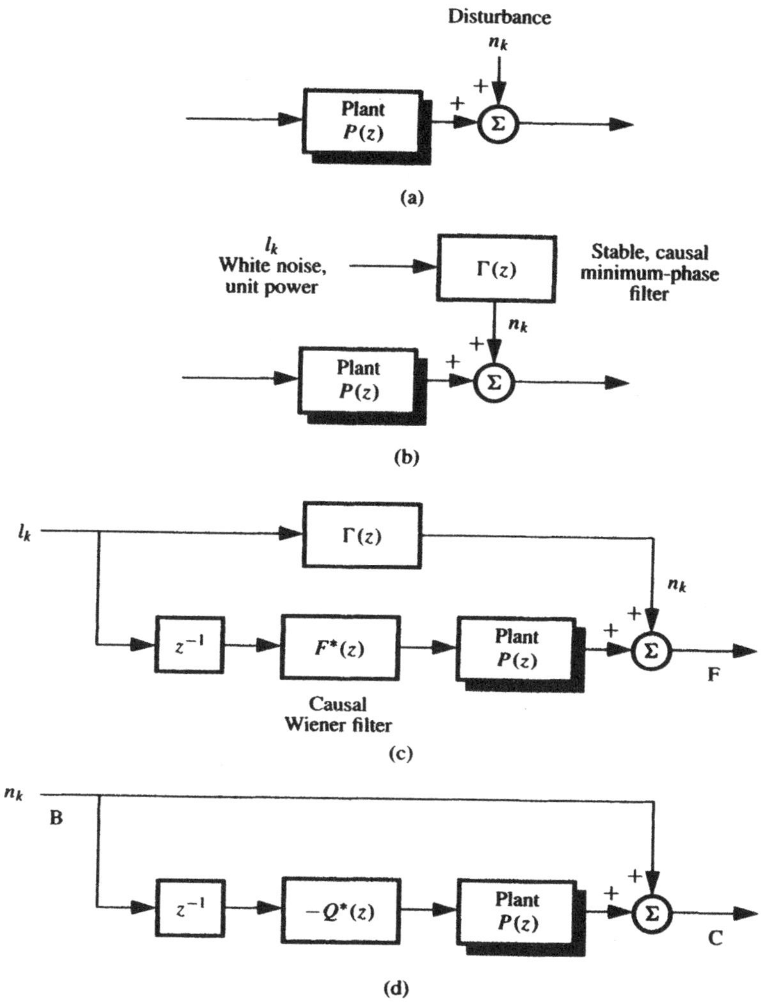

Figure 8.2 Comparative study of optimal and practical plant disturbance canceling schemes.

of canceling plant disturbance at the plant output, point F. The best filter would be a causal Wiener filter whose transfer function would be

$$F^*(z) = \frac{-1}{[P(z)P(z^{-1})]^+}\left[\frac{z\Gamma(z)P(z^{-1})}{[P(z)P(z^{-1})]^-}\right]_+ . \tag{8.4}$$

The Shannon-Bode method was used in deriving this result after interchanging the positions of $P(z)$ and $F^*(z)$ in the cascade.

Realizing that the optimal noise canceling scheme of Fig. 8.2(c) is merely an idealization, the question is, what can one do in reality? The noise canceling system of Fig. 8.2(d) uses the actual plant output disturbance rather than the unavailable source noise. A study of this system will lead us to a practical result.

The noise canceling scheme of Fig. 8.2(d) turns out to be identical in function to the optimal canceler of Fig. 8.2(c). This can be demonstrated as follows. Referring to Fig. 8.2(d) and using the Shannon-Bode method once again, the optimal causal form of $-Q_k(z)$ that minimizes plant disturbance power at point C is

$$-Q^*(z) = \frac{-1}{[P(z)P(z^{-1})]^+\Gamma(z)}\left[\frac{z\Gamma(z)P(z^{-1})}{[P(z)P(z^{-1})]^-}\right]_+ . \tag{8.5}$$

Using (8.4) and (8.5), we can relate $-Q^*(z)$ and $F^*(z)$:

$$-Q^*(z) = \frac{F^*(z)}{\Gamma(z)}. \tag{8.6}$$

Referring to Fig. 8.2(c), it is apparent that the z-transform of the plant disturbance at point F is

$$L(z)[\Gamma(z) + z^{-1}F^*(z)\cdot P(z)], \tag{8.7}$$

where $L(z)$ is the z-transform of the noise l_k. Now referring to Fig. 8.2(d), it is apparent that the z-transform of the plant disturbance at point C is

$$\begin{aligned} N(z)\left[1 - z^{-1}Q^*(z)P(z)\right] &= N(z)\left[1 + z^{-1}\frac{F^*(z)}{\Gamma(z)}P(z)\right] \\ &= L(z)\Gamma(z)\left[1 + z^{-1}\frac{F^*(z)}{\Gamma(z)}P(z)\right] \\ &= L(z)\left[\Gamma(z) + z^{-1}F^*(z)P(z)\right], \end{aligned} \tag{8.8}$$

where $N(z)$ is the z-transform of the plant disturbance n_k. We note that the plant disturbance at point C, given by (8.8) is identical to the plant disturbance at point F, given by (8.7). The conclusion is that the Wiener canceler of Fig. 8.2(d) performs equivalently to the Wiener canceler of Fig. 8.2(c). They are both optimal linear least squares noise cancelers. No other linear canceler can give better performance.

We'll show next that the disturbance canceler of Fig. 8.1 is equivalent to that of Fig. 8.2(d) and is thereby an optimal canceler, assuming that $\widehat{P}_k(z) = P(z)$. The adaptive process for finding $Q_k(z)$ of Fig. 8.1 is driven by the plant disturbance n_k. The same is true for the Wiener canceler of Fig. 8.2(d). The optimal $Q^*(z)$ in Fig. 8.2(d) is the asymptotic

values of plant noise. If the discretized plant $P(z)$ has at least one unit of delay in its impulse response, the delays of z^{-1} can and should be eliminated.

Wiener solution for the adaptive $Q_k(z)$ in Fig. 8.1 (because of like configurations of filters and error signals). The adaptive filter $Q_k(z)$ therefore converges to $Q^*(z)$ given by (8.5). The z-transform of the plant output disturbance in Fig. 8.1, after convergence, can be calculated using (8.2) and (8.6). It is

$$\begin{aligned} N(z)H_{BC}(z) &= L(z)\Gamma(z)H_{BC}(z) \\ &= L(z)\Gamma(z)\frac{1 - z^{-1}\widehat{P}_k(z)\cdot Q_k(z)}{1 + z^{-1}P(z)\cdot Q_k(z) - z^{-1}\widehat{P}_k(z)\cdot Q_k(z)} \\ &= L(z)\Gamma(z)\frac{1 - z^{-1}P(z)\cdot Q^*(z)}{1} \\ &= L(z)[\Gamma(z) + z^{-1}P(z)\cdot F^*(z)]. \end{aligned} \tag{8.9}$$

This is the same result as (8.7). Therefore, when $\widehat{P}(z) = P(z)$, the disturbance canceler of Fig. 8.1 performs equivalently to that of Fig. 8.2(d) (which is optimal).

8.3 POWER OF UNCANCELED PLANT DISTURBANCE

Even the best disturbance canceler cannot cancel perfectly and will leave a residue of uncanceled plant disturbance. This occurs because the minimum mean square error of the Wiener filter is generally not zero. An expression for the power of the uncanceled plant disturbance can be derived from (8.7) as follows. Since $L(z)$ corresponds to white noise of unit power, the transform of the autocorrelation of the uncanceled disturbance at the plant output (at point C, Fig. 8.1) will be

$$\begin{aligned} \Phi_{CC}(z) &= (\Gamma(z) + P(z)\cdot z^{-1}F^*(z))(\Gamma(z^{-1}) + P(z^{-1})\cdot z^{-1}F^*(z^{-1})) \\ &= \cdots + z\phi_{CC}(-1) + \phi_{CC}(0) + z^{-1}\phi_{CC}(1) + \cdots. \end{aligned} \tag{8.10}$$

The uncanceled plant output disturbance power will be $\phi_{CC}(0)$. This can often be evaluated when some knowledge of $P(z)$ and $\Gamma(z)$ is available. One can factor the plant disturbance spectrum $\Phi_{nn}(z)$ to obtain $\Gamma(z)$, in accord with

$$\Phi_{nn}(z) = \Gamma(z)\Gamma(z^{-1}). \tag{8.11}$$

Furthermore, $F^*(z)$ can be obtained from (8.4) with knowledge of $P(z)$ and $\Gamma(z)$. Having all of these ingredients, $\phi_{CC}(0)$ can be obtained by making use of (8.10).

8.4 OFFLINE COMPUTATION OF $Q_k(z)$

If $Q_k(z)$ were computed by an adaptive process as in Fig. 8.1, very slow adaptation would be required in order for the effects of misadjustment due to noise in the adaptive weights of $Q_k(z)$ to be negligible. On the other hand, fast adaptation might be required to maintain an inverse of $\widehat{P}_k(z)$ if $\widehat{P}_k(z)$ were tracking a time variable plant $P(z)$. If this were the case, it would be a good idea to calculate $Q_k(z)$ by an offline process as illustrated in Fig. 8.3. One would need synthetic noise whose statistical character would be the same as that of the original plant disturbance. The offline process could operate much faster than real time[4] so

[4]Hundreds or possibly thousands of iterations could be undergone offline in the calculation of $Q_k(z)$ in the time of one sampling period of the physical system.

that, given a particular $\widehat{P}_k(z)$, the offline process would deliver the optimal $Q_k(z)$ for it immediately. The offline process could be an adaptive algorithm based on the LMS algorithm or based on any other optimal least squares algorithm. Knowing the disturbance statistics and knowing $\widehat{P}_k(z)$, another approach would be to compute $Q_k(z)$ directly from the Wiener solution. This was discussed previously in connection with the computation of the inverse controller $\widehat{C}_k(z)$ in Chapter 6.

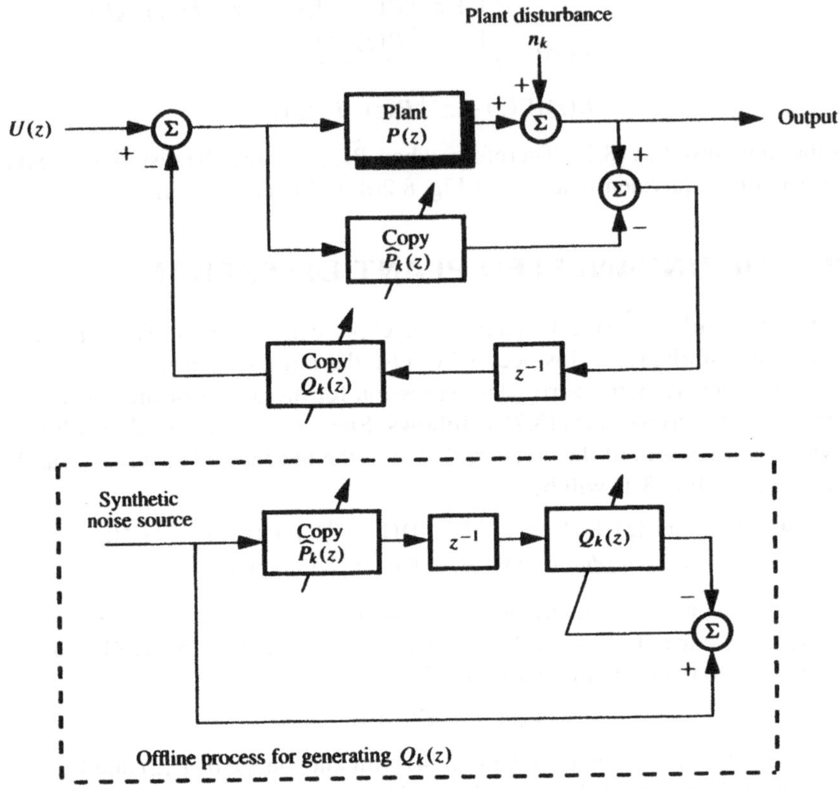

Figure 8.3 Offline computation of $Q_k(z)$.

8.5 SIMULTANEOUS PLANT MODELING AND PLANT DISTURBANCE CANCELING

A plant modeling and plant disturbance canceling system is shown in Fig. 8.4. This system incorporates the plant disturbance canceling scheme of Fig. 8.1 and the plant modeling scheme of Fig. 4.1. An offline inverse modeling process to obtain $Q_k(z)$ is shown. The purpose of this system is clearly twofold: to obtain $\widehat{P}_k(z)$, and to cancel the plant disturbance as well as possible in the least squares sense. Copies of the plant model $\widehat{P}_k(z)$ are needed for the disturbance canceler and for the offline inverse modeling process to get $Q_k(z)$. An additional copy of $\widehat{P}_k(z)$ will be needed to find the plant controller $\widehat{C}_k(z)$, but this is not shown.

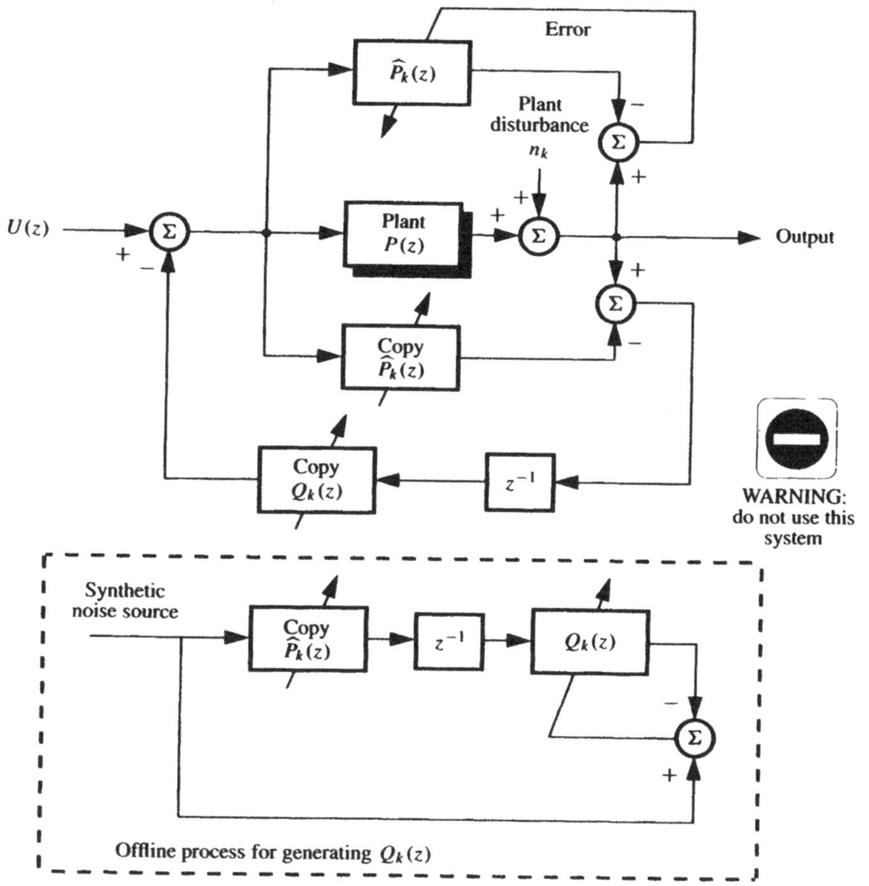

Figure 8.4 A plant modeling and disturbance-canceling system that does not work.

It is clear that the system of Fig. 8.4 is a straightforward combination of simple and robust subsystems. This entire system was built in software and tested in our laboratory, and it did not work! Much to our surprise, it was discovered that the system was fatally flawed. When $\widehat{P}_k(z)$ converged, it did not converge on $P(z)$, that is, it did not become a best match to $P(z)$. Instead it developed a bias. We soon learned to be very careful when building "simple systems."

In connection with the basic process of Fig. 4.1 for modeling a disturbed plant, Eqs. (4.6)–(4.13) show that the Wiener solution for $\widehat{P}_k(z)$ will be a best match to $P(z)$ in spite of the presence of plant disturbance n_k as long as this plant disturbance is uncorrelated with the plant input. But in the system of Fig. 8.4 the plant disturbance is fed back via $Q_k(z)$ for purposes of disturbance cancelation. The result is that the input to the adaptive filter that forms $\widehat{P}_k(z)$ has components related to and correlated with the plant disturbance. Once one realizes this, it is no longer surprising that the presence of plant disturbance caused a biased solution for $\widehat{P}_k(z)$.

To avoid this difficulty, a means of performing the intended function of the system of Fig. 8.4 is needed that does not allow plant disturbance components into the input of the adaptive filter used to find $\widehat{P}_k(z)$. Two ways to accomplish this are described here, that of Fig. 8.5 using no dither, and that of Fig. 8.7 using dither. The system of Fig. 8.5 will be discussed next. The system of Fig. 8.7 will be discussed in Section 8.6.

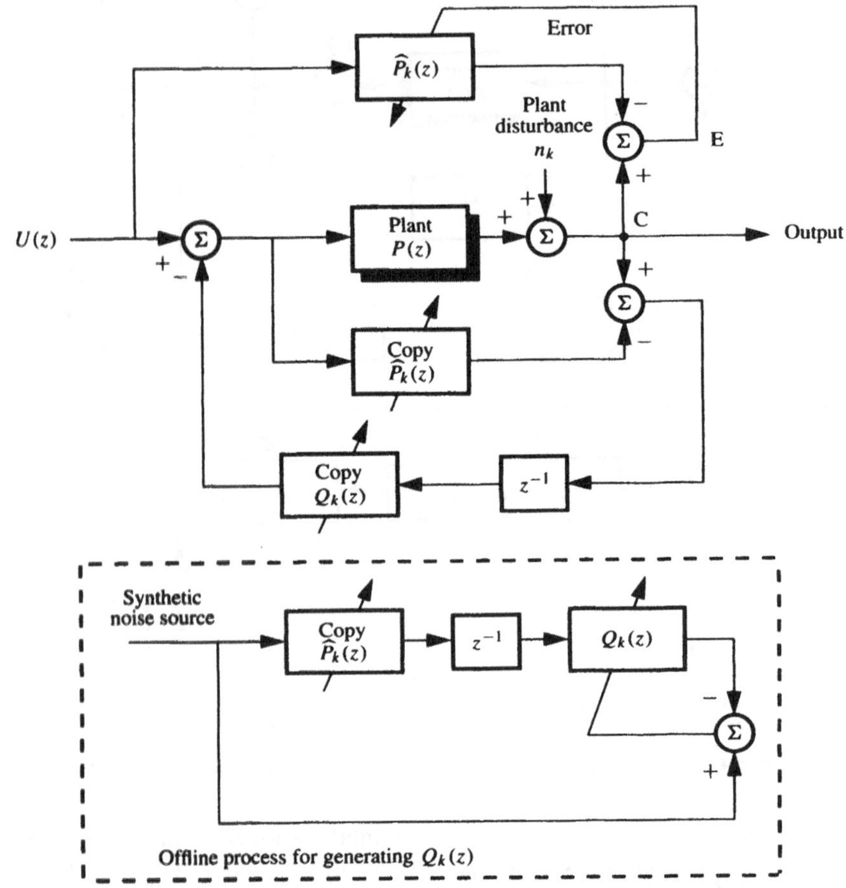

Figure 8.5 A plant modeling and disturbance-canceling system that works.

To demonstrate that the system of Fig. 8.5 does work, that is, that it causes $\widehat{P}(z)$ to adapt and converge on $P(z)$ without bias and that $z^{-1}Q(z)$ is properly obtained from $\widehat{P}(z)$ for purposes of disturbance canceling, a simulation of the system of Fig. 8.5 was done. For comparison, a simulation of the system of Fig. 8.4 was also done.

For both simulations, the plant was chosen to be

$$P(z) = \frac{1 - \frac{1}{3}z^{-1}}{1 - \frac{1}{2}z^{-1}}.$$

The command input was a zero-mean, uniformly distributed white signal of unit power. The plant noise was generated by passing a zero-mean white noise through a moving average low-pass filter having 20 equal weights. The plant disturbance was set to have a variance of $E[n_k^2] = 0.2$. The number of weights of both $P(z)$ and $Q(z)$ was chosen to be 10. Adaptation was started assuming letting the initial value for $\widehat{P}(z)$ be

$$\widehat{P}(z) = P(z) + 0.01(1 + z^{-1} + z^{-2} + \cdots + z^{-9}).$$

Both systems, those of Fig. 8.4 and of Fig. 8.5, were run for 5,000 samples before taking data, ensuring that steady-state conditions had been reached.

Initially, the value of the adaptation constant was set at $\mu = 0.003$ for both systems. The results of the experiments are shown in Table 8.1. The Wiener solution, the weights of $P(z)$, were compared with snapshot values of the weights of $\widehat{P}_k(z)$ for both systems. Clearly, the system of Fig. 8.5 provided a good estimate of the Wiener solution (estimation error, the average of the sum of the squares of the errors in all of the weights, was 0.004). However, as expected, the system of Fig. 8.4 developed quite a large estimation error (0.033). The bias is evident from the numbers displayed in Table 8.1.

The adaptation constant was then reduced threefold to $\mu = 0.001$ for both systems. The same experiment was repeated. The results are given in Table 8.2.

For the system of Fig. 8.5, noise in the weights of the plant model was reduced from a modeling error of 0.004 to the very low level of 0.0013. This reduction is consistent with the expectation that for adaptive LMS processes, noise variance in the weights should be proportional to the adaptation constant. However, for the system of Fig. 8.4, there were no significant changes in plant modeling error. In fact, the error grew from 0.033 to 0.051. The reason, of course, is that the plant estimation bias does not depend on the speed of adaptation. Hence, in this case, changes in μ had little effect on the modeling error.

The system of Fig. 8.5 is a very good one. If $\widehat{P}_k(z)$ has the flexibility to match $P(z)$, it will do so and the system dynamic response from point A to point C will have the transfer function of $P(z)$, and the plant disturbance will be optimally canceled.

It is useful to show that if, at some initial time, $\widehat{P}_k(z)$ differed by a small amount from $P(z)$, then the adaptive process of Fig. 8.5 would work to bring $\widehat{P}_k(z)$ toward $P(z)$. The following discussion is not a proof, but a heuristic argument. Refer to Fig. 8.6, where simple plant modeling is taking place. Convergence of $\widehat{P}_k(z)$ toward $P(z)$ for the system of Fig. 8.6 is assured (refer to Chapter 3 and Appendix A). It is useful to note that the transfer function from point A to point E, the transfer function from the input to the error point, is

$$H_{AE}(z) = P(z) - \widehat{P}_k(z) = -\Delta P_k(z). \tag{8.12}$$

It is clear that making the error smaller and smaller requires that $\Delta P_k(z)$ become closer and closer to zero. For a given plant input $X(z)$, the simple proportionality between the z-transform of the error signal $E(z)$ and $-\Delta P_k(z)$, the plant modeling error, corresponds to the mean square error being a quadratic function of the components of the vector $\Delta\mathbf{P}_k$. Steepest descent makes the error smaller and makes $\widehat{P}_k(z)$ relax toward $P(z)$. The quadratic mean square error function was discussed in Chapter 3.

The question is, does this also occur with the system of Fig. 8.5? Referring to this figure, we obtain the transfer function from the input at point A to the error at point E as

$$H_{AE}(z) = -\widehat{P}_k(z) + \frac{P(z)}{1 - \Delta P_k(z) \cdot Q_k(z) \cdot z^{-1}}$$

TABLE 8.1 COMPARISON OF WEIGHTS OF $P(z)$ WITH THOSE OF $\widehat{P}(z)$ FOR THE SYSTEMS OF FIGS. 8.4 AND 8.5, $\mu = 0.003$

Weights of $\widehat{P}(z)$ for syst. of Fig. 8.4	0.906	0.085	0.01	-0.013	-0.017	-0.037	-0.037	-0.026	-0.023	-0.043
Weights of $P(z)$ (Wiener Solution)	1	0.167	0.083	0.042	0.021	0.01	0.005	0.003	0.001	0.001
Weights of $\widehat{P}(z)$ for syst. of Fig. 8.5	1.005	0.172	0.074	0.05	0.007	-0.007	-0.039	-0.033	-0.009	-0.005

TABLE 8.2 COMPARISON OF WEIGHTS OF $P(z)$ WITH THOSE OF $\widehat{P}(z)$ FOR THE SYSTEMS OF FIGS. 8.4 AND 8.5, $\mu = 0.001$

Weights of $\widehat{P}(z)$ for syst. of Fig. 8.4	0.915	0.082	0.003	-0.040	-0.049	-0.052	-0.065	-0.057	-0.058	-0.051
Weights of $P(z)$ (Wiener Solution)	1	0.167	0.083	0.042	0.021	0.01	0.005	0.003	0.001	0.001
Weights of $\widehat{P}(z)$ for syst. of Fig. 8.5	0.999	0.164	0.082	0.039	0.009	-0.003	-0.004	0.004	0.042	0.012

TABLE 8.3 COMPARISON OF WEIGHTS OF $P(z)$ WITH SNAPSHOTS OF THOSE OF $\widehat{P}(z)$ FOR THE SYSTEMS OF FIG. 8.7

$P(z)$ Wiener Solution	1	0.167	0.083	0.042	0.021	0.01	0.005	0.003	0.001	0.001
Impulse response of $\widehat{P}(z)$, $\mu = 0.003$	1.01	0.182	0.107	0.07	0.038	0.027	0.014	0.022	0.007	-0.003
Impulse response of $\widehat{P}(z)$, $\mu = 0.001$	1	0.166	0.083	0.041	0.018	0.007	0.006	0.003	0	-0.001

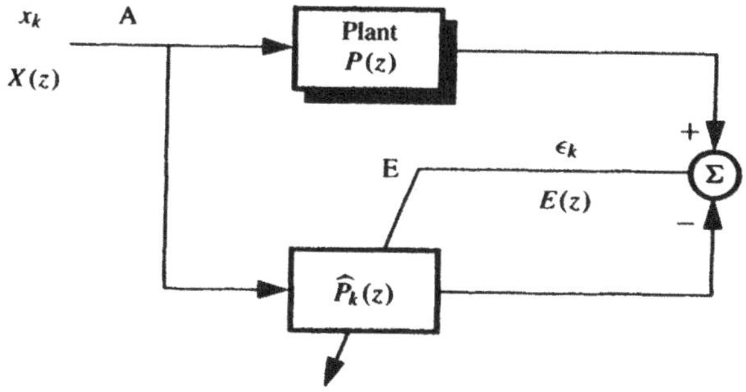

Figure 8.6 Plant modeling.

$$= -\Delta P_k(z)\frac{(1 - P(z) \cdot Q_k(z) \cdot z^{-1})}{1 - \Delta P_k(z) \cdot Q_k(z) \cdot z^{-1}}$$
$$\approx -\Delta P_k(z)(1 - P(z) \cdot Q_k(z) \cdot z^{-1}). \tag{8.13}$$

In this case, it is clear that the z-transform of the error is not simply proportional to $-\Delta P_k(z)$. There are complicating factors. When $\Delta P_k(z)$ is small, the above approximation works well. Assuming this to be the case, the error transform is proportional to $-\Delta P_k(z)$ and to $(1 - P(z) \cdot Q_k(z) \cdot z^{-1})$. When $z^{-1}Q_k(z)$ is very close to being a true inverse of $P(z)$, the plant noise will be almost perfectly canceled and the factor $(1 - P(z) \cdot Q_k(z) \cdot z^{-1})$ will be very small. This works like a low-gain factor in the error feedback loop. The result is slow convergence of $\hat{P}_k(z)$. The proper choice of the constant μ is difficult to determine a priori.

The system of Fig. 8.5 does work well, however. It does not require external dither as long as the input signal is adequate for the task of adapting $\hat{P}_k(z)$, and this is an advantage. It has the disadvantage of being more difficult than usual to work with, because of the uncertainty in the relations between μ and the adaptive time constants of $\hat{P}_k(z)$ and the uncertainty in one's knowledge of the stable range of μ. A suitable value of μ could always be found empirically however.

The system of Fig. 8.7 uses external dither, and as a result, these uncertainties are for the most part alleviated. The plant input signal needs no longer to be relied upon to provide the proper stimulus for the plant modeling process. The disadvantage in the use of dither is that an additional disturbance is introduced into the system. The effects of this disturbance can be reduced, however, if one is willing to reduce the speed of convergence of $\hat{P}_k(z)$. In any event, the behavior of the system of Fig. 8.7 is solid and predictable, as will be demonstrated in the next section.

The system of Fig. 8.7 is like the one of Fig. 8.3, obtaining the required $\hat{P}_k(z)$ by using the dither modeling scheme C introduced in Fig. 4.3(c). Although the system of Fig. 8.7 might seem complicated, it is easily implemented with modern day digital hardware and software. An analysis of this system is our next objective.

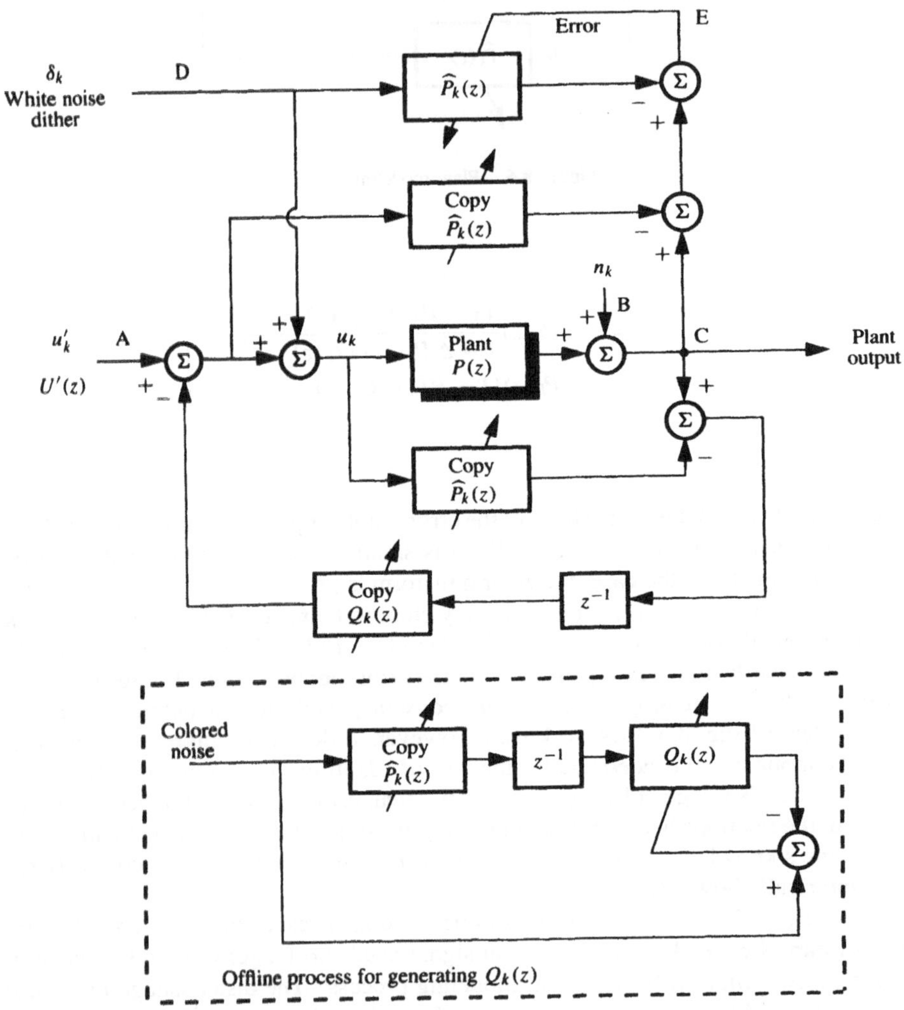

Figure 8.7 A plant modeling and disturbance-canceling system that uses dither scheme C.

8.6 HEURISTIC ANALYSIS OF STABILITY OF A PLANT MODELING AND DISTURBANCE CANCELING SYSTEM

With space limitations, it is not possible to present analyses of all workable combinations of plant modeling and plant disturbance canceling schemes. The system of Fig. 8.7 is robust and efficient and relatively easy to analyze, so we have selected this system for discussion and for analysis of its convergence, stability, time constants, and error. Similar methods of analysis can be applied to other systems configurations.

When examining the system of Fig. 8.7, one should first observe that if $\widehat{P}_k(z)$ properly matches $P(z)$, the transfer function from the input at point A to the output at point C is simply $P(z)$, that of the plant itself. Therefore, when this system is working ideally, the plant dynamics are unaltered.

Next we show that if $\widehat{P}_k(z)$ departs from the value $P(z)$, the adaptive process will tend to pull $\widehat{P}(z)$ back to $P(z)$. The approach used above will be used again here. We start by finding the transfer function from point D, where the dither input to the adaptive filter is injected, to point E, the error point in the adaptive process for finding $\widehat{P}_k(z)$:

$$\begin{aligned} H_{DE}(z) &= \frac{\{P(z) - \widehat{P}_k(z) \cdot \Delta P_k(z) \cdot Q_k(z) \cdot z^{-1}\}}{1 - \Delta P_k(z) \cdot Q_k(z) \cdot z^{-1}} - \widehat{P}_k(z) \\ &= \frac{-\Delta P_k(z)}{1 - \Delta P_k(z) \cdot Q_k(z) \cdot z^{-1}} \\ &\approx -\Delta P_k(z). \end{aligned} \tag{8.14}$$

When $\Delta P_k(z)$ is small, Eq. (8.14) holds and the z-transform of the error of the adaptive process is proportional to $-\Delta P_k(z)$, just as it was for the simple modeling scheme of Fig. 8.6. Accordingly, the mean square error will be a quadratic function of the components of the vector $\Delta \mathbf{P}_k$ for small $\Delta \mathbf{P}_k$. Steepest descent will make the error smaller by making $\widehat{P}_k(z)$ relax toward $P(z)$.

From this heuristic argument we conclude that for small $\Delta P_k(z)$, the dynamic behavior of the adaptive process in Fig. 8.7 is essentially identical to that in Fig. 8.6. For a given value of μ, both systems will have the same weight noise statistics and the same time constants. Accordingly, the pth time constant will be

$$\tau_p = \frac{1}{2\mu\lambda_p}. \tag{8.15}$$

In order that $\widehat{P}_k(z)$ be a good broadband model of the plant, a white dither should be used. As a result, all eigenvalues will be equal and the single time constant will be given by

$$\tau = \frac{1}{2\mu E[\delta_k^2]}, \tag{8.16}$$

where $E[\delta_k^2]$ is the dither power.

When close to convergence, the weight noise covariance of $\widehat{P}_k(z)$ will be

$$\begin{bmatrix} \text{weight noise} \\ \text{covariance matrix} \end{bmatrix} = \text{cov}\,[\Delta \mathbf{P}_k] = \beta \mathbf{I} = \mu E[n_k^2]\mathbf{I}, \tag{8.17}$$

where $E[n_k^2]$, the power of the plant output disturbance (before canceling), is the minimum mean square error that can be achieved by the adaptive process for finding $\widehat{P}_k(z)$. This can be verified by careful inspection of Fig. 8.7, letting $\widehat{P}_k(z) = P(z)$.

The issue of stability for this system is a complicated one, although it is not hard to experimentally determine a stable range for the parameter μ when working with a physical system or with a computer simulated system. There are two aspects to the stability problem. One involves the stability of the adaptive algorithm for finding $\widehat{P}_k(z)$. The other involves the stability of the disturbance-canceling feedback loop. We begin with a study of the latter issue.

Referring to Fig. 8.7, the transfer function from input point A to plant output point C is

$$H_{AC}(z) = \frac{P(z)}{1 - \Delta P_k(z) \cdot Q_k(z) \cdot z^{-1}}. \tag{8.18}$$

Of importance is the denominator of this transfer function. This denominator is the characteristic polynomial of the system, and it determines stability of the disturbance-canceling loop. Difficulty is introduced from the randomness of this polynomial because the vector $\Delta\mathbf{P}_k$ has random components. The components of $\Delta\mathbf{P}_k$ are mutually uncorrelated, but individually they are correlated over time and they will vary slowly if μ, as usual, is made small. We will assume that $\Delta P_k(z)$ is *quasi-constant* over time, and that the concept of transfer function can be used. The characteristic polynomial will therefore be used for stability considerations. The characteristic equation is

$$1 - \Delta P_k(z) \cdot Q_k(z) \cdot z^{-1} = 0. \tag{8.19}$$

Assuming that $\widehat{P}_k(z)$ has been constructed with n weights, n being large, then $\Delta\mathbf{P}_k$ has essentially n components. Since these components are mutually uncorrelated and there are a lot of them, the transform $\Delta P_k(z)$ will be approximately flat for real frequencies (on the unit circle in the z-plane) and will be approximately constant over time although the components of $\Delta\mathbf{P}_k$ will all be time variable. The value of the transform of $\Delta P_k(z)$ on the unit circle can be obtained as follows. The Fourier transform of the vector $\Delta\mathbf{P}_k$ is defined as

$$\left[1 \;\; e^{-j\omega} \;\; e^{-2j\omega} \;\; \ldots \;\; e^{-(n-1)j\omega}\right] \cdot \Delta\mathbf{P}_k \triangleq \Delta P_k(z)|_{z=e^{j\omega}} = \Delta P_k(e^{j\omega}). \tag{8.20}$$

Next consider the following expectation, in light of Eq. (8.17).

$$\begin{aligned} E\left[|\Delta P_k(e^{j\omega})|^2\right] &= E\left[[1 \;\; e^{-j\omega} \;\; \ldots \;\; e^{-(n-1)j\omega}] \cdot \Delta\mathbf{P}_k \cdot \Delta\mathbf{P}_k^T [1 \;\; e^{j\omega} \;\; \ldots \;\; e^{(n-1)j\omega}]^T\right\} \\ &= [1 \;\; e^{-j\omega} \;\; \ldots \;\; e^{-(n-1)j\omega}][\text{cov}\Delta\mathbf{P}_k][1 \;\; e^{j\omega} \;\; \ldots \;\; e^{(n-1)j\omega}]^T \\ &= \beta[1 \;\; e^{-j\omega} \;\; \ldots \;\; e^{-(n-1)j\omega}][1 \;\; e^{j\omega} \;\; \ldots \;\; e^{(n-1)j\omega}]^T \\ &= n\beta \\ &= n\mu E[n_k^2]. \end{aligned} \tag{8.21}$$

We have indicated above that $|\Delta P_k(e^{j\omega})|$ is essentially constant over frequency and time. Accordingly, it can be approximated by its expected value. Therefore

$$\begin{aligned} E\left[|\Delta P_k(e^{j\omega})|^2\right] &= n\beta \\ E\left[|\Delta P_k(e^{j\omega})|\right] &= \sqrt{n\beta}. \end{aligned} \tag{8.22}$$

Equation (8.22) gives an approximate constant value for the transform magnitude for $\Delta P_k(z)$ on the unit circle in the z-plane. The phase would be a slowly randomly changing function of frequency. The phase function can be represented by $\phi(w)$.

With very small μ, $\Delta P_k(z)$ will be very small and the roots of the characteristic equation will all be inside the unit circle. As μ is made larger, the roots will spread until μ is made large enough for a first root to just cross the unit circle. $\Delta P_k(z)$ can be represented in the characteristic equation (8.19) in terms of its magnitude and phase. Accordingly, on the unit circle, the characteristic equation can be written as

$$1 - e^{j\phi(w)} \cdot \sqrt{n\beta} \cdot Q_k(e^{j\omega}) \cdot e^{-j\omega} = 0. \tag{8.23}$$

Under worst case conditions, the phase factor $e^{j\phi(w)} \cdot e^{-jw}$ will manage at random to perfectly compensate $Q_k(e^{j\omega})$ at the frequency w that maximizes the magnitude of $Q_k(e^{j\omega})$ to cause satisfaction of (8.23). Let the maximum magnitude of $Q_k(e^{j\omega})$ be designated by $|Q|_{\text{max}}$. The condition enabling the first characteristic-equation zero to reach the unit circle by increasing μ (which causes β to increase) is therefore

$$1 - \sqrt{n\beta}|Q|_{\text{max}} = 0, \quad \text{or}$$

$$\beta = \frac{1}{n|Q|^2_{\text{max}}}. \tag{8.24}$$

Substituting (8.17) into (8.24) gives

$$\mu = \frac{1}{n|Q|^2_{\text{max}} E[n_k^2]}. \tag{8.25}$$

Our conclusion is that the disturbance canceling loop could become unstable if μ is made larger than specified by (8.25). To keep this feedback loop stable, μ should be chosen within the following range:

$$\frac{1}{n|Q|^2_{\text{max}} E[n_k^2]} > \mu > 0. \tag{8.26}$$

It is possible to keep μ within this range in practice. One knows the value of n, the number of weights of $\widehat{P}_k(z)$, and one would generally know the value of $E[n_k^2]$, the variance of the original plant output disturbance (before canceling), but getting a preliminary estimate of $|Q|^2_{\text{max}}$ takes a special effort. $Q_k(z)$ is one way or another related to $P(z)$, and it can be obtained by the offline process of Fig. 8.7. An approximate knowledge of $P(z)$ would provide something to use in place of $\widehat{P}_k(z)$ in an offline process for generating $Q_k(z)$. A nominal value for $Q_k(z)$ can be obtained this way. An estimate of $|Q|_{\text{max}}$ can be obtained from the nominal $Q_k(z)$ by finding the maximum value for $z = e^{j\omega}$, scanning over frequency w from zero to half the sampling frequency. Accordingly,

$$|Q|^2_{\text{max}} \approx \max_{0<w<\frac{\Omega}{2}} \left\{ |Q_k(e^{j\omega})|^2_{\text{nominal}} \right\}. \tag{8.27}$$

After estimating $|Q|^2_{\text{max}}$ from approximate knowledge of the plant, one can then choose μ within range (8.26) to keep the disturbance canceling loop stable. A suitable value of μ could also be found empirically.

The stability of the plant modeling algorithm, the adaptive algorithm for finding $\widehat{P}_k(z)$ in the system of Fig. 8.7, is the next issue of concern. The stability issue has already been addressed for the system of Fig. 4.3(c) (scheme C) which has the same plant modeling process and dither scheme. These two systems are alike except that the system of Fig. 8.7 incorporates plant disturbance canceling and as such, the input to the plant u_k has added to it the disturbance canceling feedback signal. The stability criterion (4.29) for the system of Fig. 4.3(c) can be applied to the system of Fig. 8.7 by adding the power of the output of $Q_k(z)$ to the power of the input u'_k and the dither power in order to obtain the total power of the plant input u_k. Since the input to $Q_k(z)$ is essentially equal to n_k, and since the spectrum of n_k is generally low-pass, and since $Q_k(z) \approx 1/P(z)$ in the passband of n_k, and since $P(e^{j\omega})$ is generally flat or approximately flat at low-frequencies, the power of the output of $Q_k(z)$ is approximately given by

$$\begin{pmatrix} \text{power} \\ \text{output} \\ \text{of } Q_k(z) \end{pmatrix} \approx E[n_k^2]\left(\frac{1}{|P(e^{j\omega})|^2}\right)_{\omega\ \text{low}} \tag{8.28}$$

The stability criterion for the LMS adaptive algorithm used in the system of Fig. 8.7 for finding $\widehat{P}_k(z)$ is, from modification of (4.29), accordingly

$$\frac{1}{n\left[E[\delta_k^2] + E[(u'_k)^2] + E[n_k^2]\left(\frac{1}{|P(e^{j\omega})|^2}\right)_{\omega\ \text{low}}\right]} > \mu > 0. \tag{8.29}$$

Values of μ in this range can be chosen with knowledge of the dither power, and with approximate knowledge of the input command power, the plant disturbance power, and gain magnitude of the plant at low frequencies.

In order for the system of Fig. 8.7 to be stable, it is necessary that both the disturbance canceling loop be stable and that the adaptive process for finding $\widehat{P}_k(z)$ be stable. Sometimes the disturbance canceling loop stability will limit μ, sometimes the plant modeling stability will limit μ. It is necessary therefore that μ be chosen to satisfy both (8.26) and (8.29).

8.7 ANALYSIS OF PLANT MODELING AND DISTURBANCE CANCELING SYSTEM PERFORMANCE

We have thus far determined the time constants and weight covariance noise for $\widehat{P}_k(z)$ and have approximately determined the stable range of μ for the system of Fig. 8.7. Our next objective is to determine the components of signal distortion and noise that appear at the plant output due to dither, uncanceled plant disturbance, and the effects of adaptation noise in the weights of $\widehat{P}_k(z)$.

Referring to Fig. 8.7, the transfer function from the dither injection point D to the plant output point C is

$$\begin{aligned} H_{DC}(z) &= \frac{P(z)}{1 - \Delta P_k(z)\cdot Q_k(z)\cdot z^{-1}} \\ &\approx P(z). \end{aligned} \tag{8.30}$$

Once again assuming a white dither, the dither noise power at the plant output is

$$\begin{pmatrix} \text{plant} \\ \text{output} \\ \text{dither} \\ \text{power} \end{pmatrix} = E[\delta_k^2] \cdot \left[\sum_{i=0}^{\infty} p_i^2 \right]. \tag{8.31}$$

Recall that $E[\delta_k^2]$ is the dither input power, and that p_i is the ith impulse response value of $P(z)$. An approximate knowledge of $P(z)$ would be needed to make use of Eq. (8.31) in practice.

Next, the effects of the random $\Delta P_k(z)$ will be taken into account. The transfer function from the input $u'(z)$ at point A to the plant output at point C is

$$\begin{aligned} H_{AC}(z) &= \frac{P(z)}{1 - \Delta P_k(z) \cdot Q_k(z) \cdot z^{-1}} \\ &\approx P(z) + \Delta P_k(z) \cdot Q_k(z) \cdot P(z) \cdot z^{-1}. \end{aligned} \tag{8.32}$$

Equation (8.32) applies when $\Delta P_k(z)$ is small, which is the usual operating condition. The first term of the right-hand side is $P(z)$, and this is the normal transfer function for the input $U(z)$ to encounter in propagating through to the plant output. The second term is proportional to $\Delta P_k(z)$, which varies slowly and randomly. This component of the transfer function causes a randomly modulated component of output signal distortion. We are concerned with its total power. For ease of computation, we will assume that

$$z^{-1} \cdot Q(z) \cdot P(z) \approx 1. \tag{8.33}$$

Therefore, the transform of the output signal distortion is approximately given by

$$\begin{pmatrix} \text{transform of} \\ \text{output signal} \\ \text{distortion} \end{pmatrix} = U'(z) \cdot \Delta P_k(z). \tag{8.34}$$

This can be written in the time domain in vector form as

$$\begin{pmatrix} \text{output} \\ \text{signal} \\ \text{distortion} \end{pmatrix} = \mathbf{U}'^T \cdot \Delta\mathbf{P}_k. \tag{8.35}$$

From this we obtain the power of the output signal distortion:

$$\begin{pmatrix} \text{output} \\ \text{signal} \\ \text{distortion} \\ \text{power} \end{pmatrix} = E[\mathbf{U}'^T \cdot \Delta\mathbf{P}_k \cdot \Delta\mathbf{P}_k^T \cdot \mathbf{U}']. \tag{8.36}$$

We assume that the fluctuations of $\Delta\mathbf{P}_k$ are independent of the input $\mathbf{U}'$, and that $\mathbf{U}'$ is statistically stationary. The covariance of $\Delta\mathbf{P}_k$ is given by (8.17). Accordingly,

$$\begin{aligned} \begin{pmatrix} \text{output} \\ \text{signal} \\ \text{distortion} \\ \text{power} \end{pmatrix} &= E[\mathbf{U}'^T \cdot E[\Delta\mathbf{P}_k \cdot \Delta\mathbf{P}_k^T] \cdot \mathbf{U}'] \\ &= E[\mathbf{U}'^T \cdot \text{cov}\,[\Delta\mathbf{P}_k] \cdot \mathbf{U}'] \end{aligned} \tag{8.37}$$

$$= \beta E[\mathbf{U}'^T \cdot \mathbf{U}']$$
$$= \beta n E[(u'_k)^2].$$

Our next goal is to obtain the power of the uncanceled plant disturbance at the plant output, the power of the plant disturbance residue when using an optimal canceling system. It results in general from the nonexistence of a perfect instantaneous inverse of $z^{-1}P(z)$. Assuming that $\widehat{P}_k(z)$ is a near perfect match to $P(z)$, the spectrum of the uncanceled plant disturbance is given by (8.10). The components of (8.10) can be obtained from (8.4) and (8.11). The power of the uncanceled plant output disturbance is, in accord with (8.10),

$$\begin{pmatrix} \text{uncanceled} \\ \text{plant output} \\ \text{disturbance power} \end{pmatrix} = \phi_{CC}(0). \tag{8.38}$$

The overall output error power for the system of Fig. 8.7 is the sum of the output dither power (8.31), the output signal distortion power (8.37), and the uncanceled plant output disturbance power (8.38). Accordingly,

$$\begin{pmatrix} \text{overall} \\ \text{output} \\ \text{error} \\ \text{power} \end{pmatrix} = E[\delta_k^2] \cdot \left[\sum_{i=0}^{\infty} p_i^2\right] + \beta n E[(u'_k)^2] + \phi_{CC}(0). \tag{8.39}$$

Using Eqs. (8.16) and (8.17), the above equation may be expressed as

$$\begin{pmatrix} \text{overall} \\ \text{output} \\ \text{error} \\ \text{power} \end{pmatrix} = E[\delta_k^2]\left[\sum_{i=0}^{\infty} p_i^2\right] + \frac{nE[(u'_k)^2] \cdot E[n_j^2]}{2\tau E[\delta_k^2]} + \phi_{CC}(0). \tag{8.40}$$

After choosing a given value of time constant τ for the adaptive process for $\widehat{P}_k(z)$, the overall output error power can be minimized by best choice of the dither power. Setting the derivative of the right-hand side of (8.40) with respect to $E[\delta_k^2]$ to zero, we obtain

$$E[\delta_k^2]_{\text{OPT.}} = \sqrt{\frac{nE[(u'_k)^2] \cdot E[n_k^2]}{2\tau \sum_{i=0}^{\infty} p_i^2}}. \tag{8.41}$$

Therefore,

$$\begin{pmatrix} \text{overall} \\ \text{output} \\ \text{error} \\ \text{power} \end{pmatrix}_{\text{min}} = \sqrt{\frac{2nE[(u'_k)^2] \cdot E[n_k^2]\sum_{i=0}^{\infty} p_i^2}{\tau}} + \phi_{CC}(0). \tag{8.42}$$

The uncanceled plant output disturbance power $\phi_{CC}(0)$ will always be a fraction of the plant output noise power $E[n_k^2]$.

For the plant modeling and noise canceling system of Fig. 8.7, Eqs. (8.41) and (8.42) are useful indicators of system performance. Once an adaptive time constant τ is chosen, Eq. (8.41) indicates how to select the dither power. The value of μ is selected, with knowledge of $E[\delta_k^2]$ and τ, by using Eq. (8.16). In any event, the determination of μ and dither power $E[\delta_k^2]$ requires only a choice of the adaptive time constant τ and approximate knowledge of plant characteristics, plant disturbance power, and plant input power. For all of this to work, the entire system must be stable, with μ chosen to satisfy both stability criteria of (8.26) and (8.29).

8.8 COMPUTER SIMULATION OF PLANT MODELING AND DISTURBANCE CANCELING SYSTEM

In order to verify workability of the plant modeling and disturbance canceling system of Fig. 8.7 and its analytic treatment, a set of computer simulation experiments were performed. Of greatest importance were

(i) Verification of the convergence of $\widehat{P}_k(z)$ to $P(z)$ with time constant τ given by (8.16).
(ii) Verification of the stability conditions of (8.26) and (8.29).
(iii) Verification that output noise power and signal distortion are reduced to the greatest extent possible. Verification of Eqs. (8.41) and (8.42).
(iv) Verification that the overall transfer function remains essentially unaltered by the disturbance canceling feedback.

As a first step, however, we wanted to verify that the system of Fig. 8.7 did indeed converge to an unbiased estimate $\widehat{P}(z)$ of the plant $P(z)$. For this purpose, we repeated the same simulation that was performed in Section 8.4 for illustrating the behavior of the systems of Figs. 8.4 and 8.5. In other words, we have assumed that

$$P(z) = \frac{1 - \frac{1}{3}z^{-1}}{1 - \frac{1}{2}z^{-1}}. \tag{8.43}$$

The plant disturbance n_k was created by inputting a white noise to a 20-weight low-pass moving average filter with

$$E[n_k^2] = 0.2. \tag{8.44}$$

The plant input u'_k was zero-mean, white, and of unit power. However, in contrast to the simulations of Section 8.5, the plant modeling was performed by dithering scheme C, as depicted in Fig. 8.7.

The dither was chosen to be white, zero-mean, and of unit power. Keeping all these parameters constant, we were able to vary the adaptation constant μ. For each choice of μ, the system was run for 5,000 iterations to come to equilibrium, and then in each case, a snapshot was taken of the impulse response of $\widehat{P}(z)$, with the results presented in Table 8.3.

Referring to Table 8.3, line 1 shows the first 10 impulses of plant $P(z)$ (or the Wiener solution for the plant model $\widehat{P}(z)$ having 10 weights). In line 2 are presented snapshot weights of $\widehat{P}(z)$ while adapting with $\mu = 0.003$. Clearly the match between the adaptive plant model and the Wiener solution is quite good (square error was only 0.00276). For $\mu = 0.001$, the plant model (as depicted in line 3) was even closer to the Wiener solution (square error was only 0.0000237). We can see from this that the system of Fig. 8.7 was capable of delivering an unbiased, accurate plant model.

The next issue was stability: What was the stable range for the adaptation constant μ? In order for the system to be stable, we know that μ must obey, simultaneously, the two stability criteria of (8.26) and (8.29). In certain cases, Eq. (8.26) would limit the maximum stable value of μ, while in other cases, (8.29) would limit the maximum stable value of μ. For simulation purposes, we were interested in cases for which there would be a pronounced difference between the two criteria so that in each case, we would be able to determine if instability were caused by the plant disturbance canceling loop or by the plant modeling process.

Accordingly, for the stability study we choose a new plant,

$$P(z) = \sum_{i=0}^{9} z^{-i}. \tag{8.45}$$

For this plant, (8.26) and (8.29) gave completely different stability ranges. The simulation described above was repeated for this plant. As before, the plant disturbance power and the plant input power were kept constant. The parameter μ was increased until the system became unstable. For the first test, the plant disturbance power was set at 0.2. For the second test, the plant disturbance power was raised and set at 20. For both tests, the plant input power was 1. The results of the stability study, both measured and theoretical, are given in Table 8.4. The maximum theoretical value of μ permitting stability is in each case the smaller of the two limits obtained from (8.29) (stability of $\widehat{P}_k(z)$), and (8.26) (stability of the disturbance canceling loop). Inspection of Table 8.4 shows close correlation between the theoretical predictions and the experimental results.

TABLE 8.4 STABLE RANGE OF μ FOR THE SCHEME OF FIG. 8.7, FOR TWO VALUES OF PLANT DISTURBANCE POWER

Plant disturbance power	Maximal stable range of μ from experiment	Maximal stable range of μ from Eq. (8.26)	Maximal stable range of μ from Eq. (8.29)
0.2	0.08	0.18	0.050
20	0.003	0.0018	0.045

The theoretical predictions were obtained in the following way. In order to utilize Eq. (8.26), we needed to evaluate $|Q|^2_{\text{max}}$. Given that $\widehat{P}(z) = P(z)$ (they both had the same number of weights), the Wiener solution for $Q(z)$ was

$$Q(z) = 0.701 - 0.657z^{-1} + 0.176z^{-8} - 0.132z^{-9}. \tag{8.46}$$

The maximal value of $|Q|^2$ is obtained for $z = -1$. Then,

$$|Q|^2_{\text{max}} = (1.667)^2 = 2.78. \tag{8.47}$$

Since the number of weights of $\widehat{P}(z)$ is $n = 10$, Eq. (8.26) is

$$\frac{1}{(10)(2.78)E[n_k^2]} > \mu > 0. \tag{8.48}$$

For $E[n_k^2] = 0.2$ and for $E[n_k^2] = 20$, the maximal stable values of μ are as presented in Table 8.4.

In order to utilize Eq. (8.29), we needed to know the dither power (this was set to be unity), the input power $E[(u_k')^2]$ (which was chosen to be unity), and $|P(e^{j\omega})|^2_{\omega \text{ low}}$ (which is the plant response for low real frequencies). However, for low-frequency inputs,

$$P(e^{j\omega}) = \sum_{i=0}^{9} p_i = 10. \tag{8.49}$$

Hence,

$$\frac{1}{|P(e^{j\omega})|^2_{\omega \text{ low}}} = 0.01. \tag{8.50}$$

Accordingly, (8.29) can be expressed as

$$\frac{1}{10(1 + 1 + (0.01)E[n_k^2])} > \mu > 0. \tag{8.51}$$

This expression was used in order to compute maximal stable values of μ for $E[n_k^2] = 0.2$ and $E[n_k^2] = 20$. The results are given in Table 8.4.

Inspecting the first row of Table 8.4, we conclude that for small values of plant disturbance power, stability condition (8.29) is the dominant one. Inspecting the second row of Table 8.4, we conclude that for cases with large plant disturbance power, the criterion of (8.26) becomes dominant.

For both cases, theoretical prediction of the stable range of μ matched the experimental range to within an error of 40 percent. The theory was somewhat more conservative than necessary to ensure stability.

Next we wanted to verify Eqs. (8.16), (8.26), (8.29), and (8.42) for a nonminimum-phase plant. For these simulations, a simple one-pole plant with a unit transport delay was chosen:

$$P(z) = \frac{z^{-1}}{1 - \frac{1}{2}z^{-1}}. \tag{8.52}$$

This plant is nonminimum-phase. The plant output disturbance n_k was chosen to be low-pass Gaussian and was generated by applying white Gaussian noise to a single-pole filter. The plant input signal u_k' was a sampled low-frequency square wave. Its frequency was approximately 0.0008 of the sampling frequency. The adaptive filter $Q_k(z)$ was allowed only three weights (increasing the length of $Q_k(z)$ caused only negligible improvement in disturbance canceling performance because of the low-frequency character of the plant disturbance).

To establish convergence of $\widehat{P}_k(z)$ to $P(z)$, a plot of $||\Delta \mathbf{P}_k||$ versus number of adaptations was drawn and this is shown in Fig. 8.8. This is a form of learning curve. The theoretical time constant was $\tau = 2,500$ sample periods. The measured time constant obtained from the learning curve was approximately 2,000 sample periods. Convergence was solid, with very low residual error.

The dither power was fixed at unity for a stability checking experiment. The system was run with μ gradually increased until the adaptive process for $\widehat{P}_k(z)$ blew up. A series of runs was made, each with a different value of plant disturbance power. Both stability criteria

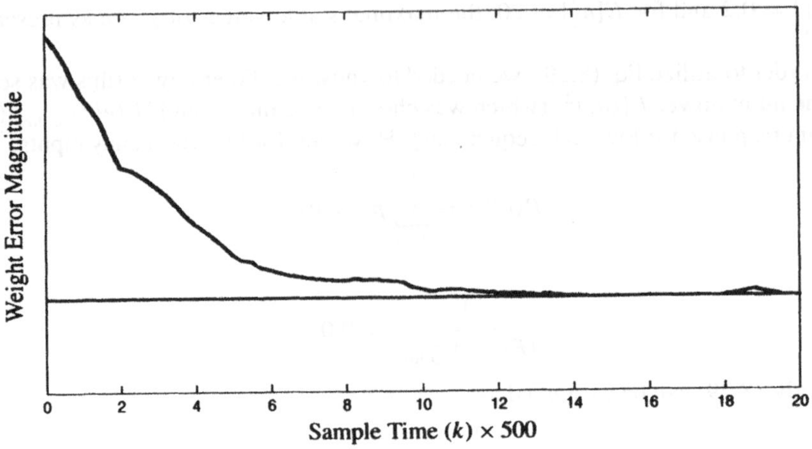

Figure 8.8 A learning curve for adaptive plant disturbance canceler.

were checked. The theoretical stability limits for μ did work, but they were not precisely confirmed, only verified within a factor of 2 or 3.

The results of a plant disturbance canceling experiment are shown in Fig. 8.9. The plant itself without disturbance was subjected to a square-wave input, and the output response, almost a square wave, is shown in Fig. 8.9(a). The same plant with the same input was then set up with low-pass plant disturbance added to its output, and the result is shown in Fig. 8.9(b). Next, this same plant with the same low-pass plant disturbance at its output was connected for disturbance cancelation in accord with Fig. 8.7. The dither was applied, and the same square wave was input as u'_k. The dither power was optimized in accord with (8.41). The plant output is shown in Fig. 8.9.c. The effects of plant disturbance cancelation are in clear evidence. The plant output disturbance power was reduced by 18 dB. Using (8.42) to calculate the overall output error power, the plant output disturbance power should have been reduced by 19.2 dB.

Many more experiments have been made in addition to this one with equally close verification of these formulas being the result. Preservation of the plant's dynamic response is evidenced by the similarity of the waveforms of Fig. 8.9(a), without disturbance canceling, and Fig. 8.9(c), with disturbance canceling in operation.

One more experiment has been performed in order to demonstrate in greater detail how the dynamic plant response is preserved when the adaptive plant disturbance canceling system of Fig. 8.7 is in operation. The results are shown in Fig. 8.10. A step input was applied to the disturbance-free plant, and the output response was plotted. The same plant with additive output disturbance was then tested. With the disturbance canceling system working and with the dither applied, the input u'_k was once again a unit step. The response was recorded. This was repeated many times. An ensemble average of step responses is plotted in Fig. 8.10. The ensemble average and the step response of the original disturbance-free plant are hardly distinguishable, indicating the preservation of the plant response in its original form even while the adaptive plant disturbance canceling system was operating.

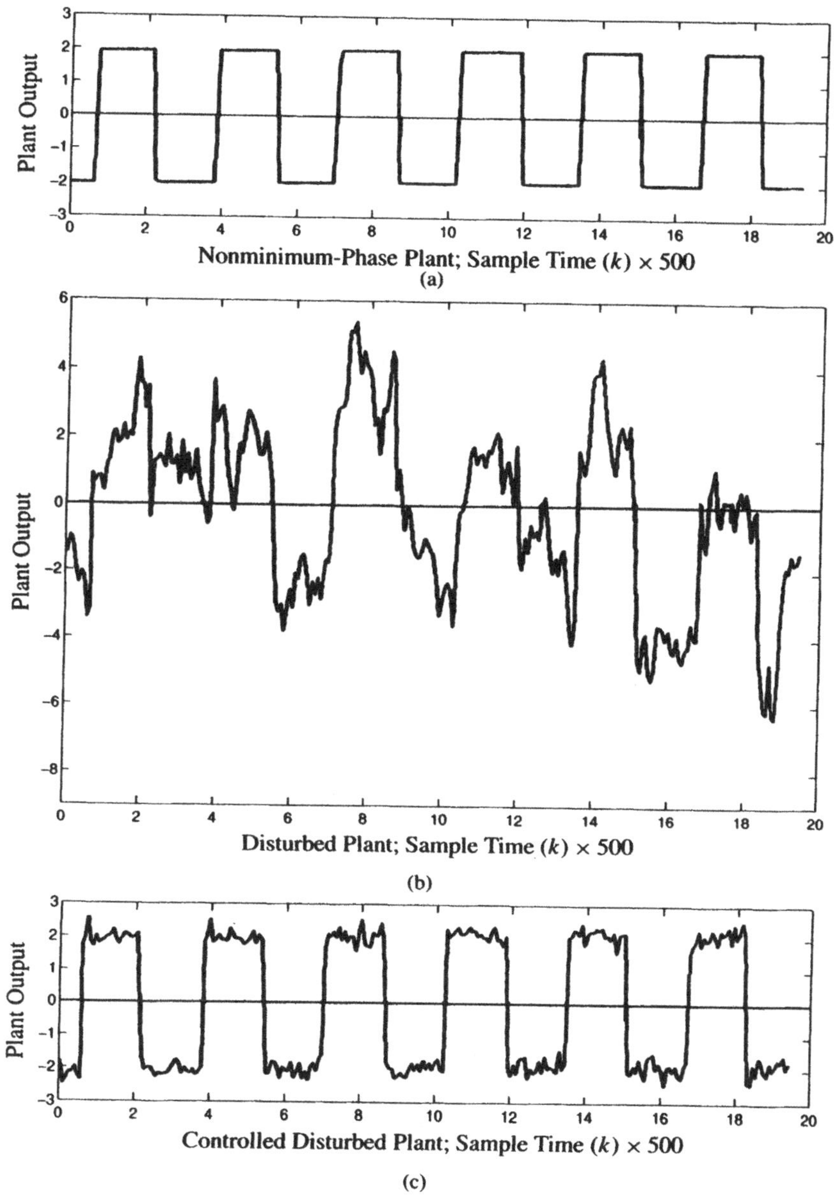

Figure 8.9 (a) Square wave response of undisturbed plant; (b) Square wave response of disturbed plant; (c) Square wave response of disturbed plant with adaptive disturbance canceler.

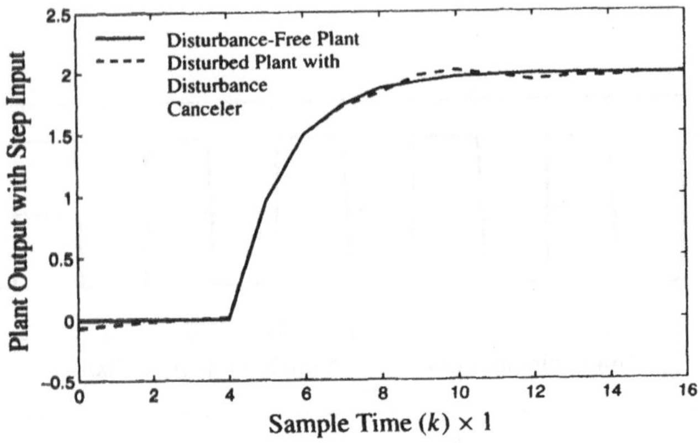

Figure 8.10 Demonstration of the preservation of plant dynamics even when canceling plant disturbance.

8.9 APPLICATION TO AIRCRAFT VIBRATIONAL CONTROL

Adaptive plant disturbance canceling can be used to alleviate a problem that sometimes afflicts passengers of aircraft, the problem of sudden changes in altitude and a bumpy ride due to vertical wind gusts. Reduction of the variance of the airplane's response to the random wind loading could greatly increase comfort and safety. An airplane subject to random updrafts and downdrafts is sketched in Fig. 8.11.

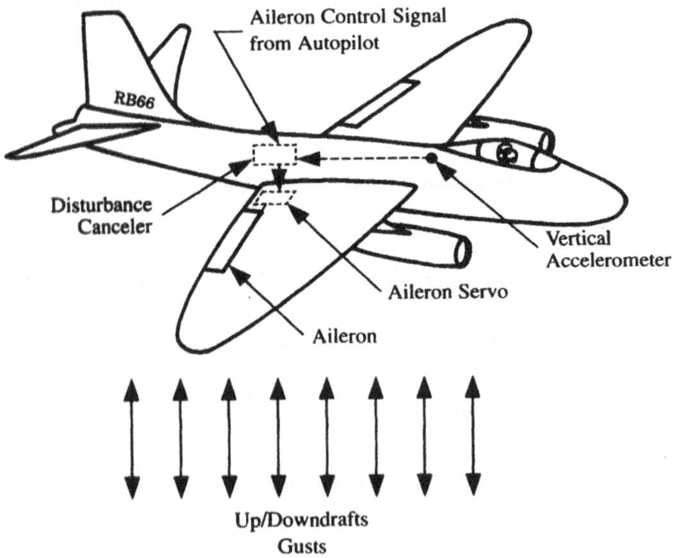

Figure 8.11 An airplane with gust loading canceler.

Corrections in altitude could be achieved in a conventional way by controlling the elevator at the tail of the airplane. By causing the tail to depress, the angle of attack changes and the airplane climbs. By raising the tail, the airplane descends. Correcting for updrafts/

downdrafts in this conventional way is slow reacting, however, and causes the airplane to operate with varying angles of attack. To overcome these disadvantages, a novel approach is hereby suggested that would apply the altitude correcting signals to the ailerons rather than to the elevator. The objective is to be able to generate wing forces to provide swift counteraction to vertical gust loading and to keep the airplane flying horizontally all the while.

Ailerons are generally not used to control altitude. They are basically used for steering. Each wing has an aileron on its trailing edge. The pair of ailerons generally operates in differential mode, that is, when one aileron goes up, the other goes down. Pushing one aileron down increases the lift on its respective wing; at the same time, the other aileron goes up and reduces the lift on its wing. The torque causes the airplane to roll, resulting in a banking turn.

We are proposing here that the linkages and servos which actuate the ailerons be modified so that the ailerons could be driven in a differential mode for steering and at the same time in a common mode for rapid variation of wing lift. The differential and common mode signals would be summed at the inputs of the two servos driving the ailerons. The differential mode signals would be the usual steering signals, while the common mode signals would be coming from an adaptive disturbance canceler rigged to suppress the effect of random updraft and downdraft wing loadings.

The random forces acting on the airplane can be treated as plant disturbance. Vertical acceleration can be sensed with a suitably placed accelerometer. The ailerons can be modulated to apply compensating wing forces to reduce vertical acceleration. For this purpose, the accelerometer output can be regarded as the plant output. The common mode input to the aileron servos can be regarded as the plant input. The plant itself is a dynamic system comprised of the ailerons, the wings, the airframe, and the surrounding air fluid. Control for only one of the aileron servos is illustrated in Fig. 8.11. Control for the other servo is identical.

Figure 8.12 shows the adaptive control system for minimizing the variance of the vertical accelerometer output. This is a more succinct form of the same adaptive plant disturbance canceler that is shown in Fig. 8.7. The dithering method is based on scheme C (of the particular configuration shown in Fig. 4.3(d)). $Q(z)$ could be obtained offline as shown in Fig. 8.7. The dynamic response of the airplane to aileron control signals remains exactly the same, with or without the adaptive canceler. The basic handling of the airplane therefore stays the same. The acceleration canceler simply reduces the bumpiness of the ride when flying through choppy air. A multi-input multi-output (MIMO) version could be devised, using appropriate sensors and actuators, to cancel random disturbances in yaw, pitch, and roll. MIMO systems are discussed in Chapter 10.

Control of vertical disturbance in the presence of vertical wind gusts has not yet been done with adaptive inverse control. The U.S. B1 bomber does have vertical disturbance control based on a more conventional servo technology applied to a pair of small *canard* wings on the forward part of the aircraft. Without this control, the ride would be so rough under certain flight conditions that the crew would not be able to physically function. Flight disturbance control needs to be developed for civilian aircraft now. Let's do it!

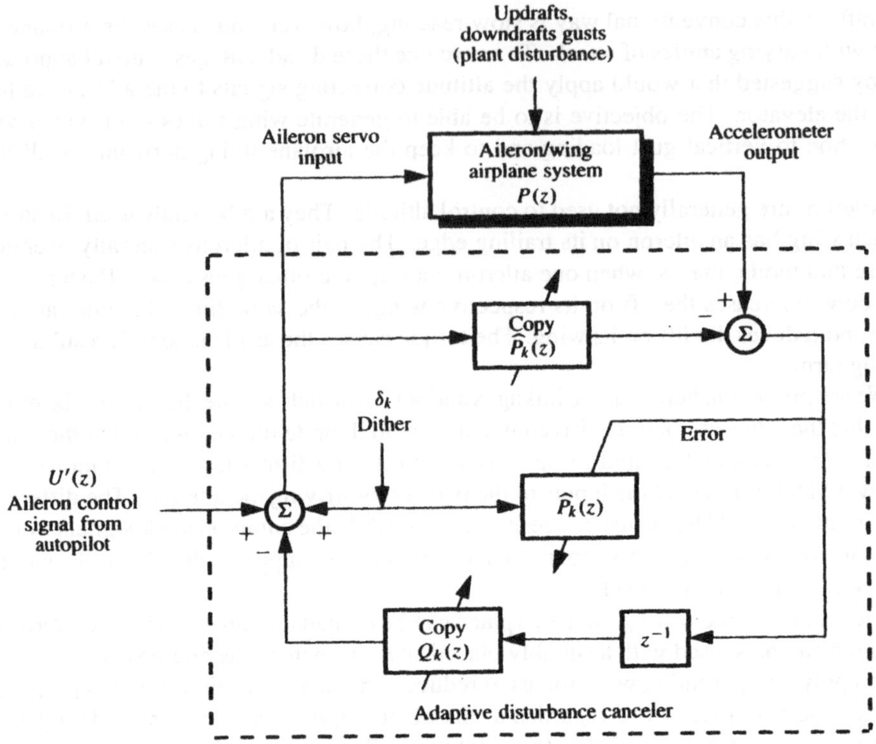

Figure 8.12 An adaptive canceler for minimizing gust loading effects.

8.10 APPLICATION TO EARPHONE NOISE SUPPRESSION

In Section 7.6 we discussed the problem of earphone noise cancelation. Another approach to the problem is illustrated in Fig. 8.13. This approach uses only one microphone, the one located between the earphone and the ear cavity. This is microphone M_2. The method to be disclosed here is based on the plant noise canceling idea embodied in the system of Fig. 8.5. A block diagram of the entire new system is shown in Fig. 8.14.

To understand how this system works, it is useful to trace through it one block at a time. The microphone signal labeled M_2 is the sum of two components, "plant disturbance" (due to (SAN), the strong ambient noise through the transfer function [M_2/SAN]), and "plant dynamic response" (the microphone output in response to the earphone signal). The "plant" here is equivalent to the transfer function [M_2/EPS]. The earphone signal is the audio input signal (AIS) minus the filtered plant disturbance, derived from the noise component of the microphone M_2 signal. The adaptive process for finding the plant model $\widehat{P}$, the offline process for finding Q, and the disturbance canceler work just like the corresponding processes of the system of Fig. 8.5. The earphone signal (EPS) connects to the earphone, and the microphone signal M_2 connects into the adaptive canceler as shown in Fig. 8.14. The audio input signal (AIS) connects to the canceler as shown in Fig. 8.14.

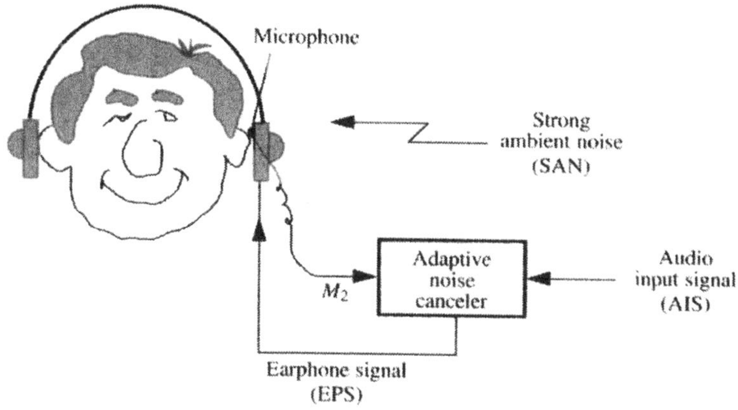

Figure 8.13 An alternative method for cancelation of earphone noise.

The system of Fig. 8.14 is the best canceler for the earphone and microphone configuration of Fig. 8.13. No linear canceler can do better.

The question is whether this is a better system than the one presented in Chapter 7. The situation is not clear, but the answer is probably no. The system of Chapter 7 uses two microphones and works on a different principle. It is probably a better system because it uses two microphone sensors instead of one, (i.e., there is more information to be used in disturbance canceling) and because the outer microphone M_1 receives the strong ambient noise earlier than M_2, making it easier to cancel the ambient noise components that leaked through the earphone.

There is no doubt that both the noise canceling schemes of Chapters 7 and 8 will work well. Systems like them have been analyzed and simulated without difficulty. Their operating characteristics can best be obtained by building the two systems in hardware and comparing them critically under real-life operational conditions. If these systems could be built cheaply enough, they could be used on commercial aircraft where they would contribute greatly to the enjoyment of in-flight movies and music.

8.11 CANCELING PLANT DISTURBANCE FOR A STABILIZED MINIMUM-PHASE PLANT

The stabilized minimum-phase plant that was studied in Chapter 7 was simulated once again to illustrate plant disturbance canceling for this case. Figure 8.15 shows a block diagram of the experiment. It is basically the same block diagram as the one in Fig. 8.5, except that it is customized for this plant of interest. It also includes a perfect $G(z)$, the discretized version of the true stabilized plant transfer function.[5] The output of this $G(z)$ is the ideal output, the discrete output of the plant if there were no disturbance and no disturbance canceling system. Comparing the plant output with the ideal output yields the residual error, due to

[5] In a real-world situation, $G(z)$ would usually not be perfectly known. $G(z)$ is assumed to be known here, for test purposes only.

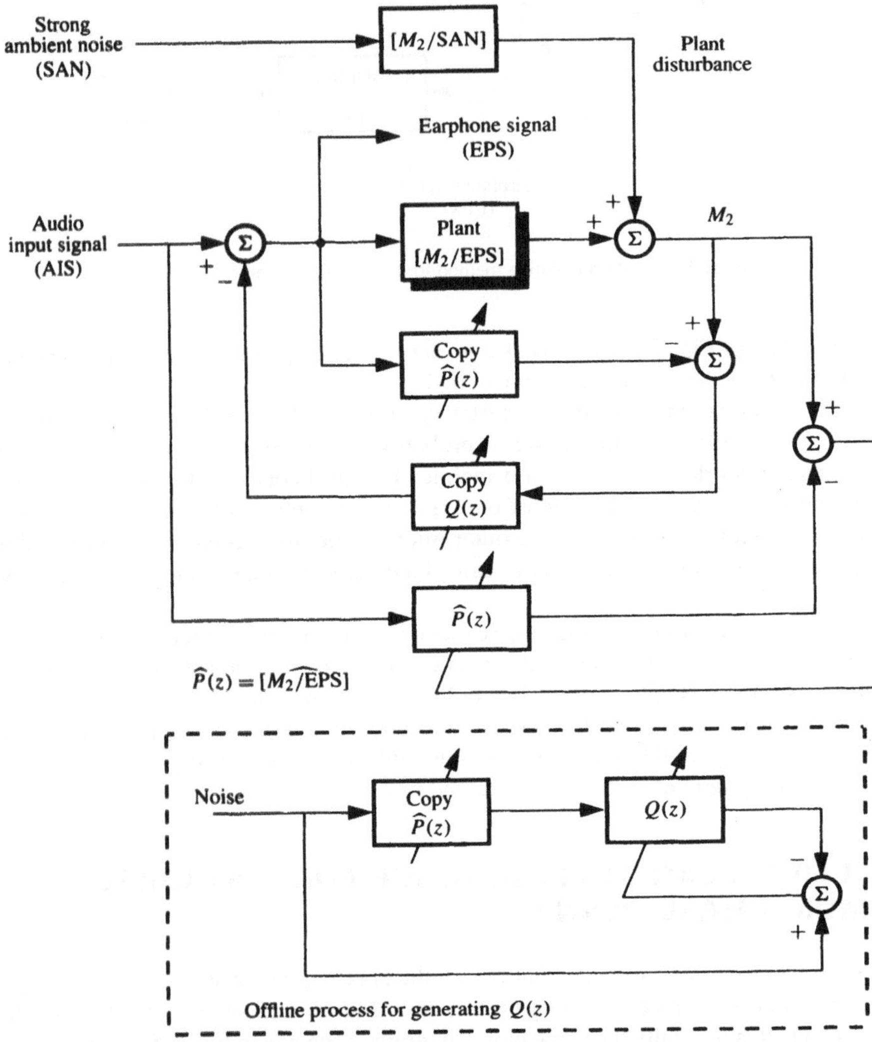

Figure 8.14 Block diagram of alternative method for cancelation of earphone noise.

uncanceled disturbance and to the effects of the canceling feedback on plant dynamics, if any. The later effect exists only when $\hat{G}(z)$ does not perfectly match $G(z)$.

Figure 8.16 shows the impulse response of the perfect $G(z)$. Using an offline process to find the corresponding optimal $Q(z)$, the resulting impulse response is shown in Fig. 8.17. Using the perfect $G(z)$ in place of an adaptive $\hat{G}(z)$, and using its almost perfect FIR $Q(z)$, we have a near optimal least squares disturbance canceler. Its performance is demonstrated in Fig. 8.18. Shown here is instantaneous squared values of the residual error plotted over time. The canceling feedback loop was closed at time sample 5,000, and disturbance canceling began then. The input driving function $U(z)$ was a zero-mean first-order Markov process, and the plant disturbance was an independent zero-mean first-order Markov process. The measured power of the residual error was reduced by a factor of 5.23 in this case. This result was very close to the theoretical optimum. Figure 8.19 shows the plant output, and the plant output when near optimally canceling the plant disturbance.

When $G(z)$ is unknown, adaptive identification is used to obtain $\hat{G}(z)$, as illustrated in Fig. 8.15. This was done by means of DCT/LMS. At one moment in time, the adaptive process was halted and the impulse response of $\hat{G}(z)$ was obtained, and it is plotted in Fig. 8.20. Corresponding to this impulse response, the offline process was used to obtain $Q(z)$, whose impulse response is shown in Fig. 8.21. These impulse responses are basically noisy versions of the ideal ones shown in Figs. 8.16 and 8.17. A plot of the square of the residual error is shown in Fig. 8.22. The canceling feedback loop was engaged at the 5,000th time sample, and the power of the residual error dropped by a factor of 5.03. This is not as good as optimal, but it is close. A comparison of the plant output with the adaptive disturbance-canceled output is shown in Fig. 8.23. Using a real-time adaptive disturbance canceler, the results are not so different from those that would be obtained with a near optimal canceler having perfect knowledge of $G(z)$ and its near optimal $Q(z)$.

A series of tests was performed next with other kinds of plant disturbances such as step and ramp functions. For one of these tests, driving function $U(z)$ was Markov as before. There was no Markov disturbance. At time sample 2,000, the canceling feedback was turned on and at time sample 3,000, a step disturbance was turned on. This simulates a plant drift with a sudden onset. The step disturbance is plotted in Fig. 8.24. The residual error is shown in Fig. 8.25. There was no residual error until the canceler was engaged. Then the error was small. At time 3,000, the error had a sharp spike because the canceler could not cancel an instantaneous disturbance, the leading edge of the step. But within one sample time, the residual error was brought down to low level and it remained at low level with a mean very close to zero. Figure 8.26 shows a comparison of the plant output with the ideal plant output in a steady-state situation, after the leading edge of the step disturbance has passed out of the time window of the plot. In many respects, the essentially zero steady-state error in the presence of a constant disturbance represents behavior like a type 1 servo.

Other tests were done with ramp disturbances, and the results were quite similar. A ramp disturbance was turned on at time sample 3,000, and allowed to plateau at time sample 4,000 as shown in Fig. 8.27. The disturbance canceling loop was closed at time sample 2,000, and a small residual error is evident in Fig. 8.28 from then on. When the ramp disturbance commenced at time sample 3,000, the effect on the residual error was almost imperceptible. The same was true at time sample 4,000 when the ramp saturated. A comparison of the actual plant output and the ideal plant output in the middle of the ramp rise up

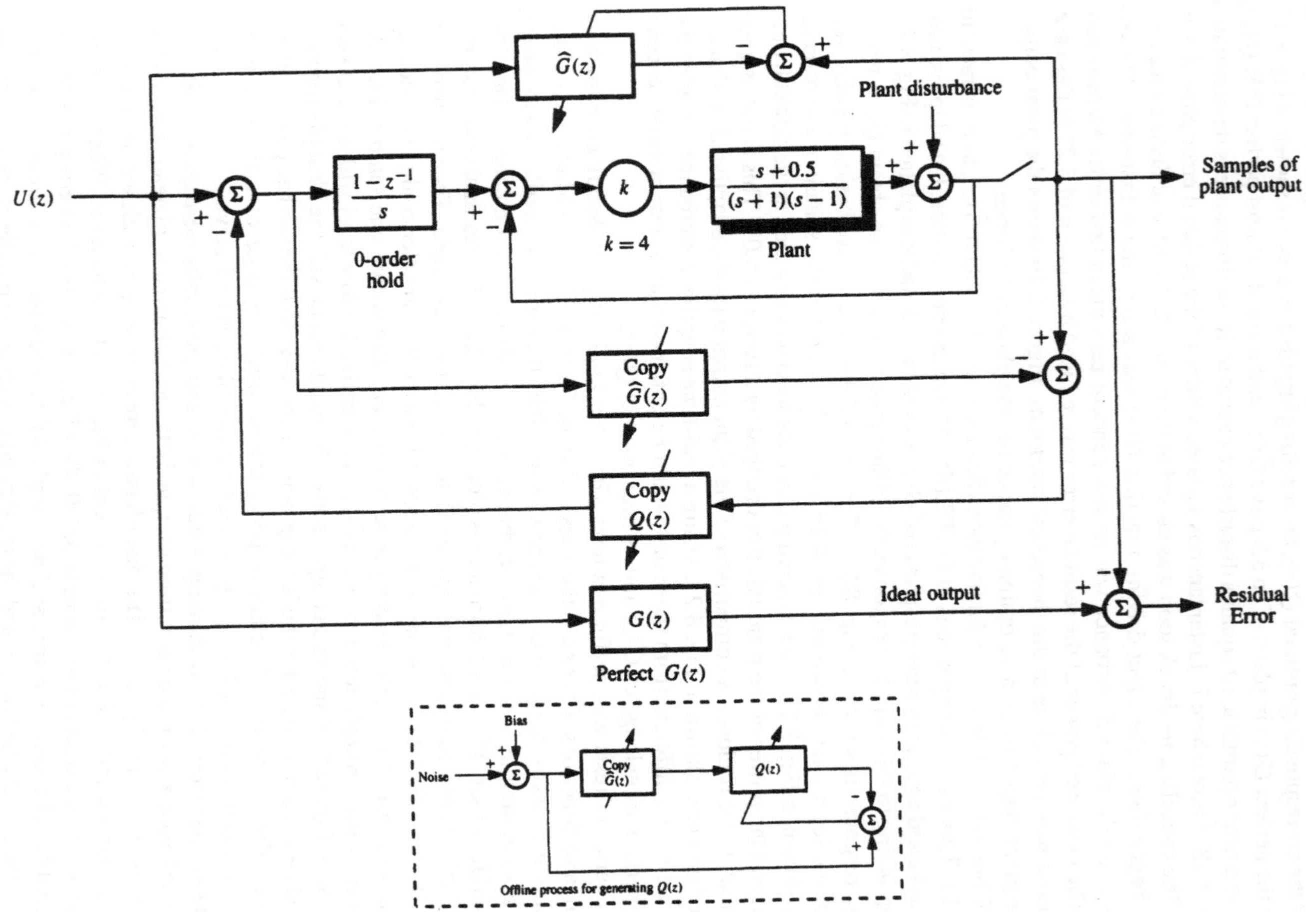

Figure 8.15 A disturbance canceling system for a stabilized minimum-phase plant.

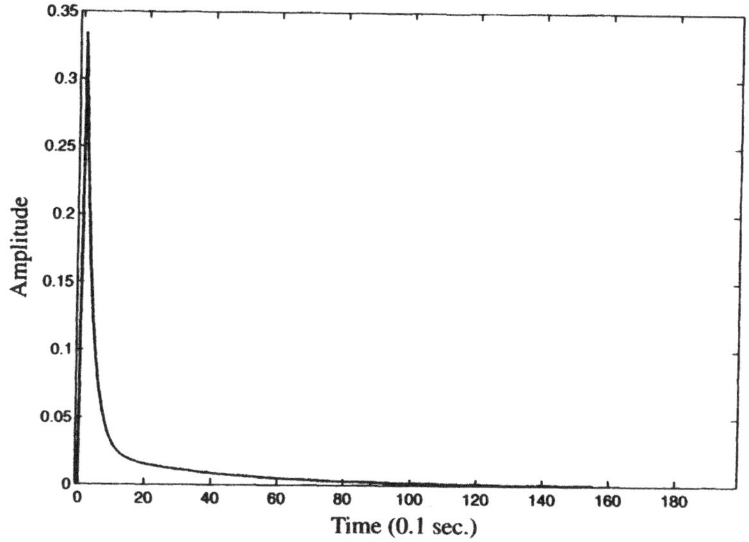

Figure 8.16 Impulse response of $G(z)$, the discretized transfer function of the stabilized minimum-phase plant.

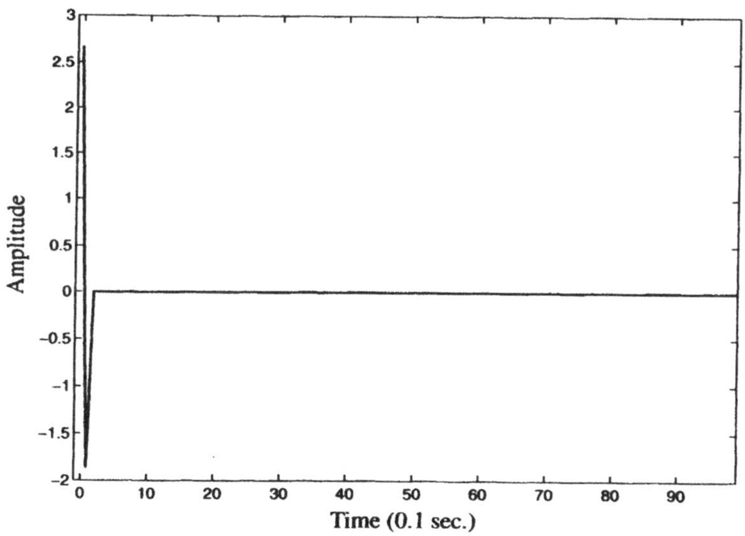

Figure 8.17 Impulse response of the near optimal $Q(z)$ for $G(z)$.

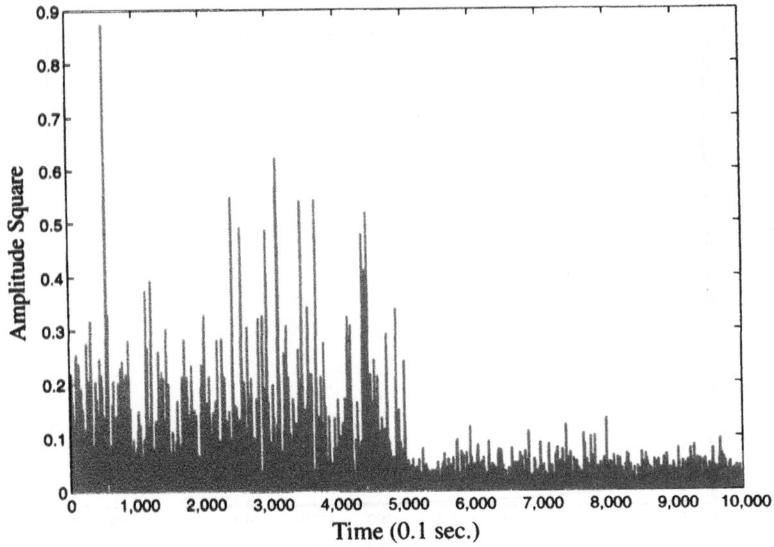

Figure 8.18 Square of the residual error plotted over time. Canceling feedback loop closed at time sample number 5,000. $Q(z)$ was obtained from $G(z)$.

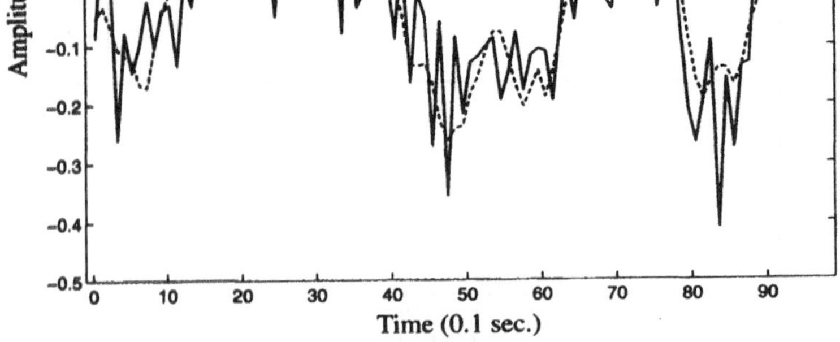

Figure 8.19 A comparison of the plant output and the output of the minimum-phase plant with near-optimal disturbance canceling.

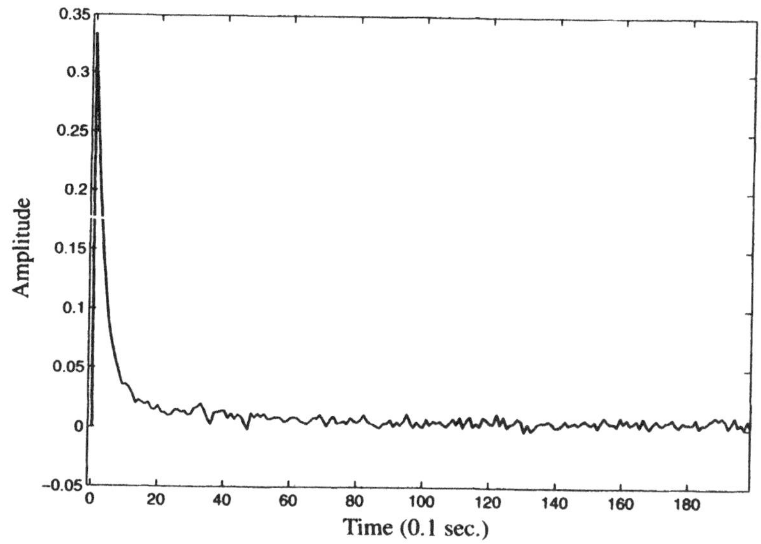

Figure 8.20 Impulse response of $\widehat{G}(z)$ for stabilized minimum-phase plant.

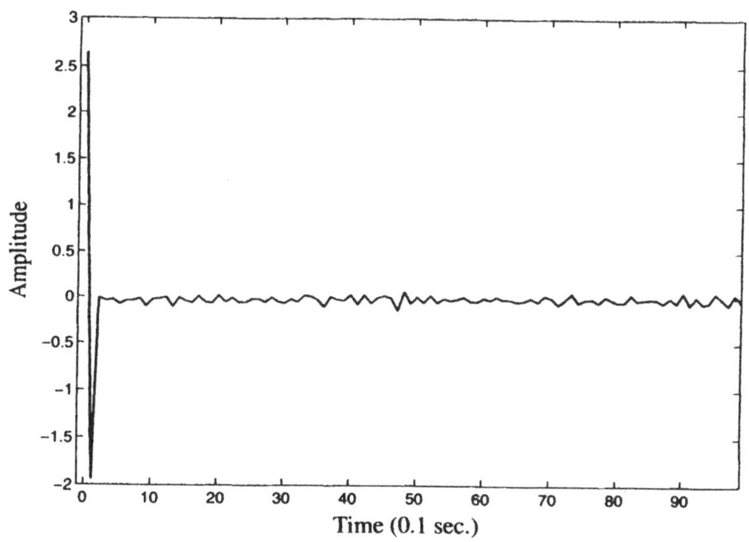

Figure 8.21 Impulse response of $Q(z)$ for $\widehat{G}(z)$.

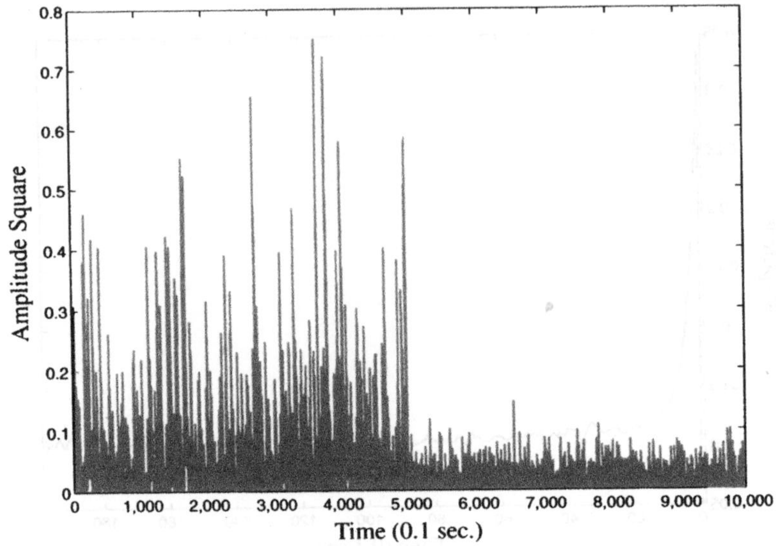

Figure 8.22 Square of the residual error. Disturbance canceling began at the 5,000th sample time. $Q(z)$ was obtained from $\widehat{G}(z)$.

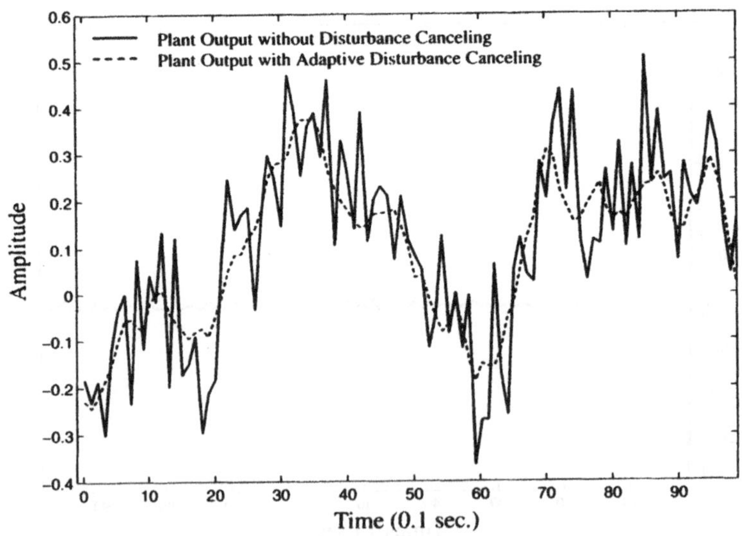

Figure 8.23 Plant output and plant output for the minimum-phase plant when using an adaptive plant disturbance canceler.

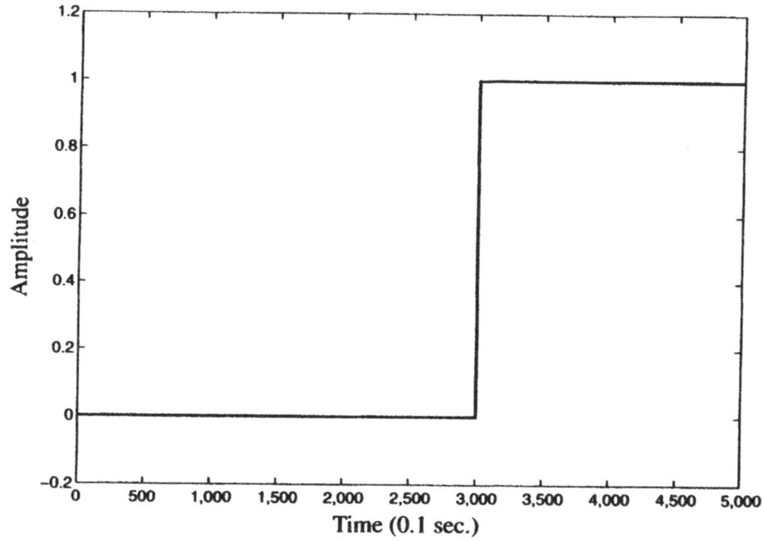

Figure 8.24 A step plant disturbance.

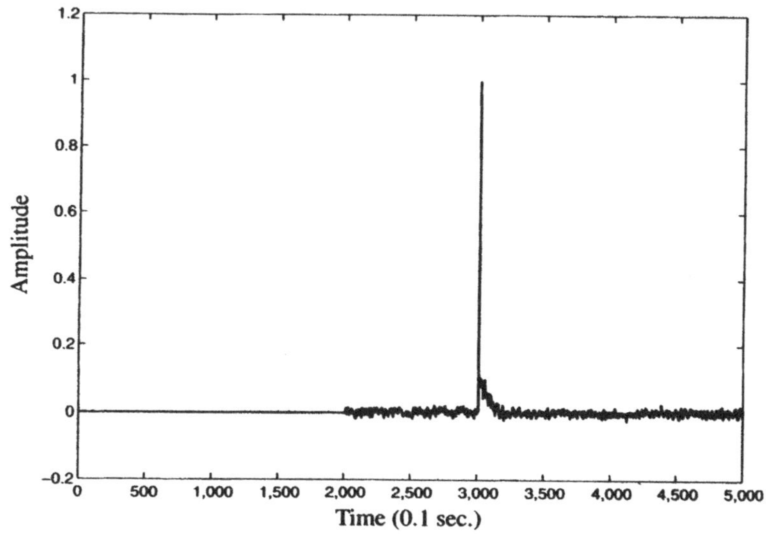

Figure 8.25 Residual error of minimum-phase system. Canceling loop turned on at time sample 2,000. Step disturbance was turned on at time sample 3,000.

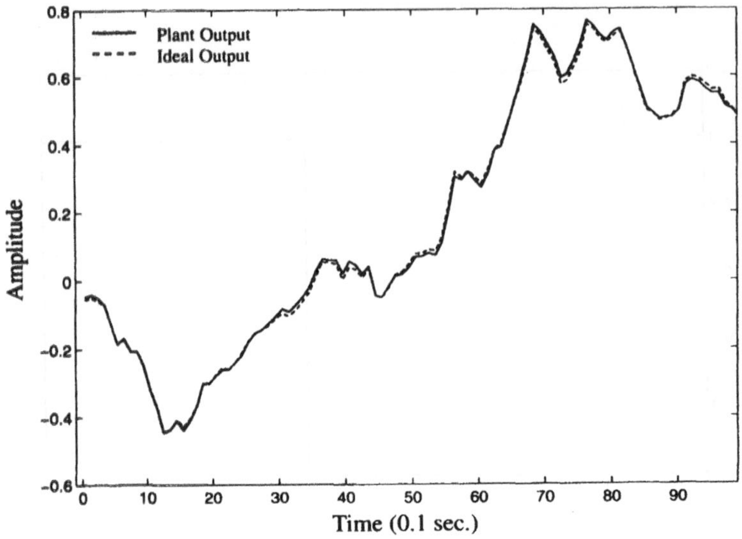

Figure 8.26 Comparison of plant output and ideal plant output in steady state with step disturbance.

is shown in Fig. 8.29. A similar comparison is shown in Fig. 8.30 in steady state, well after the ramp has plateaued. There is not much difference between the actual and ideal responses with a ramp disturbance. This behavior is like that of a type 2 servo.

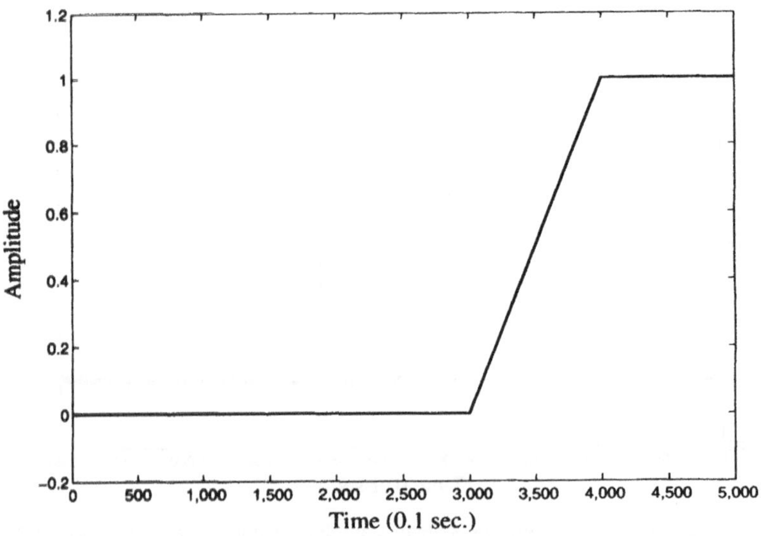

Figure 8.27 A ramp disturbance.

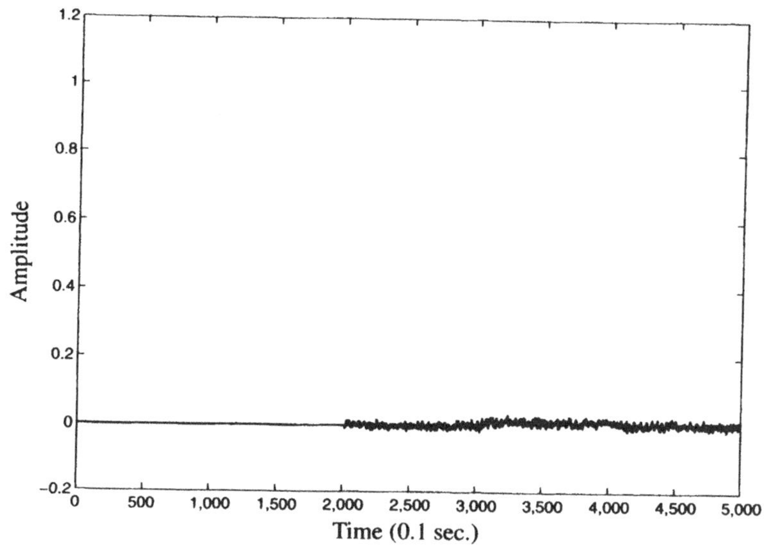

Figure 8.28 Residual error of minimum-phase system. The canceling loop was turned on at time sample 2,000. The ramp commenced at time sample 3,000 and plateaued at time sample 4,000.

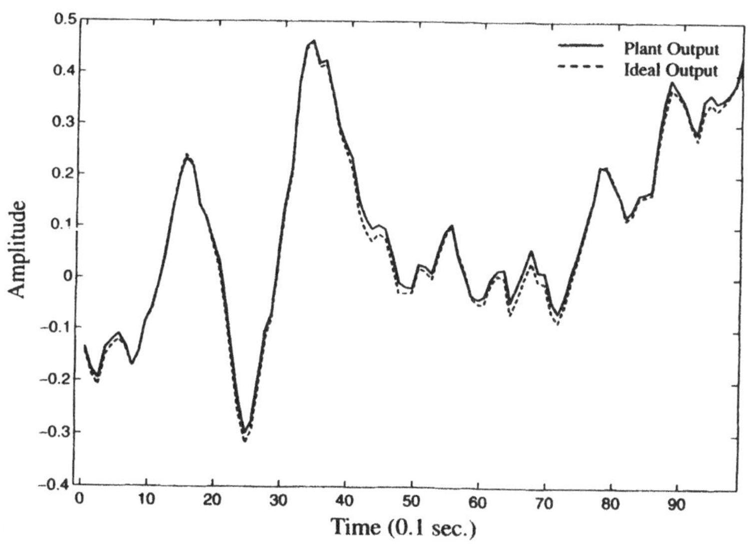

Figure 8.29 Comparison of actual and ideal plant outputs for minimum-phase system with ramp disturbance. Data were taken in the middle of the ramp rise.

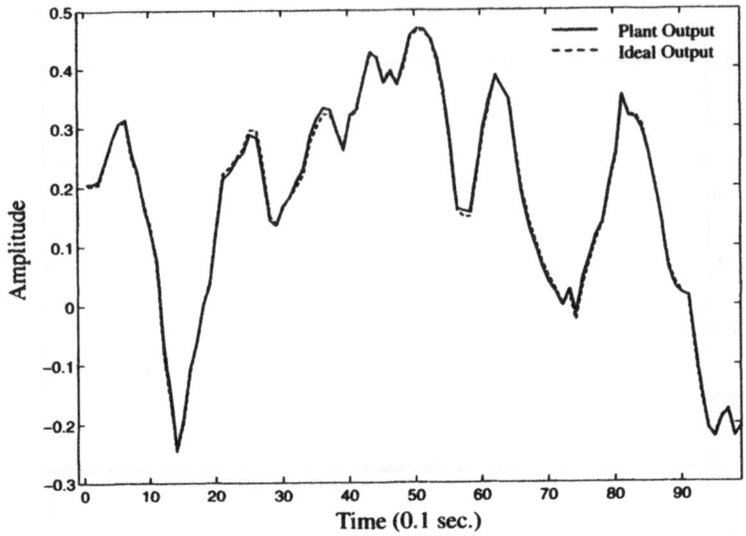

Figure 8.30 Comparison of actual and ideal plant outputs for minimum-phase system with ramp disturbance. Data were taken after the ramp saturated.

8.12 COMMENTS REGARDING THE OFFLINE PROCESS FOR FINDING Q(z)

The offline process could make use of an adaptive algorithm to find $Q(z)$, or could use Wiener theory to find a solution. The adaptive process could be based on LMS, RLS, DCT/LMS, or some other algorithm. Experience has shown, however, that a more computationally efficient approach to use offline when utilizing a computer to find $Q(z)$ is the Wiener approach. If adaptive hardware were available on the other hand, then an adaptive algorithm would be appropriate. For the above example, $Q(z)$ was calculated by conventional computer, and the Wiener approach was used.

Referring to the block diagram of Fig. 8.15, the offline process is driven by noise whose power spectrum should be similar to that of the actual plant disturbance. The choice of power spectrum is not critical. In practice, one would generally not know the disturbance spectrum exactly. One would use a best guess. For the above example, we knew the spectrum of the disturbance. Since in addition we knew $\widehat{G}(z)$, it was possible to calculate the spectrum at the input to the filter $Q(z)$ by making use of Eq. (2.22). Transforming gave the autocorrelation function. From the autocorrelation function, it was possible to compute the **R**-matrix for the inputs to the weights of the filter $Q(z)$. The **P**-vector was computed from the crosscorrelation function between the weight inputs and the desired response of $Q(z)$, that is, the driving noise. This was done by making use of (2.15) and (2.14) with knowledge of $\widehat{G}(z)$. The Wiener-Hopf equation that we used was obtained from (3.11) as

$$\mathbf{R}\mathbf{W}^* = \mathbf{P}. \tag{8.53}$$

Having **R** and **P**, Gaussian elimination was used to solve the linear equations and find $\mathbf{W}^*$, thus yielding the weights of $Q(z)$. No matrix inversion was involved.

Although the method may seem to be tedious, each step is simple, robust, and easily understood. The computation runs very fast and gives a precise result. This was our method of choice.

Referring once again to Fig. 8.15, we notice that a bias is added to the noise driving the offline process. The purpose of the bias is to cause the zero-frequency gain (the DC gain) of the filter $Q(z)$ to be the reciprocal of that of $\widehat{G}(z)$, or to be very close to this. The objective is to have close to perfect cancelation of very low-frequency or DC plant disturbance or plant drift. The bias was used in getting the results shown in Figs. 8.26, 8.29, and 8.30, where the disturbance was step and ramp. The amplitude of the bias is not critical, but should at least be comparable in amplitude to the standard deviation of the noise. Using both noise and bias to drive the offline process, the resulting canceling system works well when the plant disturbance is random with a finite mean.

When comparing the block diagrams of Figs. 8.15 and 8.3, the offline process and canceling loop of Fig. 8.3 contain a unit delay between the plant model and Q, while the offline process and canceling loop of Fig. 8.15 contain no corresponding delay. The delay is only necessary when the plant impulse response has a finite component at zero time, that is, responds instantly. For the minimum-phase example, the plant could not respond instantly because it had more poles than zeros in the s-plane. The stabilizing feedback does not alter this. The stabilized plant, when discretized, had a unit delay in its impulse response. An additional delay in the offline process was not necessary and its inclusion would have caused a loss in performance for the disturbance canceler. Therefore, the offline process and the disturbance canceling loop of Fig. 8.15 did not incorporate an additional unit delay.

8.13 CANCELING PLANT DISTURBANCE FOR A STABILIZED NONMINIMUM-PHASE PLANT

The stabilized nonminimum-phase plant that was studied in Chapter 7 is once again the object of study. We will demonstrate disturbance canceling for this system. A block diagram of the adaptive canceler is shown in Fig. 8.31. The exact impulse response of the stabilized plant is shown in Fig. 8.32 with $k = 24$. Its transform is $G(z)$. For this $G(z)$, the offline process is used to obtain $Q(z)$. Its impulse response is shown in Fig. 8.33. When used in the canceling system, a reduction in the residual error takes place. The extent of the reduction depends on the characteristics of $G(z)$, $Q(z)$, and on the spectrum of the plant disturbance. With a nonminimum-phase $G(z)$, low-frequency disturbances are much easier to cancel than high-frequency ones.

Results of a disturbance canceling experiment are shown in Fig. 8.34 when using the ideal $G(z)$ and $Q(z)$. The theoretical reduction in residual error power is 5.65. Measured results came very close to this. Figure 8.34 shows a time plot of the square of the residual error. The canceling loop was closed at time sample number 5,000. Figure 8.35 shows the actual plant output and the ideal plant output after closure of the plant disturbance canceling loop. The difference between these plots is the residual error.

An adaptive canceler was tested next. The DCT/LMS algorithm was used to obtain $\widehat{G}(z)$. An instantaneous impulse response is shown in Fig. 8.36. This is a slightly noisy version of the ideal impulse response shown in Fig. 8.32. The corresponding impulse response of $Q(z)$, obtained by the offline process block-diagrammed in Fig. 8.31, is shown

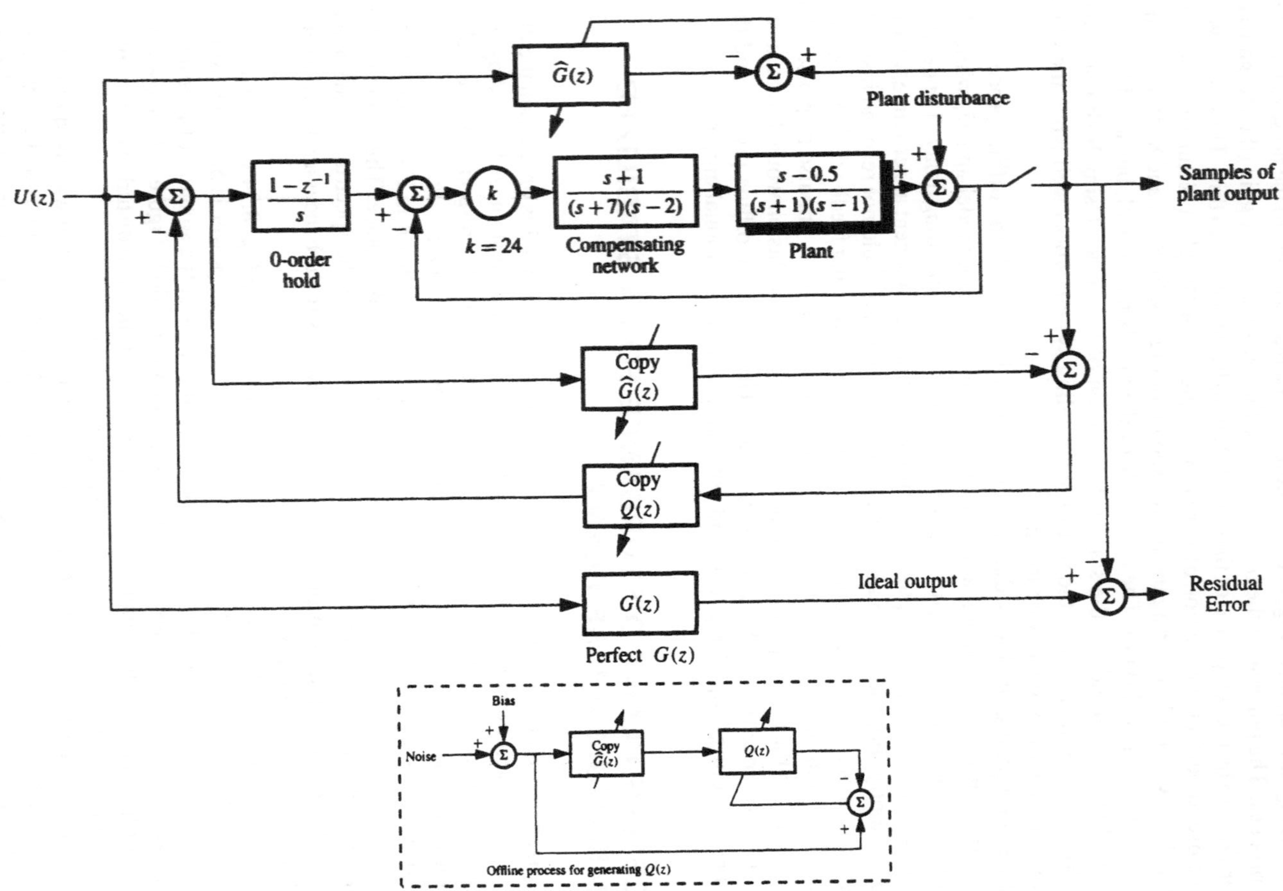

Figure 8.31 A disturbance canceling system for a stabilized nonminimum-phase plant.

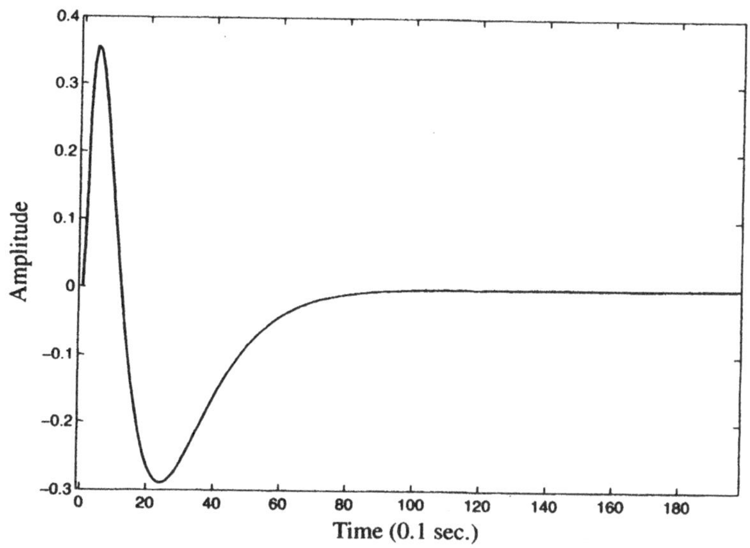

Figure 8.32 Impulse response of $G(z)$, the discretized transfer function of the stabilized nonminimum-phase plant. $k = 24$.

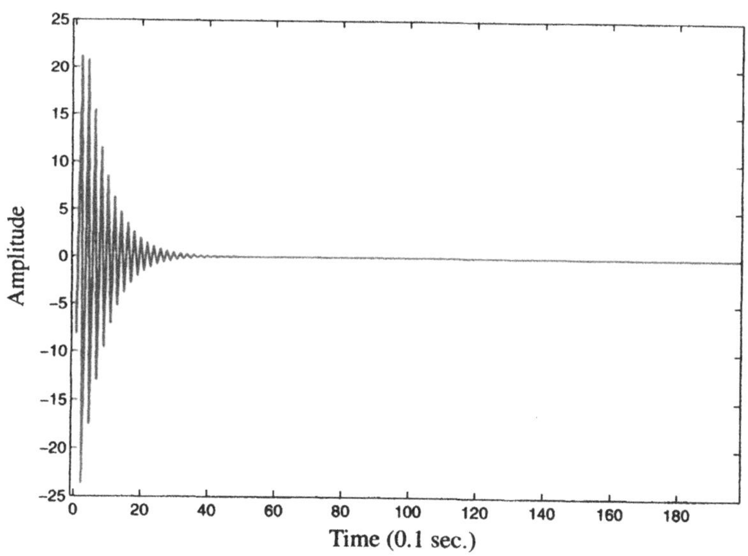

Figure 8.33 Impulse response of $Q(z)$, optimized for $G(z)$, the exact discretized transfer function of the stabilized nonminimum-phase plant with $k = 24$.

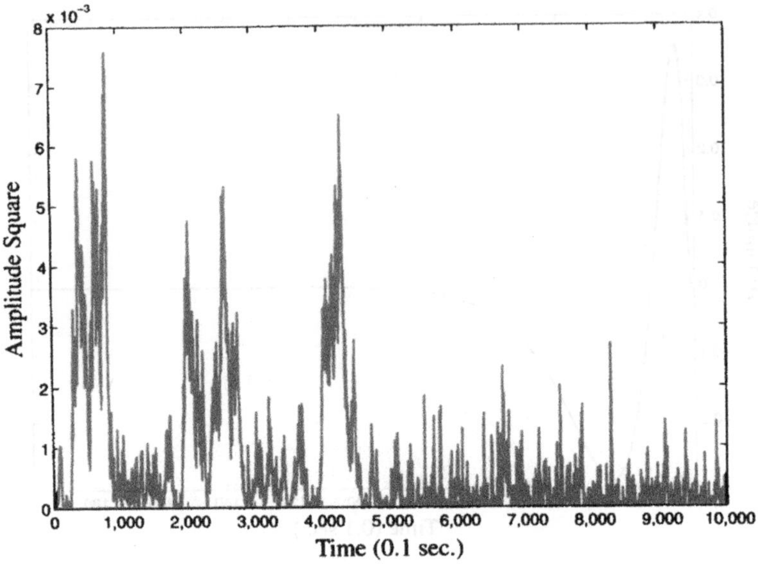

Figure 8.34 A time plot of the square of the residual error. The canceling loop was closed at sample 5,000. Ideal $G(z)$ and $Q(z)$ were used. Nonminimum-phase plant, stabilized with $k = 24$.

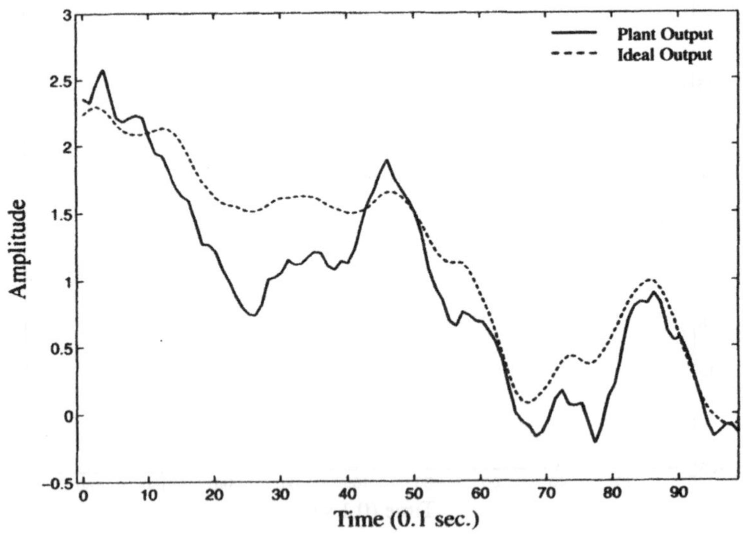

Figure 8.35 Actual plant output and ideal plant output after closure of disturbance-canceling loop. Ideal $G(z)$ and $Q(z)$ were used. Nonminimum-phase, stabilized with $k = 24$.

in Fig. 8.37. Although this is a noisy version of the ideal, shown in Fig. 8.33, it is substantially different. The extent of this difference is surprising, since the difference between the impulse responses of $G(z)$ and $\widehat{G}(z)$ is not very great.

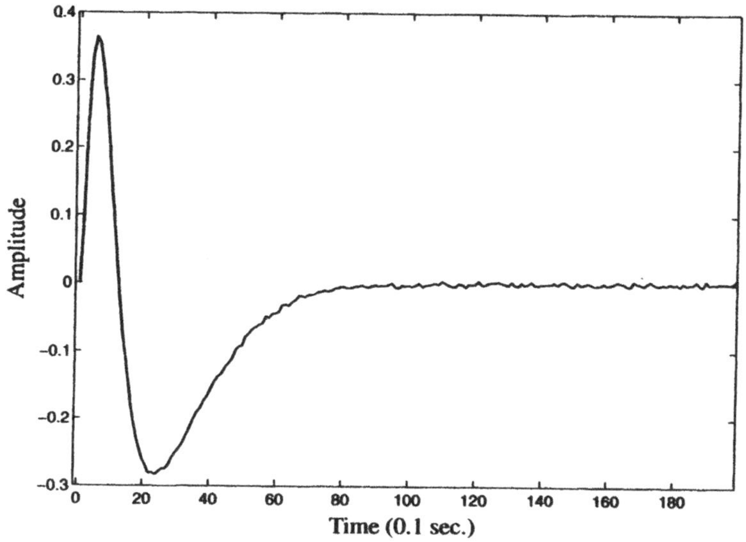

Figure 8.36 An impulse response corresponding to $\widehat{G}(z)$ obtained by DCT/LMS. Nonminimum-phase plant, stabilized with $k = 24$.

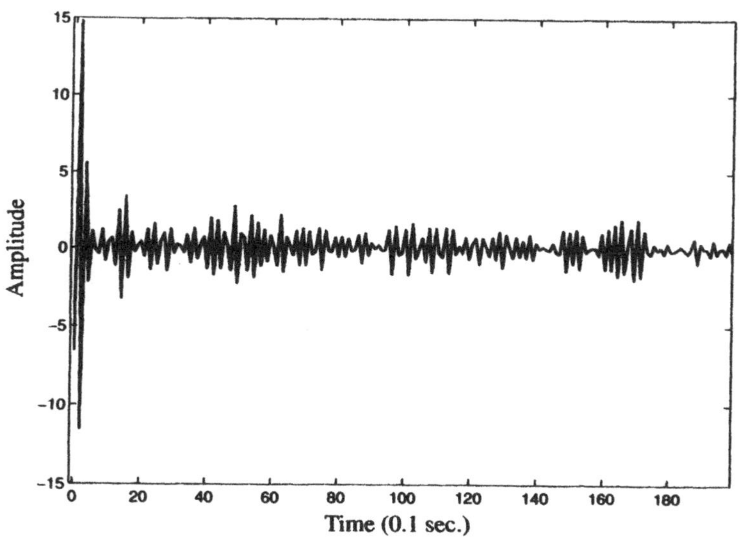

Figure 8.37 Impulse response of $Q(z)$ corresponding to $\widehat{G}(z)$ obtained by DCT/LMS. Nonminimum-phase plant, stabilized with $k = 24$.

In any event, the resulting $Q(z)$ was used in the adaptive disturbance canceler with good results. Figure 8.38 shows the square of the residual error plotted over time. The canceling loop was closed just after the 5,000th time sample. The power of the residual error was reduced by a factor of 4.44. This result is not as good as the optimal factor of 5.65, but it is not bad. The actual plant output and the ideal plant output are plotted in Fig. 8.39 after closure of the canceling loop. Once again, the difference between these plots is the residual error.

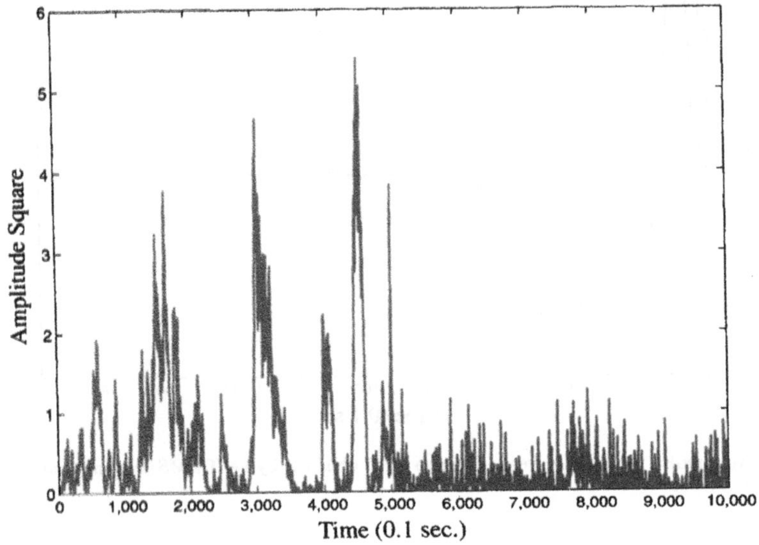

Figure 8.38 The square of the residual error. The loop was closed at sample 5,000. Adaptive $\widehat{G}(z)$ and $Q(z)$ were used. Nonminimum-phase stabilized with $k = 24$.

This experiment convinces one of the importance of having $\widehat{G}(z)$ be as close as possible to $G(z)$. Slow adaptation is the way to achieve this, and one should make the number of weights of $\widehat{G}(z)$ be no greater than needed to make a good fit.

8.14 INSENSITIVITY OF PERFORMANCE OF ADAPTIVE DISTURBANCE CANCELER TO DESIGN OF FEEDBACK STABILIZATION

In Sections 8.11 and 8.13 we analyzed two cases of plant disturbance canceling in unstable plants. However, since inverse control and disturbance canceling cannot be applied directly to an unstable plant, conventional feedback stabilization has been utilized first. The question is: How has the overall system performance been affected by the particular choice of the stabilization feedback? In other words, is it possible to have further reduction in the plant output disturbance power if different stabilization feedback is chosen?

This question is answered in Section D.2. It is shown there that, as long as the stabilization feedback itself is stable, the plant output disturbance power is independent of the

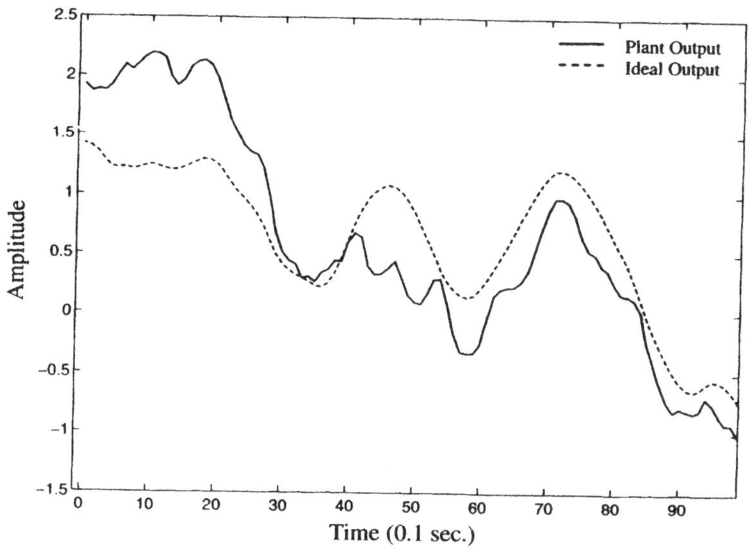

Figure 8.39 Ideal plant output and actual plant output with canceling loop closed. Adaptive $\widehat{G}(z)$ and $Q(z)$ were used. Nonminimum-phase plant was stabilized with gain set at $k = 24$.

choice of stabilization feedback. This result is very fortunate. It means that using an adaptive disturbance canceler for control of unstable plants does not create difficult problems of optimal feedback design.

SUMMARY

This chapter discussed methods for optimally reducing plant disturbance without changing plant dynamics. Feedback is used to bring plant output disturbance back to the plant input in order to effect plant disturbance canceling. The best linear least squares plant disturbance canceler may be configured this way. Although the disturbance canceler is a feedback system, the transfer function around the loop is essentially zero. For this reason, the dynamic response of the plant with or without disturbance canceling feedback is the same.

When using the system of Fig. 8.7 with white dither, the learning process for $\widehat{P}(z)$ has only one time constant, given by

$$\tau = \frac{1}{2\mu E[\delta_k^2]}, \tag{8.16}$$

where $E[\delta_k^2]$ is the dither power. When using an offline process for finding $Q(z)$, both the speed of convergence of the plant modeling algorithm and of the plant disturbance canceler are the same and this speed is determined by (8.16).

The disturbance canceling feedback has effect on both the stability of the adaptive modeling process for finding $\widehat{P}(z)$ and of the disturbance canceling loop itself. For the system of Fig. 8.7 to be stable, both processes must be stable. Stability of the disturbance canceling loop is achieved by keeping μ within the range

$$\frac{1}{n|Q|^2_{\max} E[n_k^2]} > \mu > 0, \tag{8.26}$$

where $E[n_k^2]$ is the plant output disturbance power without canceling, n is the number of weights of $\widehat{P}(z)$, and $|Q|^2_{\max}$ is the square of the largest magnitude of $Q(z)$ for values of z on the unit circle in the z-phase for real frequencies between zero and half the sampling frequency.

Stability of the plant modeling process of Fig. 8.7 is assured by keeping μ within the range

$$\frac{1}{n\left[E[\delta_k^2] + E[(u_k')^2] + E[n_k^2]\left(\frac{1}{|P(e^{j\omega})|^2}\right)_{\omega\ \text{low}}\right]} > \mu > 0, \tag{8.29}$$

where $E[(u_k')^2]$ is the power of the plant input signal and $|P(e^{j\omega})|^2_{\omega\ \text{low}}$ is the square of magnitude of the plant transfer function at low real frequencies.

By means of a heuristic proof, it was demonstrated that the equilibrium position of the vector $\widehat{\mathbf{P}}$ is the Wiener solution $\mathbf{P}$. The equilibrium position and the Wiener solution are not affected by the presence of the disturbance canceling feedback loop.

Concluding the analysis of the system of Fig. 8.7, the *overall output error power* was defined as the sum of the dither power at the plant output, the output signal distortion power caused by the modulation of the feedback loop signal due to gradient noise in the weights of $\widehat{\mathbf{P}}$, and the uncanceled plant output disturbance power. The optimal value of dither power is the value that minimizes the overall output error power. This is given by

$$E[\delta_k^2]_{\text{OPT}} = \sqrt{\frac{nE[(u_k')^2]\cdot E[n_k^2]}{2\tau\sum_{i=0}^{\infty} p_i^2}}, \tag{8.41}$$

where $E[(u_k')^2]$ is the power of u_k', the input to the plant and its disturbance canceler, as in Fig. 8.7, and $\sum_{i=0}^{\infty} p_i^2$ is the sum of squares of the impulses of the impulse response of $\mathbf{P}$. When using the optimal dither of (8.41), the minimum value of the overall output error power is

$$\begin{pmatrix}\text{overall}\\ \text{output}\\ \text{error}\\ \text{power}\end{pmatrix}_{\min} = \sqrt{\frac{2nE[(u')^2]\cdot E[n_k^2]\cdot\sum_{i=0}^{\infty} p_i^2}{\tau}} + \phi_{CC}(0), \tag{8.42}$$

where $\phi_{CC}(0)$ is the power of the uncanceled plant disturbance, defined by (8.10).

These formulas were challenged with a number of simulated experiments and were found to work well for all of the cases tested.

Adaptive canceling was applied to stabilized minimum-phase and nonminimum-phase plants. The depth of disturbance canceling was generally much greater with minimum-phase plants. In any event, the linear adaptive disturbance canceling method described in this chapter was optimal in the least squares sense. It was shown that the ability to cancel disturbance was not affected by the design of the plant stabilizer.

The next chapter presents analyses and simulation results for entire control systems with plants, plant disturbance canceling systems, and inverse controllers. The subject is called *system integration*.

Chapter 9

System Integration

9.0 INTRODUCTION

Many different control systems can be configured from the parts that have been developed in the previous chapters. Each system would have an adaptive inverse controller connected to a plant to be controlled. The plant may or may not have an attached adaptive disturbance canceler. Among these systems, one has been selected for discussion here and its block diagram is shown in Fig. 9.1. In this system, the plant uses an adaptive disturbance canceler like that of Fig. 8.7. The controller and the offline method for obtaining it are the same as those of Fig. 6.4. The adaptive process for finding $\widehat{P}_k(z)$ incorporates dithering scheme C.

This system was built in software and tested extensively. Although it has many parts, it turns out to be a solid, robust system. A straightforward start-up procedure is the following: Use the panic button to break the disturbance canceling feedback loop until $\widehat{P}_k(z)$ converges to $P(z)$, then close this loop. In the event of a sudden catastrophic change in $P(z)$ that would cause the entire system to go unstable, push the panic button until $\widehat{P}_k(z)$ converges to the new $P(z)$. Without the disturbance canceler, stability is much less of a consideration. Basically, the system will be stable as long as the plant $P(z)$ is stable.

Several questions remain about this system: the choice of dither power, the choice of μ, and the resulting time constant of the adaptive process. Finally, the output error power needs to be evaluated.

9.1 OUTPUT ERROR AND SPEED OF CONVERGENCE

There are three sources of output error or output noise: (a) uncanceled plant disturbance, (b) dither noise, and (c) effects of noisy $\Delta P_k(z)$ and $\Delta C_k(z)$ modulating the command signal $I(z)$ while it is propagating through the entire system. These errors have already been analyzed for the various parts of the system. The goal now is to put the pieces together to determine overall system performance.

We begin with a consideration of item (c) above. There are some surprises about the effects of $\Delta P_k(z)$ and $\Delta C_k(z)$ on the random modulation of the input $I(z)$ while propagating through the system. The transfer function of the overall system from input to output is

$$H_{IO}(z) = \widehat{C}_k(z)\frac{P(z)}{1 - \Delta P_k(z) \cdot z^{-1}Q_k(z)}. \tag{9.1}$$

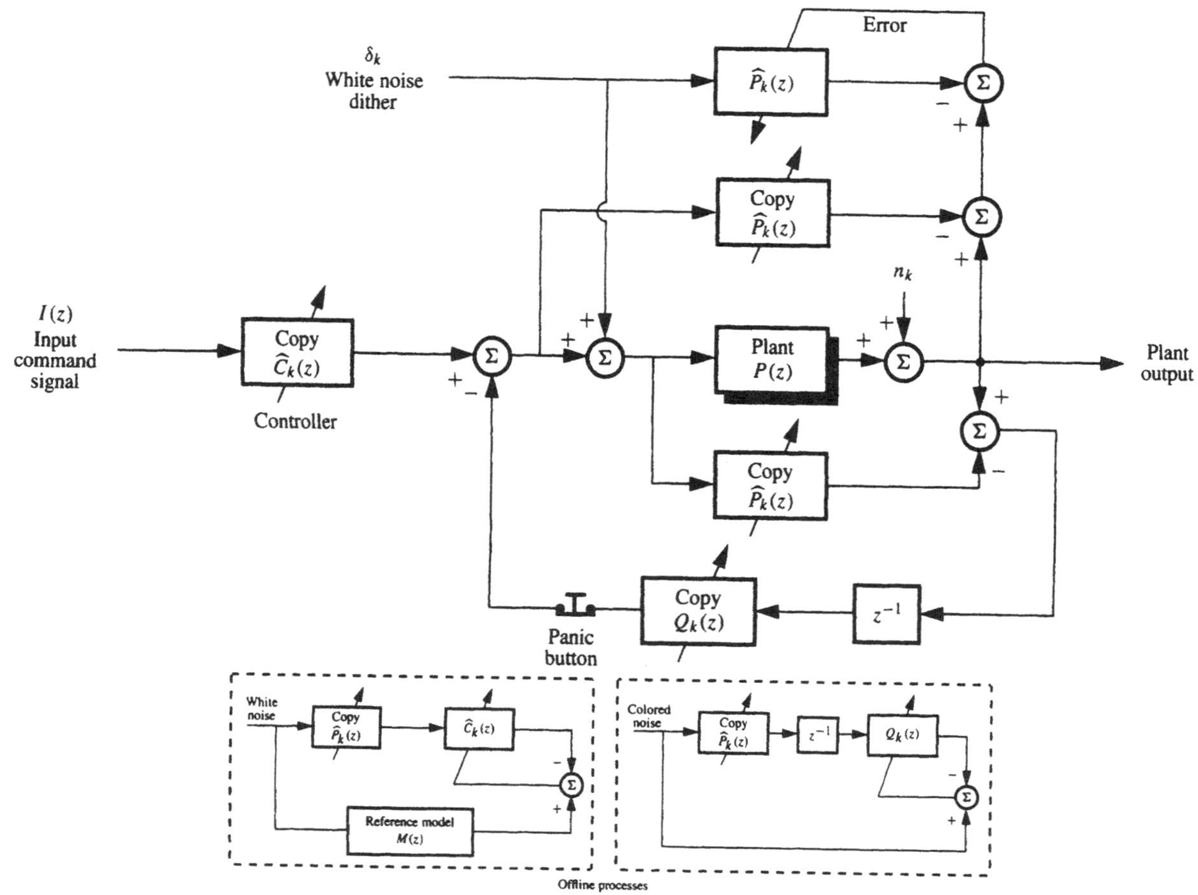

Figure 9.1 A completely integrated adaptive inverse control system using dither.

To a very good approximation,

$$\widehat{C}_k(z) \approx \frac{M(z)}{\widehat{P}_k(z)} = \frac{M(z)}{P(z) + \Delta P_k(z)}. \tag{9.2}$$

The error in $\widehat{C}_k(z)$ is caused by the error in $\widehat{P}_k(z)$, *i.e.* $\Delta P_k(z)$. The effects of this error upon $H_{IO}(z)$ can be expressed in terms of $\Delta P_k(z)$. Substituting (9.2) into (9.1) yields

$$\begin{aligned} H_{IO}(z) &\approx \frac{M(z)P(z)}{[P(z) + \Delta P_k(z)][1 - \Delta P_k(z) \cdot z^{-1} Q_k(z)]} \\ &= \frac{M(z)P(z)}{P(z) + \Delta P_k(z) - \Delta P_k(z) \cdot P(z) \cdot z^{-1} Q_k(z) - [\Delta P_k(z)]^2 z^{-1} Q_k(z)}. \end{aligned} \tag{9.3}$$

The effects of both $\Delta P_k(z)$ and $\Delta C_k(z)$ are incorporated in this expression. For small $\Delta P_k(z)$, we can neglect the last term in the denominator.

$$H_{IO}(z) \approx \frac{M(z)P(z)}{P(z) + \Delta P_k(z)[1 - P(z) \cdot z^{-1} Q_k(z)]}. \tag{9.4}$$

Assuming that

$$P(z) \cdot z^{-1} Q_k(z) \simeq 1, \tag{9.5}$$

then

$$H_{IO}(z) \approx \frac{M(z)}{1 + \frac{\Delta P_k(z)}{P(z)} \cdot [\text{small}]} \approx M(z). \tag{9.6}$$

From Eq. (9.6), one concludes that a change in $\Delta P_k(z)$ causes only a second-order change in $H_{IO}(z)$ and in turn, only a second-order effect on the plant output signal. Cancelation of the first-order effects of $\Delta P_k(z)$ and $\Delta C_k(z)$ takes place as a byproduct of the action of the plant disturbance canceling system. This conclusion, from Eq. (9.6), is based on Eq. (9.2), which is easily met in practice, and on Eq. (9.5) which can be closely met with a minimum-phase plant and approximately met with a nonminimum-phase plant.

The output noise is therefore the sum of the uncanceled plant disturbance and the dither noise. The power of the uncanceled plant disturbance is $\phi_{CC}(0)$, in accord with Eqs. (8.10) and (8.38). The various factors that enter into this residue of the plant disturbance have been described in Chapter 8. The dither noise at the plant output was derived in Chapter 8, for scheme C as (8.31):

$$\begin{pmatrix} \text{plant} \\ \text{output} \\ \text{dither} \\ \text{power} \end{pmatrix} \equiv E[\delta_k^2] \cdot \left[\sum_{i=0}^{\infty} p_i^2 \right]. \tag{9.7}$$

Therefore, the total plant output noise is

$$\begin{pmatrix} \text{total} \\ \text{plant} \\ \text{output} \\ \text{noise} \\ \text{power} \end{pmatrix} = \phi_{CC}(0) + E[\delta_k^2] \cdot \left[\sum_{i=0}^{\infty} p_i^2 \right]. \tag{9.8}$$

An expression for the time constant of the adaptive process for obtaining $\widehat{P}_k(z)$ has been derived for the system of Fig. 8.7 and is given by Eq. (8.16) as

$$\tau = \frac{1}{2\mu E[\delta_k^2]}. \tag{9.9}$$

This same formula also applies to the system of Fig. 9.1. The presence of the time-variable controller $\widehat{C}_k(z)$ does not affect the adaptive time constant of $\widehat{P}_k(z)$. To achieve rapid adaptation, either μ must be large or the dither power must be large. Making μ large may not be possible because of stability considerations, but making $E[\delta_k^2]$ large causes large noise at the plant output. One would generally like to be able to adapt rapidly and at the same time have low plant output noise power. Stability is the most important consideration, however.

Stability criteria have been obtained as (8.26) and (8.29) for the system of Fig. 8.7. The same criteria apply to the system of Fig. 9.1. The presence of $\widehat{C}_k(z)$ as the controller in Fig. 9.1 has an effect on stability in light of criterion (8.29), since the power of the plant input u_k is important to this criterion. The various factors need to be properly estimated to use (8.26) and (8.29) to calculate the stable range of μ. Once the maximum stable μ is estimated, a practical value of μ can be selected by cutting down from this maximum by a suitable factor of safety.

With μ selected, the dither power may then be selected. It is clear from (8.16) that a small dither power will result in a large time constant, and that by increasing dither power, the adaptive process will converge more rapidly. Reducing the dither power will reduce the output dither noise but slow the adaptation process. Since the uncanceled plant disturbance is irreducible, there is no point in making the output dither noise power substantially less than the uncanceled plant disturbance power. A reasonable compromise between the objectives of fast adaptation and low plant output noise would be to make the dither power at the plant output equal to the uncanceled plant disturbance power, that is, to make the two terms of (9.8) equal. As such,

$$E[\delta_k^2] = \frac{\phi_{CC}(0)}{\sum_{i=0}^{\infty} p_i^2}. \tag{9.10}$$

Now having selected μ and the dither power, the time constant of the adaptive process is given by (8.16).

The effects of dither power on system performance have already been tested experimentally for the system of Fig. 8.7. The addition of the controller in Fig. 9.1 would have only negligible effect on these results and the experiments will not be repeated here. The same is true of the previous stability studies.

9.2 SIMULATION OF AN ADAPTIVE INVERSE CONTROL SYSTEM

A number of simulated experiments were performed to test the behavior of the system of Fig. 9.1 and to compare its behavior with theoretical predictions. The plant chosen for these experiments was FIR and minimum-phase, and it had the transfer function of

$$P(z) = \sum_{i=0}^{n-1} \left(\frac{1}{2}z^{-1}\right)^i. \tag{9.11}$$

The chosen reference model was simply

$$M(z) = 1. \tag{9.12}$$

The number of weights of $\widehat{P}(z)$ was set equal to the number of weights of $P(z)$, which was chosen to be $n = 10$. The number of weights of the controller was also chosen to be $m = 10$.

The command input was a square wave with amplitude ± 1, having a period of 100 samples. The plant disturbance was simulated by taking zero-mean white noise and filtering it through a moving-average low-pass filter with 50 equal weights.

Utilizing the system of Fig. 9.1, we were able to achieve, simultaneously, cancelation of the plant disturbance and excellent control of plant dynamics.

At the beginning, we wanted to simulate the effects of imperfect plant modeling. Accordingly, we set $\widehat{P}(z)$ to be

$$\widehat{P}(z) = P(z) + 0.01(1 + z^{-1} + \cdots + z^{-9}), \tag{9.13}$$

and ceased plant model adaptation. The dither was shut off, and the plant disturbance was shut off.

Clearly, as a result of this deliberate bias in the plant model, an erroneous controller was computed. The error in the controller, in turn, might cause some excess error at the plant output.

In order to validate our prediction that the plant disturbance canceler will reduce or eliminate distortion caused by error in the weights of $\widehat{P}(z)$ and $\widehat{C}(z)$, we ran the system of Fig. 9.1 sometimes with the disturbance canceling loop open, that is,

$$Q_k(z) = 0 \quad \text{for all } k, \tag{9.14}$$

and at other times with the disturbance canceling loop closed. We have considered three cases:

a. The plant $P(z)$ was controlled by a fixed 10-weight Wiener controller based on perfect knowledge of the plant $P(z)$. The plant disturbance canceler was disconnected.

b. The plant $P(z)$ was controlled by a fixed 10-weight Wiener controller based on the biased plant estimate $\widehat{P}$ given by (9.13). The plant disturbance canceler was disconnected.

c. The plant $P(z)$ was controlled by the same fixed 10-weight Wiener controller as was used in (b), based on the biased plant estimate. Although there was no plant disturbance, the plant disturbance canceler was connected. $Q(z)$ was fixed, based on the same biased plant estimate.

In all three cases, the overall system impulse response (the convolution of plant and its controller) was found and compared with the impulse response of the reference model $M(z)$. The resulting impulse responses (truncated to their first 11 impulses) are presented in Table 9.1. Also, the output distortion power (i.e., the output error power that exists when both plant disturbance power and dither power are equal to zero) was measured for all three cases.

The first line of Table 9.1 gives the impulse response of the reference model, in accord with (9.2). The second line gives the overall impulse response obtained in case (a) (i.e., for the best 10-weight controller based on perfect knowledge of $P(z)$). The match with the

TABLE 9.1 IMPULSE RESPONSES OF ADAPTIVE INVERSE CONTROL SYSTEMS WITH BIASED PLANT MODEL, WITH AND WITHOUT PLANT DISTURBANCE CANCELING

	Impulse Number										
	1	2	3	4	5	6	7	8	9	10	11
Impulse Response of Reference Model	1	0	0	0	0	0	0	0	0	0	0
Impulse Response for Ideal Controller	1	0	0	0	0	0	0	0	0	0	-0.00097
Impulse Response for Biased Plant Model, no Disturbance Canceling	0.99	-0.005	-0.005	-0.005	-0.005	-0.005	-0.005	-0.005	-0.005	-0.005	-0.003
Impulse Response for Biased Plant Model, with Disturbance Canceling	0.99	0.005	0	0	0	0	0	0	0	0	0.001

reference model was almost perfect (imperfect due to the fact that the controller was allowed to have only 10 weights). The error was very small, close to zero:

$$\begin{pmatrix} \text{variance of output} \\ \text{error for controller} \\ \text{based on perfect } \widehat{P}(z) \end{pmatrix} = 9.64(10)^{-7}. \tag{9.15}$$

When the controller was computed for case (b) based on the biased plant model (9.13), the error power increased considerably, to the value

$$\begin{pmatrix} \text{variance of output} \\ \text{error for controller} \\ \text{based on biased } \widehat{P}(z) \end{pmatrix} = 2.82(10)^{-3}. \tag{9.16}$$

Line 3 of Table 9.1 shows the overall impulse response for case (b), and it differs more strongly from the reference model impulse response than does the impulse response for case (a).

Finally, the disturbance canceling loop was closed. Line 4 of Table 9.1 shows the overall impulse response for case (c). By comparing lines 3 and 4, we can see that the introduction of plant noise canceling indeed reduced the output error. The new error variance was

$$\begin{pmatrix} \text{variance of output} \\ \text{error for controller} \\ \text{based on biased } \widehat{P}(z), \\ \text{with plant disturbance} \\ \text{canceler on} \end{pmatrix} = 1.42(10)^{-4}. \tag{9.17}$$

Comparing (9.17) with (9.16) confirms the fact that the effects of the errors in the controller, due to plant model errors, tend to be canceled and mitigated by corresponding errors in the plant disturbance canceling loop. The variance of the output error was reduced by more than an order of magnitude by the disturbance canceling loop.

Note that error cancelation was not perfect, however. The output error variance for case (c), given by (9.17), was higher than that for case (a), given by (9.15). In case (a), the controller was almost ideal. There are three reasons that limit the capability of the disturbance canceler to cause the effects of $\Delta\mathbf{P}$ and $\Delta\mathbf{C}$ to cancel. These are

1. $\Delta\mathbf{P}$ is supposed to be small. If it is not so small, second-order effects could begin to cause output distortion;

2. The feedback filter Q has only a finite length, chosen by the system designer;

3. The disturbance canceling feedback includes, by its nature, one unit of delay. As a result, the disturbance canceling loop cannot correct any error in the first impulse of the impulse response.

In order to improve the disturbance canceler and enhance the cancelation of dynamic system error, one could adapt $\widehat{P}$ with smaller μ to reduce weight noise, one could use a longer Q where this is indicated to be advantageous, and one could run with a higher sampling rate so that error in the first samples of the impulse response would be less significant. Doing all this would make the variance of the dynamic system error extremely small.

Additional experiments were run with the plant disturbance source on, the dither on, and everything adapting in a normal way. The entire system of Fig. 9.1 was operational. The characteristics of the plant, plant disturbance, and dither were as described above. The command input was the square wave, also described above. The reference model was $M(z) = 1$. With this set of experiments, $\widehat{\mathbf{P}}$ was not fixed and its error $\Delta\mathbf{P}$ varied randomly over time k. The variations in $\widehat{\mathbf{P}}$ were caused by gradient noise.

The derivation of Eq. (9.6), the equation that led us to the conclusion that the disturbance canceling loop would work to remove the effects of $\Delta\mathbf{P}$ and $\Delta\mathbf{C}$ upon the dynamic system error, is based on the concept of transfer function. For the transfer function of a filter or of a system to be algebraically meaningful, it is necessary for that system to be linear and time invariant. While adapting, it is clear that $\widehat{\mathbf{P}}$ is not time invariant. Although the derivation of Eq. (9.6) for the time variable $\widehat{\mathbf{P}}$ and $\widehat{\mathbf{C}}$ is not strictly valid, if one were to operate with a small μ, this would cause $\Delta\mathbf{P}$ and $\Delta\mathbf{C}$ to change slowly and would make the idea of transfer function be reasonable. In any event, experimental evidence seems to indicate that the effects of $\Delta\mathbf{P}$ and $\Delta\mathbf{C}$ upon the dynamic system error cancel while adapting $\widehat{\mathbf{P}}$ at slow or at even moderate speeds.

With the basic system of Fig. 9.1 operating and adapting normally, experiments were done in the following manner. In addition to the basic system, several forms of auxiliary systems were simulated having the same architecture as the basic system but with their filters $\widehat{\mathbf{P}}_k$, $\mathbf{Q}_k$, and $\widehat{\mathbf{C}}$ not adapting normally. Their weights were obtained by copying the respective weights from the basic system. Each auxiliary system was used to explore one of the critical output components of the basic system. For example, one of the auxiliary systems was used to determine and measure the plant disturbance at the control system output. This was done by turning on the plant disturbance in the auxiliary system but leaving the dither and the command input signal off. The output of the auxiliary system contained the component of interest. Another auxiliary system was used to observe and measure the dither component of the control system output. This was done by turning off the command input signal and the plant disturbance, and turning on the dither. Another auxiliary system was used to explore the dynamic response of the control system to the command input. Everything was turned off except the command input, and the output of the auxiliary system was the dynamic response being sought. This dynamic response was able to be compared and checked against the output of the reference model.

The distortion components in this dynamic response which are due to $\Delta\mathbf{P}$ and $\Delta\mathbf{C}$ are supposed to cancel from the action of the plant disturbance canceler. An additional check of this idea was accomplished by comparing the dynamic responses at the output of the auxiliary system when the disturbance canceling loop in the auxiliary system was connected, and when it was disconnected.

We have done all of these experiments in order to check the ability of the derived formulas to predict distortion in the control system output due to uncanceled plant disturbance, dither noise, and dynamic response distortion.

Running the basic system of Fig. 9.1 in a normal way and running an auxiliary system at the same time with the dither and command input off, the power of the uncanceled plant disturbance was measured to be

$$0.0395. \tag{9.18}$$

The formula for the power of the uncanceled plant disturbance is given by (8.10). To use this formula, we need to know something of the spectrum of the plant disturbance and the transfer function of the plant. For this experiment, we knew both of these parameters. The plant transfer function is given by (9.11). The plant disturbance has unit power and was generated by filtering white noise through a 50-weight filter, equal weights. Plugging into (8.10), we calculated the theoretical power of the uncanceled plant disturbance to be

$$0.0396. \tag{9.19}$$

With the command input and plant disturbance turned off in the auxiliary system and the white dither with unit power turned on, the power of the dither in the output of the auxiliary system was measured to be

$$1.2818. \tag{9.20}$$

A formula for the dither noise in the plant output is Eq. (9.7). Since the parameters in this equation were known perfectly, the dither output power was calculated theoretically to be

$$1.3333. \tag{9.21}$$

To study dynamic response distortion, the auxiliary system was operated with plant disturbance and dither turned off, while the command input was turned on. In the basic system, the value of μ was chosen to be 0.0001, the plant disturbance was white and filtered by 50 equal weights of an FIR filter while scaled to have unit output power, and the dither was chosen to be white with unit power. The auxiliary system was first operated with its disturbance canceler on. The difference between the plant output and the output of $M(z)$ was very small, having a power of

$$1.29(10)^{-5}. \tag{9.22}$$

Theoretically, the dynamic response distortion was zero with the disturbance canceler operating. With the disturbance canceler turned off, the difference between the plant output and the output of $M(z)$ was much larger, having a power of

$$2.97(10)^{-4}. \tag{9.23}$$

The difference between (9.23) and (9.22) indicates that the disturbance canceler was able to reduce the dynamic response distortion by a factor of 23, or a reduction of 13.6 dB!

9.3 SIMULATION OF ADAPTIVE INVERSE CONTROL SYSTEMS FOR MINIMUM-PHASE AND NONMINIMUM-PHASE PLANTS

Adaptive inverse control systems have been simulated to control both the minimum-phase and nonminimum-phase plants that have been experimented with in the previous chapters. These simulations did not use dither to obtain $\widehat{G}(z)$. The controller in each case was driven by the command input signal. The controller output was used by the adaptive process to find $\widehat{G}(z)$.

A block diagram of the adaptive inverse control system for the nonminimum-phase plant is shown in Fig. 9.2. The block diagram for the minimum-phase plant is the same except that the compensating network is omitted (it is not needed to stabilize the minimum-phase plant).

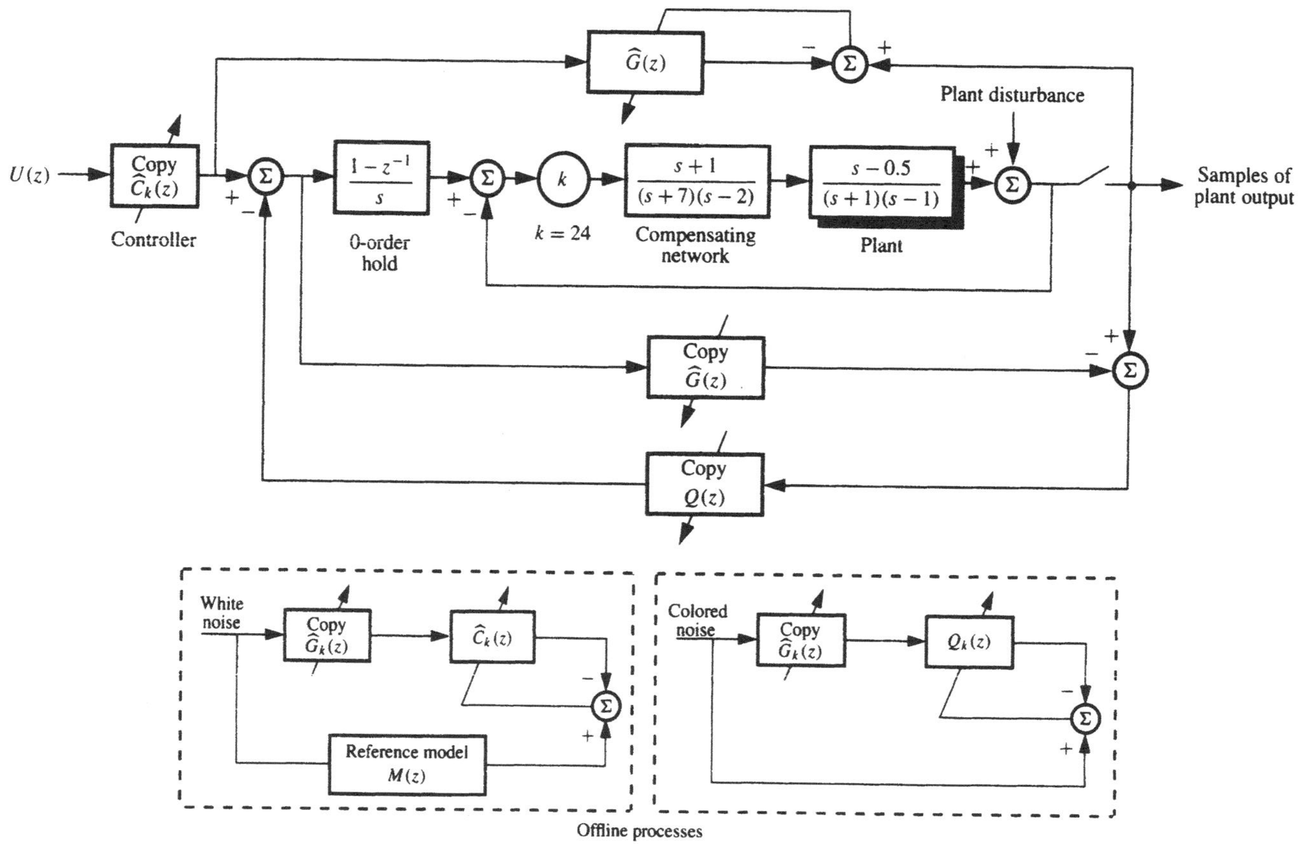

Figure 9.2 An integrated adaptive inverse control system without dither for nonminimum-phase plant.

The minimum-phase and nonminimum-phase plants were introduced in Chapter 1. Experimental results were given in Section 1.2. These results demonstrated dynamic control of the plants and showed cancelation of plant disturbances. We suggest that Section 1.2 be reviewed at this time. The results of Section 1.2 were obtained with the system block-diagrammed in Fig. 9.2. These conclusions are a culmination of our study of adaptive inverse control for linear SISO plants.

9.4 SUMMARY

One fine result obtained in this chapter is the following: The effects of $\Delta\mathbf{P}$ and $\Delta\mathbf{C}$ cancel with regard to distortion in the dynamic system response of the adaptive inverse control system of Fig. 9.1. This result is surprising and remarkable.

In this chapter, we studied the control system of Fig. 9.1. Distortion in its output is caused by dither noise, uncanceled plant disturbance, and the dynamic effects of $\Delta\mathbf{P}$ and $\Delta\mathbf{C}$. Since the latter cancel, the total plant output noise power is the sum of the dither power at the plant output and the power of the uncanceled plant disturbance.

$$\begin{pmatrix} \text{total} \\ \text{plant} \\ \text{output} \\ \text{noise} \\ \text{power} \end{pmatrix} = \phi_{CC}(0) + E[\delta_k^2] \cdot \left[\sum_{i=0}^{\infty} p_i^2 \right]. \tag{9.8}$$

The dither power at the plant output is

$$\begin{pmatrix} \text{plant} \\ \text{output} \\ \text{dither} \\ \text{power} \end{pmatrix} = E[\delta_k^2] \cdot \left[\sum_{i=0}^{\infty} p_i^2 \right]. \tag{9.7}$$

To use these formulas, one needs to have some idea of the plant impulse response in order to be able to estimate $\sum_{i=0}^{\infty} p_i^2$. In addition, the power of the uncanceled plant disturbance is $\phi_{CC}(0)$, given by (8.10). A discussion of how to estimate $\phi_{CC}(0)$ is given in Chapter 8, where Eq. (8.10) is introduced.

In Chapter 8, we obtained Eq. (8.41) which gave us a formula for optimal dither power. This formula should be used when there is dynamic system error due to $\Delta\mathbf{P}$. Generally, stronger dither gives smaller $\Delta\mathbf{P}$ but at the same time, more noise at the plant output. For the system of Fig. 9.1, we are not greatly concerned about $\Delta\mathbf{P}$ since the dynamic system error due to $\Delta\mathbf{P}$ is canceled by the error $\Delta\mathbf{C}$ in the controller. Therefore, Eq. (8.41) is irrelevant to the system of Fig. 9.1.

To choose an appropriate level of dither power for the system of Fig. 9.1, we recommend that the dither noise power at the plant output be made equal to the power of the uncanceled plant disturbance, which is irreducible. There is no point in making the dither power much smaller. Accordingly,

$$E[\delta_k^2] = \frac{\phi_{CC}(0)}{\sum_{i=0}^{\infty} p_i^2}. \tag{9.10}$$

The same stability criteria established for the system of Fig. 8.7 apply to the system of Fig. 9.1. The parameter μ must be chosen to assure the stability of the plant modeling process and stability of the plant noise canceling loop. The stability criteria for the system of Fig. 9.1 are therefore

$$\frac{1}{n|Q|^2_{\max} E[n_k^2]} > \mu > 0, \quad \text{and} \tag{8.26}$$

$$\frac{1}{n\left[E[\delta_k^2] + E[(u_k')^2] + E[n_k^2]\left(\frac{1}{P(e^{j\omega})}\right)_{\omega \text{ low}}\right]} > \mu > 0. \tag{8.29}$$

The terms are defined in Chapter 8 where these expressions are derived. How to estimate the parameters is explained in the summary of Chapter 8.

An expression for the time constant of the system of Fig. 8.7 has been obtained in Chapter 8. This same formula gives the proper time constant for estimation of $\widehat{\mathbf{P}}_k$ in the system of Fig. 9.1. The time constant is

$$\tau = \frac{1}{2\mu E[\delta_k^2]}. \tag{8.16}$$

This time constant determines the learning rate for the entire system because $\widehat{\mathbf{C}}_k$ and $\mathbf{Q}_k$, the other variable parts of the system, are obtained essentially instantly from $\widehat{\mathbf{P}}_k$ by offline processes.

We now have completed our study of linear single-input single-outputi adaptive inverse control systems. In the next chapter, we shall consider multiple-input multiple-output adaptive inverse control systems.

Chapter 10

Multiple-Input Multiple-Output (MIMO) Adaptive Inverse Control Systems

10.0 INTRODUCTION

All the systems described thus far have been of the single-input single-output (SISO) type. The purpose of this chapter is to show how the basic ideas of adaptive inverse control can be extended to apply to MIMO systems. The building blocks of such systems have transfer functions that are matrices of transfer functions. Commutability of transfer functions that is possible with linear time-invariant SISO systems is not possible with the more general MIMO systems, since the order of matrix multiplication cannot in general be commuted. The offline methods used in previous chapters to find $[\widehat{C}_k(z)]$ and $[Q_k(z)]$ depend on commutability of transfer functions, and therefore cannot be used with MIMO systems. New methods for finding $[\widehat{C}_k(z)]$ and $[Q_k(z)]$ will be required.

Before developing the techniques of adaptive inverse control for MIMO systems, we begin with a brief introduction to the basic concepts of linear MIMO systems and their representation.

10.1 REPRESENTATION AND ANALYSIS OF MIMO SYSTEMS

Figure 10.1 shows a linear dynamic MIMO filter. Its array of K-inputs, after z-transforming, can be represented by the column vector $[F(z)]$. Its array of outputs, having the same number of elements as the input array, is represented by the column vector $[G(z)]$. The transfer function of the MIMO filter is represented by a square $K \times K$ matrix of transfer functions:

$$[H(z)] = \begin{bmatrix} H_{11}(z) & H_{12}(z) & \cdots & H_{1K}(z) \\ H_{21}(z) & H_{22}(z) & \cdots & H_{2K}(z) \\ & \vdots & & \\ H_{K1}(z) & H_{K2}(z) & \cdots & H_{KK}(z) \end{bmatrix}. \qquad (10.1)$$

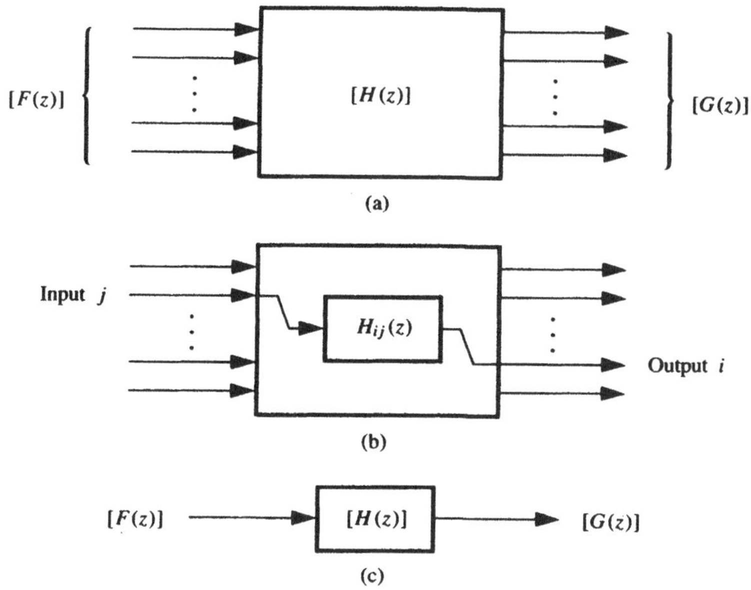

Figure 10.1 A linear MIMO filter.

The output vector can be expressed as

$$[G(z)] = [H(z)][F(z)]. \tag{10.2}$$

Each output is a linear combination of filtered versions of all the inputs. The transfer function from input j to output i is $H_{ij}(z)$.

A schematic diagram of $[H(z)]$ is shown in Fig. 10.1(a). The signal path from input line j to output line i is illustrated in Fig. 10.1(b). A block diagram of the MIMO filter is shown in Fig. 10.1(c). The input vector is $[F(z)]$. The output vector $[G(z)]$ is equal to $[H(z)][F(z)]$. The overall transfer function of the system is $[H(z)]$.

Other configurations of MIMO filters are shown in Fig. 10.2. Filters $[H_1(z)]$ and $[H_2(z)]$ are in parallel in Fig. 10.2(a). The input signal for this system is $[F(z)]$. The output signal is the sum of two signals, $[H_1(z)][F(z)]$ and $[H_2(z)][F(z)]$. The output is $[H_1(z) + H_2(z)][F(z)]$. The transfer function of this system is therefore equal to $[H(z)] = [H_1(z) + H_2(z)]$. In Fig. 10.2(b), $[H_1(z)]$ and $[H_2(z)]$ are in cascade. Following signals through this system yields certain facts. The signal at the input node A is $[F(z)]$, and the signal at node B is $[H_1(z)][F(z)]$. The signal at the output node C is $[H_2(z)][H_1(z)][F(z)]$. The transfer function of this system is therefore equal to $[H(z)] = [H_2(z)][H_1(z)]$. This is *the product of the matrix transfer functions, in reverse order to the signal flow*.

In Fig. 10.3, a system with a feedback self-loop is shown. To find the transfer function of this system, we note that the output $[G(z)]$ can be expressed as

$$[G(z)] = [F(z)] + [H_1(z)][G(z)]. \tag{10.3}$$

Combining terms,

$$[\mathbf{I} - H_1(z)][G(z)] = [F(z)]. \tag{10.4}$$

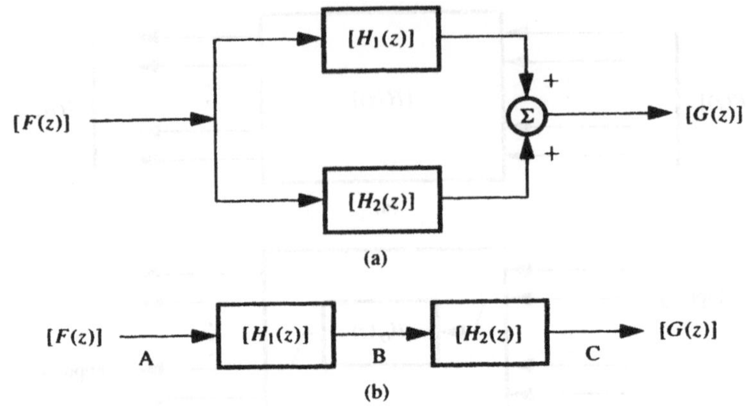

Figure 10.2 MIMO filters in parallel and cascade.

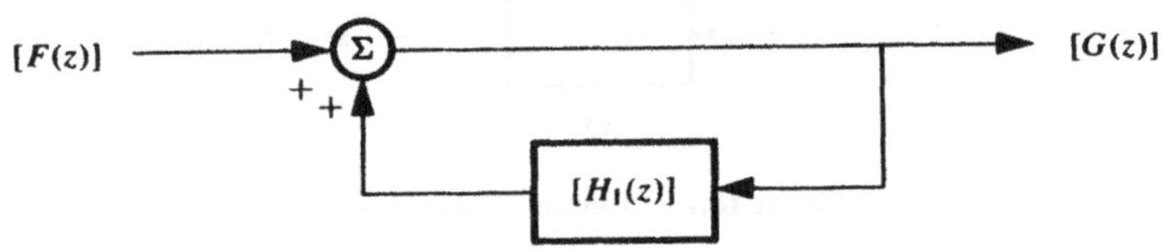

Figure 10.3 A MIMO feedback loop.

Premultiplying both sides by the inverse of $[\mathbf{I} - H_1(z)]$ gives

$$[G(z)] = [\mathbf{I} - H_1(z)]^{-1}[F(z)]. \tag{10.5}$$

The transfer function of the system is therefore

$$[H(z)] = [\mathbf{I} - H_1(z)]^{-1}. \tag{10.6}$$

Another MIMO feedback system is shown in Fig. 10.4. This system can be reduced to find the transfer function by following a number of steps, illustrated in Figs. 10.4(a), 10.4(b), and 10.4(c). The original system is shown in Fig. 10.4(a). The input is $[F(z)]$ and the output is $[G(z)]$. The diagram is redrawn in Fig. 10.4(b), with the same input causing the same output. The self-loop is the cascade of two branches whose transfer functions are combined in reverse order in Fig. 10.4(c). The overall transfer function may now be obtained by inspection, since the system is reduced to a cascade of a self-loop and a branch $[H_1(z)]$. Taking transfer functions of this cascade in reverse order, the overall transfer function is

$$[H(z)] = [H_1(z)][\mathbf{I} - H_2(z)H_1(z)]^{-1}. \tag{10.7}$$

The same system can be reduced in another way, as illustrated in Fig. 10.5. The original system is shown in Fig. 10.5(a). A step in the reduction is shown in Fig. 10.5(b). Once again, the input $[F(z)]$ produces the output $[G(z)]$. The self-loop is simplified in Fig. 10.5(c), and the transfer function can be written by inspection as

$$[H(z)] = [\mathbf{I} - H_1(z)H_2(z)]^{-1}[H_1(z)]. \tag{10.8}$$

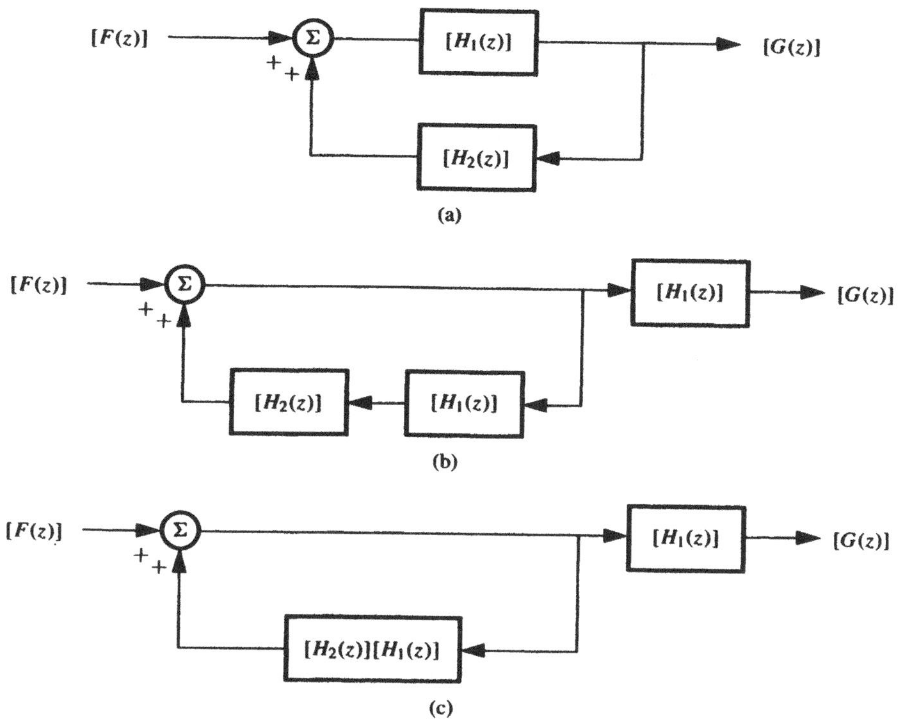

Figure 10.4 Reduction of a MIMO feedback system: (a) Original system; (b) Equivalent system; (c) Simplified equivalent system.

This result should be the same as (10.7), that is,

$$[H_1(z)][\mathbf{I} - H_2(z)H_1(z)]^{-1} = [\mathbf{I} - H_1(z)H_2(z)]^{-1}[H_1(z)]. \tag{10.9}$$

Equation (10.9) is an identity, as the reader can easily verify.

One more example will help to solidify our understanding of MIMO systems. Figure 10.6 shows a system with two feedback loops. The original system is shown in Fig. 10.6(a). The two feedback loops become self-loops in Fig. 10.6(b) without any changes to the input-output transfer function. The self-loops are simplified in Fig. 10.6(c), and their transfer functions are summed in Fig. 10.6(d). From here, the transfer function can be written by inspection as

$$[H(z)] = [H_3(z)][\mathbf{I} - H_1(z)H_2(z) - H_4(z)H_3(z)]^{-1}[H_1(z)]. \tag{10.10}$$

With this brief introduction, we are now prepared to develop adaptive inverse controls for MIMO systems.

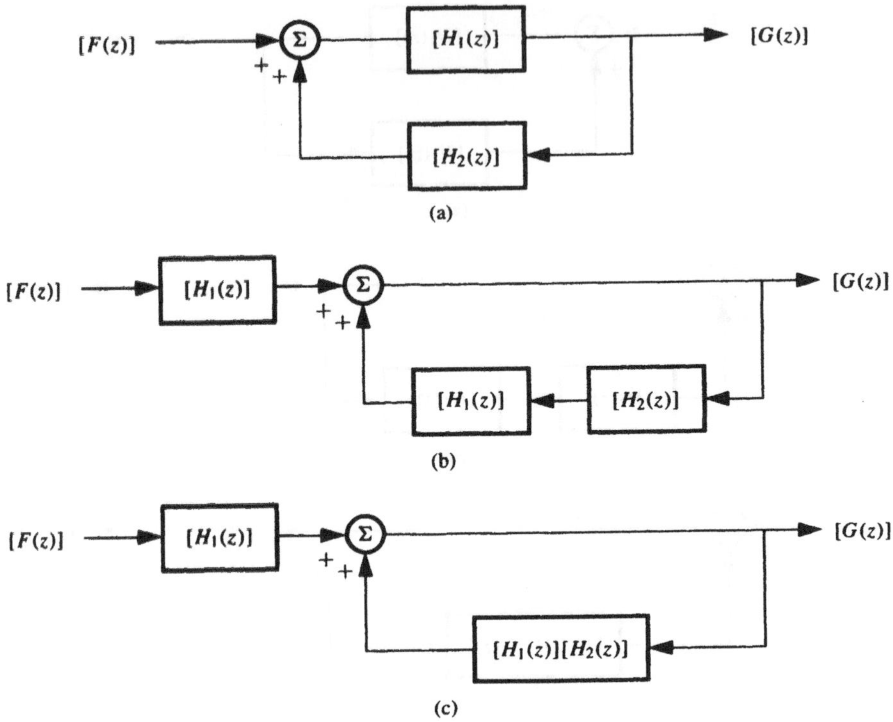

Figure 10.5 Reduction of a MIMO feedback system by another approach: (a) Original system; (b) Equivalent system; (c) Simplified equivalent system.

10.2 ADAPTIVE MODELING OF MIMO SYSTEMS

A MIMO plant to be modeled may be represented by the following transfer function:

$$[P(z)] = \begin{bmatrix} P_{11}(z) & P_{12}(z) & \cdots & P_{1K}(z) \\ P_{21}(z) & P_{22}(z) & \cdots & P_{2K}(z) \\ \vdots & \vdots & & \vdots \\ P_{K1}(z) & P_{K2}(z) & \cdots & P_{KK}(z) \end{bmatrix}. \tag{10.11}$$

The plant input is an array of signals whose transform is represented by the vector

$$[U(z)] = \begin{bmatrix} U_1(z) \\ U_2(z) \\ \vdots \\ U_K(z) \end{bmatrix}. \tag{10.12}$$

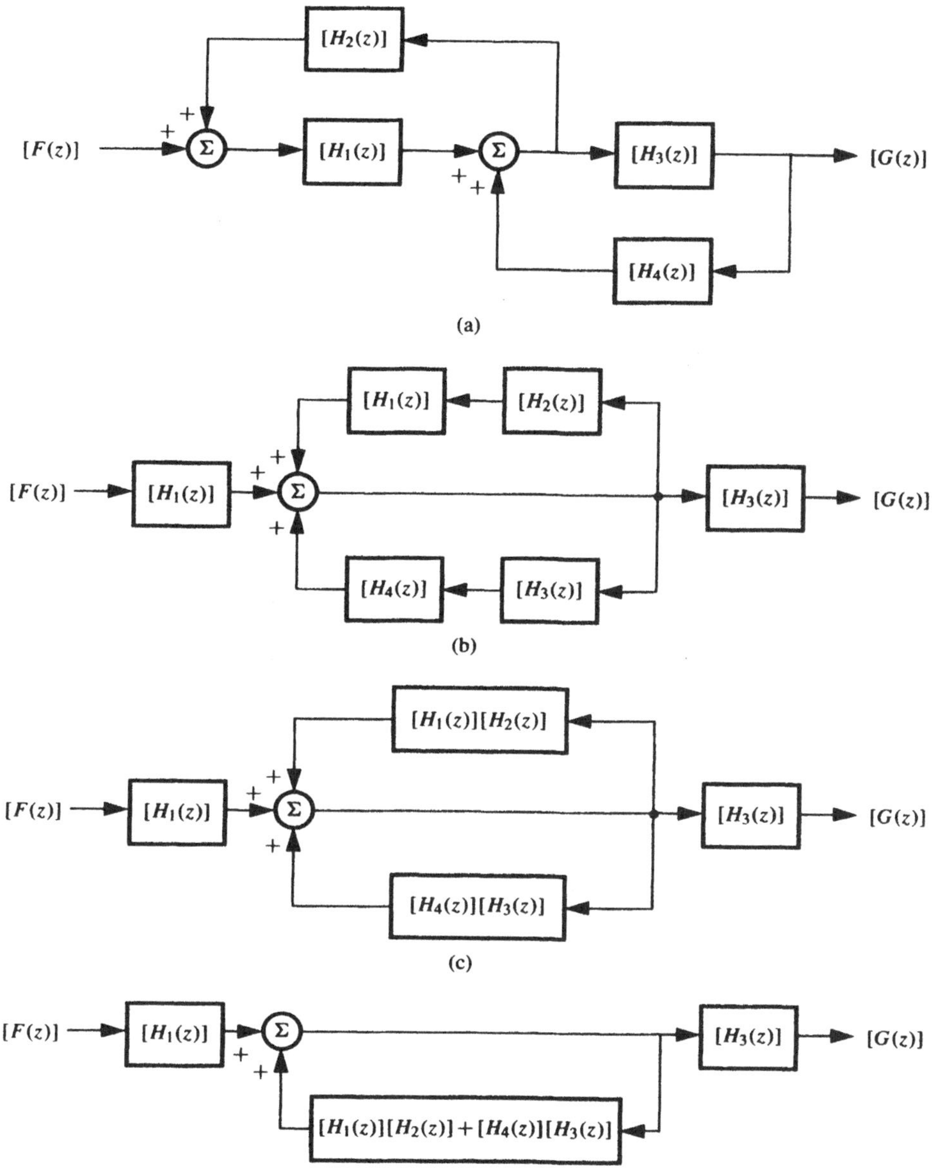

Figure 10.6 Reduction of a MIMO feedback system with two loops: (a) Original system; (b) Equivalent system; (c) Simplified equivalent system; (d) Further simplified equivalent system.

The plant output is an array of signals whose transform is represented by the vector

$$[Y(z)] = \begin{bmatrix} Y_1(z) \\ Y_2(z) \\ \vdots \\ Y_K(z) \end{bmatrix}. \tag{10.13}$$

The plant output and plant input are related by the plant transfer function:

$$[Y(z)] = [P(z)][U(z)]. \tag{10.14}$$

10.2.1 Adaptive MIMO Modeling Using Scheme B

The plant can be adaptively modeled, using independent dither inputs, as shown for the two-input two-output case of Fig. 10.7. Use of uncorrelated dither in the modeling process is necessary when the plant inputs alone cannot be used, that is, when they are correlated with each other or when they are not persistently exciting. The dither scheme illustrated in Fig. 10.7 is a MIMO version of scheme B, which is shown in Fig. 4.3(b) for the SISO case. A more succinct picture of the dither scheme is shown in Fig. 10.8, making use of a vector block diagram.

The modeling process illustrated in Fig. 10.7 involves the adaptation of four separate filters, $\widehat{P}_{11}$, $\widehat{P}_{12}$, $\widehat{P}_{21}$, and $\widehat{P}_{22}$. As these filters converge, their transfer functions should closely approximate the respective plant transfer functions P_{11}, P_{12}, P_{21}, and P_{22}. Note that a common error signal is used in the adaptation of $\widehat{P}_{11}$ and $\widehat{P}_{12}$, and that another common error signal is used in the adaptation of $\widehat{P}_{21}$ and $\widehat{P}_{22}$. It seems that the adaptation process for these filters should be affected by their interconnections, and a simple analysis shows this is indeed the case. Of concern are issues such as stability of the adaptive process, speed of convergence, misadjustment, and noise in the weights.

Figures 10.9, 10.10, 10.11, and 10.12 have been devised as an aid in addressing these issues. Figure 10.9 shows one-half of the adaptive process illustrated in Fig. 10.7. This half operates independently of the other half and can be studied all by itself. The natural driving signal, the *controller output*, is not used for plant identification by scheme B. The dither signal serves this purpose. In fact, the natural driving signal acts as noise to the adaptive identification process. Figure 10.10 is equivalent to Fig. 10.9, except that here the interference due to the controller output is accounted for by an additive noise component at the plant output, and it appears in addition to the plant disturbance.

Assume that all of the filters of Fig. 10.10 are FIR and that they all are of length n time samples. Assume that the two dither signals of Fig. 10.10 are white and of equal power. Figure 10.11 shows how a single white dither input could be used under these circumstances instead of the two dither inputs of Fig. 10.10. Two dither signals are created from a single one by using a delay of n time samples. In this way, all of the signals at the taps of the tapped delay lines of $\widehat{P}_{11}$ and $\widehat{P}_{12}$ will be mutually uncorrelated in Fig. 10.11 just as they are in Fig. 10.10. The results of adaptation in both systems will therefore be the same.

The system of Fig. 10.11 can be redrawn as shown in Fig. 10.12. The two adaptive filters $\widehat{P}_{11}$ and $\widehat{P}_{12}$ are now combined to comprise a single adaptive filter of length $2n$ time samples. Assume for simplicity that the values of μ are set to be the same for $\widehat{P}_{11}$ and $\widehat{P}_{12}$.

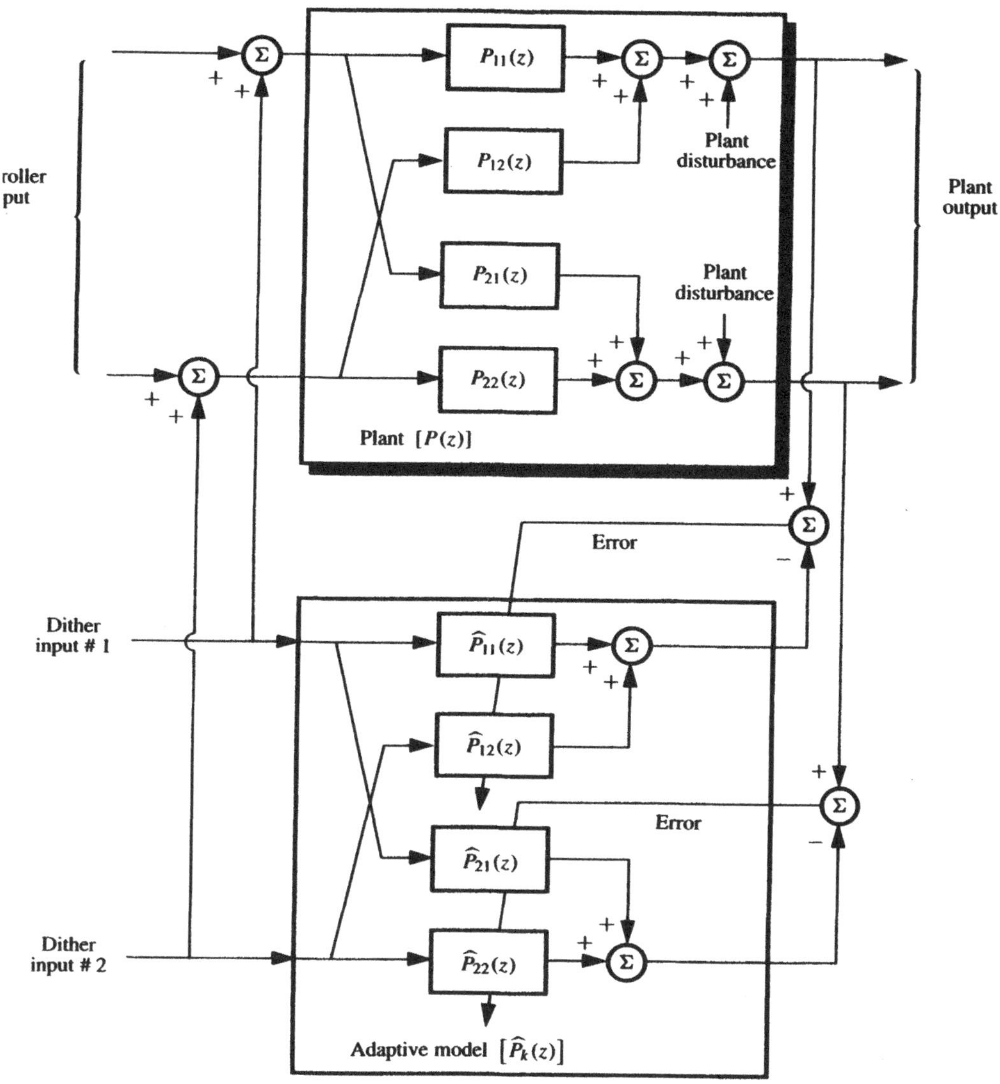

Figure 10.7 Modeling a two-input two-output system in accord with scheme B.

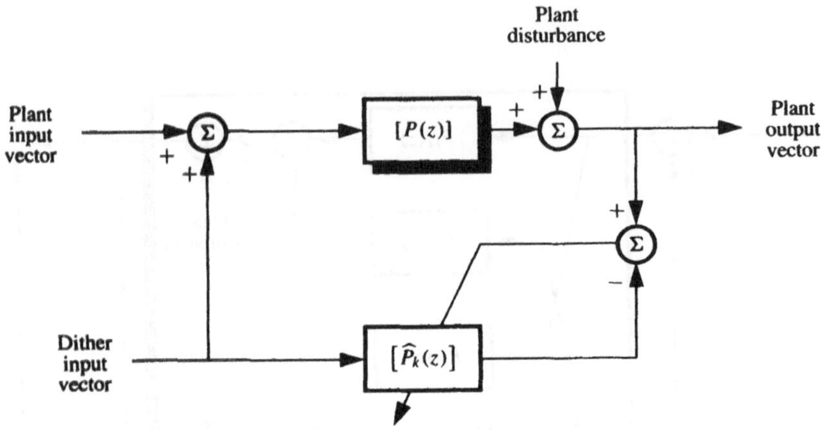

Figure 10.8 Simplified block diagram for modeling a MIMO plant based on scheme B.

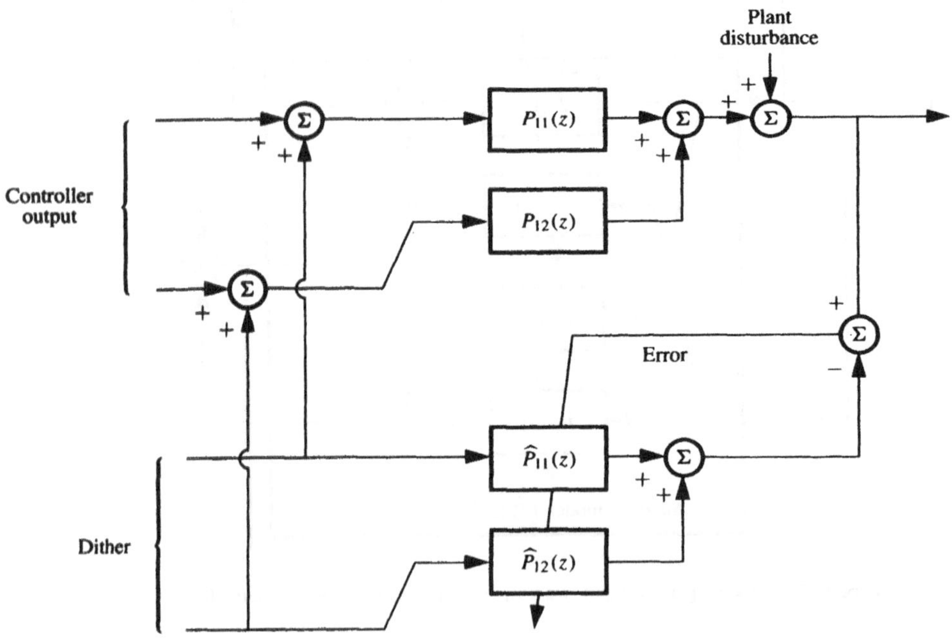

Figure 10.9 A part of the two-channel MIMO modeling process.

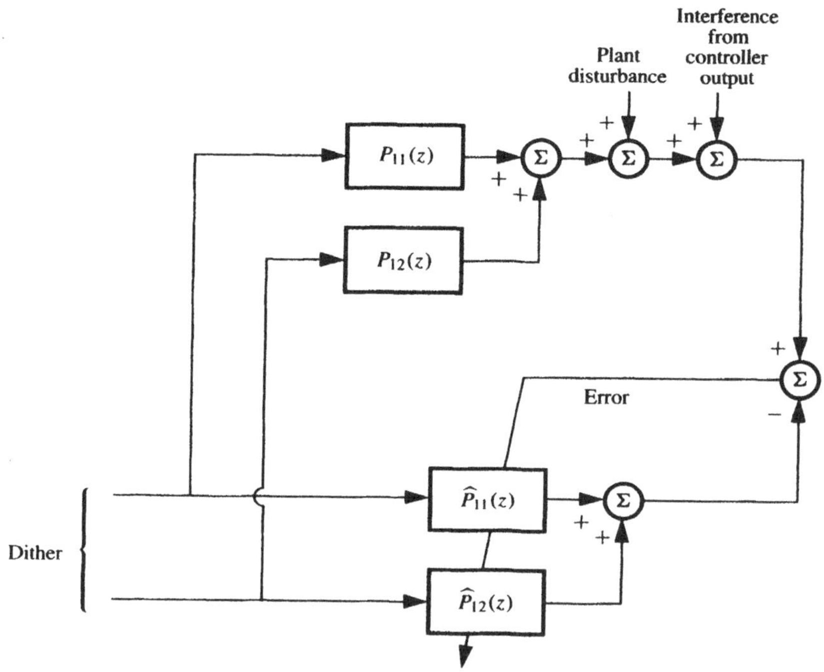

Figure 10.10 A part of the two-channel MIMO modeling process, simplified.

The issues of stability, and convergence rate, and misadjustment and noise in the weights are the same for the system of Fig. 10.12 as they are for the system of Fig. 10.9, under the conditions assumed above. Note that the adaptive behavior in Fig. 10.9 is the same as that in Fig. 10.7. So, to understand the adaptive behavior of the system of Fig. 10.7, one needs only to analyze the behavior of the system of Fig. 10.12. This can be done using existing methodology. Formulas for time constant, stable range of μ, and misadjustment are given in Appendix B. Since the length of each of the adaptive filters $\widehat{P}_{11}$, $\widehat{P}_{12}$, $\widehat{P}_{21}$, $\widehat{P}_{22}$ in Fig. 10.7 is n, then the length of the equivalent adaptive filter in Fig. 10.12 is twice that. If more channels are involved, and the number of MIMO channels is represented by K, then the equivalent filter length is n multiplied by K.

Using formulas (B.20) for time constant, (B.22) for stable range of μ, and (B.27) for misadjustment, the corresponding formulas for scheme B applied to a multichannel MIMO system are

$$\tau = \frac{1}{2\mu E[\delta_k^2]}. \tag{10.15}$$

$$0 < \mu < \frac{1}{nK\,E[\delta_k^2]}. \tag{10.16}$$

$$M = \frac{\mu n K\,E[\delta_k^2]}{1 - \mu n K\,E[\delta_k^2]}. \tag{10.17}$$

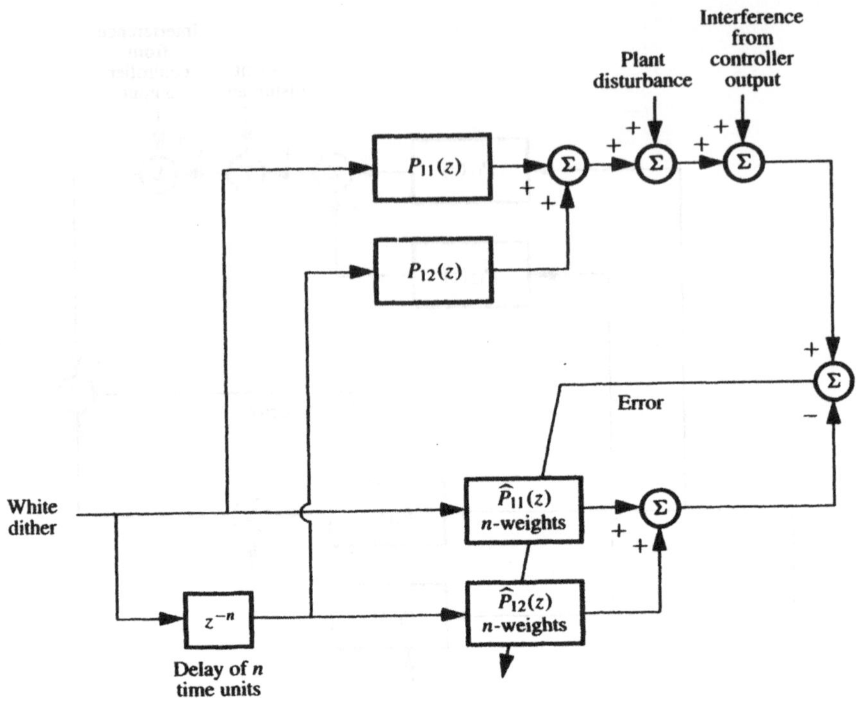

Figure 10.11 Alternative dither for two-channel MIMO modeling process.

Everything stays the same, except that the equivalent filter is long for MIMO. To keep misadjustment down at a low level, it is necessary to adapt more slowly, roughly by a factor equal to K.

10.2.2 Adaptive MIMO Modeling Using Scheme C

As scheme C was often the preferable choice for SISO systems, it is of course important to consider its application to MIMO systems. Figure 10.13 shows a detailed plan for a two-channel MIMO application of scheme C. Figure 10.14 gives an overview vector diagram for general use of scheme C for plant modeling in a multichannel MIMO system.

Figure 10.15 shows half of the two-channel modeling modeling process of Fig. 10.13. Since it and the other half act independently, adaptive behavior can be determined by its study.

We attempted to relate the behavior of the system of Fig. 10.15 to that of the one-dimensional SISO system of scheme C (shown in Fig. 4.3(c)), but this did not work out. Such an approach worked out well for scheme B above, but failed for scheme C because of the presence of $\widehat{P}_{11}$ COPY and $\widehat{P}_{12}$ COPY. The various filters cannot be stacked in time to make longer filters in this case.

To analyze the system of Fig. 10.15 in order to gain an understanding of the system of Fig. 10.13, it is necessary to go back to fundamentals. We will use the analytical techniques developed in Section B.4.

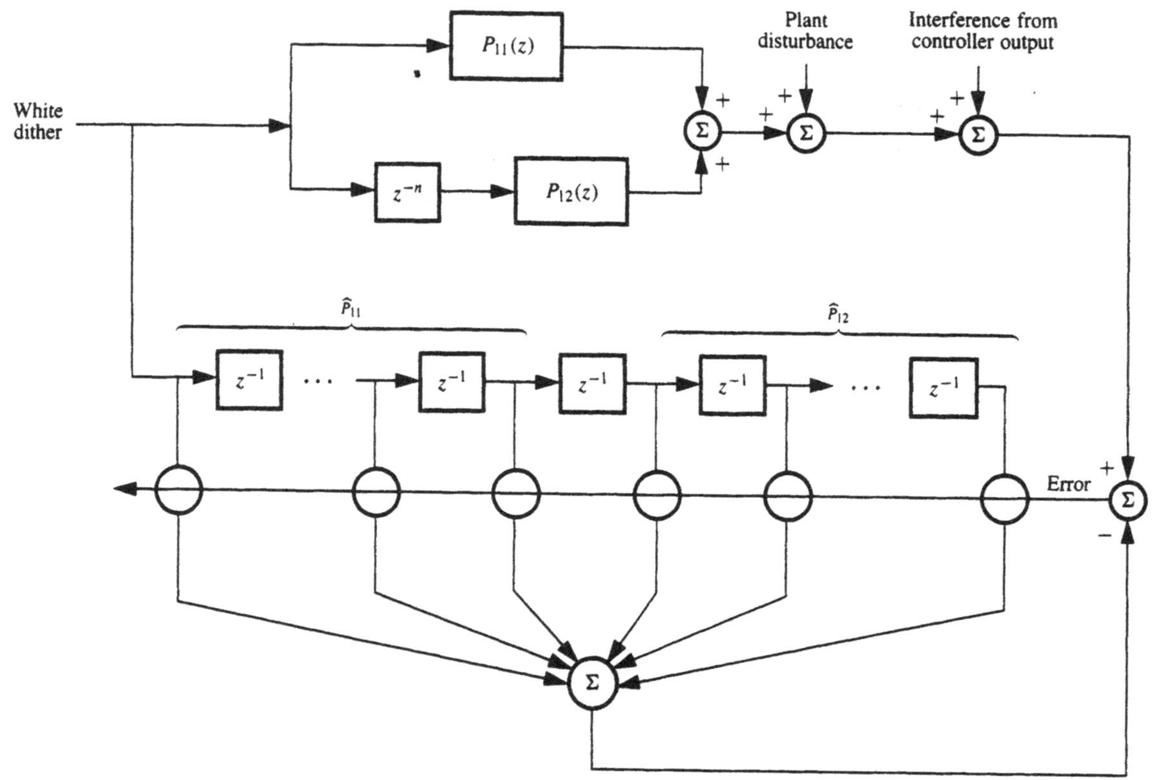

Figure 10.12 Detailed diagram of two-channel MIMO modeling with alternative dither.

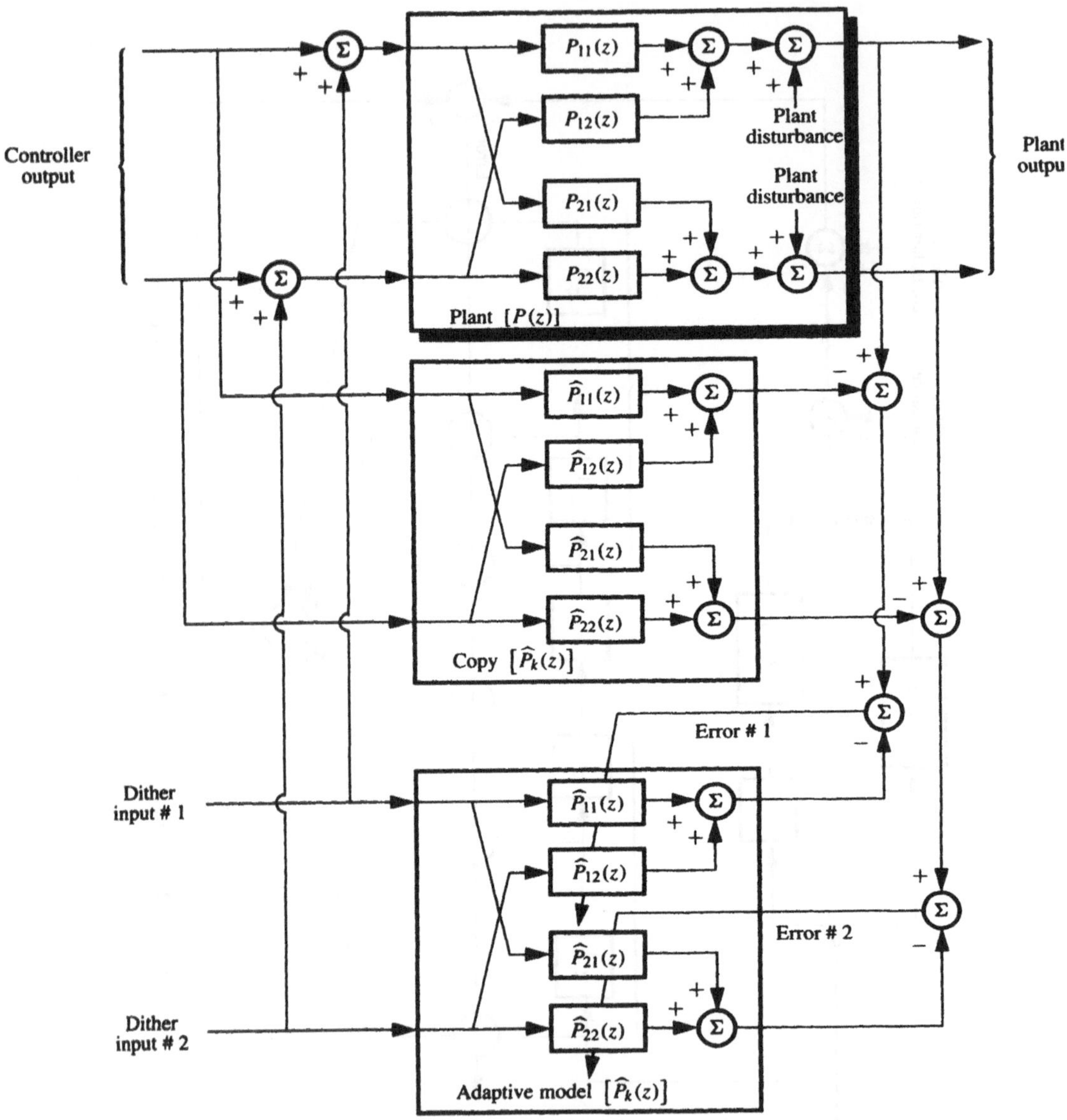

Figure 10.13 Scheme C for two-input two-output plant modeling.

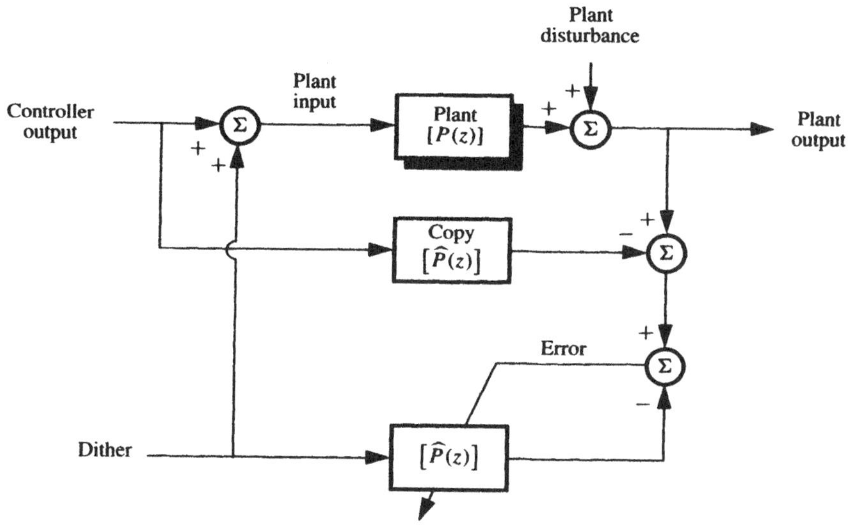

Figure 10.14 A vector signal diagram of scheme C for MIMO plant modeling.

Referring to Fig. 10.15, we observe that the average mean square error is the minimum mean square error, equal to the plant disturbance power, plus the average excess mean square error caused by noise in the weights of $\widehat{P}_{11}$, $\widehat{P}_{12}$, and $\widehat{P}_{11}$ COPY and $\widehat{P}_{12}$ COPY.

The average mean square error can be written as

$$\begin{aligned}\bar{\xi} &= \xi_{\min} + (\text{average excess MSE}) \\ &= \xi_{\min} + \mu n \bar{\xi} E[\delta_{1k}^2] + \mu n \bar{\xi} E[\delta_{2k}^2] \\ &\quad + \mu n \bar{\xi} E[(u'_{1k})^2] + \mu n \bar{\xi} E[(u'_{2k})^2].\end{aligned} \tag{10.18}$$

If we assume that the dither power is the same on both channels, and if we assume that the controller output power is the same on both channels,

$$\bar{\xi} = \xi_{\min} + 2\mu n \bar{\xi} \left[E[\delta_k^2] + E[(u'_k)^2]\right]. \tag{10.19}$$

Note that $E[\delta_k^2]$ is the dither power on one channel, and $E[(u'_k)^2]$ is the controller output power on one channel. The average mean square error can be written as

$$\begin{aligned}\bar{\xi} &= \frac{\xi_{\min}}{1 - 2\mu n\left[E[\delta_k^2] + E[(u'_k)^2]\right]} \\ &= \frac{\xi_{\min}}{1 - 2\mu n E[u_k^2]}.\end{aligned} \tag{10.20}$$

The expectation $E[u_k^2]$ is the power level on a single-input channel to the plant.

These expressions can be generalized for the multi-input case. Assume again that the dither powers are equal from channel to channel, and that the controller output powers are equal from channel to channel. The number of channels is designated by K. Accordingly,

$$\bar{\xi} = \frac{\xi_{\min}}{1 - \mu n K\left[E[\delta_k^2] + E[(u'_k)^2]\right]} \tag{10.21}$$

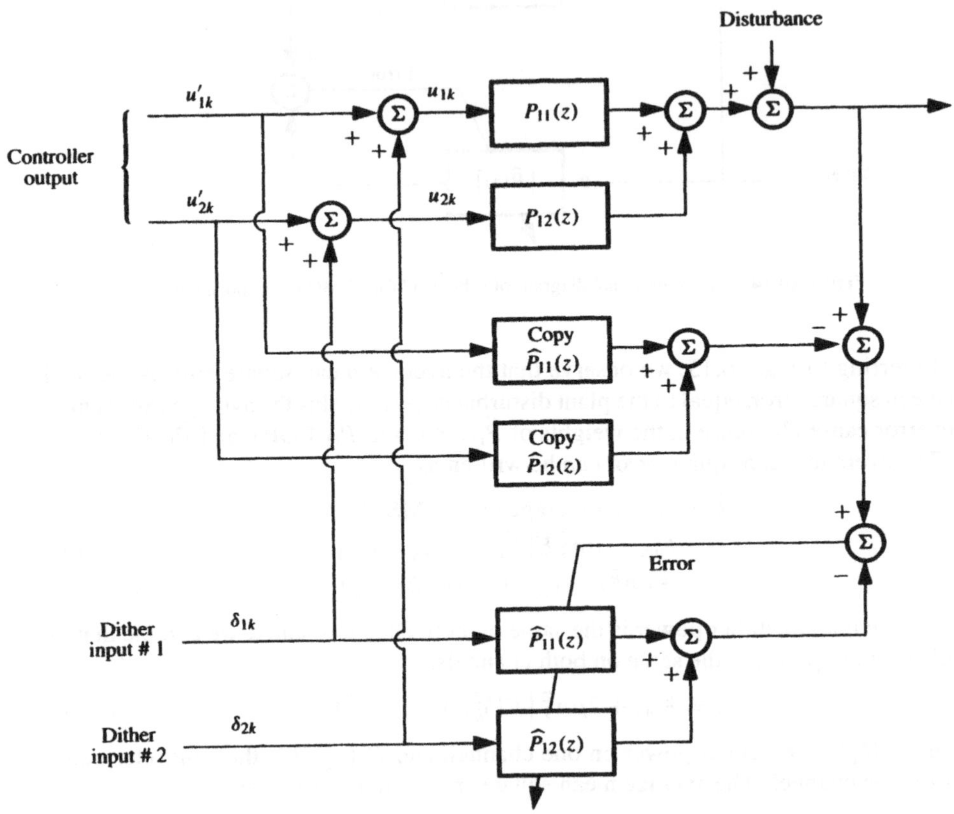

Figure 10.15 A part of the two-channel MIMO modeling process in accord with scheme C.

$$= \frac{\xi_{\min}}{1 - \mu n K\, E[u_k^2]}.$$

The misadjustment can be obtained by combining (10.18) with (10.21):

$$M = \frac{\text{(average excess MSE)}}{\xi_{\min}} = \frac{K \mu n E[u_k^2]}{1 - K \mu n E[u_k^2]}. \tag{10.22}$$

For stability, μ must be positive and small enough so that M remains finite. The stable range of μ for MIMO scheme C is

$$\frac{1}{K n E[u_k^2]} > \mu > 0. \tag{10.23}$$

If the controller output power differed from channel to channel, and/or the dither power differed from channel to channel, new expressions for misadjustment and stable range could be derived using similar analytical techniques.

10.2.3 Conclusions on MIMO Plant Modeling

Comparing the expressions for misadjustment and stable range of μ obtained for SISO systems and given in Appendix B with the corresponding expressions for MIMO systems given in this chapter, one can draw a simple conclusion. The SISO expressions can be generalized directly to apply to the MIMO case by simply multiplying the number of weights per adaptive filter n by the number of channels K provided that, for the MIMO system,

a. all filters have the same number of weights;

b. all filters have the same input power levels;

c. dither powers are equal on all channels;

d. the same value of μ is used for all adaptive filters.

This generalization works for systems based on both scheme B and scheme C.

It is interesting to note that for a given level of misadjustment, convergence time increases only linearly with the number of MIMO channels. Before doing this analysis, we were concerned that learning time would be proportional to the square of the number of channels (since the total number of weights grows with the square of the number of channels). Fortunately, learning time is only proportional to the product of the number of channels and the number of weights per adaptive filter. This is a surprising result.

10.3 ADAPTIVE INVERSE CONTROL FOR MIMO SYSTEMS

So far, several ways for adaptively modeling a MIMO plant have been described in this chapter. Having a plant model $[\widehat{P}(z)]$, one can use it to develop an inverse controller for the plant $[P(z)]$. Two ways of finding the inverse controller will be considered. The first way is primarily an algebraic technique, and the second way is a technique based on the filtered-ϵ approach to adaptive inverse control.

10.3.1 The Algebraic Approach

An algebraic technique for finding the adaptive inverse controller $[\widehat{C}_k]$ is described next. Referring to Fig. 6.1, it is clear that for the SISO case, a proper $\widehat{C}_k(z)$ would be such that:

$$\widehat{C}_k(z)\widehat{P}(z) = M(z). \tag{10.24}$$

The same idea pertains to the MIMO case. Taking into account that the transfer function matrix of the cascade of two transfer functions is the matrix product in reverse order to that of the signal flow, a proper $[\widehat{C}_k(z)]$ for the MIMO system would be

$$[\widehat{P}(z)][\widehat{C}_k(z)] = [M(z)]. \tag{10.25}$$

An algebraic technique for obtaining $[\widehat{C}_k(z)]$ is the following. Refer to Fig. 10.16. The matrix $[\widehat{P}_k]$ COPY can be obtained by the methods illustrated in Figs. 10.8 or 10.14. A new matrix transfer function $[V_k(z)]$ is introduced here for mathematical purposes. An adaptive process for finding $[V_k(z)]$ is indicated in Fig. 10.16. From $[V_k(z)]$, an algebraic process can be used to find the controller $[C_k(z)]$.

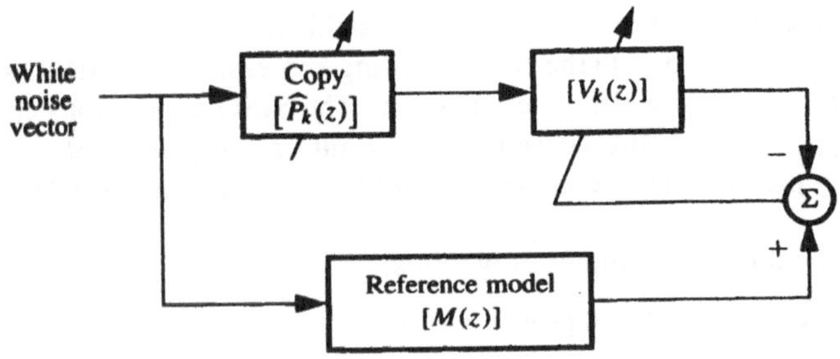

Figure 10.16 A step in the calculation of $[\widehat{C}(z)]$.

Assume that the reference model $[M(z)]$ is chosen to give good system operation and that its inverse is stable. Then, let $[\widehat{C}_k(z)]$ be

$$[\widehat{C}_k(z)] = [M(z)]^{-1}[V_k(z)][M(z)]. \tag{10.26}$$

Justification for Eq. (10.26) will be given below.

Using Eq. (10.26) to obtain $[\widehat{C}_k(z)]$, we next show that (10.25) will be satisfied. Refer to Fig. 10.16. Assuming that the adaptive process has converged and that the errors are small. Then

$$[V_k(z)][\widehat{P}_k(z)] = [M(z)]. \tag{10.27}$$

Postmultiplying both sides by $[\widehat{P}_k(z)]^{-1}$ yields

$$[V_k(z)] = [M(z)][\widehat{P}_k(z)]^{-1}. \tag{10.28}$$

Substituting into (10.26) yields

$$\begin{aligned}[\widehat{C}_k(z)] &= [M(z)]^{-1}[M(z)][\widehat{P}_k(z)]^{-1}[M(z)] \\ &= [\widehat{P}_k(z)]^{-1}[M(z)].\end{aligned} \tag{10.29}$$

Premultiplying both sides by $[\widehat{P}_k(z)]$ yields

$$[\widehat{P}_k(z)][\widehat{C}_k(z)] = [M(z)]. \tag{10.30}$$

Obtaining $[V(z)]$ from the adaptive process of Fig. 10.16 and obtaining $[\widehat{C}(z)]$ from the algebraic process of Eq. (10.26) gives the required MIMO controller. Although this technique may seem to be a bit roundabout, it does work!

In order to obtain $[\widehat{C}_k(z)]$ with Eq. (10.26), an inverse of $[M(z)]$ is needed. Since this inverse is to become a factor of $[\widehat{C}_k(z)]$, it is necessary that $[M(z)]^{-1}$ be stable. Since $[M(z)]$ is chosen by the system designer, it should be chosen so that $[M(z)]^{-1}$ is stable.

This restriction can be relaxed if a sufficiently delayed stable inverse of $[M(z)]$ is used.[1] Referring to Eq. (10.26), the delay incorporated in $[M(z)]^{-1}$ could be compensated for by using a correspondingly time advanced form of $[M(z)]$. This can be readily obtained if the reference model $[M(z)]$ has sufficient transport delay in all of its impulse responses. Without sufficient transport delay in $[M(z)]$, one could simply redefine $[M(z)]$ to have this delay. The end result would be an overall system response that would be like a delayed form of the original $[M(z)]$.

Analysis of the performance of adaptive inverse control for MIMO systems proceeds in like manner to that for SISO systems as explained in Chapter 6. The overall system error consists of a sum of four components:

i. Plant output disturbance
ii. Dither noise filtered through the plant
iii. System error due to truncation of $[\widehat{P}(z)]$ and/or $[\widehat{C}(z)]$
iv. Dynamic system error

Calculation of noise at the plant output due to plant disturbance and dither noise is straightforward. The effects of system error due to truncation of $[\widehat{P}(z)]$ or $[\widehat{C}(z)]$ can be made small by making the involved adaptive filters sufficiently long. We assume that truncation effects are negligible. The dynamic system error is due to the effects of noise in the weights of $[\widehat{C}(z)]$, originating from noise in the weights of $[\widehat{P}(z)]$.

This component of error can be calculated in an analogous fashion to the SISO case given by Eqs. (5.26)–(5.41) and Eqs. (6.6)–(6.11). From Eq. (10.30), one can write

$$[\widehat{C}(z)] = [\widehat{P}(z)]^{-1}[M(z)]. \tag{10.31}$$

It is required that $[\widehat{P}(z)]^{-1}$ be stable. For this to be, it may be necessary to make $[\widehat{P}(z)]^{-1}$ a delayed inverse. Continuing,

$$\begin{aligned}
[\widehat{C}(z)] &= [\widehat{P}(z)]^{-1}[M(z)] \\
&= [P(z) + \Delta P(z)]^{-1}[M(z)] \\
&= \left[[P(z)](\mathbf{I} + [P(z)]^{-1}[\Delta P(z)])\right]^{-1}[M(z)] \\
&= \left(\mathbf{I} + [P(z)]^{-1}[\Delta P(z)]\right)^{-1}[P(z)]^{-1}[M(z)] \\
&\approx \left(\mathbf{I} - [P(z)]^{-1}[\Delta P(z)]\right)[P(z)]^{-1}[M(z)].
\end{aligned} \tag{10.32}$$

[1] It can be shown that the delayed inverse of any $[M(z)]$ will be stable, with enough delay.

An error vector can be defined as follows

$$[E(z)] \triangleq ([M(z)] - [P(z)][\widehat{C}(z)])\, I(z), \tag{10.33}$$

where $I(z)$ is the z-transform of the command input vector. Combining (10.32) with (10.33), we obtain

$$\begin{aligned}[E(z)] &= ([M(z)] - [P(z)](\mathbf{I} - [P(z)]^{-1}[\Delta P(z)])[P(z)]^{-1}[M(z)])\, I(z) \\ &= ([M(z)] - [M(z)] + [\Delta P(z)][P(z)]^{-1}[M(z)])\, I(z) \\ &= ([\Delta P(z)][P(z)]^{-1}[M(z)])\, I(z).\end{aligned} \tag{10.34}$$

Equation (10.34) is completely analogous to the SISO Eq. (5.37). Figure 5.15 shows, for the SISO case, how the dynamic system error forms as the command input signal propagates through the system and encounters the weight noise $\Delta P(z)$. For the MIMO case, Fig. 10.17 shows the formation of the dynamic system error.

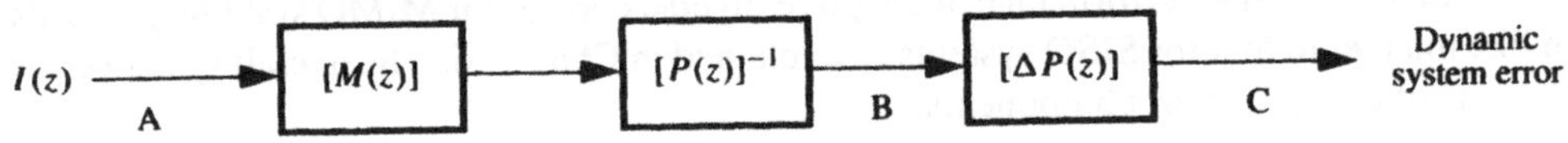

Figure 10.17 MIMO dynamic system error due to fluctuation in $[\widehat{P}(z)]$.

The variance of all components of the dynamic system error vector can be computed in an analogous way with a method like the SISO calculation. A simple formula analogous to the SISO formula in Eq. (5.40) can be obtained if special conditions exist, such as

a. The power levels on all channels at point B in Fig. 10.17 are the same. This is equivalent to the requirement that the power levels on all output channels of the controller be equal.

b. All filters of $[\widehat{P}(z)]$ have the same number of weights n. All filters of $[\widehat{C}(z)]$ have the same number of weights n.

c. The values of μ for the adaptive filters of $[\widehat{P}(z)]$ are equal.

d. The plant disturbance power levels are the same on all channels.

At the plant output, all channels have the same dynamic system error variance. For a single plant output channel,

$$\begin{pmatrix}\text{variance of}\\ \text{dynamic}\\ \text{system}\\ \text{error}\end{pmatrix} = \phi_{BB}(0)\cdot \min(m,n)\cdot \beta \cdot K. \tag{10.35}$$

Using scheme C to obtain $[\widehat{P}(z)]$, optimal dither levels can be determined for the special case described above. Referring to the SISO derivations of Chapter 6, the overall system error power (including plant output disturbance, dither output noise, and dynamic system error) on a single-output channel of the MIMO system can be expressed as

$$\begin{pmatrix}\text{overall}\\ \text{system}\\ \text{error}\\ \text{power}\end{pmatrix} = E[n_k^2] + E[\delta_k^2]\cdot\left[\sum_{i=0}^{\infty} p_i^2\right]\cdot K + \min(m,n)\cdot\beta\cdot\phi_{BB}(0)\cdot K. \quad (10.36)$$

This relation is analogous to Eq. (6.6), and all definitions made in connections with (6.6) are relevant here. Other special assumptions are necessary for (10.36) to apply, and they are

e. Dither power is equal on all channels.

f. All the filters of $[\widehat{P}(z)]$ have essentially equal sum of squares of the impulses of their impulse responses.

For the special case above, simple relations can be obtained for optimal dither power. With a given choice of time constant τ for adaptation of $[\widehat{P}(z)]$, the overall system error power for a single-output channel can be expressed as

$$\begin{pmatrix}\text{overall}\\ \text{system}\\ \text{error}\\ \text{power}\end{pmatrix} = E[n_k^2] + E[\delta_k^2]\cdot\left[\sum_{i=0}^{\infty} p_i^2\right]\cdot K \quad (10.37)$$
$$+\min(m,n)[\phi_{BB}(0)]\frac{E[n_k^2]}{2\tau E[\delta_k^2]\cdot K}.$$

This equation is based on the SISO formula of Eq. (6.6), and all definitions made relative to (6.6) apply here. Adaptation of $[\widehat{P}(z)]$ is done by scheme C. Differentiating (10.37) with respect to $E[\delta_k^2]$ and setting the derivative to zero yields the optimal dither power:

$$E[\delta_k^2]_{\text{OPT}} = \sqrt{\frac{E[n_k^2]\cdot\phi_{BB}(0)\cdot\min(m,n)}{2\tau K^2\sum_{i=0}^{\infty} p_i^2}}. \quad (10.38)$$

This equation is analogous to the SISO relation of Eq. (6.9).

The minimum overall system error power can be obtained by analogy to Eq. (6.12). Using the optimal dither with scheme C, the result for a single-output channel is

$$\begin{pmatrix}\text{overall}\\ \text{system}\\ \text{error}\\ \text{power}\end{pmatrix}_{\min} = E[n_k^2] + \sqrt{\frac{2}{\tau}E[n_k^2]\cdot\phi_{BB}(0)\cdot\left[\sum_{i=0}^{\infty} p_i^2\right]\cdot\min(m,n)}. \quad (10.39)$$

10.3.2 The Filtered-ϵ Approach

Adaptive inverse control of MIMO systems can also be done by making use of the filtered-ϵ method of adaptive inverse control introduced in Chapter 7. None of the other adaptive inverse control techniques taught above, including the filtered-X algorithm, can be used with

MIMO systems because they all require impermissible commutation of matrix operators. The ordering of matrix operations and the ordering of signal flow through matrix filters occurs in a natural way with the filtered-ϵ approach, leaving it a viable and useful technique for MIMO operation.

A filtered-ϵ adaptation scheme for a MIMO plant is shown in Fig. 10.18. In order to adapt the controller $[\widehat{C}(z)]$ COPY, an error vector is required. Ideally this would be ϵ', shown in Fig. 10.18. Since ϵ' is not available, a filtered error vector is used in its place. The filter is $[P_{\Delta}^{-1}(z)]$, a delayed plant inverse, estimated from $[\widehat{P}(z)]$. Both online and offline means for finding $[P_{\Delta}^{-1}(z)]$, the error filter, are indicated in Fig. 10.18. The plant model, $[\widehat{P}(z)]$, is obtained using scheme C. The method is analogous to that described in Chapter 7, Figs. 7.4 and 7.5, for SISO systems.

A few details about finding the delayed plant inverse filter, especially pertinent for MIMO systems, need to be discussed. Suppose that online inverse modeling is being done (switch left in Fig. 10.18). The inverse modeling signal is then the filtered error vector, as indicated in Fig. 10.18. The reason for using the filtered error for this purpose is to cause the input for $[\widehat{P}_{\Delta}^{-1}]$ to have the right spectral character. Since a copy of this filter will be having ϵ_k, the overall system error vector, as its input while generating the filtered error, it is reasonable to have the same or approximately the same input signal or a signal with an equivalent spectrum for an input as this filter is created by adaptation. Having the right input spectrum during adaptation is not critical, but taking the trouble to "do it right" could yield a better error filter. Offline adaptation (switch right in Fig. 10.18) could be accomplished by using synthetic noise generated for inverse modeling whose spectrum should be chosen to match that of the filtered error as well as can be done. Offline adaptation would normally be done at the outset, to initialize online adaptation. On the other hand, offline adaptation is an excellent idea and it can be used in its own right.

Analysis of the system of Fig. 10.18 has been attempted, but because of the noncommutability of the matrix operator, no simple expressions for misadjustment, noise in the weight vector of $[\widehat{C}(z)]$, or range of μ for stability have been obtained so far. The time constants for the adaptation of $[\widehat{C}(z)]$ are the same as if $[\widehat{C}(z)]$ were isolated from the rest of the system. The time constants for the adaptation of the plant model $[\widehat{P}(z)]$ by scheme C are determined by the methods of Section 10.2.2. The time constants for the adaptation of the inverse plant model $[\widehat{P}_{\Delta}^{-1}(z)]$ are the same as if the subsystem for calculating it were isolated from the rest of the system.

10.4 PLANT DISTURBANCE CANCELING IN MIMO SYSTEMS

Plant disturbance canceling in SISO systems was described in Chapter 8. The same techniques can be utilized, with some modification, to cancel disturbance in MIMO plants. Modification is required in some places in order to ensure that the ordering of signal flow in the disturbance canceling feedback is not commuted.

Figure 8.1 shows a SISO plant disturbance canceler using online adaptation to form $Q(z)$, and Fig. 8.3 shows a SISO plant disturbance canceler using an offline process to form $Q(z)$. Disturbance cancelers such as these should generally not be used for MIMO applications. The reason is that the plant disturbance must first be filtered by $Q(z)$ and then applied to $P(z)$ to achieve disturbance canceling. But $Q(z)$ is formed in the systems of Figs. 8.1 and

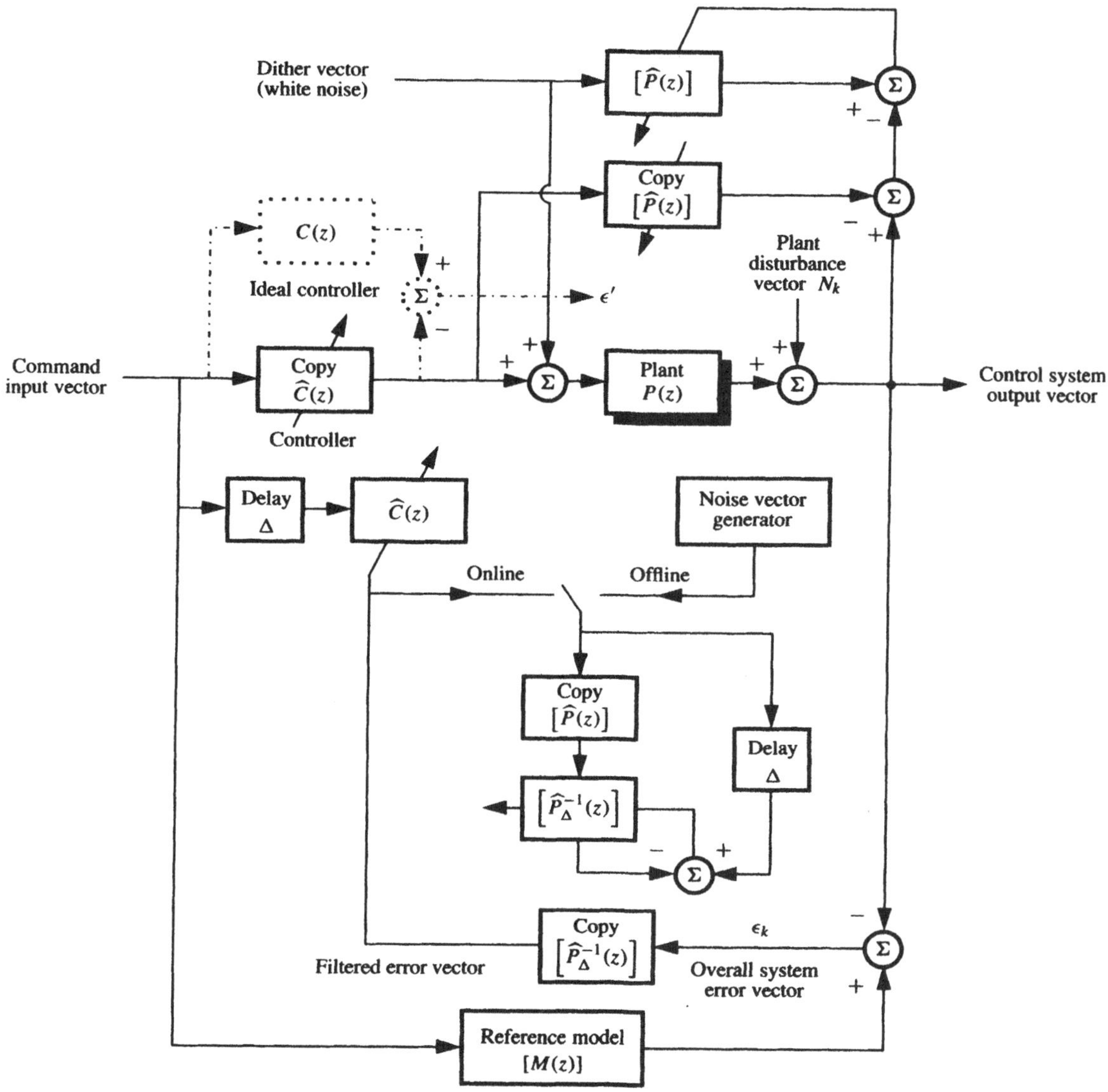

Figure 10.18 Filtered-ϵ adaptation for a MIMO plant.

8.3 as an inverse of $\widehat{P}(z)$, following $\widehat{P}(z)$ rather than leading $\widehat{P}(z)$. The proper way to do this for MIMO control is to adapt $Q(z)$ as an inverse of $\widehat{P}(z)$ with $Q(z)$ leading $P(z)$ in the same order of signal flow as in the disturbance canceling loop itself. The idea is illustrated in Fig. 10.19.

The filtered-ϵ algorithm is used to find $[Q(z)]$ in Fig. 10.19. In order to filter the error vector, it is necessary to have an inverse of the plant $[P(z)]$. A delayed inverse $[\widehat{P}_{\Delta}^{-1}]$ may be found by an offline process, as illustrated in the figure. Also, the process for generating $[Q(z)]$ may be offline, as illustrated. The system of Fig. 10.19 is therefore analogous to that of Fig. 8.3. Online processes can be devised alternatively, and the resulting system would be analogous to that of Fig. 8.1.

It may be noted that the use of a delayed inverse in Fig. 10.19 often makes a better inverse and otherwise causes no additional problems. The delay must be accounted for, however, when adapting $Q(z)$. This has been done by delaying the input of $[Q(z)]$ by $\Delta + 1$ units of time, Δ units for the delay in the inverse of $[\widehat{P}]$, plus one unit for the delay in the cascade $[Q(z)]$ COPY, $z^{-1}\mathbf{I}$, and $[\widehat{P}(z)]$ COPY.

Referring again to Fig. 10.19, we should note that the synthetic noise vector number one used to obtain $[Q(z)]$ should, in principle, be spectrally like the plant disturbance. The synthetic noise vector number two used to obtain $[\widehat{P}_{\Delta}^{-1}]$ should be such that the input to $[\widehat{P}_{\Delta}^{-1}]$ would be spectrally like the error signal of the process for generating $[Q(z)]$. Noise having this kind of spectrum is not so easy to generate. White noise should suffice, however, since errors in $[\widehat{P}_{\Delta}^{-1}]$ are not critical for finding the best $[Q(z)]$, as we learned in the SISO case.

It is useful to note that if $[P(z)]$ is minimum-phase and if the sampling frequency is high so that z^{-1} is a very small delay, then the error in generating $[Q(z)]$ will be very small and $[Q(z)]$ will be a close approximation to an exact inverse of $[\widehat{P}(z)]$. When this is the case, $[Q(z)]$ and $[\widehat{P}(z)]$ are commutable and the methods of both Figs. 8.1 and 8.3 would be directly applicable as alternatives to the methods developed in this chapter if one wished to use them.

10.5 SYSTEM INTEGRATION FOR CONTROL OF THE MIMO PLANT

Figure 9.1 shows an integrated control system for a SISO plant. A modified version of this system is shown in Fig. 10.20 for a MIMO plant.

The disturbance canceling techniques of Fig. 10.19 have been incorporated into Fig. 10.20. What is also added is an offline process for finding the controller, and this is based on the filtered error algorithm. Notice that in all cases, the adaptive offline processes are configured so that the training signal flows occur in the same sequence as in the actual system. This is necessary for MIMO.

Another method for finding the controller based on filtered error is shown in Fig. 10.21. Offline processes are used for finding $[\widehat{P}_{\Delta}^{-1}]$ and $[Q(z)]$. The process for finding the controller is online. The result is a system that learns a little more slowly than the system of Fig. 10.20 but would provide the correct controller for the plant and its noise canceling system even if $[\widehat{P}(z)]$ does not perfectly represent $[P(z)]$.

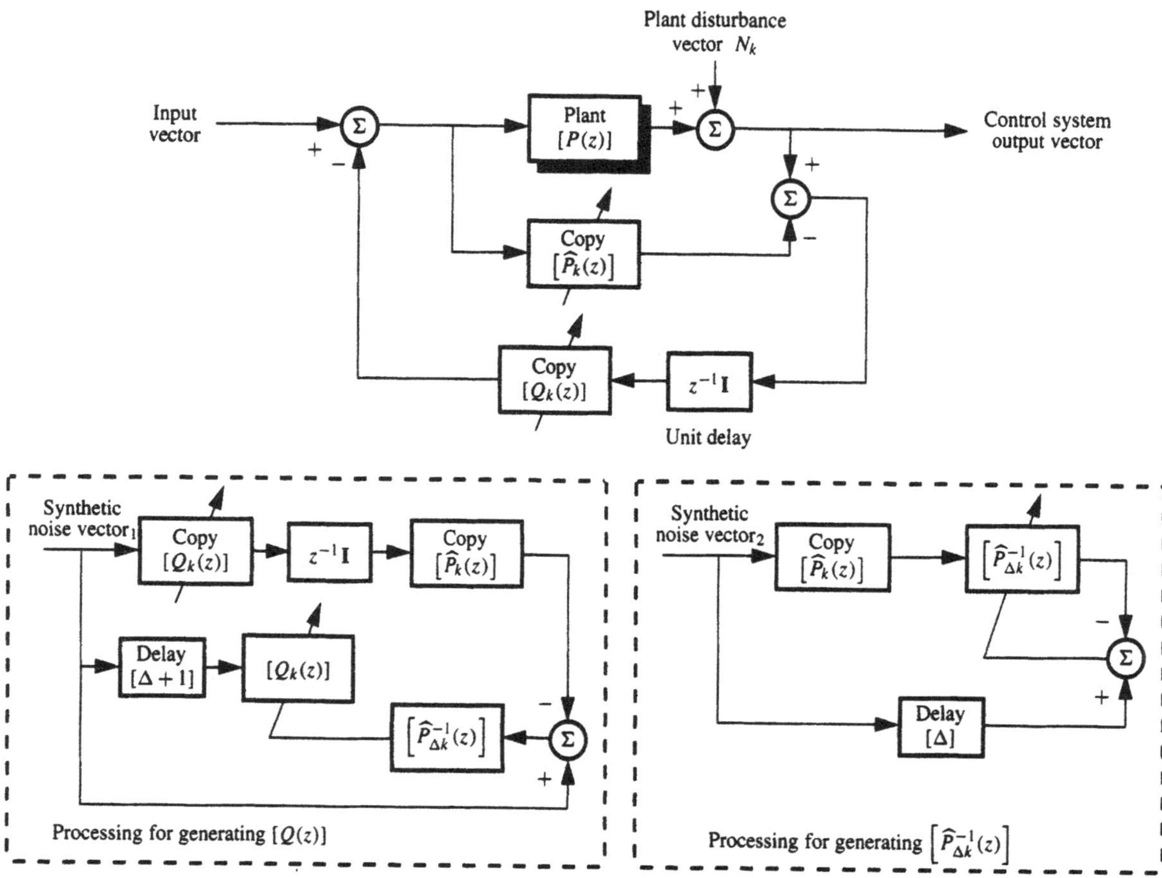

Figure 10.19 MIMO plant disturbance canceling using offline formation of $[\hat{P}_\Delta^{-1}]$ and $[Q(z)]$.

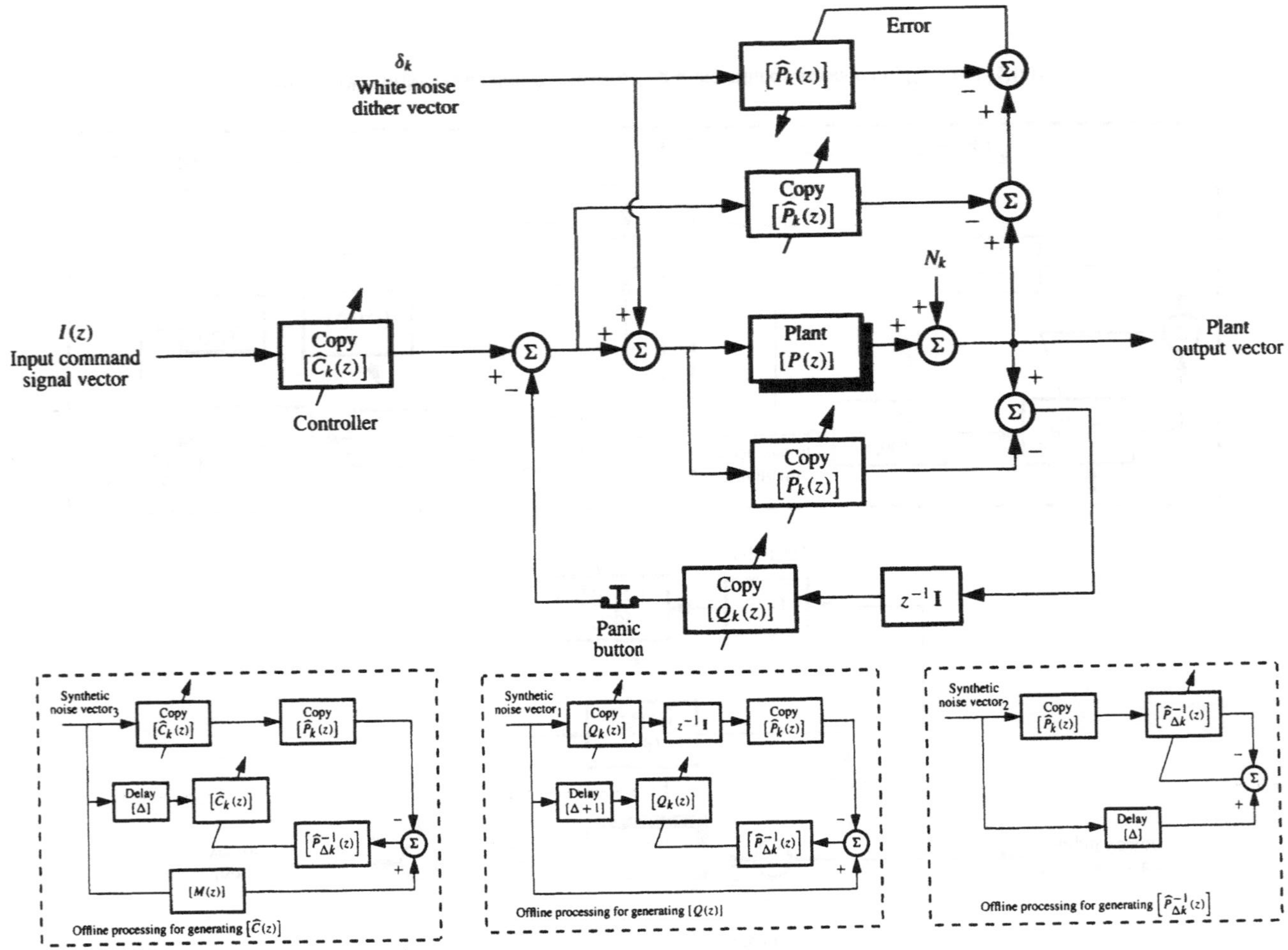

Figure 10.20 An integrated MIMO system.

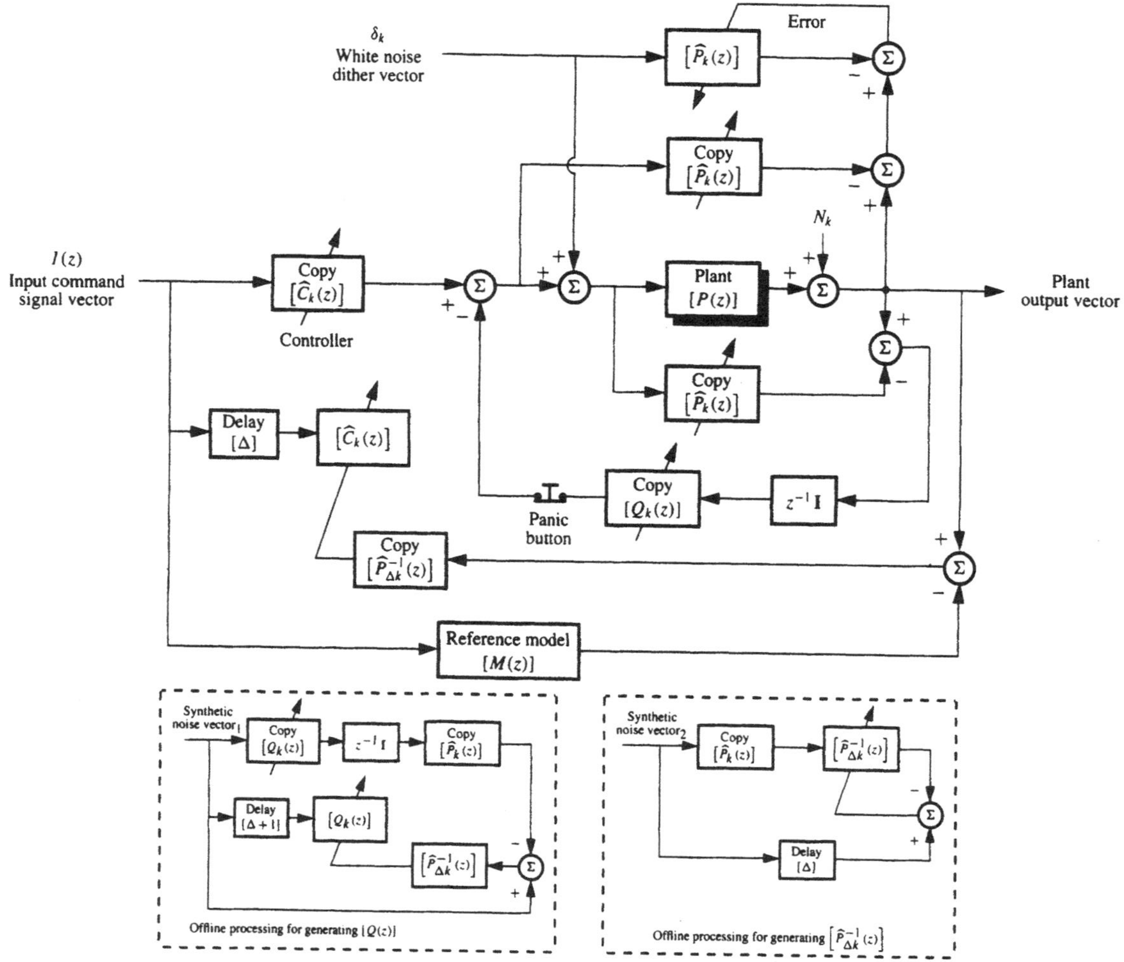

Figure 10.21 Another integrated MIMO system.

We have not yet done a detailed convergence and error analysis of the MIMO system shown in Figs. 10.20 and 10.21. We are confident that students of this subject will be able to develop such analysis.

10.6 A MIMO CONTROL AND SIGNAL PROCESSING EXAMPLE

An extraordinary problem of great current interest is that of canceling noise in the passenger compartment of commercial aircraft. A new generation of airplanes is being developed using a combination of turbojet engines and propellers that promises to be 30 to 40 percent more fuel efficient than the best turbojet airplanes flying in the mid 1990s. The engines and pusher propellers will most likely be mounted at the tail. A serious drawback is the increase in cabin noise resulting from the use of propellers. We propose to utilize adaptive inverse control systems of the MIMO type to control and cancel noise due to the turbojet engines and the propellers.

Referring to Fig. 10.22(a), we see the aircraft with a turbine-driven propeller at the rear. We assume that there will be two engines and two propellers, and that they will be synchronized. Thus, obtaining a single reference signal from a sensor on the shaft of one of the engines should be sufficient. We will need to obtain signals corresponding to the fundamental and harmonics of the turbine blades, and the fundamental and harmonics of the propeller blades. A single reference signal containing a sum of all those components would be satisfactory. The shaft sensor signal might need to undergo nonlinear processing to generate all the important harmonics.

The proposed approach to the problem of canceling the noise is presented in Fig. 10.22(b). Each passenger seat has an internal microphone located approximately at the passenger's head level. Loudspeakers are placed inside the aircraft cabin at some distance away from each other and from the microphones. The engine reference signal is fed to an adaptive noise canceler. The microphone signals are also fed to the adaptive canceler. The output of the canceler is a set of loudspeaker signals, obtained by optimally filtering the engine reference signal and intended to drive the loudspeakers so that their acoustic outputs will cancel the ambient engine and propeller noise in the vicinity of the microphones, near the heads of the passengers. The system illustrated uses two microphones and two loudspeakers. The number of microphones and loudspeakers could be increased as required.

The objective for the adaptive system is to generate the loudspeaker signals in order to minimize the sum of the powers of the microphone outputs. The microphone outputs are the error signals of the system. Figure 10.23 is a schematic diagram showing the origin and the propagation paths of the engine and propeller noise and the loudspeaker sound output. Each microphone senses the sound, a sum of these components, and outputs an electrical signal in accord with its acoustic to electrical transfer function. Adaptive filters drive the loudspeakers. Their inputs come from the common engine reference signal. The question is: How should one adapt the adaptive filters?

An answer to this question comes from the block diagram of Fig. 10.24. The microphone signals, the error vector of the system, are minimized in the mean square sense (minimize the mean of the sum of the squares of its components) by using the filtered-ϵ algorithm of the MIMO type to adapt the controller $[\widehat{C}]$. To filter the error, we need a delayed *plant inverse*. The way to get this inverse is shown in Fig. 10.25.

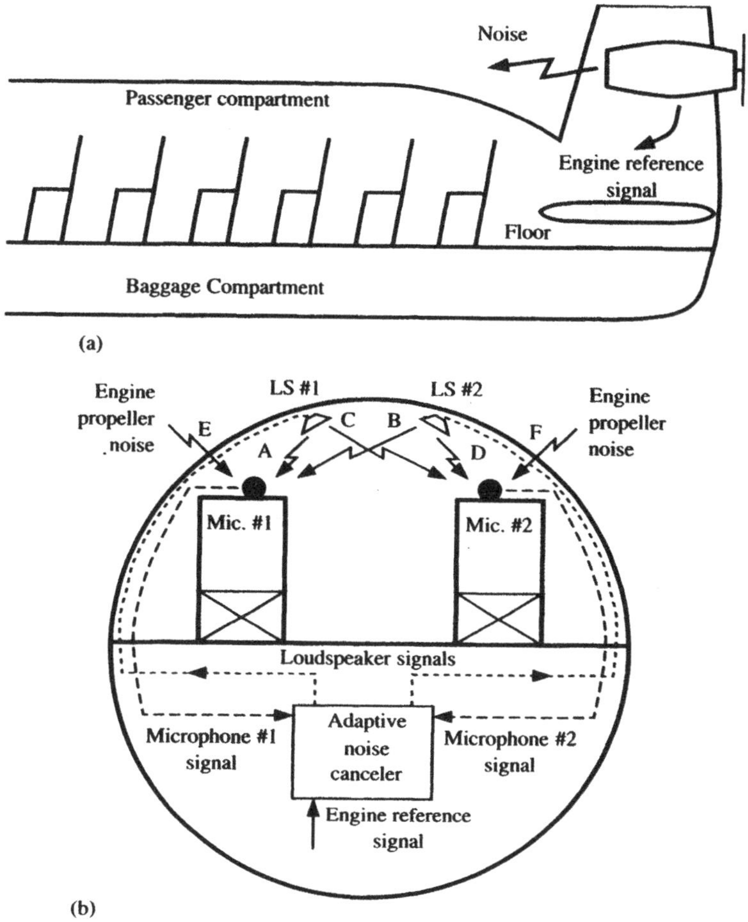

Figure 10.22 Aircraft noise problem and adaptive system for its mitigation.

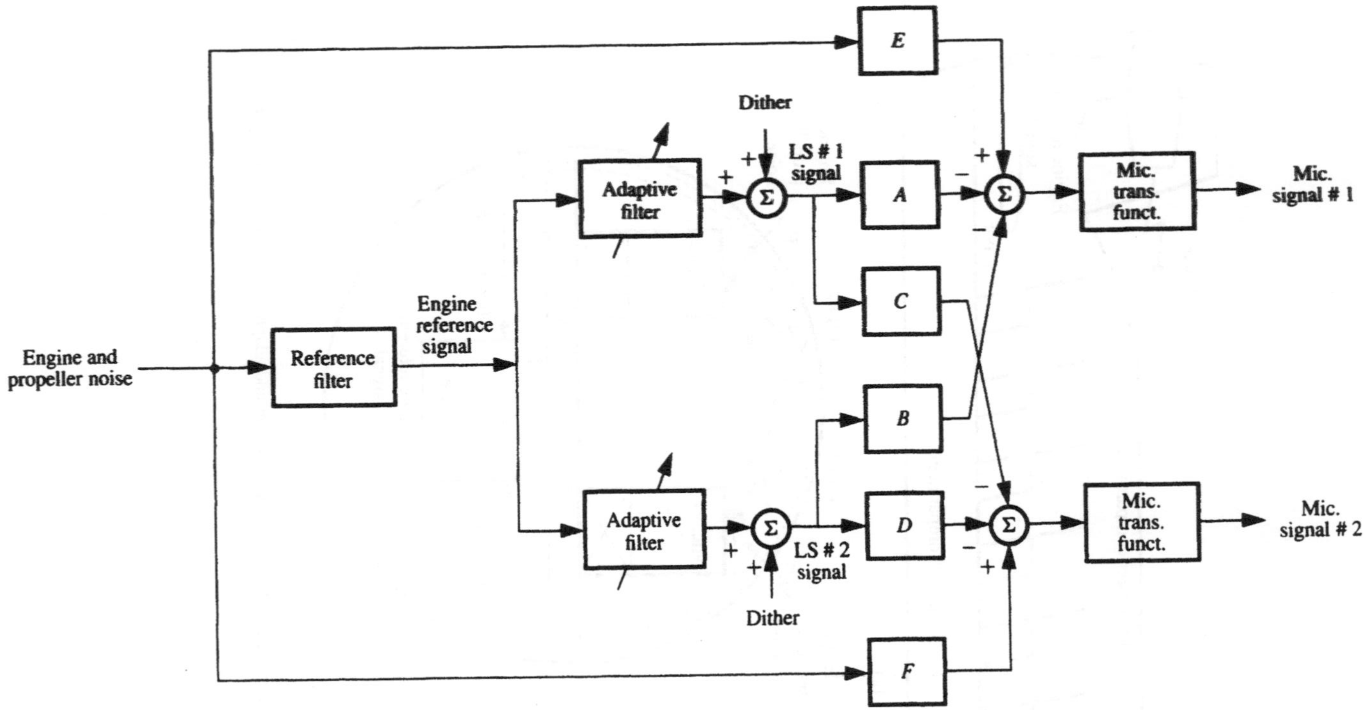

Figure 10.23 Schematic of noise propagation paths and noise canceling system.

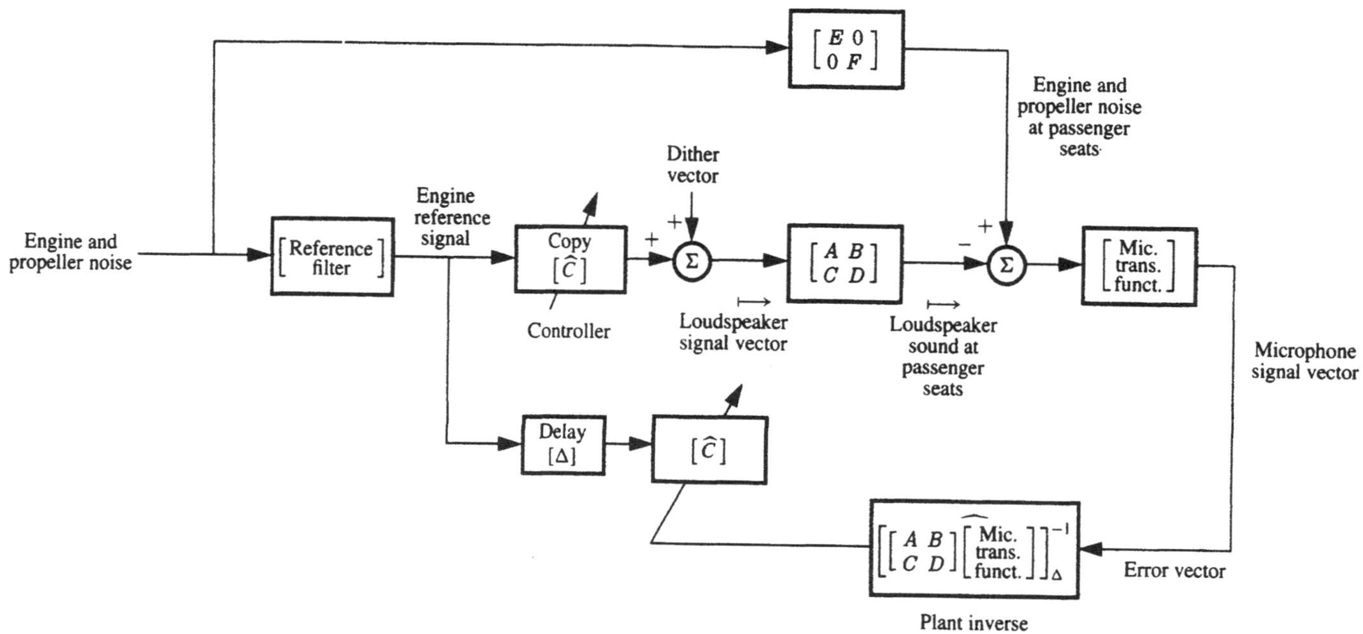

Figure 10.24 Vector block diagram of adaptive aircraft noise-canceling system.

In Fig. 10.25, scheme C is used to obtain a plant model. The plant is the matrix transfer function of the signal flow path from the loudspeaker electrical inputs to the microphone electrical outputs. An offline process is used to obtain, by MIMO adaptation, a delayed inverse of the plant model. The inverse is then able to be used in the filtered-ϵ algorithm shown in Fig. 10.24.

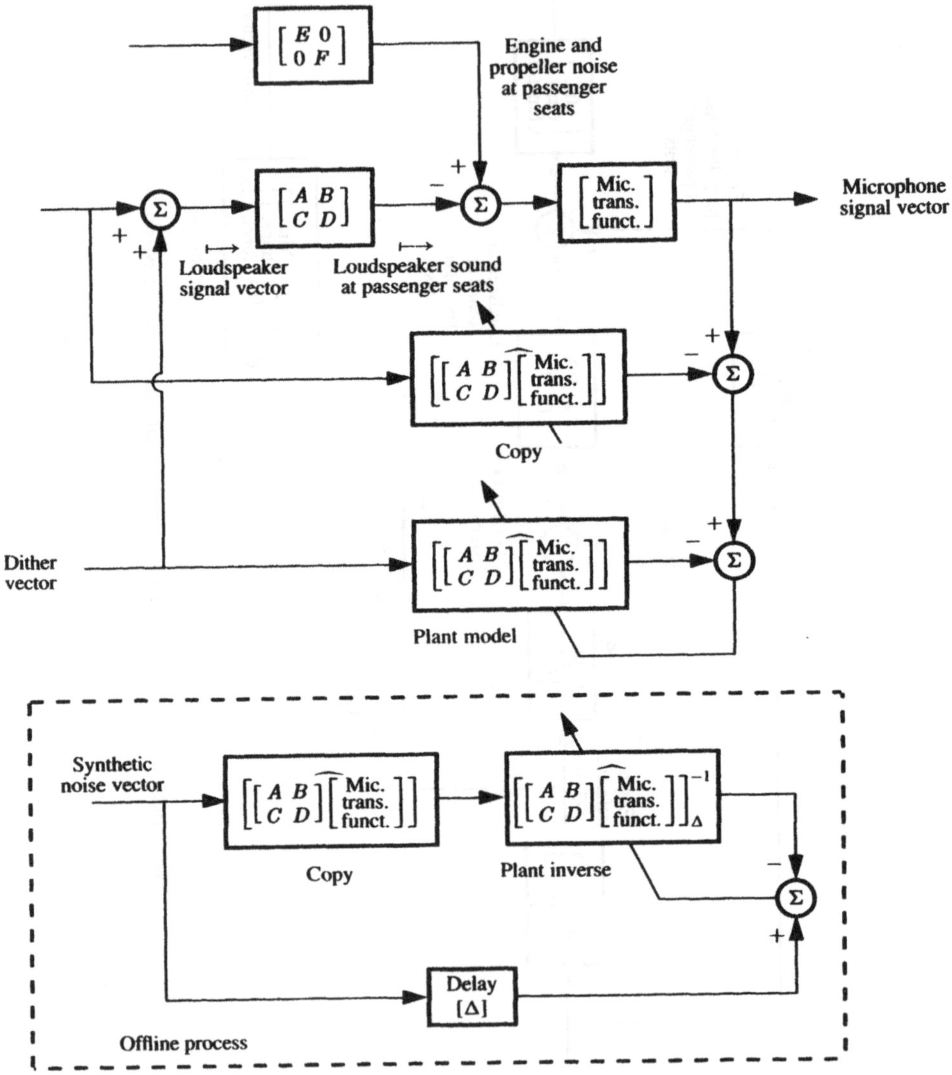

Figure 10.25 Plant modeling and inverse modeling for the aircraft noise-canceling system.

The noise canceling system diagrammed in Figs. 10.22–10.25 uses adaptive inverse control techniques to control noise for the passengers in an aircraft. Once again, one can see the close relationship that exists between control problems and signal processing problems. In many cases, they are really quite the same.

The noise canceling system described above can be used to simultaneously cancel noise at more than two microphones. The number of canceling loudspeakers should be equal to the number of microphones. Other ways using adaptive MIMO techniques could be employed to solve this problem. For example, without using an engine reference signal, the microphone signals alone could be adaptively filtered making a MIMO version of the earphone noise canceling system illustrated in Figs. 8.13–8.14. The noise coming from the microphones would be treated like plant noise, and the loudspeaker outputs would be used to cancel this noise. Many other ways can be thought of to approach and solve this important noise canceling problem.

The First International Conference on Active Control of Sound and Vibration was held at Virginia Polytechnic Institute, April 15–17, 1991. The meeting was sponsored by NASA Langley Research Center, Office of Naval Research, and the U.S. Army Aerostructures Directorate. The *Conference Proceedings* were published by Technomic Publishing Co., Inc. (851 New Holland Ave., Box 3535, Lancaster, PA, 17604, USA, Fax: (717) 295-4538). The *Proceedings* were edited by Professors C.A. Rogers and C.R. Fuller of Virginia Polytechnic Institute and State University, Blacksburg, Virginia. Papers were presented describing work on the above problem by British Aerospace and others. Many papers reported on canceling fan noise in air conditioning ducts, canceling auto exhaust noise, and canceling road noise in cars. Most of these projects make use of the filtered-X LMS algorithm. Adaptive noise and vibration canceling has become a major new field.

10.7 SUMMARY

In this chapter, means of describing linear multiple-input multiple-output (MIMO) systems have been developed. Block diagrams and flow graphs are useful for this purpose, as are algebraic methods. Adaptive techniques for modeling and inverse modeling were introduced, and they turned out to be very similar to those used with single-input single-output (SISO) systems except that care is exercised not to commute matrix transfer function operators.

Formulas for misadjustment and time constant of the adaptive MIMO plant modeling process have been obtained when using dither schemes B or C. For a given level of misadjustment, learning time is the same as for SISO, multiplied by the number MIMO channels, K. It is a surprise that learning time goes up only linearly with K, not with K^2 for example.

Inverse controls for MIMO plants were devised. One approach was based on an algebraic technique. A second approach was based on the filtered-ϵ LMS algorithm. Both methods work quite well.

Cancelation of plant disturbance in MIMO systems is possible. Several methods were explained for this, offline and online. The filtered-ϵ algorithm proved to be quite useful in finding the disturbance-canceling feedback transfer function $[Q(z)]$ from the plant model $[\widehat{P}(z)]$. Care was taken in all these developments to ensure that the ordering of matrix transfer functions was not commuted.

Adaptive inverse control systems were described for MIMO plants. Two different approaches for finding the inverse controller $[\widehat{C}(z)]$ were demonstrated, both based on the filtered-ϵ algorithm, one offline, the other online.

A practical application of adaptive disturbance canceling in a MIMO system is described in Appendix F by Dr. Thomas Himel of the Stanford Linear Accelerator Center. An eight-input, eight-output adaptive canceler is used 24 hours a day for beam control with a two mile long high-power linear accelerator. Beam position is controlled to within a micron. This is a fascinating application.

Many of the rules that are invoked in dealing with MIMO systems are applicable to nonlinear systems, such as noncommutability of operators. The next chapter deals with adaptive *inverse* control of nonlinear plants. The idea of inverse control for nonlinear systems is a strange one, because nonlinear systems do not generally have inverses. The nonlinearity invokes even more rules. In the next chapter, we develop techniques like adaptive inverse control for application to nonlinear SISO and MIMO systems.

Chapter 11

Nonlinear Adaptive Inverse Control

.0 INTRODUCTION

The principles of adaptive inverse control can be applied to the control of nonlinear plants. The purpose of this chapter is to show how to do this using SISO plant control as an example. Nonlinear MIMO plants can also be controlled with systems of greater complexity.

Many of the rules of MIMO systems apply to nonlinear systems, such as noncommutability of filtering and signal processing operations. Additional rules apply to nonlinear systems, such as exploring or modeling plant behavior only with input power level and input signal characteristics set to correspond to that of the actual plant input. Scaling and linearity do not work.

Strictly speaking, nonlinear systems do not have inverses. Nevertheless, the methods of adaptive inverse control can be made to apply. Inverse control of nonlinear plants can be done and canceling disturbance in nonlinear plants can also be done by making use of nonlinear adaptive filters.

For inverse control, the command input is applied to a nonlinear controller whose adjustable parameters are adapted so that when the output of the controller drives the plant input, the plant output becomes a best least squares match to the reference model's output. The resulting controller would be a good inverse only for the particular input command signal, not in general. If the characteristics of the command input signal were to change, it would be necessary for the controller to adapt rapidly and keep up with the changes. As long as this is feasible, nonlinear inverse control will work. Let this be the case. Our first step then is to study nonlinear adaptive filtering.

1 NONLINEAR ADAPTIVE FILTERS

One of the simplest of nonlinear adaptive filters is shown in Fig. 11.1. The input signal is applied to a tapped delay line. Signals at the taps are weighted, squared and weighted, and so forth, with all weighted signals summed and output. The output signal is therefore a linear combination of the signals at the taps plus a linear combination of squares of these signals. The desired response input is compared with the filter output signal. The difference, the error signal, is used to set the weights. The mean square error is a quadratic function of the weights. The weights can be adapted using the LMS algorithm. Adaptation is swift and sure, since there are no local optima.

This filter can be further generalized by making the output, in addition to the above, contain linear combinations of the tap signals raised to higher powers, and linear combinations of cross products of the tap signals raised to various powers. The output is therefore the sum of a Volterra series in the tap signals. The filter is often called an adaptive Volterra filter. Its weights are the coefficients of the Volterra series.

Another way to construct an adaptive filter is illustrated in Fig. 11.2. The input signal is applied to a tapped delay line, and the tap signals are in turn applied to an adaptive neural network. A three-layer network is shown in the figure.

Each neural element, labeled "AD," is an independent nonlinear adaptive device called Adaline. Connected into a network, these devices allow the realization of complex nonlinear adaptive systems. The neural device is an adaptive threshold element, and it was studied by Rosenblatt [1] and by Widrow and Hoff [2] in the early 1960s. An adaptive linear bypass is shown in Fig. 11.2, which contributes to an output which is a combination of linear and nonlinear functions of the tap signals.

A method for adapting multilayered feedforward networks as in Fig. 11.2 was developed in the early 1970s and 1980s. The algorithm is called *backpropagation* [3–5]. A simple explanation and derivation is given by Widrow and Lehr [6] in the September 1990 issue of the *Proceedings of the IEEE.*[1] The September and October issues of these *Proceedings* are special issues devoted to neural networks. General references on the subject of neural networks are the *IEEE Transactions on Neural Networks*, the journal *Neural Networks* published by the International Neural Network Society (INNS), and the *Journal of Neural Computation* published by M.I.T. Press. Many international conferences on neural networks are held in the United States, Europe, and Asia each year. Their published proceedings are always a source of the latest information on the subject. Many books have also been written on neural networks. A representative sample is given by references [5,7–13]. Use of neural networks for nonlinear control has been proposed by Kawato, Uno, Isobe, and Suzuki [14], and by Psaltis, Lideris, and Yamamura [15]. Their works relate well to the subject matter of this chapter. A collection of papers on this subject has been compiled by Miller, Sutton, and Werbos [16].

Backpropagation is a very sophisticated generalization of the LMS algorithm, using an instantaneous gradient with the method of steepest descent to minimize mean square error. The mean square error is not a quadratic function of the weights, however. The unknown nature of the mean square error function makes prediction of rate of convergence very difficult. Furthermore, local optima exist and often the backpropagation algorithm will hang up on local optima. This does not happen with the Volterra filter, whose behavior is simpler and in many respects, more predictable.

How to configure nonlinear adaptive filters cannot yet be determined by analytic procedures. Present day methods are empirical. There is no simple way to decide if a Volterra filter or a neural network filter would be the best for a given application. If possible, both

[1] This paper is reproduced here as Appendix G.

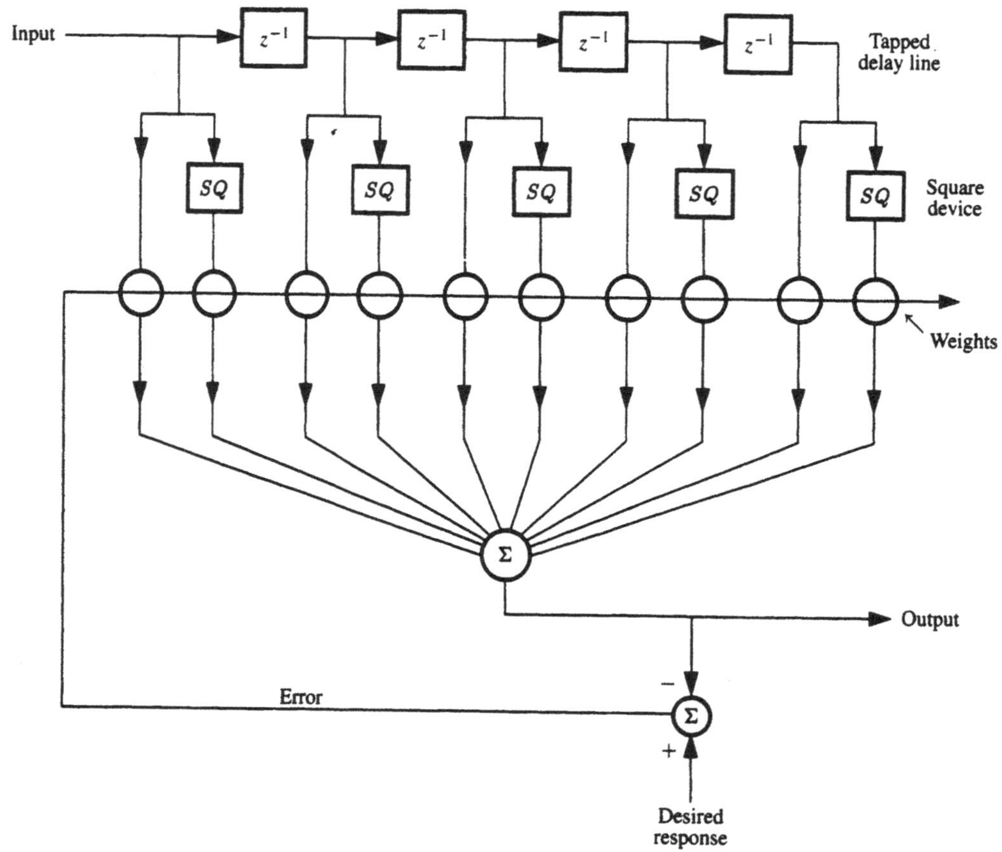

Figure 11.1 A simple nonlinear adaptive filter.

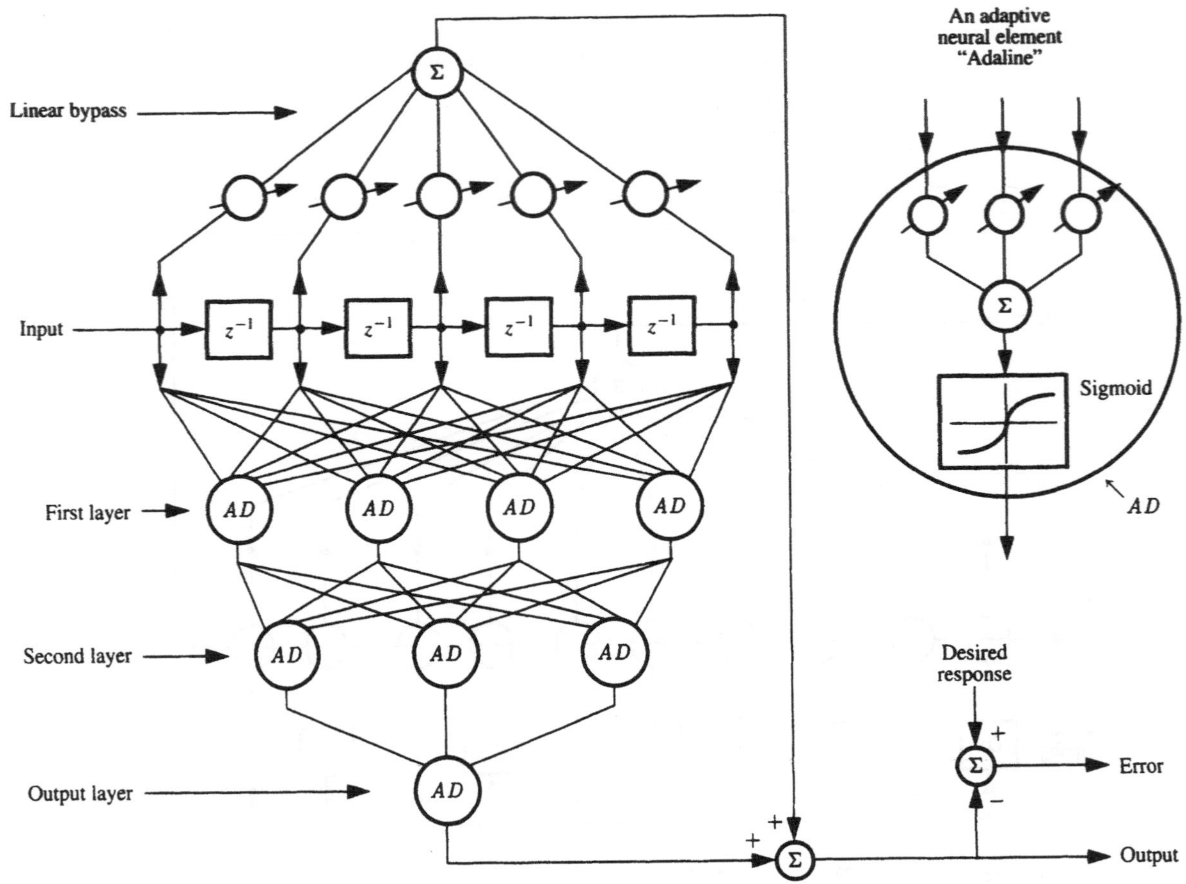

Figure 11.2 An adaptive nonlinear filter based on neural networks.

should be tried. Recent experience suggests that more often than not, better results are obtained with the neural network realization.

11.2 MODELING A NONLINEAR PLANT

Figure 11.3 shows a method for modeling an unknown nonlinear plant P. If the plant input alone can be used for modeling, then the dither can be omitted. Otherwise, it should be added to the plant input signal. This is like scheme A of Chapter 4, and since we are doing nonlinear modeling, let this process be called scheme A_{NL}. The input to the adaptive filter $\widehat{P}$ is the same as that of the plant. This is as it should be when dealing with a nonlinear plant. Modeling schemes B and C of Chapter 4 should be avoided since when using them, the dither signal inputs to the adaptive modeling filters are typically small in amplitude and are therefore quite different in character from those going into the respective plants. Proper nonlinear modeling algorithms like schemes B and C will be described below.

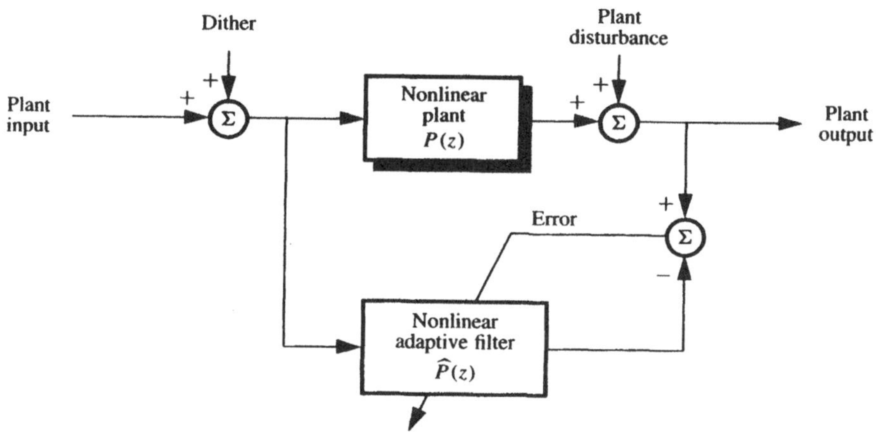

Figure 11.3 Modeling a nonlinear dynamic system using scheme A_{NL}.

When doing modeling in accord with scheme A_{NL} of Fig. 11.3, care must be taken in utilizing the dither signal. If the plant input is stationary, the dither should not be used. If the plant input is nonstationary, dither can be used having the intensity and character to cause the adaptive filter to model the plant under input conditions of the greatest interest. At the same time, one should keep the dither as small as possible to minimally disturb the plant. Another technique that could be used to achieve these objectives would turn the adaptive algorithm off and on to cause the adaptive filter to model the plant under selected conditions.

11.2.1 A Nonlinear Modeling Scheme like Scheme B

Using the simple adaptive Volterra filter that is illustrated in Fig. 11.1, one can develop a means for modeling a nonlinear plant P that works very much like scheme B of Chapter 4. Refer to Fig. 11.4. The nonlinear plant is represented for this illustration as a fixed Volterra filter with only linear and square terms. The adaptive modeling filter has an unusual structure which will be explained. It has in this case the same number of weights as the plant. The scheme will work well in practice as long as the adaptive filter has a number of weights equal to or greater than the number of weights of the plant.

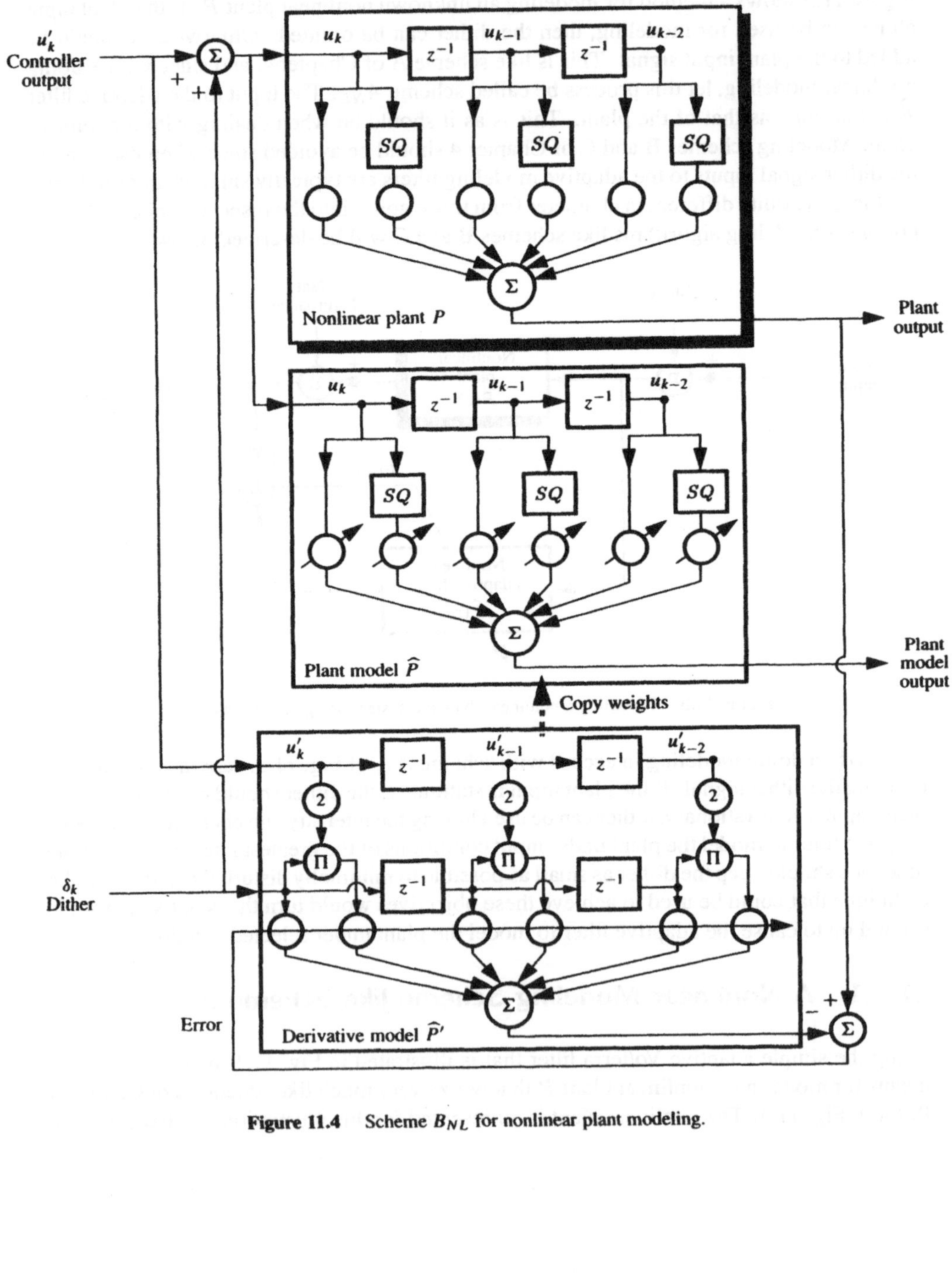

Figure 11.4 Scheme B_{NL} for nonlinear plant modeling.

The adaptive model $\widehat{P}'$ shown in Fig. 11.4, called the *derivative model*, does not resemble the Volterra plant but, when its weights converge, they are supposed to take values that correspond exactly to the weights of the plant.

Referring to Fig. 11.4, the input to the plant is u_k, the sum of the dither δ_k and the controller output u'_k. The dither power should be small compared to the power of u_k. The output of the plant will be a sum of the responses to u'_k and to δ_k. The response to u'_k will not be significantly affected by the presence of the dither, but the response to δ_k will be greatly affected by the presence of the controller output u'_k.

The adaptive derivative model $\widehat{P}'$ has δ_k as its input, but to produce an output equal to the response of the plant to δ_k, the adaptive derivative model must have access to u'_k. Accordingly, the adaptive derivative model has both u'_k and δ_k as its inputs. Its output does not match the total plant output, just the component of the plant output due to δ_k. The difference between the plant and adaptive derivative model outputs is used as an error signal for LMS adaptation of the weights of the adaptive derivative model. This is the same situation as is encountered with scheme B.

When the weights of the adaptive derivative model $\widehat{P}'$ correspond exactly to the weights of the plant, the output of the model corresponds almost exactly to the component of the plant output response due to the dither δ_k. The correspondence of these output components becomes exact in the limit as dither power is made small. An explanation of how this works can be developed by referring to Fig. 11.4.

The effects of small dither on the plant output at any moment of time can be obtained by perturbing one at a time the signals at the taps of the plant's tapped delay line, u_k, u_{k-1} and u_{k-2}. The derivatives of the output with respect to these signals multiplied by the corresponding dither components δ_k, δ_{k-1}, and δ_{k-2} are summed to yield the plant output signal component due to the dither.

The adaptive derivative model $\widehat{P}'$ is structured to make an analogous computation. It does so when its weights correspond to the weights of the plant. Then, the output of the derivative model is identical to the dither output of the plant. Making the weights of the derivative model equal to those of the plant causes the mean square of the error between the outputs of the plant and the derivative model to be minimized. In fact, when the adaptive process adjusts the weights of the derivative model to minimize mean square error, the adaptive process causes the derivative model weights to converge on the plant weights. We call this adaptive algorithm scheme B_{NL}. The derivative model $\widehat{P}'$ is not a model of the plant. It is merely used to find the weights of the actual plant, shown in Fig. 11.4. The plant model $\widehat{P}$ is a true Volterra model resembling the plant.

Scheme B_{NL} works well for higher order Volterra plants and adaptive plant models. The simple Volterra form illustrated in Fig. 11.4 can be generalized for plants having higher-degree nonlinearities and crossproducts of higher degree. A simplified block diagram that provides an overview of scheme B_{NL} for modeling higher-order Volterra plants is shown in Fig. 11.5. Scheme B_{NL} can also be used with adaptive plant models based on neural networks, utilizing adaptive filters of the type shown in Fig. 11.2.

11.2.2 A Nonlinear Modeling Scheme like Scheme C

Scheme B_{NL} adapts by matching the output of the adaptive derivative model to the output of the plant P. The output of the adaptive derivative model, at convergence, contains only

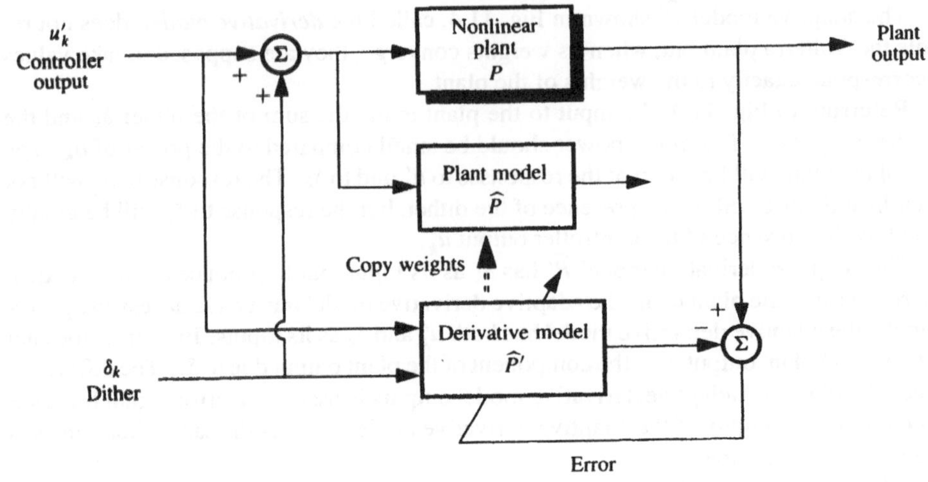

Figure 11.5 Scheme B_{NL}.

the plant output component due to the dither input and does not contain the plant's response to u'_k. The error will therefore be large, even at convergence. The original scheme B suffers from the same problem, and this motivated the development of the original scheme C. A scheme for nonlinear plant modeling which is analogous to scheme C is illustrated in Fig. 11.6. We call this scheme C_{NL}.

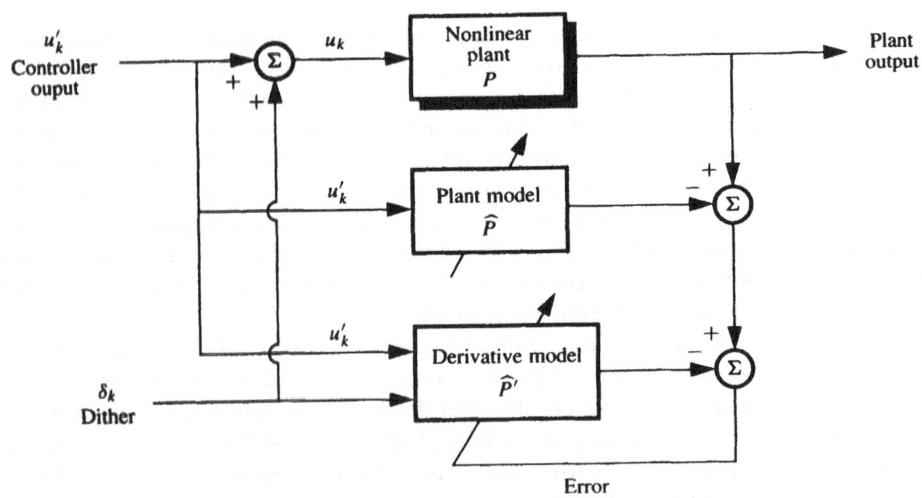

Figure 11.6 Scheme C_{NL}.

In Fig. 11.6, the nonlinear plant P can be considered to be of a high-order Volterra type. The plant model $\widehat{P}$ is a Volterra filter with adjustable weights. The adaptive derivative model is the same as the one shown in Fig. 11.5, and it is represented in more detail in Fig. 11.4. The weights for the plant model are copied from the weights of the nonlinear adaptive derivative model.

Schemes B_{NL} and C_{NL} do not depend on the stationarity and persistence of the controller output u'_k in order to achieve satisfactory plant modeling. They depend on the well-chosen characteristics of the dither instead. Scheme C_{NL} is advantageous over scheme B_{NL} in that its adaptive error signal is much smaller and tends to zero when the adaptive derivative model approaches convergence. It should be noted that if there were plant disturbance (not shown in Figs. 11.4, 11.5, and 11.6), the error signal would be equal to the plant disturbance signal at convergence.

It should further be noted that both schemes B_{NL} and C_{NL} converge on the correct solution even if there is plant disturbance and this disturbance is correlated with the controller output u'_k. The same is true for schemes B and C. This would not be true for scheme A_{NL} or for scheme A. This issue becomes important when developing methods for canceling internal plant disturbance.

Although not illustrated here, nonlinear adaptive plant models based on neural networks can be implemented by means of the adaptive filter of Fig. 11.2 and its appropriate derivative model. Schemes A_{NL}, B_{NL}, and C_{NL} can all be realized with neural networks.

.3 NONLINEAR ADAPTIVE INVERSE CONTROL

The inverse controller for a linear plant has a transfer function which is a close approximation to the reciprocal of the plant transfer function. In the nonlinear case, transfer functions do not exist. So what does *inverse control* mean for nonlinear plants?

Control of a nonlinear plant can be achieved by filtering the command input signal with a nonlinear filter and by applying the filtered command signal as the input to the plant. The idea is the same as for control of the linear plant. We define this as inverse control, for both linear and nonlinear plants.

A model-reference adaptive inverse control system for a nonlinear plant is shown in Fig. 11.7. The plant is represented by P and the adaptive plant model by $\widehat{P}$. The reference model M could be linear or nonlinear. A simple plant modeling process is shown in the figure. In its place, schemes A_{NL}, B_{NL}, or C_{NL} could be used if dither is required for plant modeling.

A copy of the plant model $\widehat{P}$ is used in an offline process to find the best $\widehat{C}$. Since $\widehat{C}$ and $\widehat{P}$ are nonlinear, their ordering cannot be commuted. The noise driving the process should be constructed to be as statistically alike the command input signal as possible, having the same power density spectrum and the same probability density function (first order and higher orders).

The parameters of $\widehat{C}$ need to be adjusted to minimize the mean square of the error. The LMS algorithm cannot be used because error at the output of $\widehat{C}$ is not available. This can be done, however, by using linear random search (LRS) or differential steepest descent (DSD), described in Chapter 6. Once the weights of $\widehat{C}$ are obtained, they can be copied into $\widehat{C}$ COPY to form the controller.

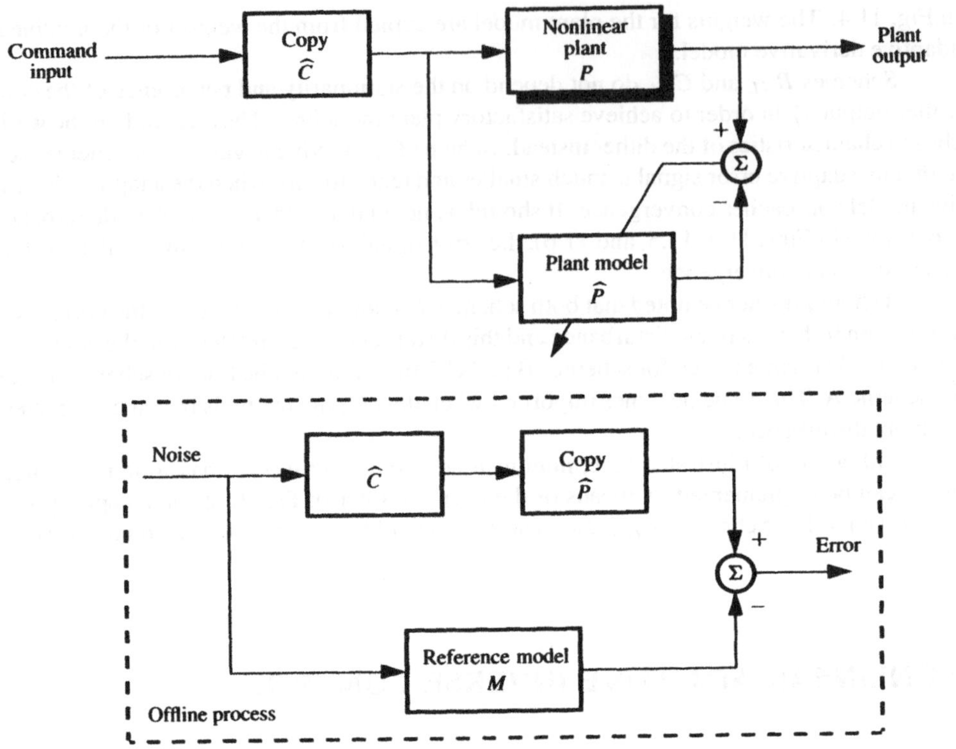

Figure 11.7 A nonlinear model-reference adaptive inverse control system.

11.3.1 A Filtered-ϵ Technique for Nonlinear Adaptive Inverse Control

The LRS and DSD algorithms are easy to apply and work well, but they are very slow to converge and use much more data than LMS to achieve convergence with a given level of misadjustment. Accordingly, there is good reason to attempt to develop a learning scheme based on LMS to obtain $\hat{C}$. It turns out that the filtered error algorithm can be adapted to this purpose.

The adaptive inverse control techniques described in previous chapters, including the filtered-X algorithm, cannot be used with nonlinear systems because they require impermissible commutation of filtering operations. The ordering of filtering operations occurs in a natural way with the filtered-ϵ approach, and as long as signal levels and statistical characteristics are maintained appropriately, the filtered-ϵ technique is applicable to the problem of finding $\hat{C}$.

The basic approach to the filtered-ϵ algorithm is illustrated in Figs. 7.4 and 7.5 for linear SISO control systems, and in Fig. 10.18 for linear MIMO control systems. Figure 11.8 incorporates many of the features of these block diagrams. The operation of this control system is straightforward and needs no special description here. The system in Fig. 11.8 is suitable for linear SISO control only. Our job is to adapt this approach to nonlinear SISO control.

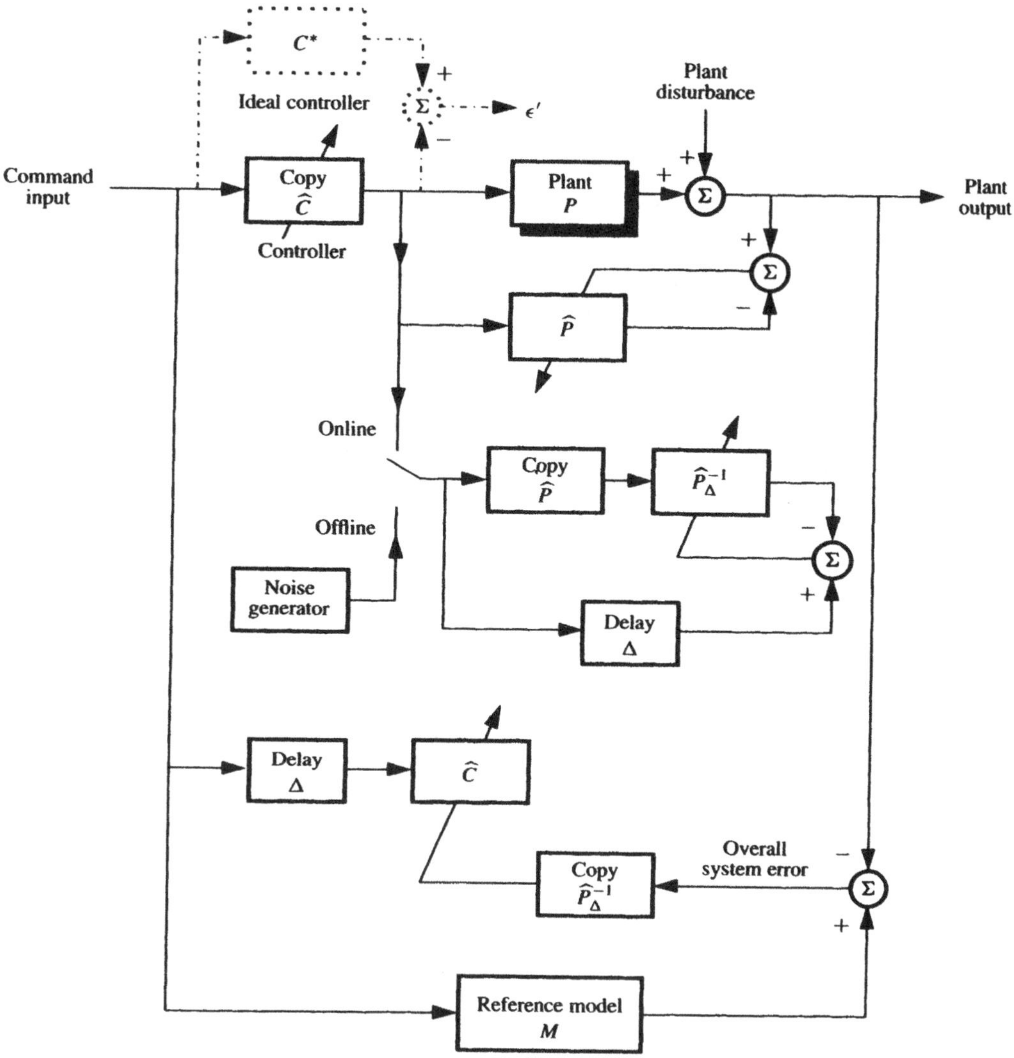

Figure 11.8 The filtered-ϵ LMS system, suitable for linear control.

The system of Fig. 11.9 is very similar to that of Fig. 11.8, except that with its minor modification, it is now capable of doing nonlinear control. If dither is required for the plant modeling process, schemes A_{NL}, B_{NL}, or C_{NL}could be used. The adaptive process for finding the nonlinear $\widehat{P}_\Delta^{-1}$ works in the same way as that for finding $\widehat{P}_\Delta^{-1}(z)$ of the linear case. How well does the nonlinear inverse of the nonlinear plant function? This is a good question. A lot of experience and theoretical development (not yet available) will be needed to provide a definitive answer. Preliminary work seems to indicate that it is generally possible to find a satisfactory nonlinear inverse, given nonlinearities and a sufficient number of variable weights in the inverse filter.

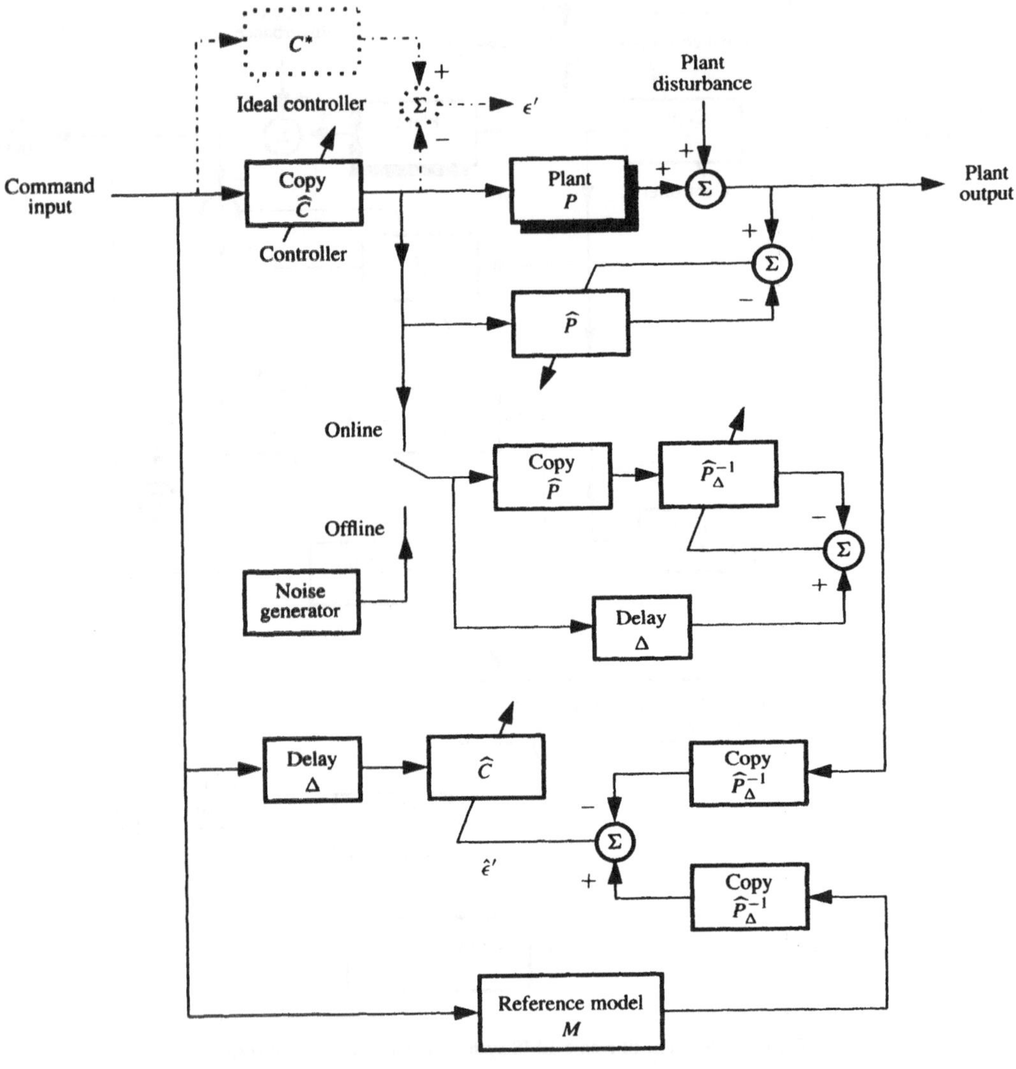

Figure 11.9 A filtered-ϵ LMS system capable of nonlinear control.

It should be noted that when one is finding $\widehat{P}_\Delta^{-1}$ offline, the noise generator needs to have the same dynamic range and statistical character as the output of the controller. Having the right dynamic range is critical to nonlinear inverse modeling, just as it is for direct modeling.

When filtering the error ϵ, the inverse plant model $\widehat{P}_\Delta^{-1}$ works properly when its input signal has statistical character and dynamic range comparable to that of the plant output. For the system of Fig. 11.8, the overall system error is expected to be very small in amplitude compared to the plant output, so $\widehat{P}_\Delta^{-1}$ COPY will not serve well. In order to achieve proper nonlinear filtering of error ϵ, the system of Fig. 11.9 uses two identical nonlinear filters, $\widehat{P}_\Delta^{-1}$ COPY. Each carries input signals having approximately correct amplitude, dynamic range, and statistical character. Their outputs are subtracted to provide an error signal for the adaptive process that finds $\widehat{C}$. Small errors in $\widehat{P}_\Delta^{-1}$ are quite tolerable as long as both of the $\widehat{P}_\Delta^{-1}$ COPY filters are identical to each other.

An alternative to the system of Fig. 11.9 is shown in Fig. 11.10. Here the output of $\widehat{P}$ is used in generating the error signal for the adaptive process for finding $\widehat{C}$. The advantage is that the output of $\widehat{P}$ does not contain plant disturbance. The disadvantage is that differences between $\widehat{P}$ and P could bias the adaptive process for finding $\widehat{C}$. Thus, the controller could differ from the ideal. The control engineer must trade advantages against disadvantages in each practical case.

Yet another alternative, this one an alternative to the system of Fig. 11.10, is shown in Fig. 11.11. The plant input is used here instead of the plant output to obtain a portion of the error signal for the adaptive process for finding $\widehat{C}$. Instead of running the controller output through P or $\widehat{P}$ and then through $\widehat{P}_\Delta^{-1}$, one simply takes the plant input and uses it directly as shown. A delay Δ must be included in the signal path. The behavior of the system of Fig. 11.11 is similar to that of the system of Fig. 11.10. More work will be needed to determine which system is better and when. They both appear to work quite well when $\widehat{P}_\Delta^{-1}$ is an accurate plant inverse.

11.3.2 Adaptive Inverse Control for Nonlinear MIMO Systems

The block diagram of Fig. 11.9 represents a system capable of adaptive nonlinear control for SISO plants. It is able to be generalized, however, to represent an adaptive inverse control system for MIMO applications. To do MIMO, the form and structure of the system stays the same. Within the blocks, the plant model $\widehat{P}$ would be structured like the adaptive MIMO filter of Fig. 10.7, except that each component adaptive filter would be a nonlinear adaptive filter. The same remarks apply to $\widehat{P}_\Delta^{-1}$, used to filter the components of the overall system error, and to $\widehat{C}$, the controller. The reference model M would also be MIMO, and it too could be nonlinear.

An algebraic technique has been devised in order to demonstrate that the system of Fig. 11.9 will perform correctly for adaptive inverse control of nonlinear MIMO systems. This simple form of algebra will be introduced here and it will prove to be helpful in the analysis of nonlinear signal processing systems that are either SISO or MIMO.

Refer once again to Fig. 11.9. We shall represent the input to the control system as I. This is meant to be a symbolic representation, not necessarily a z-transform of the input vector. If we neglect plant disturbance, we can represent the plant output as

$$Z = P\widehat{C}I. \tag{11.1}$$

The notation is similar to that of finding the z-transform of the plant output, as if P and $\widehat{C}$ were linear transfer functions. But in this use, both P and $\widehat{C}$ are nonlinear. Symbolically, (11.1) represents that input I is first applied to $\widehat{C}$, whose output is then applied to P, whose

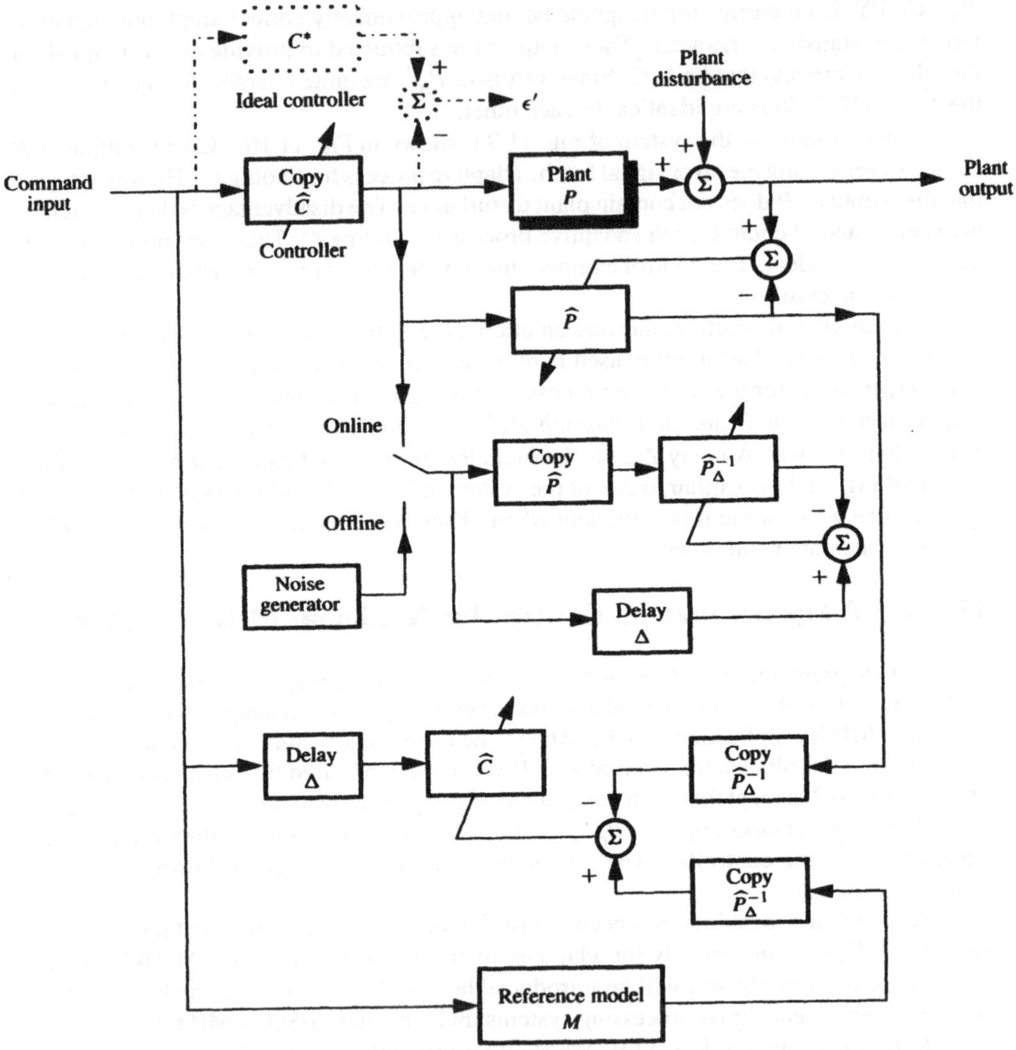

Figure 11.10 An alternative filtered-ϵ LMS system capable of nonlinear control.

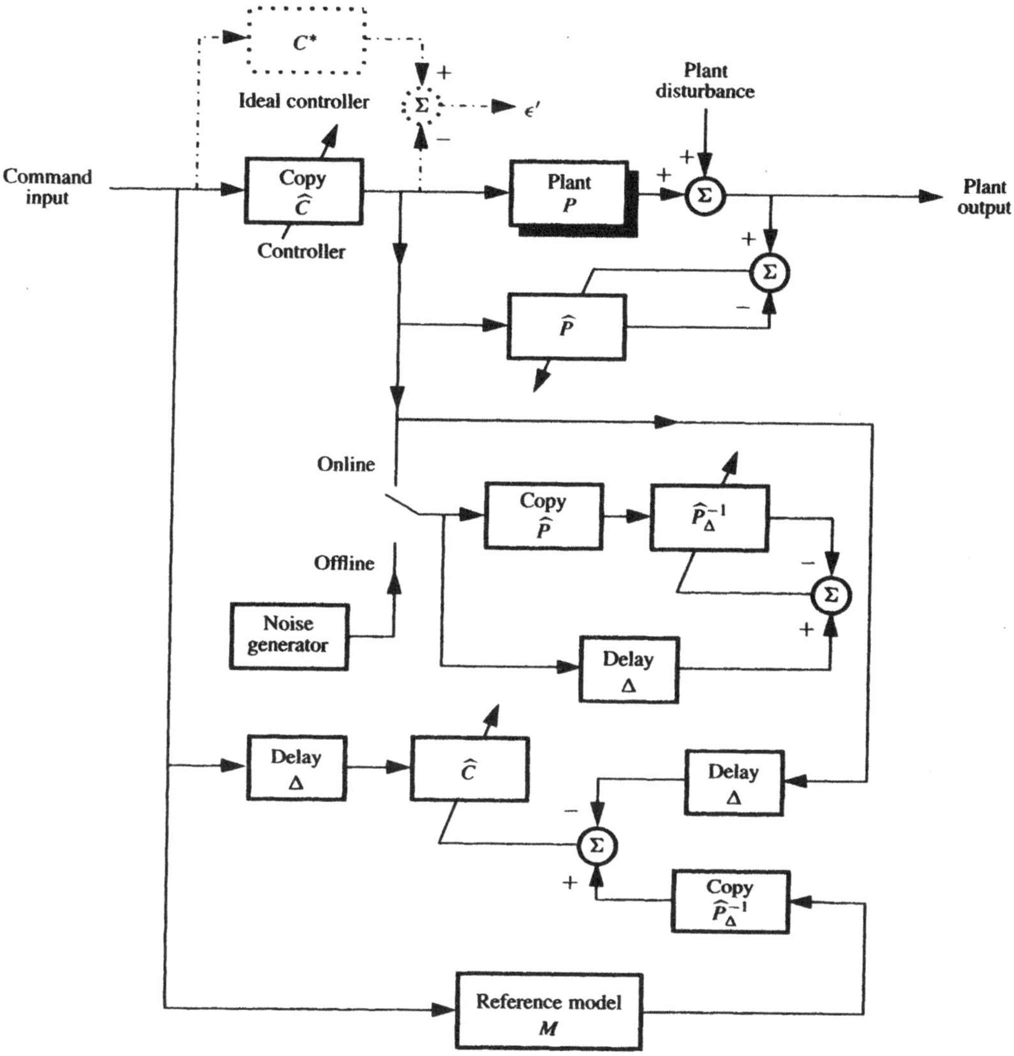

Figure 11.11 Another filtered-ϵ LMS system capable of nonlinear control.

output is the plant output Z. What looks like a product in (11.1) is not meant to be multiplication.

The ideal controller in Fig. 11.9 is designated by C^*. This controller is imaginary and nonexistent, but if one imagined applying input I to C^* and allowing its output to drive the plant P, this would produce a plant output which would be equal to the output of the reference model, designated by M, when driven by input I. Thus,

$$PC^*I = MI. \tag{11.2}$$

Comparing the output of the actual controller with that of the ideal controller, the controller error ϵ' would be expressed as

$$\epsilon' = C^*I - \widehat{C}I. \tag{11.3}$$

The error signal used in the adaptation of $\widehat{C}$ is designated in Fig. 11.9 by $\hat{\epsilon}'$. Following the block diagram and neglecting plant disturbance, this error signal can be expressed as

$$\hat{\epsilon}' = \widehat{P}_{\Delta}^{-1}MI - \widehat{P}_{\Delta}^{-1}P\widehat{C}I. \tag{11.4}$$

Substituting (11.2) into (11.4) yields

$$\hat{\epsilon}' = \widehat{P}_{\Delta}^{-1}PC^*I - \widehat{P}_{\Delta}^{-1}P\widehat{C}I. \tag{11.5}$$

The adaptive process for the formation of $\widehat{P}_{\Delta}^{-1}$ is shown in Fig. 11.9. By inspection, it is clear that if the error of this process is very small, then the cascade of $\widehat{P}$ and $\widehat{P}_{\Delta}^{-1}$ will be equivalent to a delay Δ, given that the input to $\widehat{P}$ is I through $\widehat{C}$ or what is approximately the same, I through C^*. (It is interesting to note that the cascade of $\widehat{P}$ and $\widehat{P}_{\Delta}^{-1}$ will be equivalent to the delay Δ, but because of nonlinearity, the cascade of $\widehat{P}_{\Delta}^{-1}$ and $\widehat{P}$ would not necessarily be equivalent to the simple delay Δ.)

Refer back to (11.5). Assume that $\widehat{P}$ is a close enough fit to P that $\widehat{P}$ can be substituted for P. Accordingly,

$$\hat{\epsilon}' = \widehat{P}_{\Delta}^{-1}\widehat{P}C^*I - \widehat{P}_{\Delta}^{-1}\widehat{P}\widehat{C}I. \tag{11.6}$$

Replacing $\widehat{P}_{\Delta}^{-1}\widehat{P}$ by Δ yields

$$\hat{\epsilon}' = \Delta C^*I - \Delta\widehat{C}I = \Delta(C^*I - \widehat{C}I). \tag{11.7}$$

In light of (11.3), we notice that

$$\hat{\epsilon}' = \Delta(\epsilon'). \tag{11.8}$$

Accordingly, when adapting $\widehat{C}$ to minimize the mean square of $\hat{\epsilon}'$, we are minimizing the mean square of ϵ' and are thereby choosing a controller that is as close as possible to the ideal controller. In obtaining this result, we have assumed that $\widehat{P}$ closely fits P and that $\widehat{P}_{\Delta}^{-1}$ is close to a perfect delayed inverse of $\widehat{P}$. It should be noted that Eqs. (11.1)–(11.8) are not real equations. They are relations that help us keep track of the flow of signals through systems operators and delays, whether linear or nonlinear, SISO or MIMO.

Similar algebraic techniques can be used to show that the adaptive processes illustrated in Figs. 11.10 and 11.11 develop controllers that are close to the ideal for linear or nonlinear plants, and for SISO or MIMO plants. On the other hand, the adaptive process of Fig. 11.8 is not necessarily ideal for nonlinear applications.

To understand why the system of Fig. 11.8 is not generally useful for controlling nonlinear plants, we use the following algebra. The error signal for the adaptive process for finding $\widehat{C}$ in Fig. 11.8 is

$$\begin{pmatrix} \text{error} \\ \text{signal} \\ \text{of } \widehat{C} \end{pmatrix} = \widehat{P}_{\Delta}^{-1}(MI - P\widehat{C}I). \tag{11.9}$$

The error ϵ' is

$$\epsilon' = C^*I - \widehat{C}I. \tag{11.10}$$

The question is: Are these two error signals the same?

Regarding Fig. 11.8, we note that

$$MI = PC^*I. \tag{11.11}$$

This is really a definition of C^*. Substitution of (11.11) into (11.9) yields

$$\begin{pmatrix} \text{error} \\ \text{signal} \\ \text{of } \widehat{C} \end{pmatrix} = \widehat{P}_{\Delta}^{-1}(PC^*I - P\widehat{C}I). \tag{11.12}$$

Assume once again that $\widehat{P}$ is a very close fit to P, so that

$$\begin{pmatrix} \text{error} \\ \text{signal} \\ \text{of } \widehat{C} \end{pmatrix} = \widehat{P}_{\Delta}^{-1}(\widehat{P}C^*I - \widehat{P}\widehat{C}I). \tag{11.13}$$

Since $\widehat{P}$ and $\widehat{P}_{\Delta}^{-1}$ are nonlinear operators, we note that

$$\widehat{P}_{\Delta}^{-1}(\widehat{P}C^*I - \widehat{P}\widehat{C}I) \neq \Delta(C^*I - \widehat{C}I). \tag{11.14}$$

The reason is that for the left-hand side of (11.14), there are two signal components input to $\widehat{P}_{\Delta}^{-1}$. If each component were acting separately, then the combination of $\widehat{P}_{\Delta}^{-1}$ and $\widehat{P}$ would be equivalent to Δ. The two components acting simultaneously cause this equivalence to be invalid. Using (11.14), (11.13), and (11.10), we may conclude that

$$\begin{pmatrix} \text{error} \\ \text{signal} \\ \text{of } \widehat{C} \end{pmatrix} \neq \Delta(\epsilon'). \tag{11.15}$$

Accordingly, adapting $\widehat{C}$ in the system of Fig. 11.8 to minimize mean square error will not necessarily provide a controller that performs as closely as possible to the ideal controller.

4 NONLINEAR PLANT DISTURBANCE CANCELING

Disturbance can be canceled for both the SISO and MIMO cases, even when the plant is nonlinear. Since MIMO is more general, a scheme for doing this with nonlinear MIMO plants is shown in Fig. 11.12. The methodology involved is based on the linear MIMO plant noise-canceling scheme of Fig. 10.19 but adapted to work with nonlinear MIMO plants by utilizing concepts developed for the systems of Figs. 11.4 and 11.9.

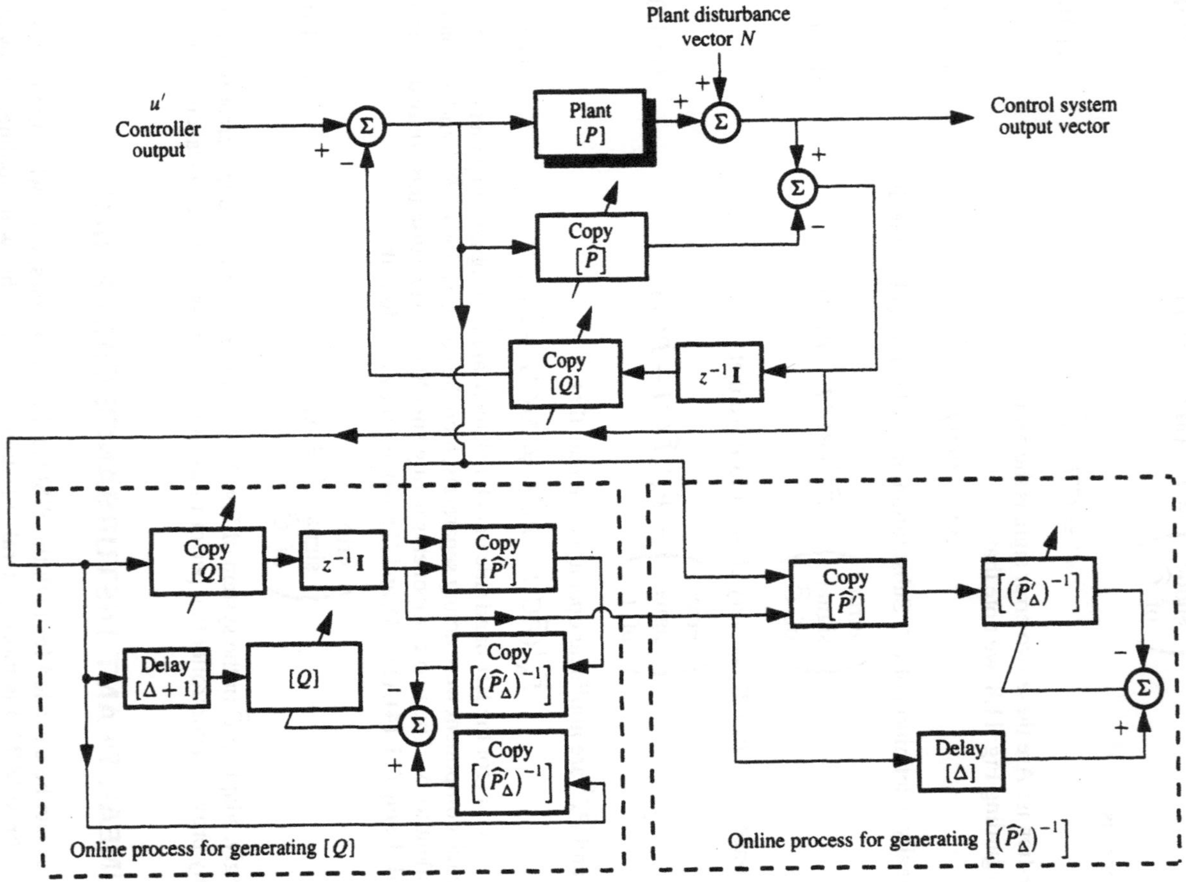

Figure 11.12 Nonlinear MIMO plant disturbance canceling with online formation of $(\widehat{P}'_\Delta)^{-1}$ and Q.

Assume that either scheme B_{NL} or C_{NL} (not shown in Fig. 11.12) has been used to make the plant model $\widehat{P}$. A byproduct of modeling is the derivative model $\widehat{P}'$. This will be used in the processes for generating $(\widehat{P}'_\Delta)^{-1}$ and Q. We assume that the plant disturbance is small compared to the plant input signals, and that the plant responses to plant disturbance can be obtained by appropriately using the derivative model $\widehat{P}'$.

Referring to Fig. 11.12, the online process for generating Q is driven by the input to Q in the actual plant disturbance canceling circuit. This is the correct driving function, having the right dynamic range and statistics. For this subsystem to work properly, however, the correct plant model and inverse derivative model $\widehat{P}'$ must be available (see Fig. 11.4). The input level for this derivative model is set by the actual plant input signal, as shown in Fig. 11.12. The inverse derivative model $(\widehat{P}'_\Delta)^{-1}$ is obtained by another online adaptive process, shown in Fig. 11.12. The driving function for the plant derivative model is the same as that used in the subsystem for generating Q, and the signal level is set by the actual plant input signal. Adapting $(\widehat{P}'_\Delta)^{-1}$ as shown provides the appropriate $(\widehat{P}'_\Delta)^{-1}$ for the process for generating Q.

An inspection of Fig. 11.12 allows one to verify that after convergence of $\widehat{P}$, $\widehat{P}'$, Q, and $(\widehat{P}'_\Delta)^{-1}$, correct signal levels and signal statistics will be present at the inputs of Q, $\widehat{P}'$, and $(\widehat{P}'_\Delta)^{-1}$ in the subsystem for generating Q, and the levels and statistics will be correct at the inputs of $\widehat{P}'$ and $(\widehat{P}'_\Delta)^{-1}$ in the subsystem for generating $(\widehat{P}'_\Delta)^{-1}$. The entire system will provide the capability for the optimal nonlinear plant disturbance canceling. Optimal means that for the chosen configuration of the filter Q and of the filters $\widehat{P}'$ and $(\widehat{P}')^{-1}$, best weights will be found to minimize mean square error and to minimize plant output disturbance power.

11.5 AN INTEGRATED NONLINEAR MIMO INVERSE CONTROL SYSTEM INCORPORATING PLANT DISTURBANCE CANCELING

Figure 11.13 is a nonlinear MIMO system incorporating adaptive inverse dynamic control and plant disturbance canceling. Scheme C_{NL} is used to obtain $\widehat{P}$ and $\widehat{P}'$. Online processes for generating Q and $(\widehat{P}'_\Delta)^{-1}$, which are used in the plant disturbance canceling circuit, are identical to the corresponding processes in Fig. 11.12.

The controller $\widehat{C}$ is generated by an offline process driven by synthetic noise having the same dynamic range and statistics as the input command signal vector. Alternatively, this process could be online with the driving signal being the input command signal vector itself. In order to generate $\widehat{C}$, $(\widehat{P}'_\Delta)^{-1}$ is needed. A process to generate it is shown in Fig. 11.13. This process is driven by the input signal to $\widehat{P}$ in the subsystem for generating $\widehat{C}$, providing the correct dynamic range and statistics for finding $(\widehat{P}'_\Delta)^{-1}$.

When all the adaptive processes in Fig. 11.13 converge (finding $\widehat{P}$ and $\widehat{P}'$, $(\widehat{P}'_\Delta)^{-1}$, Q, $\widehat{P}'_\Delta$, and $\widehat{C}$), the entire system will respond like M as best possible in the least squares sense, with optimal canceling of the plant disturbance.

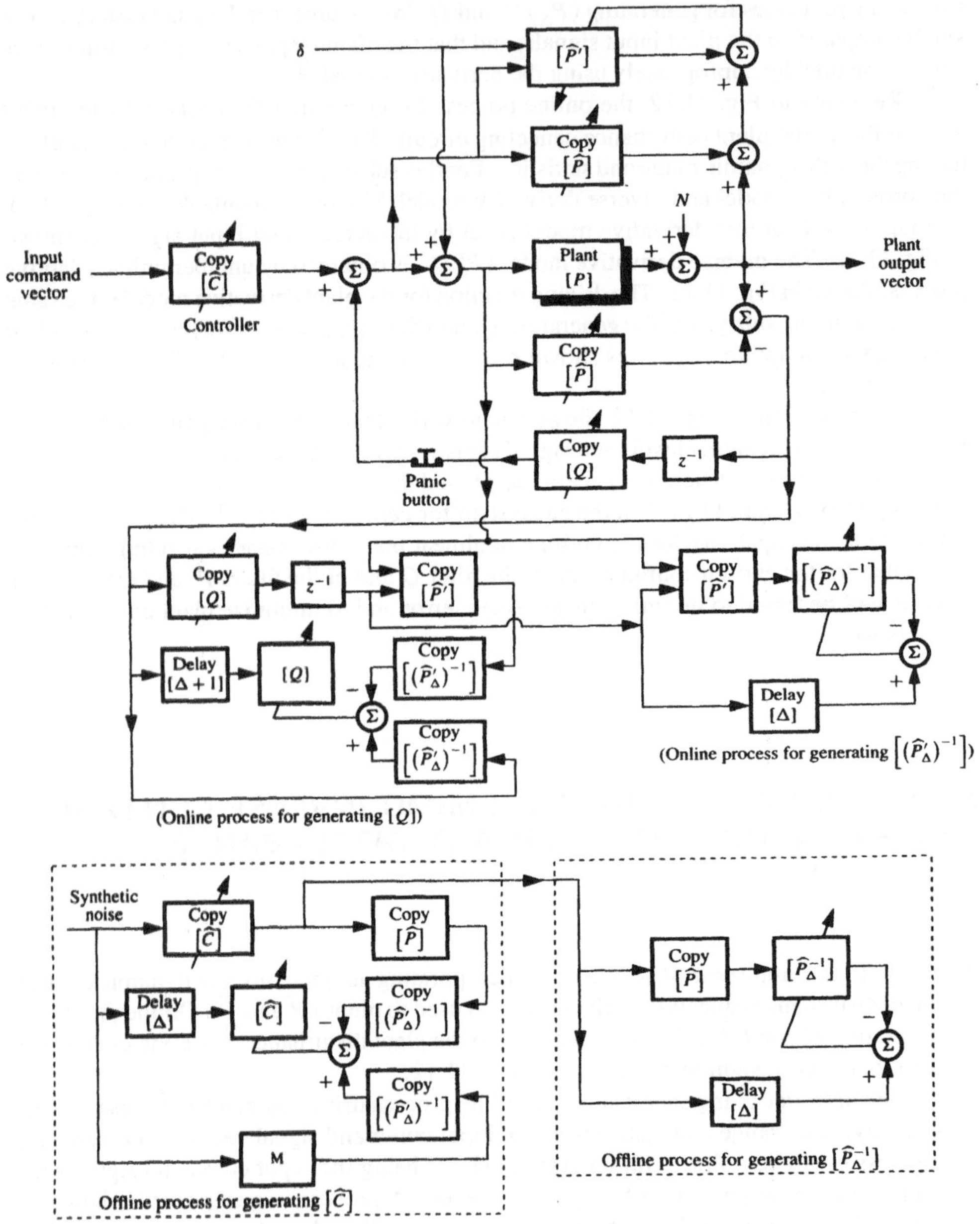

Figure 11.13 An integrated nonlinear MIMO system.

11.6 EXPERIMENTS WITH ADAPTIVE NONLINEAR PLANT MODELING

Practical applications of adaptive inverse control for nonlinear plants are coming along. At this moment, analytical studies and simulation studies of portions of these systems are being made. In this section, we review results of simulation studies of adaptive modeling of an unknown nonlinear plant.

To provide background information for these studies, Appendix G contains a reprint of [6] by Widrow and Lehr which describes the backpropagation algorithm used in adapting neural networks and explains how the LMS algorithm is at the foundation of backpropagation. Appendix H shows alternative ways of structuring nonlinear adaptive filters based on tapped delay lines and neural networks and shows how adaptive inverse control can be done for MIMO plants using these structures and adapting with the filtered-ϵ method.

In this section, we will use an adaptive filter of the type shown in Fig. 11.14. The input signal is fed to a tapped delay line. The tap signals are used to drive the weights of a linear combiner to make an adaptive linear filter. The same tap signals are also used to drive a two-layer neural network comprising a nonlinear adaptive filter. The outputs of the linear adaptive filter and the nonlinear adaptive filter are summed to make the output signal. The weights of the linear adaptive filter are adapted by LMS. The weights of the nonlinear adaptive filter are adapted by backpropagation. The output could be compared to a desired response to obtain an error signal. All of the weights are thus adapted by steepest descent to minimize mean square error. This filter is like the one shown in Fig. 11.2.

The filter of Fig. 11.4 can be used for nonlinear plant identification. The idea is illustrated in Fig. 11.15. The nonlinear dynamic plant was constructed for demonstration purposes by combining two linear one-pole digital filters in a cascade with a nonlinear sigmoid. The dynamic parts are linear. The nonlinear part is memoryless. This plant was not intended to be a generic nonlinear dynamic system, but merely a good example of such a system.

The adaptive system, containing linear and nonlinear parts, has the challenging problem of modeling the plant and learning to imitate its behavior. The architecture of the adaptive model is completely different from that of the plant. At the outset, there is no assurance that values of the adaptive weights would exist to allow equivalent behavior for the plant and the adaptive model, given a wide variety of plant input signals.

The first experiment in plant modeling was done by setting and fixing the weights of the nonlinear adaptive filter to zero. The linear adaptive filter was adjusted to minimize mean square error (refer to Fig. 11.15). After training, a sample of the nonlinear plant output was plotted simultaneously with the linear adaptive filter output. In Fig. 11.16, the solid curve is the plant output, and the dashed curve is the adaptive filter output. The error is the difference between the two curves and, not surprisingly, it is significant. One cannot expect to be able to model a nonlinear plant with a linear adaptive filter.

The next experiment with the systems of Fig. 11.15 was done by adapting both the linear and nonlinear filters simultaneously, making possible a nonlinear model of the nonlinear plant. After convergence, segments of plant output signal and model output signal were plotted. In Fig. 11.17, the solid curve is the plant output and the dashed curve is the nonlinear model output. The error between them is very small, indicating that the model is a good fit to the plant, at least for the given random input.

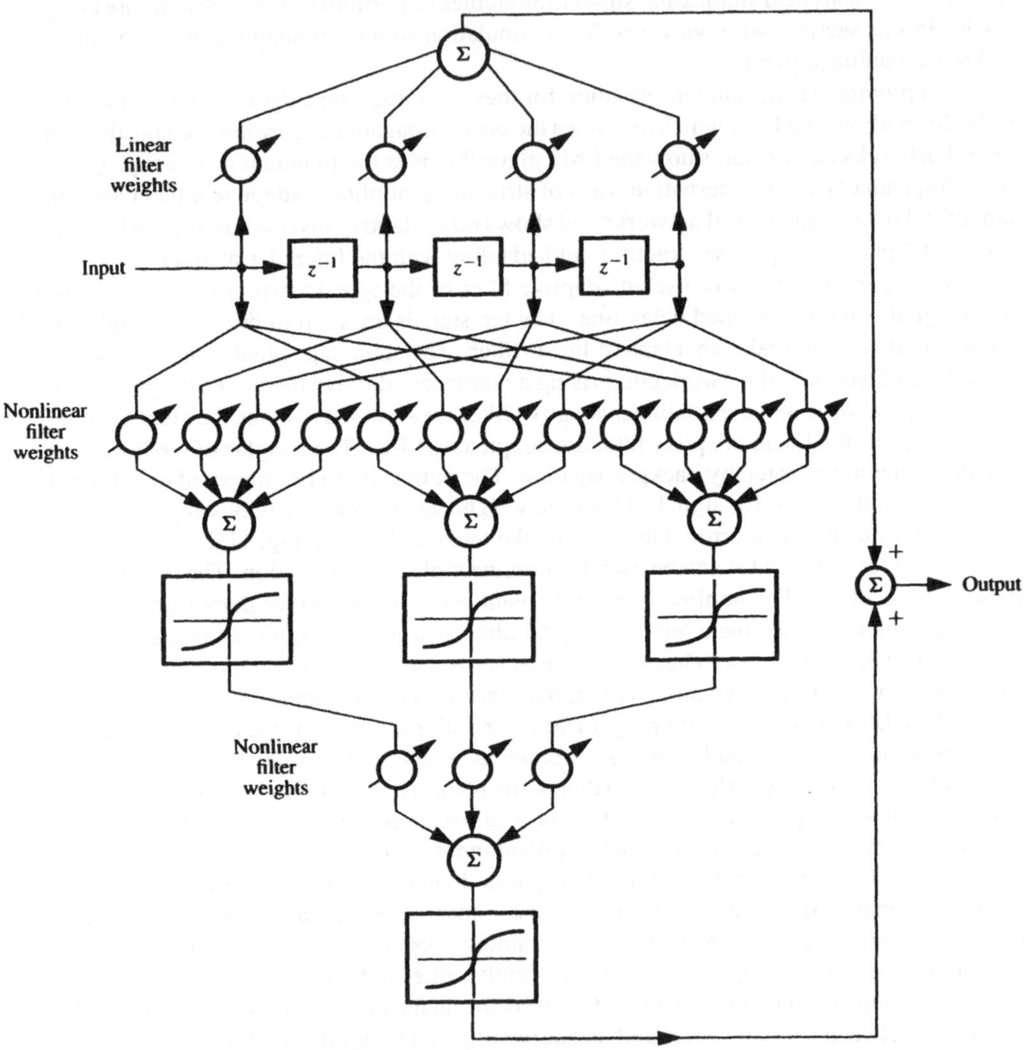

Figure 11.14 A linear/nonlinear adaptive filter.

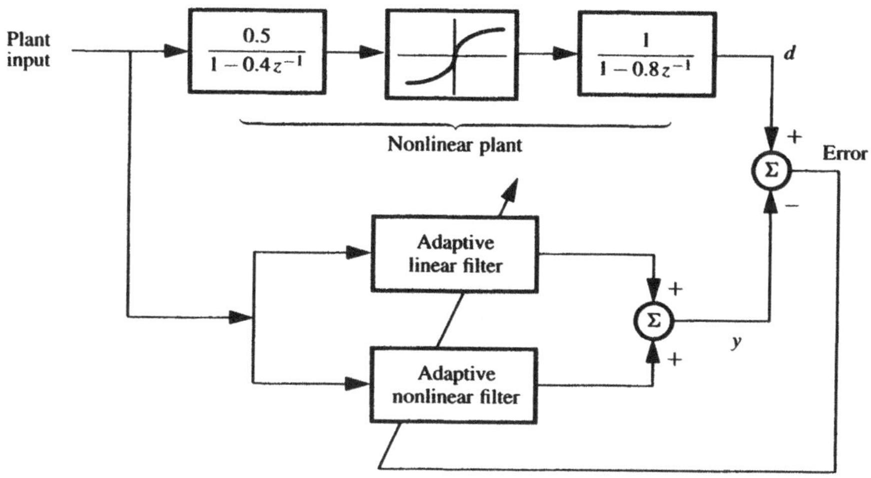

Figure 11.15 Nonlinear plant identification.

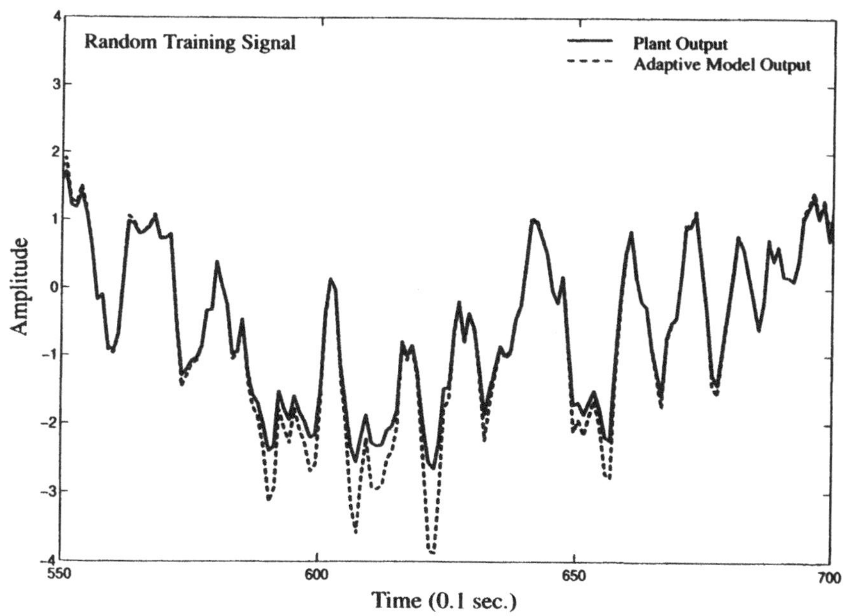

Figure 11.16 Plant output and adaptive linear model output.

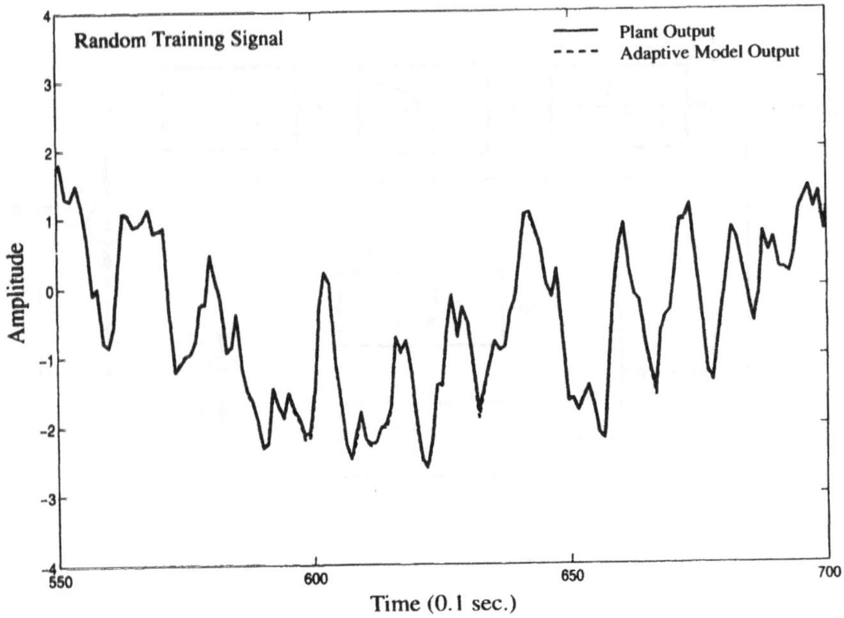

Figure 11.17 Plant output and adaptive nonlinear model output.

When the adaptive model was only linear, the converged weights were stored after training with the random input signal. The weights of the nonlinear filter were set to zero. A sinusoidal test input signal was applied to the plant and the linear model and typical outputs are plotted in Fig. 11.18. The solid curve is the plant output and the dashed curve is the output of the linear model. The error between them is considerable, and this is no surprise.

Another experiment was done, using the converged weights of the nonlinear model. The linear and nonlinear filters were simultaneously trained with a random input, and the weights converged and were fixed. A sinusoidal test input signal was applied to both the plant and the nonlinear converged model, and the outputs are plotted in Fig. 11.19. The solid curve is the plant output and the dashed curve is the output of the nonlinear model. The curves do not match perfectly, but they are very close.

These experiments were done by Bradley Smith. They are interesting, and they suggest that it will be possible to make good direct and inverse models for nonlinear dynamic plants. But there is no doubt that much work remains to be done before this kind of modeling becomes routine.

11.7 SUMMARY

This chapter develops a set of fundamental techniques for using adaptive inverse control with nonlinear plants, both SISO and MIMO. The chapter begins with block diagrams for nonlinear tapped delay-line filters based on Volterra series and on neural networks. The weights of the Volterra filter are adapted by the LMS algorithm. The weights of the neu-

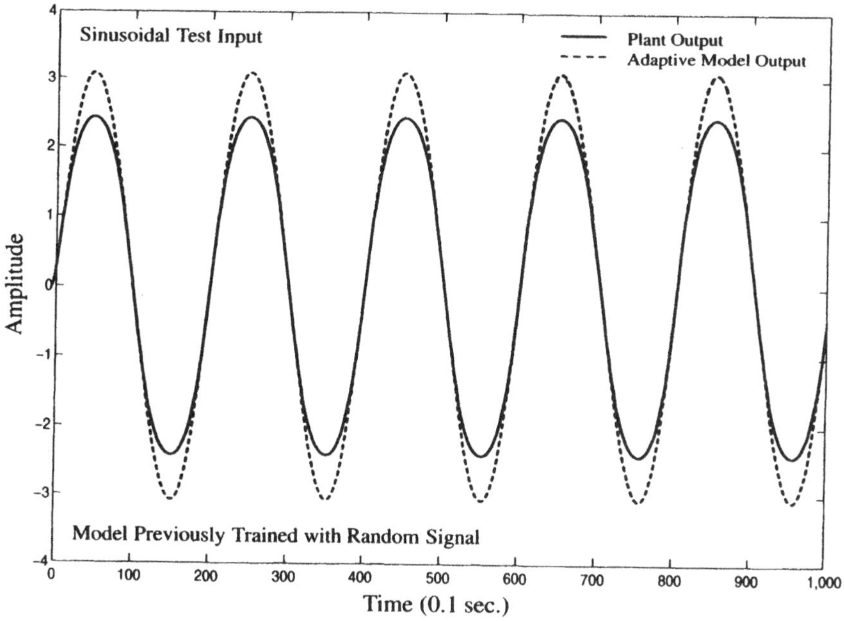

Figure 11.18 Plant output and linear model output with sinusoidal test input signal. Linear model converged and fixed after training on random input.

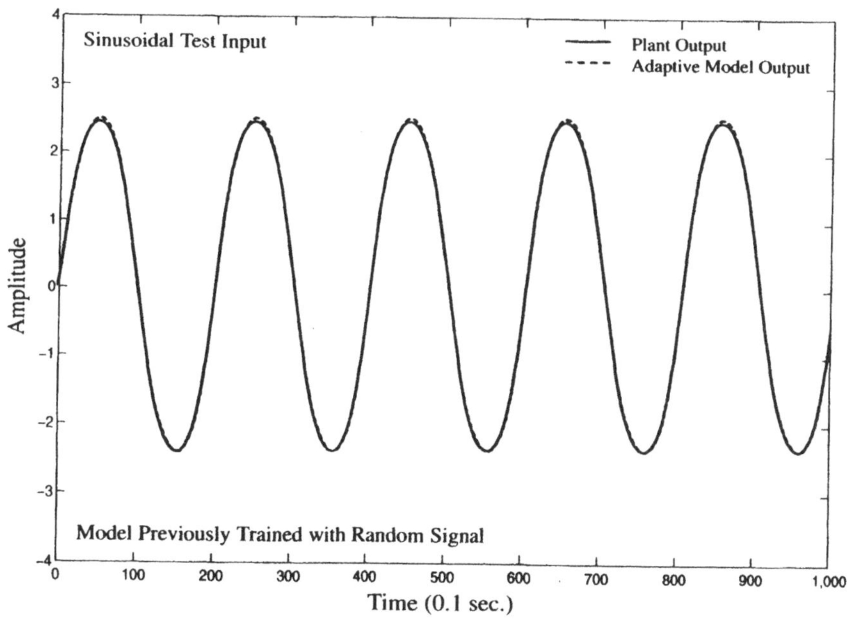

Figure 11.19 Plant output and nonlinear model output with sinusoidal test input signal. Nonlinear model converged and fixed after training on random input.

ral network filter are adapted by the backpropagation algorithm. Backpropagation is a substantial generalization of the LMS algorithm [2, 6]. For the development of the chapter, the simpler Volterra filter was chosen for illustration. In general, experience indicates that the neural network filter yields better performance, however.

Ordinary plant modeling without dither can be done in a conventional way. Plant modeling with dither added to the plant input, like scheme A, can also be done in a conventional way. But where the modeling action is primarily based on the dither alone, as with schemes B and C, a whole new approach is needed. The reason is that the dither, typically small in amplitude compared with the plant input signal, would excite only a small signal response on top of the large plant input response. Proper modeling of the plant can be done by making use of a derivative model rather than an actual model. The weights of the derivative model are in turn used to obtain the weights of the actual model. Using the derivative model idea, nonlinear versions of schemes B and C are developed.

Finding the inverse of a nonlinear plant is described next. Although nonlinear plants do not have exact inverses, it is possible to determine weights for a nonlinear filter so that, for a given input signal, the cascade of the nonlinear filter and the nonlinear plant will produce an output that is a best least squares match to the output of a reference model driven by the same input signal. The weights of the nonlinear filter were adapted by the filtered-ϵ LMS algorithm.

Methods for plant disturbance canceling are developed next. With small but significant differences, they are quite similar to those used with linear plants. Both offline and online techniques may be used to find Q.

Having all the parts, an entire nonlinear adaptive inverse control system is assembled, featuring separate control of plant disturbance and separate control of nonlinear plant dynamics.

MIMO as well as SISO designs are presented in this chapter. Since nonlinear operators are not commutable and MIMO systems components are also not commutable, making the systems MIMO did not significantly add to the conceptual complexity. At least the block diagrams for nonlinear MIMO systems are not much different from those of nonlinear SISO systems.

In working with nonlinear plants, care was exercised not to commute nonlinear operators. The same was true for MIMO operators, even if they were linear. An algebra was developed for nonlinear and MIMO signal flow to help in the design and analysis of such systems.

Whenever one is adapting a nonlinear filter, SISO or MIMO, the input signal levels and statistical characteristics must be the same as will be encountered in actual practice in the final system.

We believe that the proposed methods for finding the weights of a nonlinear inverse controller and the weights of an adaptive plant disturbance canceler are optimal and lead to the best least squares solutions possible for the chosen configurations of adaptive filters. Much work needs to be done with nonlinear techniques for adaptive control. Some of the questions are beginning to emerge. Surprisingly, even some of the answers are beginning to emerge as well. There are great opportunities here.

Bibliography for Chapter 11

[1] F. ROSENBLATT, "On the convergence of reinforcement procedures in simple perceptrons," Cornell Aeronautical Laboratory Report, VG-1196-G-4, February 1960, Buffalo, NY.

[2] B. WIDROW, and M.E. HOFF, "Adaptive switching circuits," in *IRE WESCON Conv. Rec.*, Pt. 4, 1960, pp. 96–104.

[3] P. WERBOS, "Beyond regression: New tools for prediction and analysis in the behavioral sciences," Ph.D. diss., Harvard University, August 1974.

[4] D.B. PARKER, "Learning logic," Tech. Rep. TR-47, Center for Computational Research in Economics and Management Science, M.I.T., April 1985.

[5] D.E. RUMELHART, and J.L. MCCLELLAND, *Parallel distributed processing*, Vols. I and II (Cambridge, MA: M.I.T. Press, 1986).

[6] B. WIDROW, and M.A. LEHR, "30 Years of adaptive neural networks: Perceptron, Madaline and backpropagation," *Proc. IEEE* Vol. 78, No. 9 (September 1990), pp. 1415–1441.

[7] S.I. GALLANT, *Neural network learning* (Cambridge, MA: M.I.T. Press, 1993).

[8] S. HAYKIN, *Neural networks* (New York: Macmillan, 1994).

[9] R. HECHT-NIELSEN, *Neurocomputing* (Reading, MA: Addison-Wesley, 1990).

[10] J.A. HERTZ, A. KROGH, and R.G. PALMER, *Introduction to the theory of neural computation* (Reading, MA: Addison-Wesley, 1991).

[11] B. KOSKO, *Neural networks and fuzzy systems* (Englewood Cliffs, NJ: Prentice Hall, 1992).

[12] P. MEHRA, and B.W. WAH, Ed., *Artificial neural networks: Concepts and theory* (Los Alamitos, CA: IEEE Computer Society Press, 1992).

[13] P.K. SIMPSON, *Artificial neural systems* (New York: Pergamon Press, 1990).

[14] M. KAWATO, Y. UNO, M. ISOBE, and R. SUZUKI, "Hierarchical neural network model for voluntary movement with application to robotics," *IEEE Control Systems Magazine*, Vol. 8, No. 2 (April 1988), pp. 8–16.

[15] D. PSALTIS, A. SIDERIS, and A.A. YAMAMURA, "A multilayered neural network controller," *IEEE Control Systems Magazine*, Vol. 8, No. 2 (April 1988), pp. 17–21.

[16] W.T. MILLER, R.S. SUTTON, and P. Werbos, Eds., *Neural Networks for Control* (Cambridge, MA: M.I.T. Press, 1990).

Chapter 12

Pleasant Surprises

While developing the principal mathematical results of adaptive inverse control, we would come upon a result from time to time that seemed to us to be a pleasant surprise. Many of these results were desired and were anticipated intuitively. Some were easy to prove and some required a great deal of algebra. We were hoping that the desired results would be true because if they were, it would make the theory simple and the applications easy. Whenever one of the desired results was proven to be true, we said to ourselves: "This is amazing. There must be something right about this approach to adaptive control." The purpose of this brief chapter is to review the pleasant surprises and to summarize the findings of this book.

1. Precise inverse controllers can be devised for minimum-phase plants and, with somewhat delayed response, for nonminimum-phase plants too.
2. The effect of closed-loop response can be obtained in an open-loop feedforward control system by using the feedback inherent in adaptive filtering to find the inverse adaptive controller.
3. Plant disturbance can be optimally canceled using feedback with zero gain around the loop. Best linear least squares plant disturbance canceling can be accomplished without altering the plant transfer function.
4. Plant disturbance canceling can be done independently of plant dynamic control. The optimization of one of these processes is not compromised by the optimization of the other.
5. Achieving an overall system response which is a best least squares estimate of a reference model's response is generally straightforward and natural with adaptive inverse control.
6. If the plant is unstable, it must first be stabilized by feedback. Then the plant and its feedback are subject to adaptive inverse control, treating the plant and its stabilizing feedback as an equivalent plant. The ability to cancel plant disturbance is unaffected by the choice of stabilizing feedback. The ability to achieve a desired overall system dynamic response is also unaffected by the choice of the stabilizing feedback. If the inverse needs delay for its realization, the required delay will not depend on the choice of the stabilization feedback.

7. When creating an inverse controller by the filtered-ϵ or the filtered-X algorithm, small errors in the plant model cause no errors in the controller. Feedback in the adaptive process causes the overall system dynamic response to hover about an equilibrium condition corresponding to an overall dynamic response that is a best least squares match to the reference model response.
8. When adapting an inverse controller with an algorithm other than filtered-ϵ or filtered-X, errors in the plant model cause errors in the controller. However, these errors are compensated for by second-order errors in the plant dynamics caused by the feedback of the plant disturbance canceler. Thus, the interaction of the multiple adaptive processes in the integrated system cause its overall dynamic response to hover about an equilibrium condition which is a best least squares match to the reference model response. The adaptive feedback provides robust behavior for the overall system.
9. The use of adaptive feedback does not create stability problems, except during start-up or during a sudden catastrophic change in plant dynamics when the plant disturbance canceling loop could get out of balance and go unstable. The remedy is to temporarily abstain from plant disturbance canceling by breaking the disturbance canceling loop until the plant model regains a response close to that of the plant.
10. Adaptive inverse control applies readily to the control of MIMO systems. Learning time in a MIMO system goes up only linearly with the number of channels, rather than with the square of the number of channels.
11. Adaptive inverse control applies readily to the control of nonlinear systems, whether SISO or MIMO.
12. Dynamic control of either a minimum-phase plant or a nonminimum-phase plant can certainly be accomplished with adaptive inverse control. But what of a plant having a zero exactly on the unit circle in the z-plane? The inverse of such a plant would need to have a pole on the unit circle, and it would be unstable with either a left-handed or a right-handed impulse response. Here is a case where inverse control should fail. But it does not seem to, and this is surprising. We finish this chapter with a set of experiments making inverse controllers for plants with zeros which are very near the unit circle and in some cases, exactly on it.

Figure 12.1 shows the impulse response of a stable, discrete, disturbance-free plant having two poles and one zero. The poles are complex conjugates which correspond to a damped oscillatory response. The transfer function of the plant is given by

$$P(z) = \frac{(z - 1.05)}{[z - (0.9 + 0.3j)][z - (0.9 - 0.3j)]}. \tag{12.1}$$

The plant is nonminimum-phase since its zero lies close to but just outside the unit circle. The impulse response of a delayed inverse filter is shown in Fig. 12.2. The delay was chosen arbitrarily to be 100 sample periods, although it was really not necessary to have such a large delay in order to get an excellent inverse.

Figure 12.3 shows the convolution of the impulse response of the plant and the delayed inverse. It is almost a perfect delayed impulse. Figure 12.4 is a plot of the plant output when driven by the inverse filter as a controller. The command input was white noise. Superimposed is a plot of the output of a pure delay of 100 sample times. It is driven by the

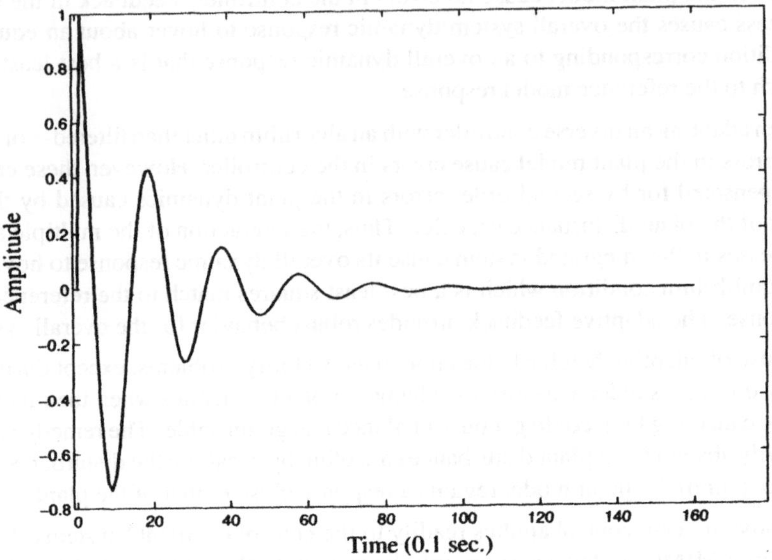

Figure 12.1 Impulse response of a discrete damped oscillatory nonminimum-phase plant.

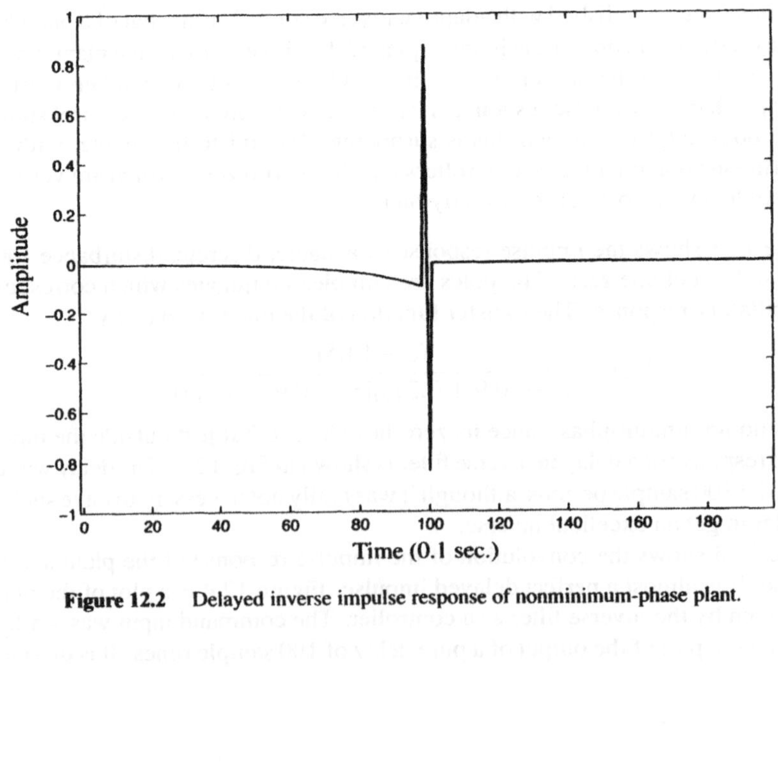

Figure 12.2 Delayed inverse impulse response of nonminimum-phase plant.

same command input as the controller and plant. The two plots are almost identical. There is nothing surprising about this.

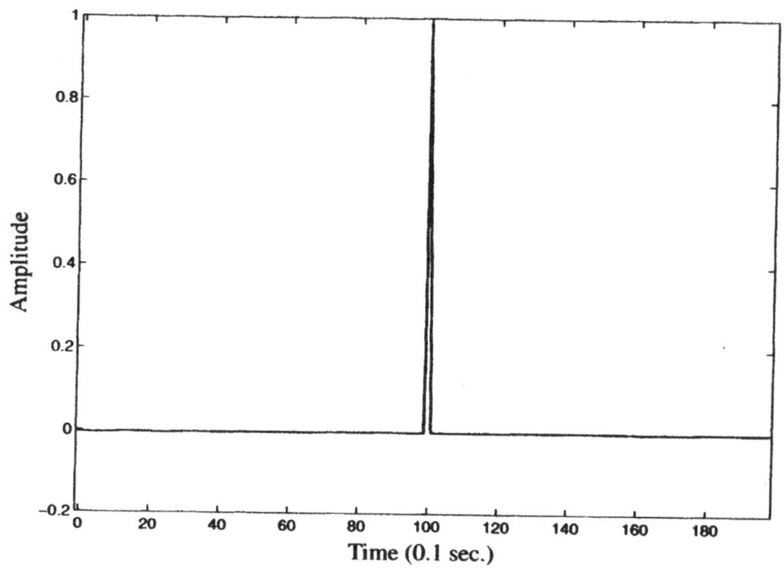

Figure 12.3 Convolution of impulse response of nonminimum-phase plant and its delayed inverse controller.

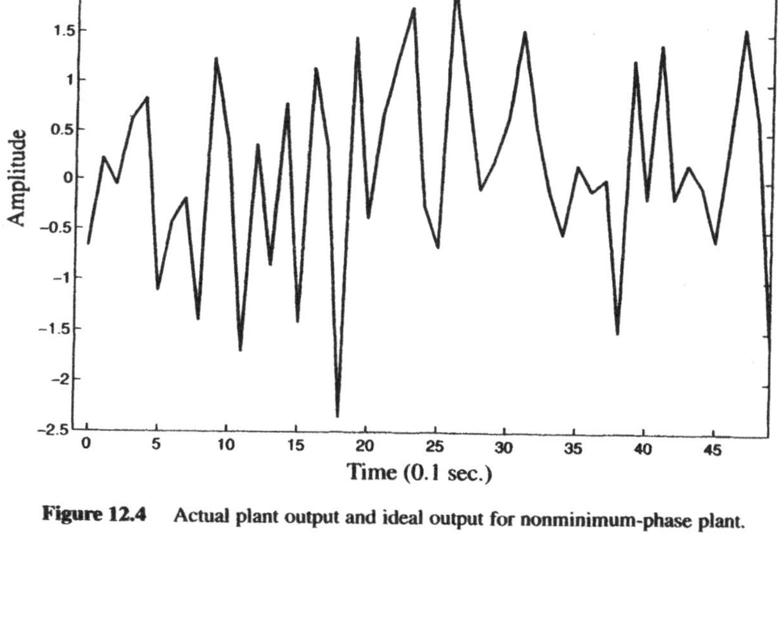

Figure 12.4 Actual plant output and ideal output for nonminimum-phase plant.

Figure 12.5 shows the impulse response of a similar plant, one that has the same poles but has a single zero close to but inside the unit circle. Its transfer function is

$$P(z) = \frac{(z - 0.95)}{[z - (0.9 + 0.3j)][z - (0.9 - 0.3j)]}. \tag{12.2}$$

This plant is minimum-phase. A delayed inverse impulse response is shown in Fig. 12.6. The delay was chosen to be 100 sample times to be like the previous case. A perfect inverse could have been obtained with no delay at all, however.

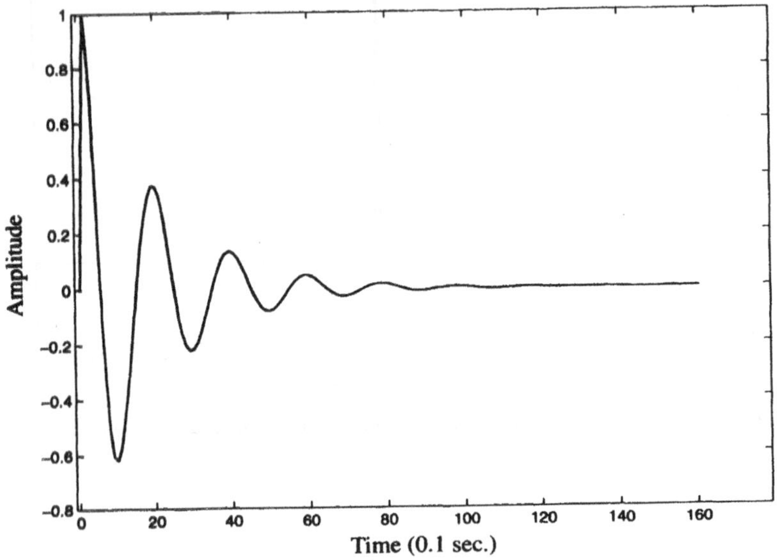

Figure 12.5 Impulse response of a discrete damped oscillatory minimum-phase plant.

The convolution of the impulse responses of the plant and its delayed inverse are shown in Fig. 12.7. The output of the plant driven by the inverse filter as a controller and the output of a delay unit of 100 sample periods are plotted in Fig. 12.8, both driven by the same command input. The two plots superimpose with almost perfect registration. Again, there is no surprise here.

Figure 12.9 shows the impulse response of another plant, similar to the previous two plants. In this case, the poles are the same but the zero is exactly on the unit circle. The transfer function is

$$P(z) = \frac{(z - 1)}{[z - (0.9 + 0.3j)][z - (0.9 - 0.3j)]}. \tag{12.3}$$

This plant is not minimum-phase or nonminimum-phase. Its inverse transfer function contains the factor $1/(1 - z^{-1})$. Either to the right or to the left in time, the corresponding impulse response could not be approximated by an FIR impulse response. If we were to allow an IIR inverse, it would not be stable. We should not be able to construct an inverse for the plant represented by (12.3). But we did it! Figure 12.10 shows the impulse response of a delayed FIR inverse, with the delay set to 100 sample periods. Figure 12.11 shows the

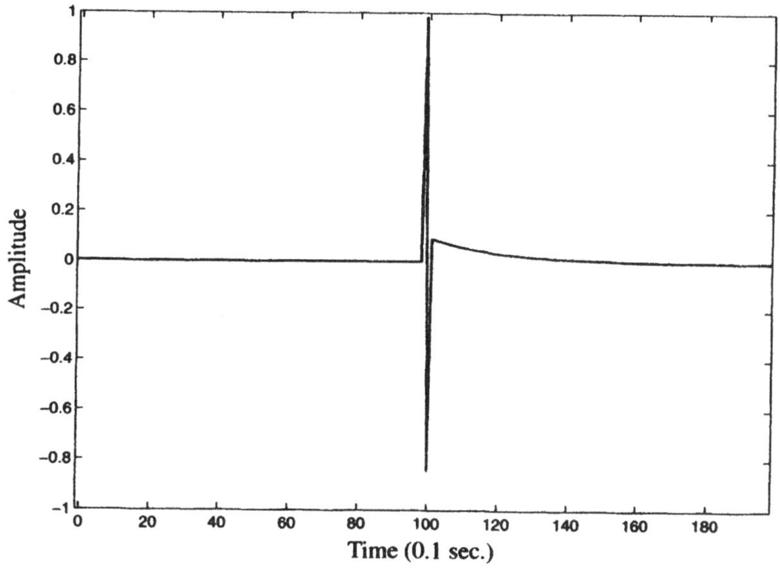

Figure 12.6 Delayed inverse impulse response of minimum-phase plant.

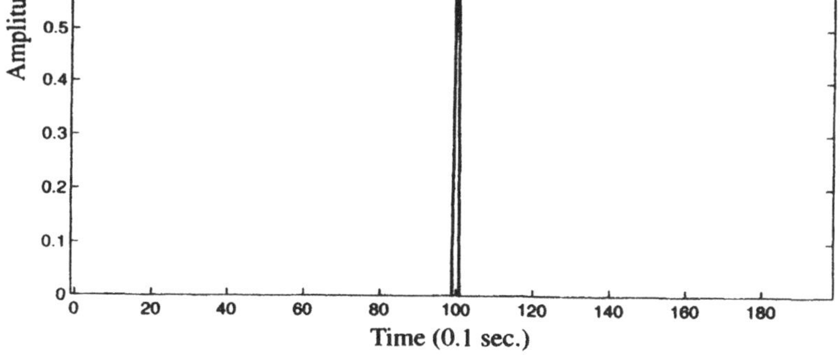

Figure 12.7 Convolution of impulse response of minimum-phase plant and its delayed inverse controller.

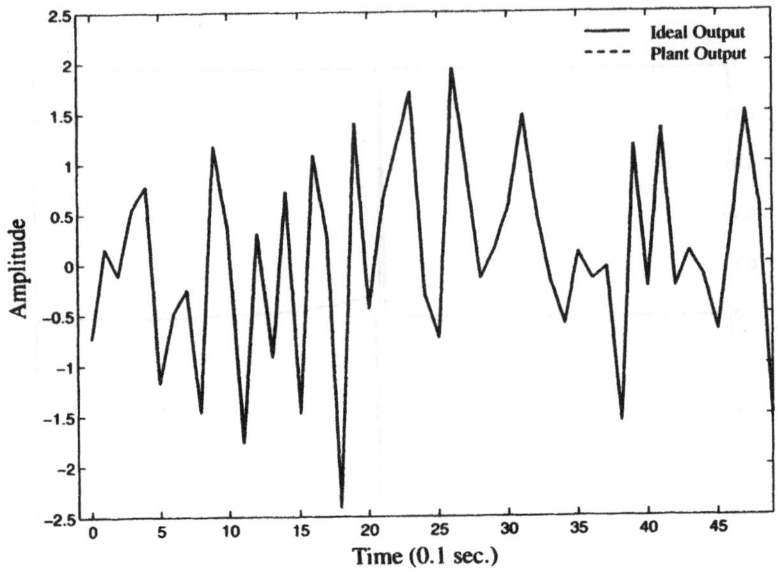

Figure 12.8 Actual plant output and ideal output for minimum-phase plant.

convolution of the impulse responses of the plant and its delayed inverse. This is not a bad delayed impulse, not far from perfect. Using the delayed inverse filter as a controller to drive the plant, the plant output is plotted in Fig. 12.12. Superimposed is a plot of the output of a delay of 100 sample periods, driven by the same command input as the controller of the plant. The two plots are very close, although they do not register perfectly.

These experiments exhibit a continuum of behavior as the plant zero is moved from outside the unit circle to the unit circle itself, to inside the unit circle. Inverse control works easily whether the plant is nonminimum-phase or minimum-phase, with the zero in question close to the unit circle or far from it. It also appears to work reasonably well even with a zero exactly on the unit circle. These are surprising results. We don't know why this works. It is not clear that this will work with all possible zero placements on the unit circle. Our knowledge of inverse control of dynamic systems is based on the theory of two-sided Laplace transforms. There may be some other way to analyze and explain these things that we do not yet understand.

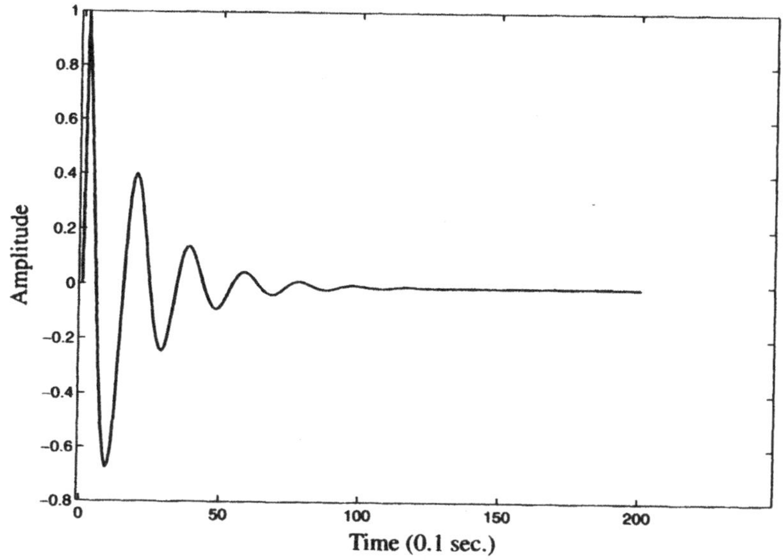

Figure 12.9 Impulse response of a discrete damped oscillatory plant with a zero exactly on the unit circle.

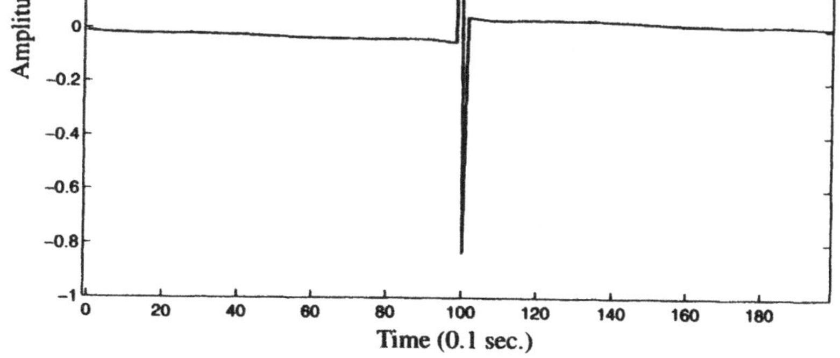

Figure 12.10 Delayed inverse impulse response of plant with zero on the unit circle.

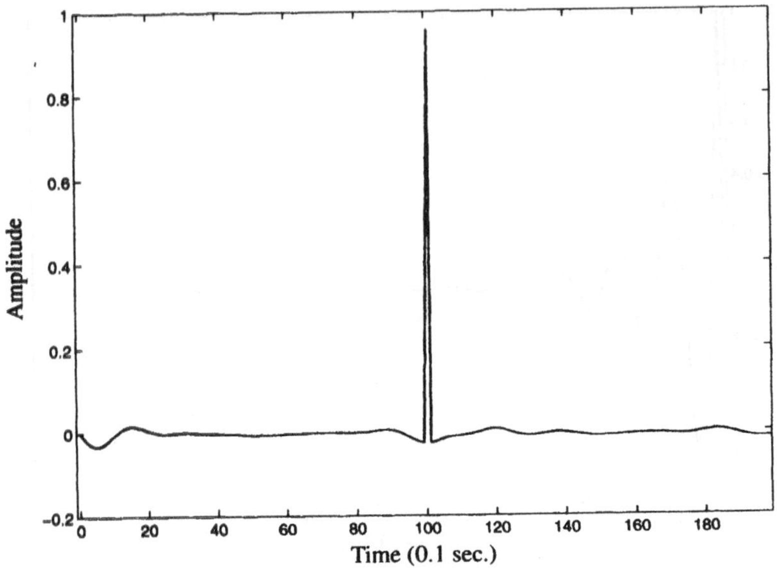

Figure 12.11 Convolution of impulse responses of plant with zero on unit circle and its delayed inverse controller.

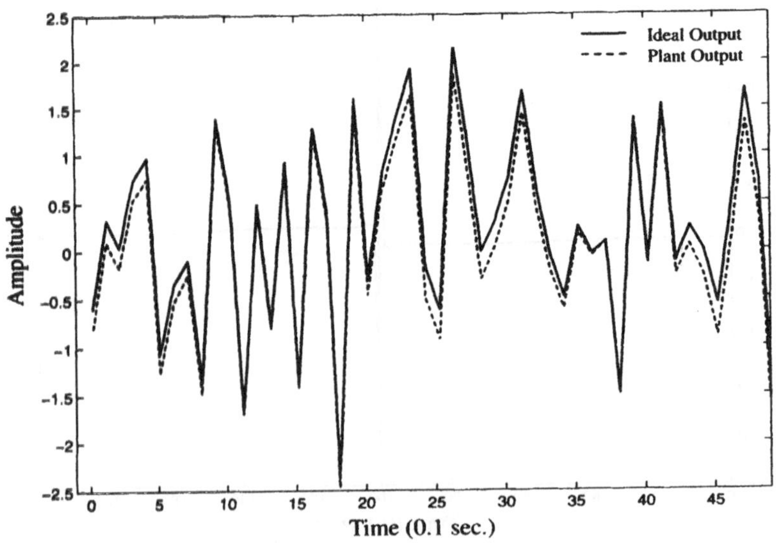

Figure 12.12 Actual plant output and ideal output for plant with zero on unit circle.

Appendix A

Stability and Misadjustment of the LMS Adaptive Filter

In this appendix, we will analyze stability and misadjustment of the adaptive transversal filter whose weights are controlled by the Widrow-Hoff LMS algorithm. The adaptive filter is comprised of a tapped delay line, shown in Fig. 3.1, whose tap signals are applied to an adaptive linear combiner in order to generate the filter output.

A.1 TIME CONSTANTS AND STABILITY OF THE MEAN OF THE WEIGHT VECTOR

At each moment of time k, the input vector to the weights is $\mathbf{X}_k$ defined by (3.1), the filter output is y_k defined by (3.3), the weight vector is $\mathbf{W}_k$ defined by (3.2), and the error ϵ_k is defined by (3.4). The Wiener solution is $\mathbf{W}^*$, defined by (3.10), and the difference between the weight vector $\mathbf{W}_k$ and the Wiener solution is defined as $\mathbf{V}_k$ and is given by (3.14).

We will analyze stability of the LMS adaptive filter by examining it in the application of modeling an unknown system in the configuration of Fig. A.1. A common input signal x_k is applied to both the adaptive filter and the fixed unknown filter. The adaptive filter is modeling the fixed filter. The fixed filter is assumed to have an independent internal noise, which is represented by an additive zero-mean noise n_k at the filter output. The difference between the noisy output and the adaptive filter output is the error signal ϵ_k which is used by the LMS algorithm to adapt $\mathbf{W}_k$. We assume that the adaptive filter has enough weights to perfectly match the dynamic behavior of the fixed filter. Therefore, the impulse response of the fixed filter is the Wiener solution for the adaptive filter. If the adaptive weights were set to their Wiener values, the error ϵ_k would be equal to the noise n_k. The minimum mean square error of the adaptive filter is accordingly,

$$\xi_{\min} = E[n_k^2]. \tag{A.1}$$

A study of stability of the LMS algorithm as used in the context of Fig. A.1 will not be perfectly general, but it will be applicable to a wide variety of practical cases. The resulting analysis leads to a simple stability criterion that works not only for systems configurations like in Fig. A.1, but it is known to work for many other configurations not covered by the following derivation [2].

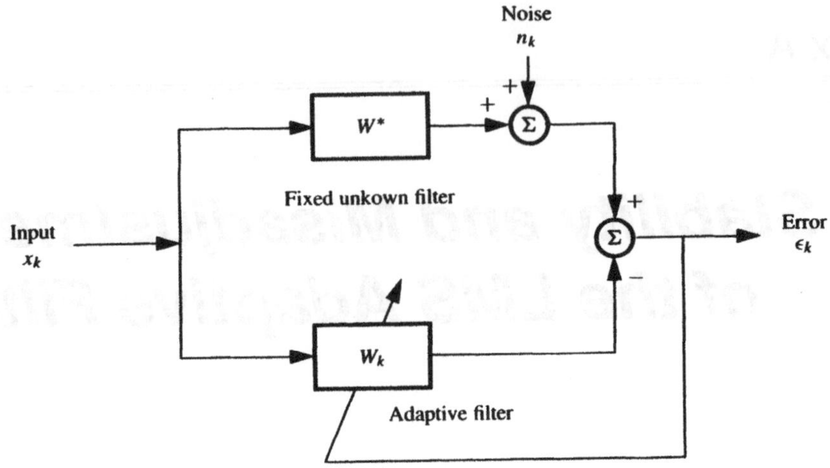

Figure A.1 Adaptive modeling configuration used in study of LMS stability.

Inspection of Fig. A.1 shows that the error ϵ_k can be expressed as

$$\begin{aligned}\epsilon_k &= (\mathbf{W}^* - \mathbf{W}_k)^T\mathbf{X}_k + n_k \\ &= -\mathbf{V}_k^T\mathbf{X}_k + n_k \\ &= -\mathbf{X}_k^T\mathbf{V}_k + n_k.\end{aligned} \tag{A.2}$$

The LMS adaptation rule (3.35) requires that

$$\mathbf{W}_{k+1} = \mathbf{W}_k + 2\mu\epsilon_k\mathbf{X}_k.$$

Subtracting $\mathbf{W}^*$ from both sides yields

$$\mathbf{V}_{k+1} = \mathbf{V}_k + 2\mu\epsilon_k\mathbf{X}_k. \tag{A.3}$$

Substituting (A.2) into this expression gives

$$\begin{aligned}\mathbf{V}_{k+1} &= \mathbf{V}_k + 2\mu(-\mathbf{X}_k^T\mathbf{V}_k + n_k)\mathbf{X}_k, \text{ or} \\ \mathbf{V}_{k+1} &= \mathbf{V}_k + 2\mu\mathbf{X}_k(-\mathbf{X}_k^T\mathbf{V}_k + n_k), \text{ or} \\ \mathbf{V}_{k+1} &= (\mathbf{I} - 2\mu\mathbf{X}_k\mathbf{X}_k^T)\mathbf{V}_k + 2\mu n_k\mathbf{X}_k.\end{aligned} \tag{A.4}$$

Now, we take expectation of both sides to get

$$E[\mathbf{V}_{k+1}] = E\left[(\mathbf{I} - 2\mu\mathbf{X}_k\mathbf{X}_k^T)\mathbf{V}_k\right]. \tag{A.5}$$

The last term of (A.4) vanished under expectation since the independent noise n_k is uncorrelated with input $\mathbf{X}_k$.

Assume that μ is small. With slow adaptation, the adaptive weight vector depends on many past input samples, and its dependence on the present input vector is negligible. Accordingly, we assume that the input vector $\mathbf{X}_k$ is independent of the weight vector $\mathbf{W}_k$ and therefore it is independent of $\mathbf{V}_k$. This same assumption was first made in [1]. It follows that

$$E[\mathbf{V}_{k+1}] = E[\mathbf{V}_k] - 2\mu E[\mathbf{X}_k\mathbf{X}_k^T]E[\mathbf{V}_k]. \tag{A.6}$$

Using Eq. (3.8), we can write

$$\begin{aligned} E[\mathbf{V}_{k+1}] &= E[\mathbf{V}_k] - 2\mu \mathbf{R} E[\mathbf{V}_k] \\ &= (\mathbf{I} - 2\mu\mathbf{R}) E[\mathbf{V}_k]. \end{aligned} \tag{A.7}$$

Putting $\mathbf{R}$ into normal form (3.17), we obtain

$$E[\mathbf{V}_{k+1}] = \mathbf{Q}(\mathbf{I} - 2\mu\mathbf{\Lambda})\mathbf{Q}^{-1} E[\mathbf{V}_k]. \tag{A.8}$$

Premultiplying both sides by $\mathbf{Q}^{-1}$,

$$E[\mathbf{V}'_{k+1}] = (\mathbf{I} - 2\mu\mathbf{\Lambda}) E[\mathbf{V}'_k]. \tag{A.9}$$

This equation is in pure diagonal form, and its solution is

$$E[\mathbf{V}'_{k+1}] = (\mathbf{I} - 2\mu\mathbf{\Lambda})^k E[\mathbf{V}'_0], \tag{A.10}$$

where $\mathbf{V}'_0$ is the initial value of $\mathbf{V}_k$ expressed in the rotated (primed) coordinates. Now, we premultiply both sides by $\mathbf{Q}$ to get

$$E[\mathbf{V}_{k+1}] = \mathbf{Q}(\mathbf{I} - 2\mu\mathbf{\Lambda})^k \mathbf{Q}^{-1} E[\mathbf{V}_0]. \tag{A.11}$$

As k goes to infinity, the expected value of $\mathbf{V}_{k+1}$ will go to zero as long as all the geometric ratios of (A.11) are less than unity, that is,

$$|r_p| < 1, \quad \text{for all} \quad p, \quad \text{where} \tag{A.12}$$

$$r_p = (1 - 2\mu\lambda_p). \tag{A.12a}$$

Therefore, the condition for stability of the mean of the weight vector is

$$\frac{1}{\lambda_{\max}} > \mu > 0, \quad \text{or}$$

$$1 > \mu\lambda_{\max} > 0. \tag{A.13}$$

This is in accordance with Eq. (3.27). Since the geometric ratios given by (A.12a) are identical to (3.28), it follows from (3.30) that the pth time constant is

$$\tau_p \cong \frac{1}{2\mu\lambda p}. \tag{A.13a}$$

For the special case of the input being white, all eigenvalues of $\mathbf{R}$ could be equal, and Eq. (A.13) becomes

$$\frac{n}{\operatorname{tr} \mathbf{R}} > \mu > 0, \quad \text{or} \tag{A.14}$$

$$\frac{1}{E[x_k^2]} > \mu > 0. \tag{A.15}$$

The single time constant for this case is

$$\tau \cong \frac{1}{2\mu\lambda} = \frac{1}{2\mu E[x_k^2]}. \tag{A.15a}$$

A.2 CONVERGENCE OF THE VARIANCE OF THE WEIGHT VECTOR AND ANALYSIS OF MISADJUSTMENT

To begin the analysis, it is necessary to consider the nature of the noise in the weight vector when practicing LMS adaptation in the vicinity of the Wiener solution. Equation (3.60) shows that the weight noise covariance matrix is a scalar matrix. We will derive a more refined version of (3.60) below, but in order to do this and to establish criteria for stability of the weight variance, we shall accept as fact that the weight covariance matrix is scalar when operating with LMS in the vicinity of the Wiener solution. Accordingly,

$$\begin{bmatrix} \text{weight noise} \\ \text{covariance} \\ \text{matrix} \end{bmatrix} = E[\mathbf{V}_k\mathbf{V}_k^T] = c\mathbf{I}. \tag{A.16}$$

The variance of the weight noise, a scalar, can be defined as

$$v_k \triangleq E[\mathbf{V}_k^T\mathbf{V}_k]. \tag{A.17}$$

From (A.16),

$$v_k = nc, \quad \text{and} \tag{A.18}$$

$$E[\mathbf{V}_k\mathbf{V}_k^T] = \frac{1}{n}v_k\mathbf{I}. \tag{A.19}$$

To establish stability of the variance of the weight noise, we shall examine the behavior of v_k during LMS adaptation. Refer to Eq. (A.4). Premultiply both sides of this equation by the respective transposes to obtain

$$\mathbf{V}_{k+1}^T\mathbf{V}_{k+1} = \mathbf{V}_k^T(\mathbf{I} - 2\mu\mathbf{X}_k\mathbf{X}_k^T)^2\mathbf{V}_k + 4\mu^2n_k^2\mathbf{X}_k^T\mathbf{X}_k \tag{A.20}$$

$$+4\mu n_k\mathbf{X}_k^T(\mathbf{I} - 2\mu\mathbf{X}_k\mathbf{X}_k^T)\mathbf{V}_k.$$

Next, we take expectation of both sides. The last term of (A.20) disappears under expectation because the noise n_k has zero-mean and is independent of $\mathbf{X}_k$ and $\mathbf{V}_k$. The expectation of the next to last term of (A.20) is factorable because of these same properties of noise n_k. Thus

$$E[\mathbf{V}_{k+1}^T\mathbf{V}_{k+1}] = E[\mathbf{V}_k^T(\mathbf{I} - 2\mu\mathbf{X}_k\mathbf{X}_k^T)^2\mathbf{V}_k] \tag{A.21}$$

$$+4\mu^2E[n_k^2]E[\mathbf{X}_k^T\mathbf{X}_k].$$

Equation (A.21) can now be rewritten as

$$v_{k+1} = E[\mathbf{V}_k^T(\mathbf{I} - 2\mu\mathbf{X}_k\mathbf{X}_k^T)^2\mathbf{V}_k] \tag{A.22}$$

$$+4\mu^2E[n_k^2]\,\text{tr}\,\mathbf{R}.$$

For convenience in the utilization of (A.22), we define a scalar a as

$$a \triangleq E\left[\mathbf{V}_k^T(\mathbf{I} - 2\mu\mathbf{X}_k\mathbf{X}_k^T)^2\mathbf{V}_k\right]. \tag{A.23}$$

We can compute a by computing its trace.

$$a = \text{tr}\, a = E\left[\text{tr}\ \left(\mathbf{V}_k^T(\mathbf{I} - 2\mu\mathbf{X}_k\mathbf{X}_k^T)^2\mathbf{V}_k\right)\right]. \tag{A.24}$$

Since trace (**AB**) = trace (**BA**), trace a can be written as

$$\begin{aligned}\text{tr}\, a &= E\left[\text{tr}\left(\mathbf{V}_k\mathbf{V}_k^T(\mathbf{I} - 2\mu\mathbf{X}_k\mathbf{X}_k^T)^2\right)\right] \\ &= \text{tr}\ \left(E\left[\mathbf{V}_k\mathbf{V}_k^T(\mathbf{I} - 2\mu\mathbf{X}_k\mathbf{X}_k^T)^2\right]\right).\end{aligned} \tag{A.25}$$

Recall that μ was chosen to be small (corresponding to slow adaptation), so the weight vector $\mathbf{W}_k$ and the weight deviation from the Wiener solution $\mathbf{V}_k$ are essentially independent of the input signal vector $\mathbf{X}_k$. As such (A.25) can be expressed as

$$a = \text{tr}\, a = \text{tr}\left(E[\mathbf{V}_k\mathbf{V}_k^T]E\left[(\mathbf{I} - 2\mu\mathbf{X}_k\mathbf{X}_k^T)^2\right]\right). \tag{A.26}$$

Now,

$$E\left[(\mathbf{I} - 2\mu\mathbf{X}_k\mathbf{X}_k^T)^2\right] = E\left[\mathbf{I} - 4\mu\mathbf{X}_k\mathbf{X}_k^T + 4\mu^2\mathbf{X}_k(\mathbf{X}_k^T\mathbf{X}_k)\mathbf{X}_k^T\right]. \tag{A.27}$$

The scalar quantity $(\mathbf{X}_k^T\mathbf{X}_k)$ does fluctuate as the input vector varies. If we assume however that the number of weights is large (i.e., let n be large), the law of large numbers indicates that this scalar quantity, the magnitude square of the input vector, will be essentially constant. Accordingly,

$$(\mathbf{X}_k^T\mathbf{X}_k) = nE[x_k^2] = \text{tr}\, \mathbf{R}. \tag{A.28}$$

Equation (A.27) can now be written as

$$\begin{aligned}E\left[(\mathbf{I} - 2\mu\mathbf{X}_k\mathbf{X}_k^T)^2\right] &= E\left[\mathbf{I} - 4\mu\mathbf{X}_k\mathbf{X}_k^T + 4\mu^2\ \text{tr}\ \mathbf{R}(\mathbf{X}_k\mathbf{X}_k^T)\right] \\ &= \mathbf{I} - 4\mu\mathbf{R} + 4\mu^2\mathbf{R}\ \text{tr}\ \mathbf{R}.\end{aligned} \tag{A.29}$$

We may now substitute (A.29) and (A.19) into (A.26) to obtain

$$\begin{aligned}a &= \text{tr}\ \left(\frac{1}{n}v_k\mathbf{I}\left[\mathbf{I} - 4\mu\mathbf{R} + 4\mu^2\mathbf{R}\ \text{tr}\ \mathbf{R}\right]\right) \\ &= \frac{1}{n}v_k\ \text{tr}\ \left(\mathbf{I} - 4\mu\mathbf{R} + 4\mu^2\mathbf{R}\ \text{tr}\ \mathbf{R}\right) \\ &= \frac{1}{n}v_k\left(n - 4\mu\ \text{tr}\ \mathbf{R} + 4\mu^2(\text{tr}\ \mathbf{R})^2\right).\end{aligned} \tag{A.30}$$

Next we substitute (A.30) into (A.23), and substitute the result into (A.22) to obtain

$$\begin{aligned}v_{k+1} &= v_k\left(1 - \frac{4\mu}{n}\ \text{tr}\ \mathbf{R} + \frac{4\mu^2}{n}(\text{tr}\ \mathbf{R})^2\right) \\ &\quad +4\mu^2E[n_k^2]\ \text{tr}\ \mathbf{R}.\end{aligned} \tag{A.31}$$

This is a first-order linear difference equation that governs the behavior of the variance of the weight noise at each step of the adaptive process. Its geometric ratio is

$$\begin{pmatrix} \text{geometric} \\ \text{ratio} \end{pmatrix} = \left(1 - \frac{4\mu}{n}\text{tr}\,\mathbf{R} + \frac{4\mu^2}{n}(\text{tr}\,\mathbf{R})^2\right). \tag{A.32}$$

Stability of v_k, the variance of the weight vector, is assured by choosing μ to cause the magnitude of the geometric ratio to be less than unity, that is,

$$\left|1 - \frac{4\mu}{n}\text{tr}\,\mathbf{R} + \frac{4\mu^2}{n}(\text{tr}\,\mathbf{R})^2\right| < 1. \tag{A.33}$$

This geometric ratio is a quadratic function of μ that can be shown to never go negative. Equation (A.33) is therefore equivalent to

$$1 - \frac{4\mu}{n}\,\text{tr}\,\mathbf{R} + \frac{4\mu^2}{n}(\text{tr}\,\mathbf{R})^2 < 1. \tag{A.34}$$

This is equivalent to

$$\frac{4\mu^2}{n}(\text{tr}\,\mathbf{R})^2 < \frac{4\mu}{n}\text{tr}\,\mathbf{R}. \tag{A.35}$$

Since trace $\mathbf{R}$ is positive and since the number of weights n is positive, μ must be positive to satisfy (A.35). Dividing through both sides of the inequality by the positive quantity

$$\frac{4\mu}{n}\text{tr}\,\mathbf{R},$$

we obtain the stability condition

$$0 < \mu\text{tr}\,\mathbf{R} < 1. \tag{A.36}$$

Satisfaction of (A.36) assures stability of the variance of the weight vector. Since it is a more stringent condition than (A.13), satisfaction of (A.36) assures both stability of the variance of the weight vector and convergence of its mean to the Wiener solution. Assumptions made in the derivation include small μ, large n, a scalar weight noise covariance matrix, and adaptation for the purpose of plant modeling (in the context of Fig. A.1).

Further use of Eq. (A.31) yields a more refined derivation of misadjustment than the one obtained in Chapter 3. Having chosen μ within the stable range of (A.36) and allowing enough adaptation time for Eq. (A.31) to come to equilibrium, it is clear that

$$v_{k+1} = v_k. \tag{A.37}$$

Substituting (A.37) into (A.31) gives the variance of the weight noise as

$$v_k = \frac{n\mu E[n_k^2]}{1 - \mu\text{tr}\,\mathbf{R}}. \tag{A.38}$$

The weight noise covariance matrix can be obtained from this, using Eq. (A.19):

$$\begin{aligned} \begin{bmatrix} \text{weight noise} \\ \text{covariance matrix} \end{bmatrix} &= \frac{1}{n}v_k\mathbf{I} \\ &= \frac{\mu E[n_k^2]}{1 - \mu\text{tr}\,\mathbf{R}}\mathbf{I}. \end{aligned} \tag{A.39}$$

For the configuration of Fig. A.1, Eq. (A.1) states that the minimum mean square error equals the power of the plant noise. Accordingly,

$$\begin{bmatrix} \text{weight noise} \\ \text{covariance matrix} \end{bmatrix} = \frac{\mu\xi_{\min}}{1 - \mu \operatorname{tr} \mathbf{R}}\mathbf{I}. \tag{A.40}$$

Comparing this result with the original expression (3.60), the weight noise covariance matrix (A.40) contains a new factor

$$\frac{1}{1 - \mu \operatorname{tr} R}. \tag{A.41}$$

Approximations were made in the derivation of (3.60) that were not made in the derivation of (A.40), such as the neglect of terms dependent on μ^2 that in a small way contribute to the misadjustment.

The original formula for misadjustment based on the noise covariance matrix (3.60) is

$$M = \mu \operatorname{tr} \mathbf{R}. \tag{A.42}$$

Using the new expression for weight noise covariance provides a new formula for misadjustment that is derived simply by incorporating the factor (A.41). Accordingly,

$$M = \frac{\mu \operatorname{tr} \mathbf{R}}{1 - \mu \operatorname{tr} \mathbf{R}}. \tag{A.43}$$

This new refined formula predicts misadjustment more accurately. For small values of μ, it is identical to the original formula.

Experimental verification of (A.43) has been done in comparison with (A.42). A modeling experiment was done following the configuration of Fig. A.1. Results are given in Table A.1 for an experiment with an adaptive filter containing 20 weights. The input signal to both the adaptive filter and the fixed filter was white noise of unit power. The value of trace **R** was 20. Both uniformly distributed and Gaussian signals were used, and the same results were obtained. The new more refined formula was a more accurate predictor of misadjustment, especially for large values of μ. For misadjustment levels below 25 percent there was not much difference, but for larger levels of misadjustment, the new formula was much more accurate. Refer to Table A.1.

Horwitz and Senne [2] have presented a derivation of stability of the variance and have obtained stable bounds on μ that are somewhat more precise, but very similar to (A.36). Their earlier derivation is more complicated. However, it is based on assumptions like those that we have used. They also present a refined misadjustment formula that is almost the same as (A.43), but once again their derivation is more complicated. Their paper has always been of very high interest to researchers in the field.

It should be noted that there are more general proofs of convergence of the LMS algorithm (see [3] and [4]), which do not necessitate all the assumptions that we have made above. On the other hand, these proofs are valid for very small values of adaptation constant μ (i.e., for values of μ much smaller than the practical stability range (A.36)).

TABLE A.1 COMPARISON OF MEASURED MISADJUSTMENT WITH FORMULA PREDICTION

μ	Measured misadjustment	μ tr $\mathbf{R}$	$\frac{\mu \text{ tr } \mathbf{R}}{1-\mu \text{ tr } \mathbf{R}}$
0.0025	0.054	0.05	0.053
0.005	0.110	0.10	0.110
0.01	0.251	0.20	0.250
0.015	0.438	0.30	0.429
0.02	0.676	0.40	0.667
0.025	1.050	0.50	1.00
0.03	1.640	0.60	1.50
0.035	3.00	0.70	2.333
0.04	6.23	0.80	4.00
0.045	(*)	0.90	9.00

(*): sometimes blows up

A.3 A SIMPLIFIED HEURISTIC DERIVATION OF MISADJUSTMENT AND STABILITY CONDITIONS

The same results obtained in A.2 can be obtained in a much simpler and more intuitive manner. Assume that the LMS adaptive process has converged so that the mean of the weight vector is equal to $\mathbf{W}^*$. The minimum mean square error is $\xi_{\min}$, and for the modeling configuration of Fig. A.1, this is equal to the plant noise power $E[n_k^2]$. Because of noise in the weights, the actual mean square error will be greater than $\xi_{\min}$. The weights will be undergoing random brownian motion about the Wiener solution at the bottom of the bowl.

Excess mean square error is created by the input signal propagating through the noisy weights. The average mean square error can be expressed as

$$\bar{\xi} = \xi_{\min} + \begin{pmatrix} \text{average} \\ \text{excess} \\ \text{MSE} \end{pmatrix}. \tag{A.44}$$

The idea of average excess mean square error has been discussed in Chapter 3.

The noise in the weights is uncorrelated from weight to weight. With an adaptive filter of the FIR type, the average excess MSE will be equal to the input signal power multiplied by the sum of the variances of the weight noises. Accordingly,

$$\begin{pmatrix} \text{average} \\ \text{excess} \\ \text{MSE} \end{pmatrix} = E[x_k^2] \cdot n \cdot \begin{pmatrix} \text{variance} \\ \text{of noise of} \\ \text{single weight} \end{pmatrix}. \tag{A.45}$$

The covariance matrix of the weight noise is derived in Eq. (3.60) and is given by

$$\text{cov}\,[\mathbf{V}_k] = \mu\xi_{\min}\mathbf{I}. \tag{A.46}$$

The weight noise power is proportional to $\xi_{\min}$, according to this formula. This is a good approximation when operating near the bottom of the bowl, adapting with a small value of μ. Otherwise an appropriate heuristic formula is the following:

$$\text{cov}\,[\mathbf{V}_k] = \mu\bar{\xi}\mathbf{I}. \tag{A.47}$$

Equation (A.47) shows the weight noise power as proportional to the actual average mean square error of the operating adaptive process rather than to the minimum mean square error of the perfect Wiener solution. The variance of the noise of a single weight can now be expressed as $\mu\bar{\xi}$. Using Eqs. (A.45) and (A.47), we can write

$$\begin{aligned}\begin{pmatrix}\text{average}\\ \text{excess}\\ \text{MSE}\end{pmatrix} &= E[x_k^2]\cdot n\cdot \mu\bar{\xi} \\ &= \mu\bar{\xi}\ \text{tr}\,\mathbf{R}.\end{aligned} \tag{A.48}$$

This is very similar to Eq. (3.64) with $\bar{\xi}$ used in the expression in place of ξ_{min}. The average mean square error can be obtained from (A.44) and (A.48):

$$\begin{aligned}\bar{\xi} &= \xi_{\text{min}} + \mu\bar{\xi}\ \text{tr}\,\mathbf{R} \\ \bar{\xi}(1-\mu\ \text{tr}\,\mathbf{R}) &= \xi_{\text{min}} \\ \bar{\xi} &= \frac{\xi_{\text{min}}}{1-\mu\ \text{tr}\,\mathbf{R}}.\end{aligned} \tag{A.49}$$

The misadjustment is defined by (3.65). Combining with (A.49), we have

$$\begin{aligned}M &= \frac{\bar{\xi}-\xi_{\text{min}}}{\xi_{\text{min}}} \\ &= \frac{\frac{\xi_{\text{min}}}{1-\mu\ \text{tr}\,\mathbf{R}}-\xi_{\text{min}}}{\xi_{\text{min}}} \\ &= \frac{\mu\ \text{tr}\,\mathbf{R}}{1-\mu\ \text{tr}\,\mathbf{R}}.\end{aligned} \tag{A.50}$$

This result is the same as that of (A.43).

Stability can be determined by keeping M, given by (A.43) or (A.50), finite. In addition, we know that μ must be positive for stable adaptation. Accordingly, to have a stable adaptive process, we must choose μ from within the range

$$\frac{1}{\text{tr}\,\mathbf{R}} > \mu > 0, \quad \text{or} \tag{A.51}$$

$$\frac{1}{nE[x_k^2]} > \mu > 0.$$

This is the same as (A.36), the criterion for stability of the variance of the LMS algorithm.

Bibliography for Appendix A

[1] B. WIDROW, "Adaptive filters," in *Aspects of Network and System Theory*, ed. R.E. Kalman and N. De Claris (New York: Holt, Rinehart and Winston, 1970), pp. 563–587.

[2] L.L. HORWITZ, and K.D. SENNE, "Performance advantage of complex LMS for controlling narrow-band adaptive arrays," *IEEE Trans. on Circuits and Systems*, Vol. CAS-28, No. 6 (June 1981), pp. 562–576.

[3] R.R. BITMEAD, "Convergence in distribution of LMS-type adaptive parameter estimates," *IEEE Trans. on Auto. Control*, Vol. AC-28, No. 1 (January 1983), pp. 54–60.

[4] O. MACCHI, and E. EWEDA, "Second-order convergence analysis of stochastic adaptive linear filtering," *IEEE Trans. on Auto. Control*, Vol. AC-28, No. 1 (January 1983), pp. 76–85.

Appendix B

Comparative Analyses of Dither Modeling Schemes A, B, and C

Chapter 4 describes three plant modeling schemes involving the use of a random dither signal. The purpose of this appendix is to analyze these schemes and to compare their performances so that their relative advantages and disadvantages can be delineated.

We are concerned with convergence conditions, speed of adaptation, misadjustment, and the noise in plant estimation (the weight noise covariance matrix). For each scheme, faster adaptation causes more noise in the weights of the plant model. The schemes can be compared by evaluating their weight noise covariances for a given speed of adaptation.

Before doing these analyses, it is convenient to make certain definitions. Referring to Figs. 4.3(a), (b), and (c), we will denote the last n samples of the controller output as

$$\mathbf{U}'_k \triangleq (u'_k, u'_{k-1}, \ldots u'_{k-n+1})^T, \tag{B.1}$$

the last n samples of the dither as

$$\Delta_k \triangleq (\delta_k, \delta_{k-1}, \ldots, \delta_{k-n+1})^T, \tag{B.2}$$

and the last n samples of the plant input as

$$\mathbf{U}_k \triangleq (u_k, u_{k-1}, \ldots, u_{k-n+1})^T. \tag{B.3}$$

For these signals, we may define autocorrelation matrices and corresponding eigenvalues as

$$\mathbf{R}_{U'} \triangleq E[\mathbf{U}'_k \mathbf{U}'^T_k], \quad \text{with eigenvalues} \tag{B.4}$$

$$(\lambda_{1U'}, \lambda_{2U'}, \ldots, \lambda_{pU'}, \ldots, \lambda_{nU'}), \quad \text{and} \tag{B.5}$$

$$\mathbf{R}_\Delta \triangleq E[\Delta_k \Delta_k^T], \quad \text{with eigenvalues} \tag{B.6}$$

$$(\lambda_{1\Delta}, \lambda_{2\Delta}, \ldots, \lambda_{p\Delta}, \ldots, \lambda_{n\Delta}), \quad \text{and} \tag{B.7}$$

$$\mathbf{R}_U \triangleq E[\mathbf{U}_k\mathbf{U}_k], \text{ with eigenvalues} \tag{B.8}$$

$$(\lambda_{1U}, \lambda_{2U}, \ldots, \lambda_{pU}, \ldots, \lambda_{nU}). \tag{B.9}$$

We will assume that the plant has an infinite impulse response, but that it is representable with high accuracy as an FIR digital filter with a sufficient number of weights. We will also assume that the adaptive model has the same number of taps and weights as the plant. (If instead we were to assume that the plant model has more taps and weights than the plant, the following analysis would be unchanged. That would not be the case if the plant model has fewer taps and weights than the plant.)

B.1 ANALYSIS OF SCHEME A

The configuration of scheme A, shown in Fig. 4.3(a), is similar to Fig. A.1 except that the plant input and plant model input with scheme A consist of the sum of the dither signal and the controller output signal. We can make direct use of the analytical results of Appendix A. From (A.13a), the pth time constant for scheme A is

$$\tau_p = \frac{1}{2\mu\lambda_{pU}}. \tag{B.10}$$

The plant input autocorrelation matrix is

$$\mathbf{R}_U = \mathbf{R}_{U'} + \mathbf{R}_\Delta. \tag{B.11}$$

If the dither is white, the eigenvalues add, so that

$$\lambda_{pU} = \lambda_{pU'} + \lambda_{p\Delta}. \tag{B.12}$$

Since all the eigenvalues of the white dither are equal,

$$\lambda_{p\Delta} = \lambda_\Delta = E[\delta_k^2]. \tag{B.13}$$

The time constant can be expressed as

$$\tau_p = \frac{1}{2\mu(\lambda_{pU'} + \lambda_\Delta)} = \frac{1}{2\mu(\lambda_{pU'} + E[\delta_k^2])}. \tag{B.14}$$

The condition for stability of the variance of the weight vector with scheme A can be obtained from (A.36) as

$$\frac{1}{\operatorname{tr} \mathbf{R}_U} > \mu > 0. \tag{B.15}$$

It may be noted that

$$\begin{aligned}\operatorname{tr} \mathbf{R}_U &= \operatorname{tr} \mathbf{R}_{U'} + \operatorname{tr} \mathbf{R}_\Delta \\ &= n(E[u_k'^2] + E[\delta_k^2]).\end{aligned} \tag{B.16}$$

The stability of the variance is now conditioned by

$$\frac{1}{n(E[u_k'^2] + E[\delta_k^2])} > \mu > 0. \tag{B.17}$$

The noise covariance matrix of the weight vector can be obtained for scheme A directly from (A.39):

$$\begin{bmatrix} \text{weight noise} \\ \text{covariance matrix} \end{bmatrix} = \frac{\mu E[n_k^2]}{1-\mu \operatorname{tr} \mathbf{R}_U}\mathbf{I}.$$

$$= \frac{\mu E[n_k^2]}{1-\mu n\left(E[u_k'^2]+E[\delta_k^2]\right)}\mathbf{I}. \tag{B.18}$$

The misadjustment of the adaptive model of scheme A is obtained from (A.43) as

$$M = \frac{\mu \operatorname{tr} \mathbf{R}_U}{1-\mu \operatorname{tr} \mathbf{R}_U} = \frac{\mu n\left(E[u_k'^2]+E[\delta_k^2]\right)}{1-\mu n\left(E[u_k'^2]+E[\delta_k^2]\right)}. \tag{B.19}$$

.2 ANALYSIS OF SCHEME B

Scheme B is shown in Fig. 4.3(b). To facilitate the analysis of this system it is convenient to redraw the diagram in the form shown in Fig. B.1. The resulting adaptive model $\widehat{P}(z)$ will be the same. In Fig. B.1, the effect of the control signal u_k' driving the plant is seen as a disturbing influence from the point of view of the modeling process, adding to the plant disturbance n_k to give an equivalent plant disturbance $\tilde{n}_k$.

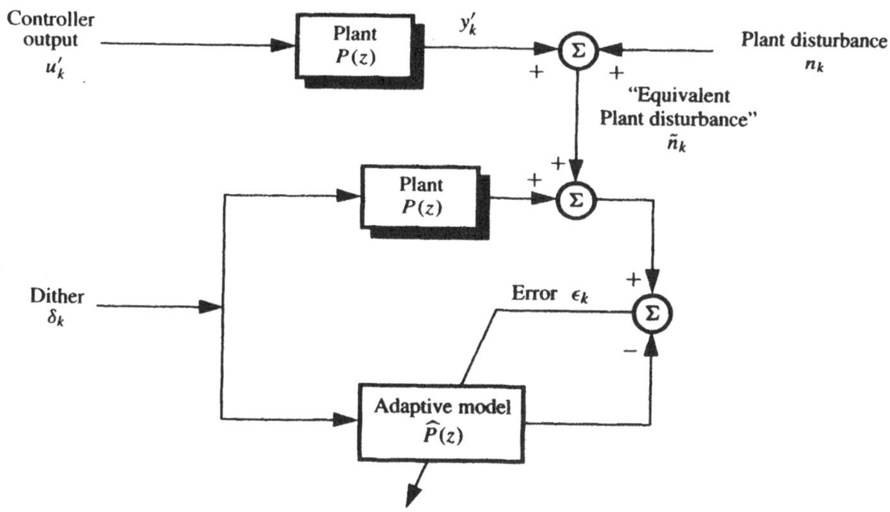

Figure B.1 Alternate representation of plant modeling scheme B.

The configuration of Fig. B.1 is very similar to that of Fig. A.1, so that the analysis of Appendix A can be directly applied to the analysis of scheme B. Assuming white dither, the time constant of scheme B is obtained from (A.13a) as

$$\tau = \frac{1}{2\mu\lambda_\Delta} = \frac{1}{2\mu E[\delta_k^2]}. \tag{B.20}$$

The range of stability for scheme B, obtained from (A.36), is

$$\frac{1}{\operatorname{tr}\mathbf{R}_\Delta} > \mu > 0, \quad \text{or} \tag{B.21}$$

$$\frac{1}{nE[\delta_k^2]} > \mu > 0. \tag{B.22}$$

The weight noise covariance matrix for scheme B, obtained from (A.39), is

$$\begin{bmatrix} \text{weight noise} \\ \text{covariance matrix} \end{bmatrix} = \frac{\mu E[\tilde{n}_k^2]}{1 - \mu \operatorname{tr}\mathbf{R}_\Delta}\mathbf{I}. \tag{B.23}$$

If the response of the plant to the controller output u'_k is represented by y'_k, then the equivalent plant noise is

$$\tilde{n}_k = y'_k + n_k. \tag{B.24}$$

Since y'_k and n_k are mutually independent,

$$E[\tilde{n}_k^2] = E[y_k'^2] + E[n_k^2]. \tag{B.25}$$

Equation (B.23) can now be rewritten as

$$\begin{bmatrix} \text{weight noise} \\ \text{covariance matrix} \end{bmatrix} = \frac{\mu\left(E[y_k'^2] + E[n_k^2]\right)}{1 - \mu n E[\delta_k^2]}\mathbf{I}. \tag{B.26}$$

The misadjustment for scheme B can be obtained from (A.43) as

$$M = \frac{\mu \operatorname{tr}\mathbf{R}_\Delta}{1 - \mu \operatorname{tr}\mathbf{R}_\Delta} = \frac{\mu n E[\delta_k^2]}{1 - \mu n E[\delta_k^2]}. \tag{B.27}$$

B.3 ANALYSIS OF SCHEME C

Scheme C is a more sophisticated adaptive process than either scheme A or scheme B. It is not in any way similar in configuration to the system of Fig. A.1. The analysis of Appendix A, although helpful in methodology, cannot be used directly to analyze scheme C. We must "start from scratch."

Refer to Fig. 4.3(c). Let $\mathbf{P}$ and $\widehat{\mathbf{P}}_k$ represent the plant impulse response vector and the kth estimation of the plant impulse response vector correspondingly. We assume that the Wiener solution $\widehat{\mathbf{P}}^*$ is equal to $\mathbf{P}$, (assuming as was done above that the impulse response $\widehat{\mathbf{P}}_k$ is long enough to match that of $\mathbf{P}$). $\Delta\mathbf{P}_k$ will denote the weight error vector at time k.

$$\Delta\mathbf{P} = \widehat{\mathbf{P}}_k - \widehat{\mathbf{P}}^* = \widehat{\mathbf{P}}_k - \mathbf{P}. \tag{B.28}$$

Using this notation, we can represent the adaptive modeling error ϵ_k as

$$\begin{aligned} \epsilon_k &= n_k + (\Delta_k + \mathbf{U}'_k)^T\mathbf{P} - (\Delta_k + \mathbf{U}'_k)^T\widehat{\mathbf{P}}_k \\ &= n_k - (\Delta_k + \mathbf{U}'_k)^T\Delta\mathbf{P}_k. \end{aligned} \tag{B.29}$$

The LMS adaptation rule can be written in accord with (3.35) as

$$\widehat{\mathbf{P}}_{k+1} = \widehat{\mathbf{P}}_k + 2\mu\epsilon_k\Delta_k. \tag{B.30}$$

Subtracting **P** from both sides of (B.30) and substituting both (B.28) and (B.29) yields

$$\Delta\mathbf{P}_{k+1} = (\mathbf{I} - 2\mu\Delta_k(\Delta_k + \mathbf{U}'_k)^T)\Delta\mathbf{P}_k + 2\mu n_k \Delta_k. \tag{B.31}$$

Next we take expectations of both sides of (B.31). Since the dither was assumed to have zero-mean and to be independent of the plant disturbance, the last term on the right side vanishes under expectation. Moreover, Δ_k and $\mathbf{U}'_k$ are independent of each other, and for small μ they are both independent of $\Delta\mathbf{P}_k$. Hence,

$$\begin{aligned} E[\Delta\mathbf{P}_{k+1}] &= E[\mathbf{I} - 2\mu\Delta_k(\Delta_k + \mathbf{U}'_k)^T]E[\Delta\mathbf{P}_k] \\ &= (\mathbf{I} - 2\mu E[\Delta_k\Delta_k^T]) \cdot E[\Delta\mathbf{P}_k]. \end{aligned} \tag{B.32}$$

Since the dither is white,

$$E[\Delta\mathbf{P}_{k+1}] = (1 - 2\mu E[\delta_k^2])E[\Delta\mathbf{P}_k]. \tag{B.32a}$$

Clearly the mean of $\Delta\mathbf{P}_k$ converges to zero if and only if the geometric ratio has a magnitude less than one, that is,

$$|1 - 2\mu E[\delta_k^2]| < 1 \Longleftrightarrow 1 < \mu < \frac{1}{E[\delta_k^2]}. \tag{B.33}$$

The time constant of this convergence process will be

$$\tau = \frac{1}{2\mu E[\delta_k^2]}, \tag{B.34}$$

in accordance with Eq. (3.31).

The convergence in the mean does not necessitate the convergence of the variance of the weights. In order to analyze the convergence of the variance, we have to evaluate first the covariance of the weight noise. According to Eq. (A.16), when the model $\widehat{\mathbf{P}}_k$ adaptively approaches the vicinity of the Wiener solution $\widehat{\mathbf{P}}^*$, the weight noise covariance matrix becomes scalar.

$$\begin{pmatrix} \text{weight noise} \\ \text{covariance} \\ \text{matrix} \end{pmatrix} = E[\Delta\mathbf{P}_k \cdot \Delta\mathbf{P}_k^T] = c\mathbf{I}. \tag{B.35}$$

The question is, once the adaptive process arrives at the Wiener solution, will the weight vector remain in the vicinity of this optimal solution? In order to answer the question, we will evaluate the impact of each iteration on the variance of the weight noise, defined as

$$v_k \triangleq E[\Delta\mathbf{P}_k^T \Delta\mathbf{P}_k]. \tag{B.36}$$

According to (B.35)

$$v_k = nc, \quad \text{and} \tag{B.37}$$

$$E[\Delta\mathbf{P}_k\Delta\mathbf{P}_k^T] = \frac{1}{n}v_k\mathbf{I}. \tag{B.38}$$

We can find a connection between v_{k+1} and v_k from (B.31). Multiplying each side of (B.31) by its transpose yields

$$\begin{aligned} \Delta\mathbf{P}_{k+1}^T\Delta\mathbf{P}_{k+1} = {} & \Delta\mathbf{P}_k^T\left\{\mathbf{I} - 2\mu(\Delta_k + \mathbf{U}'_k)\Delta_k^T\right\}\left\{\mathbf{I} - 2\mu\Delta_k(\Delta_k + \mathbf{U}'_k)^T\right\}\Delta\mathbf{P}_k \\ & + 4\mu^2 n_k^2 \Delta_k^T\Delta_k + 4\mu n_k \Delta_k^T(\mathbf{I} - 2\mu\Delta_k(\Delta_k + \mathbf{U}'_k))\Delta\mathbf{P}_k. \end{aligned} \tag{B.39}$$

Next, we take expectations of both sides of (B.39). Since n_k is zero-mean independent plant disturbance, the last term on the right side vanishes under expectation. Hence,

$$\begin{aligned} v_{k+1} &= E[\Delta\mathbf{P}_{k+1}^T \Delta\mathbf{P}_{k+1}] \\ &= a + 4\mu^2 E[n_k^2 \Delta_k^T \Delta_k] \\ &= a + 4\mu^2 n E[n_k^2] E[\delta_k^2], \end{aligned} \tag{B.40}$$

where

$$a \triangleq E[\Delta\mathbf{P}_k^T \left\{\mathbf{I} - 2\mu(\Delta_k + \mathbf{U}'_k)\Delta_k^T\right\} \left\{\mathbf{I} - 2\mu\Delta_k(\Delta_k + \mathbf{U}'_k)^T\right\} \Delta\mathbf{P}_k]. \tag{B.41}$$

Since $\text{tr}(AB) = \text{tr}(BA)$ we can rewrite (B.41) as

$$\begin{aligned} a &= \text{tr}(a) \\ &= E[\text{tr}\,(\Delta\mathbf{P}_k^T \left\{\mathbf{I} - 2\mu(\Delta_k + \mathbf{U}'_k)\Delta_k^T\right\} \left\{\mathbf{I} - 2\mu\Delta_k(\Delta_k + \mathbf{U}'_k)^T\right\} \Delta\mathbf{P}_k)] \\ &= E[\text{tr}\,(\Delta\mathbf{P}_k \Delta\mathbf{P}_k^T \left\{\mathbf{I} - 2\mu(\Delta_k + \mathbf{U}'_k)\Delta_k^T\right\} \left\{\mathbf{I} - 2\mu\Delta_k(\Delta_k + \mathbf{U}'_k)^T\right\})]. \end{aligned} \tag{B.42}$$

Utilizing, once more, the assumption of independence of $\Delta\mathbf{P}_k$, Δ_k, and $\mathbf{U}'_k$ we can evaluate the scalar a as

$$\begin{aligned} a &= \text{tr}\,(E[\Delta\mathbf{P}_k \Delta\mathbf{P}_k^T] \cdot E[(\mathbf{I} - 2\mu(\Delta_k + \mathbf{U}'_k)\Delta_k^T)(\mathbf{I} - 2\mu\Delta_k(\Delta_k + \mathbf{U}'_k)^T)]) \\ &= \text{tr}\,(E[\Delta\mathbf{P}_k \Delta\mathbf{P}_k^T] \cdot E[\mathbf{I} - 4\mu\Delta_k\Delta_k^T - 2\mu\mathbf{U}'_k\Delta_k^T - 2\mu\Delta_k\mathbf{U}_k'^T \\ &\quad + 4\mu^2(\Delta_k + \mathbf{U}'_k)\Delta_k^T\Delta_k(\Delta_k + \mathbf{U}'_k)^T]) \\ &= \text{tr}\,(E[\Delta\mathbf{P}_k \Delta\mathbf{P}_k^T] \cdot (\mathbf{I} - 4\mu E[\Delta_k\Delta_k^T] + 4\mu^2 E[(\Delta_k + \mathbf{U}'_k)\Delta_k^T\Delta_k(\Delta_k + \mathbf{U}'_k)^T]). \end{aligned} \tag{B.43}$$

Assume that the number of weights n of the plant model is large. From the law of large numbers, it follows that

$$\Delta_k^T \Delta_k \equiv \sum_{l=1}^{n} \delta_{k-l+1}^2 = \text{const.} = nE[\delta_k^2]. \tag{B.44}$$

Substitution of (B.37), (B.41), and (B.44) into (B.43) yields

$$a = \frac{v_k}{n}\text{tr}\,(\mathbf{I} - 4\mu E[\delta_k^2]\mathbf{I} + 4\mu^2 n E[\delta_k^2] \cdot E[(\Delta_k + \mathbf{U}'_k)(\Delta_k + \mathbf{U}'_k)^T]). \tag{B.45}$$

Since $\mathbf{U}'_k$ and Δ_k are independent of each other,

$$a = \frac{v_k}{n}\left\{n - 4\mu n E[\delta_k^2] + 4\mu^2 n^2 E[\delta_k^2](E[\delta_k^2] + E[(u'_k)^2])\right\}. \tag{B.46}$$

Substitution of (B.46) into (B.40) yields

$$\begin{aligned} v_{k+1} &= v_k\left\{1 - 4\mu E[\delta_k^2][1 - \mu n(E[\delta_k^2] + E[(u'_k)^2])]\right\} \\ &\quad + 4\mu^2 n E[n_k^2] E[\delta_k^2]. \end{aligned} \tag{B.47}$$

According to (B.38) and (B.47) the variance of the adaptive process converges if and only if

$$1 - 4\mu E[\delta_k^2][1 - \mu n(E[\delta_k^2] + E[(u'_k)^2])] < 1. \tag{B.48}$$

Since $E[\delta_k^2]$ is strictly positive and $E[(u'_k)^2]$ is nonnegative, (B.48) is equivalent to

$$1 - \mu n\left(E[\delta_k^2] + E[(u'_k)^2]\right) > 0, \quad \text{with } \mu > 0. \tag{B.49}$$

Hence the adaptive process will have bounded variance for

$$0 < \mu < \frac{1}{n(E[\delta_k^2] + E[(u'_k)^2])} = \frac{1}{nE[u_k^2]}. \tag{B.50}$$

This simple and easily applied condition for convergence of the variance is stronger than Eq. (B.33) for the convergence in the mean, and it is therefore the only condition which needs to be satisfied. It is easy to apply because the power of the dither is known and because one would generally have some estimate of the power of the plant driving signal u'_k. This stability condition has been verified for dither scheme C by extensive simulation experiments.

Use of Eq. (B.47) allows an accurate but simple derivation of misadjustment to be made. Choosing μ in the stable range (B.50) and allowing time for the adaptive process to come to equilibrium, we have

$$v_{k+1} = v_k. \tag{B.51}$$

Substituting (B.51) into (B.47) gives the variance of the weight noise as

$$v_k = \frac{\mu n E[n_k^2]}{1 - \mu n \left(E[\delta_k^2] + E[u_k'^2]\right)}. \tag{B.52}$$

The weight noise covariance matrix can be found from this, in accord with (B.38), as

$$\begin{bmatrix} \text{weight noise} \\ \text{covariance matrix} \end{bmatrix} = \frac{1}{n} v_k \mathbf{I} = \frac{\mu E[n_k^2]}{1 - \mu n \left(E[\delta_k^2] + E[u_k'^2]\right)} \mathbf{I}. \tag{B.53}$$

For scheme C (Fig. 4.3(c)), the minimum mean square error is the plant disturbance power. Therefore, (B.53) can be rewritten as

$$\begin{bmatrix} \text{weight noise} \\ \text{covariance matrix} \end{bmatrix} = \frac{\mu \xi_{\min}}{1 - \mu n \left(E[\delta_k^2] + E[u_k'^2]\right)} \mathbf{I}. \tag{B.54}$$

Excess mean square error in the modeling process will be caused by weight noise. The modeling error is given by (B.29). This includes the effects of plant disturbance. The error caused by weight noise, apart from the effects of plant disturbance, is given by

$$\begin{bmatrix} \text{error due to} \\ \text{weight noise} \end{bmatrix} = -(\Delta_k + \mathbf{U}'_k)^T \Delta \mathbf{P}_k. \tag{B.55}$$

Since

$$\mathbf{U}_k = \Delta_k + \mathbf{U}'_k, \tag{B.55a}$$

we can write this as

$$\begin{bmatrix} \text{error due to} \\ \text{weight noise} \end{bmatrix} = -\mathbf{U}_k^T \Delta \mathbf{P}_k. \tag{B.56}$$

The mean square of this error is

$$\begin{bmatrix} \text{excess} \\ \text{MSE} \end{bmatrix} = E[\mathbf{U}_k^T \Delta \mathbf{P}_k \Delta \mathbf{P}_k^T \mathbf{U}_k]. \tag{B.57}$$

In accord with previous assumptions, we assume that $\Delta\mathbf{P}_k$ is independent of $\mathbf{U}_k^T$. Therefore,

$$\begin{bmatrix} \text{excess} \\ \text{MSE} \end{bmatrix} = E[\mathbf{U}_k^T E[\Delta\mathbf{P}_k \Delta\mathbf{P}_k^T]\mathbf{U}_k]. \tag{B.58}$$

Substituting (B.54) into this expression yields

$$\begin{bmatrix} \text{excess} \\ \text{MSE} \end{bmatrix} = \frac{\mu\xi_{\min}}{1 - \mu n\left(E[\delta_k^2] + E[u_k'^2]\right)} E[\mathbf{U}_k^T \mathbf{U}_k]. \tag{B.59}$$

$$E[\mathbf{U}_k^T \mathbf{U}_k] = \operatorname{tr} \mathbf{R}_U, \tag{B.60}$$

and since δ_k and u_k' are independent of each other,

$$n\left(E[\delta_k^2] + E[u_k'^2]\right) = \operatorname{tr} \mathbf{R}_U. \tag{B.61}$$

Substituting (B.60) and (B.61) into (B.59) gives

$$\begin{bmatrix} \text{excess} \\ \text{MSE} \end{bmatrix} = \frac{\mu\xi_{\min} \operatorname{tr}\mathbf{R}_U}{1 - \mu \operatorname{tr} \mathbf{R}_U}. \tag{B.62}$$

The misadjustment is obtained by normalizing this with respect to the minimum MSE. Accordingly,

$$M = \frac{\mu \operatorname{tr} \mathbf{R}_U}{1 - \mu \operatorname{tr} \mathbf{R}_U}. \tag{B.63}$$

B.4 A SIMPLIFIED HEURISTIC DERIVATION OF MISADJUSTMENT AND STABILITY CONDITIONS FOR SCHEME C

It is possible to use the analytical techniques developed in Section A.3, to produce a simplified analysis of stability and misadjustment for the adaptive process of scheme C.

Refer to Fig. 4.3(c) for a block diagram of scheme C. The minimum mean square error is, as stated above, equal to the power of the plant disturbance. The average mean square error of the converged adaptive filter (when the mean of the weight vector equals the Wiener weight vector) is equal to the minimum mean square error plus the average excess mean square error due to weight noise:

$$\bar{\xi} = \xi_{\min} + \begin{pmatrix} \text{average} \\ \text{excess MSE} \end{pmatrix}. \tag{B.64}$$

The weight noise of $\widehat{P}(z)$ is, of course, identical to the weight noise of $\widehat{P}(z)$ copy. Both of these filters contribute to the average excess mean square error.

The adaptive model $\widehat{P}(z)$, which has noisy weights, contributes to average excess mean square error in the following way. Since the noise in the weights is uncorrelated from weight to weight, the average excess mean square error component at the output of $\widehat{P}(z)$ will be equal the input dither power multiplied by the sum of the variances of the weight noises. Therefore

$$\begin{pmatrix} \widehat{P}(z) \text{ component of} \\ \text{excess MSE} \end{pmatrix} = E[\delta_k^2] \cdot n \cdot \begin{pmatrix} \text{variance of} \\ \text{noise of} \\ \text{single weight} \end{pmatrix}. \tag{B.65}$$

Likewise, for $\widehat{P}(z)$ copy,

$$\begin{pmatrix} \widehat{P}(z) \text{ copy} \\ \text{component of excess MSE} \end{pmatrix} = E[(u'_k)^2] \cdot n \cdot \begin{pmatrix} \text{variance of noise of} \\ \text{single weight} \end{pmatrix}. \tag{B.66}$$

The average excess MSE is the sum of these components. Accordingly,

$$\begin{pmatrix} \text{average} \\ \text{excess MSE} \end{pmatrix} = \left[E[\delta_k^2] + E[(u'_k)^2]\right] \cdot n \cdot \begin{pmatrix} \text{variance of noise} \\ \text{of single weight} \end{pmatrix}. \tag{B.67}$$

The covariance matrix of the weight noise is

$$\text{cov}[\mathbf{V}_k] = \mu\bar{\xi}\mathbf{I}. \tag{B.68}$$

The reasoning is already given in Appendix A in connection with Eq. (A.47). The variance of the noise of a single weight is therefore equal to $\mu\bar{\xi}$. Substituting this into (B.67) yields

$$\begin{pmatrix} \text{average} \\ \text{excess MSE} \end{pmatrix} = \left[E[\delta_k^2] + E[(u'_k)^2]\right] \cdot n\mu\bar{\xi}. \tag{B.69}$$

This can also be expressed as

$$\begin{pmatrix} \text{average} \\ \text{excess MSE} \end{pmatrix} = E[(u_k)^2] \cdot n\mu\bar{\xi}. \tag{B.70}$$

Combining (B.70) with (B.64) gives

$$\begin{aligned} \bar{\xi} &= \xi_{\min} + \bar{\xi} E[u_k^2] \cdot n\mu, \quad \text{or} \\ \bar{\xi} &= \frac{\xi_{\min}}{1 - \mu n E[u_k^2]} = \frac{\xi_{\min}}{1 - \mu \operatorname{tr} \mathbf{R}_U}. \end{aligned} \tag{B.71}$$

The misadjustment can be calculated as follows,

$$M = \frac{(\text{average excess MSE})}{\xi_{\min}} = \frac{n\mu\bar{\xi} E[u_k^2]}{\xi_{\min}}. \tag{B.72}$$

Substituting (B.71) into this expression gives

$$M = \frac{\mu n E[u_k^2]}{1 - \mu n E[u_k^2]} = \frac{\mu \operatorname{tr} \mathbf{R}_U}{1 - \mu \operatorname{tr} \mathbf{R}_U}. \tag{B.73}$$

This result is identical to (B.63). Stability requires that μ be positive and small enough so that the misadjustment remains finite. Accordingly,

$$\frac{1}{\operatorname{tr} \mathbf{R}_U} > \mu > 0. \tag{B.74}$$

This is the same as criterion (B.50) for stability of the variance when using scheme C. These simple heuristic derivations of misadjustment and criteria for stability for scheme C have yielded results that are identical to those that have been obtained with more rigorous treatment in Section B.3.

B.5 A SIMULATION OF A PLANT MODELING PROCESS BASED ON SCHEME C

In order to verify Eqs. (B.34), (B.50), and (B.52), we have simulated a plant modeling process based on dithering scheme C depicted in Fig. 4.3(c). We have chosen the plant to be IIR, of the form

$$P(z) = \frac{1 - \frac{1}{3}z^{-1}}{1 - \frac{1}{2}z^{-1}}. \tag{B.75}$$

The adaptive plant model $\widehat{P}(z)$ was FIR. The number of weights of the plant model was chosen to be $n = 10$.

The system of Fig. 4.3(c) has three input signals: the controller output u'_k, the dither δ_k, and the plant noise n_k. For the sake of simplicity, we have chosen all three inputs to be zero-mean, white, of unit power, and independent of each other.

Keeping all of the above design parameters fixed, we varied the adaptation constant μ. For every choice of μ, we have run the system of Fig. 4.3(c) for a few thousand data samples starting each time with the initial condition $\widehat{P}_k(z) = 0$ for $k = 0$. We monitored the sum of squares of the differences between the weights of $\widehat{P}(z)$ and the weights of the actual plant $P(z)$. Learning curves were obtained based on the plant model error as a function of time. From these curves, we have computed the experimental time constant τ. Equation (B.34) gives a theoretical time constant formula for scheme C. For various values of μ, theoretical and experimental values of time constant were taken, and the results are shown in Table B.1. Discrepancy between theoretical and experimental time constants is of the order of 10 percent or less, over a very wide range of time constant values.

TABLE B.1 COMPARISON OF EXPERIMENTAL AND THEORETICAL VALUES OF TIME CONSTANT AND WEIGHT NOISE FOR PLANT MODELING WITH SCHEME C

Adaptation constant μ	Time constant τ		v_k, Variance of the noise in the weights of plant model	
	Estimated	Measured	Estimated	Measured
0.00025	2000	2028	0.0025	0.0024
0.00100	500	567	0.0102	0.0096
0.00200	250	278	0.0208	0.0223
0.00400	125	128	0.0435	0.0433
0.01	50	53	0.1250	0.1331
0.02	25	20	0.3333	0.3275
0.03	17	16	0.7500	0.8302
0.05			unstable	655

Next we allowed the adaptive process to run for at least four time constants to reach statistical equilibrium. In statistical steady state, the average noise in the weights of the plant model was measured. For each value of μ, a considerable amount of averaging was done to get accurate experimental values of the variance of the weight noise of $\widehat{P}(z)$. Theoretical values of this parameter were obtained by using Eq. (B.52). A comparison is shown in Table

B.1. Theoretical values and experimental measurements agree within 10 percent over a very wide range of μ.

Next, we wished to verify that the weight noise covariance matrix, as expressed by (B.53), is indeed a diagonal one. We set $\mu = 0.001$, and measured the steady-state weight noise covariance matrix. Scaling the values up by a factor of 10,000 to make things visible and rounding to the nearest single integer, the data obtained are displayed in Table B.2. The sum of all the off-diagonal entries was less than 2 percent of the sum of all entries. But the experimental eigenvalue spread was 6:1, whereas the theoretical eigenvalue spread was 1:1. By and large, Eq. (B.53) appeared to work quite well, but not perfectly.

TABLE B.2 WEIGHT NOISE AUTOCORRELATION MATRIX FOR PLANT MODELING WITH SCHEME C

9	1	0	1	−1	0	−1	0	0	−1
1	14	1	1	1	−1	0	1	−1	0
0	1	13	0	−5	1	1	−1	−2	1
1	1	0	9	−1	0	−2	0	0	2
−1	1	−5	−1	9	2	0	1	0	0
0	−1	1	0	2	12	−1	0	−2	1
−1	0	1	−2	0	−1	7	0	2	0
0	1	−1	0	1	0	0	8	1	0
0	−1	−2	0	0	−2	2	1	7	−1
−1	0	1	2	0	1	0	0	−1	9

Our final goal was to verify Eq. (B.50), which determines the stable range of the adaptation constant μ. Since in our simulation the number of weights of $\widehat{P}(z)$ was $n = 10$, and

$$E[\delta_k^2] = E[(u_k')^2] = 1, \tag{B.76}$$

the system should be stable for

$$0 < \mu < \frac{1}{n\left(E[\delta_k^2] + E[(u_k')^2]\right)} = 0.05. \tag{B.77}$$

Refer again to Table B.1. Comparing measured and theoretical values of weight noise for various values of μ, it is clear that when $\mu = 0.05$, the system was theoretically unstable but experimentally just barely stable. Although the system was still stable, it was operating with weight noise that suddenly became enormous as μ was raised to the stated level. In fact, when μ was further raised to $\mu = 0.051$, the algorithm blew up. Hence, for this particular case, our prediction of the stable range of μ was correct to within 2 percent.

This experiment gives one confidence in the formulas and gives one a feeling that the analysis of scheme C, although based on a few assumptions, is reasonably accurate and quite useful.

SUMMARY

The principal results of this appendix are summarized in Table B.3, which compares the performance characteristics of modeling schemes A, B, and C.

TABLE B.3 PERFORMANCE CHARACTERISTICS OF MODELING SCHEMES A, B, AND C

	Modeling Scheme A	Modeling Scheme B	Modeling Scheme C
Stable Range for μ	$\frac{1}{n(E[\mu_k'^2]+E[\delta_k^2])} > \mu > 0$	$\frac{1}{nE[\delta_k^2]} > \mu > 0$	$\frac{1}{n(E[\mu_k'^2]+E[\delta_k^2])} > \mu > 0$
Time Constant (White Dither)	$\tau_p = \frac{1}{2\mu(\lambda_{pU'}+E[\delta_k^2])}$	$\tau = \frac{1}{2\mu E[\delta_k^2]}$	$\tau = \frac{1}{2\mu E[\delta_k^2]}$
$\left[\begin{array}{c}\text{Weight Noise}\\ \text{Covariance Matrix}\end{array}\right]$	$\frac{\mu E[n_k^2]}{1-\mu n(E[\mu_k'^2]+E[\delta_k^2])}\mathbf{I}$	$\frac{\mu(E[y_k'^2]+E[n_k^2])}{1-\mu n E[\delta_k^2]}\mathbf{I}$	$\frac{\mu E[n_k^2]}{1-\mu n(E[\mu_k'^2]+E[\delta_k^2])}\mathbf{I}$

Scheme A is a simple one, and it can be used as a basis of comparison to scheme B and scheme C. Comparing scheme A to scheme B, we can see that the stable range of μ is larger with scheme B. But with the same choice of μ, the time constant with scheme B is larger than any of the time constants of scheme A, so convergence of scheme B is slower. How much faster scheme A will be depends on the eigenvalues of $R_{U'}$. With a large eigenvalue spread, some $\lambda_{pU'}$ would be close to zero and the slowest modes of scheme A would be about as slow as the single mode of scheme B. Again keeping μ the same for both schemes A and B, and keeping μ well within its stability range, that is, keeping μ small, the weight noise covariance matrices can be compared. The numerator of the covariance for scheme B is larger than that for scheme A. With small μ, the denominators are about the same. Therefore, the weight noise for scheme B is larger than for scheme A, while the rate of convergence for scheme A is as good or better than the rate of convergence of scheme B. Therefore, from this point of view, scheme A gives superior performance over that of scheme B, except that, if one were not concerned with weight noise, scheme B could be driven a lot harder and made to converge faster than scheme A.

Why else might one use scheme B used instead of scheme A? The answer is that the input to the adaptive model of scheme B is driven by dither, which has controllable and predictable spectral characteristics and may be much better for modeling than the adaptive model input of scheme A, which consists of dither plus the output of the controller. The output of the controller, the driving input to the plant, is potentially unreliable as a modeling signal; it may be highly nonstationary, it may at times be much stronger than the dither and have poor bandwidth characteristics, and it may not be representable as a stochastic signal at all. For example, it might consist of sporadic transient pulses, step functions, and so forth. To do a better modeling job and to ensure a wideband fit between the model and the plant, scheme B may be the better choice.

Another reason for choosing scheme B arises when the plant disturbance is correlated with the plant input. This occurs when plant disturbance obtained at the plant output is filtered and fed back into the plant input to achieve plant disturbance canceling. This subject was discussed in Chapter 8.

Scheme C was devised to be an improvement over scheme B. Like scheme B, scheme C offers a better, more reliable modeling signal than that of scheme A. Both scheme C and scheme B do their modeling based on the dither signal alone. The improvement of scheme C over scheme B comes from the elimination of the component y'_k, shown in Fig. B.1, from the error ϵ_k. This reduces the minimum mean square error and reduces weight noise. Like scheme B, scheme C can be used even when the plant disturbance is correlated with the plant input, as is the case when plant disturbance is canceled.

Studying Table B.1, scheme C may be compared with scheme A. The stable range of μ is the same for both. The weight noise covariance is the same when choosing the same value of μ for both schemes. The time constant for scheme C is larger than all the time constants for scheme A, except that with big eigenvalue spread for $R_{U'}$, the slowest modes of scheme A will be comparable in decay rate to the single mode of scheme C. The basic performance in terms of weight noise and speed of convergence is similar, comparing scheme C with scheme A, but scheme C has better modeling properties than scheme A.

Again referring to Table B.1, we observe that the stable range of μ for scheme C is not as great as that of scheme B, so that, disregarding weight noise, scheme B can be pushed much harder to achieve faster convergence than that attainable with scheme C. On the other hand, the time constants of schemes B and C are the same when using the same values of μ. Operating with small values of μ, the weight noise of scheme C could be considerably less than that of scheme B. The weight noise covariances would have similar denominators, but the corresponding numerator for scheme C could be quite a bit less than that for scheme B.

Scheme C has essentially all the good features of scheme B (except that scheme B could be pushed harder to converge faster — but high rates of convergence cause large noise levels in the modeling weights) and offers the possibility of significantly less noise in the weights with the same speed of convergence. Scheme C is generally the scheme of choice.

It should be noted, finally, that all three schemes, using dither, have the disadvantage of disturbing the plant and creating noise at the plant output. This is a "necessary evil" in many cases when the plant driving function cannot be relied upon as a suitable modeling signal. When studying Figs. 4.3(a), (b), and (c), it is clear that all three schemes result in exactly the same dither noise at the plant output. None of the schemes has an advantage in this regard.

Bibliography for Appendix B

[1] B. WIDROW, "Adaptive filters," in *Aspects of Network and System Theory*, ed. R.E. Kalman and N. De Claris (New York: Holt, Rinehart and Winston, 1970), pp. 563–587.

[2] L.L. HORWITZ, and K.D. SENNE, "Performance advantage of complex LMS for controlling narrow-band adaptive arrays," *IEEE Trans. on Circuits and Systems*, Vol. CAS-28, No. 6 (June 1981), pp. 562–576.

[3] R.R. BITMEAD, "Convergence in distribution of LMS-type adaptive parameter estimates," *IEEE Trans. on Auto. Control*, Vol. AC-28, No. 1 (January 1983), pp. 54–60.

[4] O. MACCHI, and E. EWEDA, "Second-order convergence analysis of stochastic adaptive linear filtering," *IEEE Trans. on Auto. Control*, Vol. AC-28, No. 1 (January 1983), pp. 76–85.

Appendix C

A Comparison of the Self-Tuning Regulator of Åström and Wittenmark with the Techniques of Adaptive Inverse Control

The best-known adaptive control methods are based on the self-tuning regulator of Åström and Wittenmark. Their 1973 paper [2] has had great influence worldwide in the field of adaptive control. Chapter 3 of their book entitled *Adaptive Control* [1] summarizes their work on the self-tuning regulator. Figure C.1 is a generic diagram of the self-tuning regulator, based on Fig. 3.1 of *Adaptive Control*.

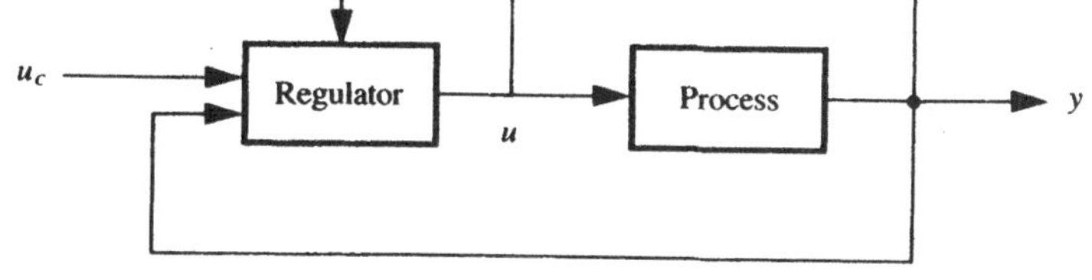

Figure C.1 The self-tuning regulator based on Fig. 3.1 of: K.J. ÅSTRÖM, and B. WITTENMARK, *Adaptive Control* (Reading, MA: Addison-Wesley, 1989).

The system of Fig. C.1 is linear and SISO, and it works in the following way. The process or plant is excited by its input u. Its output is y. This output contains a response to u, plus plant disturbance. An estimator, receiving both the signal input and signal output of the plant, estimates the plant parameters. These estimates are fed to an automatic design algorithm that sets the parameters of the regulator. This regulator could be an input controller, or a controller within the feedback loop, or both. For convenience, we have redrawn the diagram of Fig. C.1 in Fig. C.2.

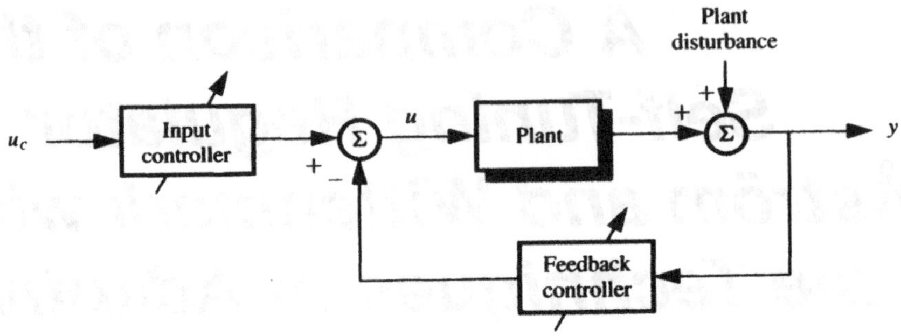

Figure C.2 An alternative representation of the self-tuning regulator.

In Chapter 8, we demonstrated that the adaptive disturbance canceler of Fig. 8.1 minimizes the plant output disturbance power. In fact, we have shown that no other linear system, regardless of its configuration, can reduce the variance of the plant output disturbance to a level lower than that of Fig. 8.1. Comparing the self-tuning regulator of Fig. C.2 with the adaptive disturbance canceler of Fig. 8.1, the question is: Can the feedback controller of the self-tuning regulator be designed to cancel the plant disturbance as well as the adaptive disturbance canceler? Another question that arises is: Can an input controller be designed for the self-tuning regulator so that when it is cascaded with the plant and its feedback controller, the entire control system will have a transfer function equal to the transfer function of a selected reference model? It is not obvious that the self-tuning regulator and the adaptive inverse control system will deliver performances that are equivalent to each other.

C.1 DESIGNING A SELF-TUNING REGULATOR TO BEHAVE LIKE AN ADAPTIVE INVERSE CONTROL SYSTEM

To address these issues, we redraw Figs. C.2 and 8.1 as Figs. C.3 and C.4, respectively, in order to simplify and bring out essential features of these block diagrams. For simplicity, we have drawn Fig. 8.1 with $\widehat{P}(z)$ approximated by $P(z)$. Also, we included a necessary unit delay z^{-1} within the feedback loop of the self-tuning regulator that will be necessary only if there is no delay either in the plant or in the feedback controller. We will assume that $P(z)$ is stable. If the plant is not really stable, let it be stabilized by a separate feedback stabilizer and let $P(z)$ represent the stabilized plant. This creates no theoretical problems for the self-tuning regulator, and as demonstrated in Appendix D, creates no theoretical problems for adaptive inverse control. To compare the two approaches, we need to first show, if possible, that the transfer function from the plant disturbance injection point to the plant output is the same for the self-tuning regulator as for the adaptive disturbance canceler. For the self-tuning regulator of Fig. C.3, this transfer function is

$$\frac{1}{1 + z^{-1} \cdot P(z) \cdot FC(z)}. \tag{C.1}$$

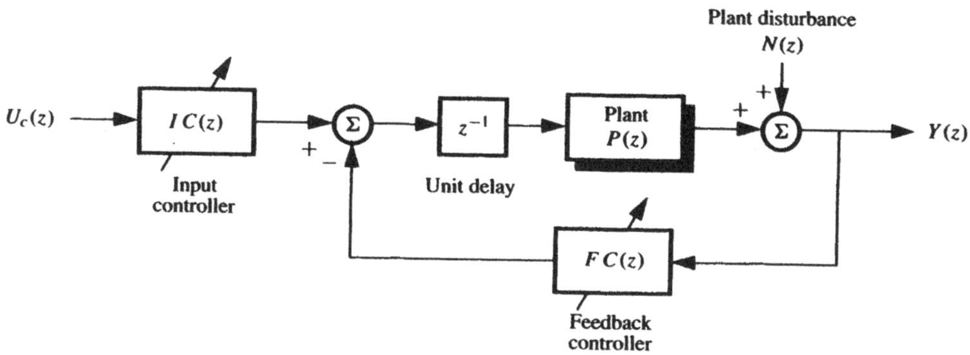

Figure C.3 Another representation of the self-tuning regulator.

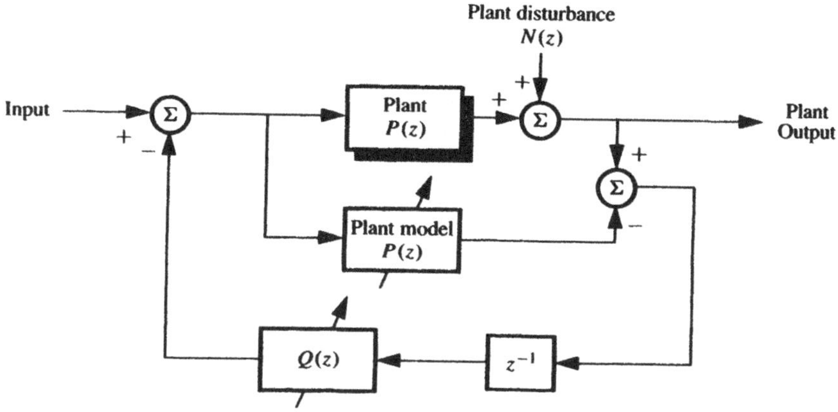

Figure C.4 Another representation of the adaptive plant disturbance canceler.

For the adaptive disturbance canceler of Fig. C.4, this transfer function is

$$1 - z^{-1} \cdot P(z) \cdot Q(z). \tag{C.2}$$

In order for these transfer functions to be equal, it is necessary that

$$FC(z) = \frac{Q(z)}{1 - z^{-1} \cdot P(z) \cdot Q(z)}. \tag{C.3}$$

To obtain $FC(z)$, we need $Q(z)$ and $P(z)$. In practice, both would be readily available (to a close approximation) from adaptive processes already described. So there would be no problem in getting a good expression for $FC(z)$.

If the feedback controller given by Eq. (C.3) were used in the self-tuning regulator, would the resulting system be stable? The answer is yes. Referring to Fig. C.3, we may note that the transfer function through the feedback loop to the plant output is

$$\frac{z^{-1}P(z)}{1+z^{-1}\cdot P(z)\cdot FC(z)} = z^{-1}\cdot P(z)\cdot(1-z^{-1}\cdot P(z)\cdot Q(z)). \tag{C.4}$$

This transfer function is stable since both $P(z)$ and $Q(z)$ are stable. So far so good.

To control the dynamic response of the system, to make it behave like the dynamic response of a selected reference model, we multiply the transfer function (C.4) by $IC(z)$ and set the product equal to $M(z)$:

$$M(z) = IC(z)\cdot z^{-1}\cdot P(z)\cdot(1-z^{-1}\cdot P(z)\cdot Q(z)). \tag{C.5}$$

Accordingly,

$$\begin{aligned} IC(z) &= \frac{M(z)}{z^{-1}\cdot P(z)\cdot(1-z^{-1}\cdot P(z)\cdot Q(z))} \\ &= \frac{M(z)\cdot FC(z)}{z^{-1}\cdot P(z)\cdot Q(z)}. \end{aligned} \tag{C.6}$$

For the entire self-tuning regulator to be stable, it is necessary that $IC(z)$ be stable. Stability has already been established for the rest of the system. $M(z)$ is stable. Since $P(z)$ and $Q(z)$ are stable, they would not cancel any unstable poles of $FC(z)$ that may occur. It is necessary, therefore, for $FC(z)$ to be stable in order for $IC(z)$ to be stable, although this is not sufficient for stability. $IC(z)$ will be unstable if either $P(z)$, $Q(z)$, or both are nonminimum-phase.

How would one build a self-tuning regulator if its feedback controller $FC(z)$ were unstable? There are two possibilities. One possibility would be to choose an $FC(z)$ that is stable but not optimal for plant disturbance canceling. The other possibility would be to use the optimal, unstable $FC(z)$ inside the feedback loop and build the input controller $IC(z)$ having the indicated poles and zeros but allowing components of $IC(z)$ to be noncausal as required for stability. The noncausal filter could be realized approximately with an appropriate delay. The entire system response would be a delayed version of the response of $M(z)$. These difficulties are not encountered when the optimal $FC(z)$ is used and the input controller $IC(z)$ is stable.

C.2 SOME EXAMPLES

Specific examples will help to clarify some of the issues. Suppose that the plant disturbance is constant, that it is a DC offset or bias. The question is: How well do the adaptive disturbance canceler and the self-tuning regulator handle this disturbance? For the adaptive disturbance canceler, the transfer function from the plant disturbance injection point to the plant output is given by (C.2). We would like this transfer function to have a value of zero at zero frequency, that is, at $z = 1$. This is easily accomplished as follows:

$$1 - P(1)\cdot Q(1) = 0,\quad \text{or}\quad Q(1) = \frac{1}{P(1)}. \tag{C.7}$$

Any form of $Q(z)$ would allow perfect canceling of the constant disturbance as long as the value of Q at $z = 1$ is the reciprocal of the value of P at $z = 1$.

So the adaptive plant disturbance canceler will perfectly eliminate constant plant disturbance. What will the self-tuning regulator do with this disturbance? Its transfer function from plant disturbance injection point to plant output point is given by (C.1). This transfer function should equal zero at zero frequency to eliminate the DC plant disturbance. Accordingly,

$$\frac{1}{1 + P(1) \cdot FC(1)} = 0. \tag{C.8}$$

Assuming that the plant transfer function is well behaved at $z = 1$, it is clear that $FC(z)$ must be infinite at $z = 1$. One way to accomplish this would be to let $FC(z)$ be a digital integrator,

$$FC(z) = \frac{1}{1 - z^{-1}}, \tag{C.9}$$

giving it a simple pole at $z = 1$. The input controller has the transfer function given by (C.6):

$$\begin{aligned} IC(z) &= \frac{M(z) \cdot FC(z)}{z^{-1} \cdot P(z) \cdot Q(z)} \\ &= \left(\frac{M(z)}{z^{-1} \cdot P(z) \cdot Q(z)}\right)\left(\frac{1}{1 - z^{-1}}\right). \end{aligned} \tag{C.10}$$

$P(z)$ and $Q(z)$ are both stable and finite at $z = 1$. If in addition $M(z)$ has a finite value at $z = 1$, $IC(z)$ will have a pole at $z = 1$ and will thereby be unstable making the entire system unstable. A noncausal realization of $IC(z)$ would not be useful in this case. The only possibility would be to choose an $FC(z)$ having its pole slightly inside the unit circle, sacrificing some disturbance canceling capability for a stable $IC(z)$.

The same kind of result would be obtained if the plant disturbance were a constant-amplitude sine wave. The adaptive disturbance canceler would adapt and learn to eliminate it perfectly. The self-tuning regulator would either be unstable or, if stable, would give somewhat less than optimal disturbance reducing performance.

3 SUMMARY

The self-tuning regulator has an easier job of disturbance reduction and dynamic control than adaptive inverse control when the plant is unstable with a pole or poles on the unit circle or outside the unit circle. Feedback used by the self-tuning regulator has the capability of moving the plant poles inside the unit circle to stabilize the plant, reduce disturbance, and control its dynamics. For adaptive inverse control, the first step would be to stabilize the plant with feedback. The choice of feedback transfer function for stabilization would not be critical and would not need to be optimized. The only requirement would be to somehow stabilize the plant. Then adaptive inverse control could be applied in the usual manner. A discussion of initial stabilization for adaptive inverse control systems is given in Appendix D.

A difficult case for both approaches occurs when the plant has one or more zeros on the unit circle. The inverse controller tries to put the FIR equivalent of a pole or poles on top of the zeros. The inverse impulse response does not die exponentially but instead persists forever. An optimal FIR inverse filter cannot be constructed.[1] The self-tuning regulator of Fig. C.3 also cannot cope with such a problem. Its feedback cannot move around the zeros. It too would try to choose $IC(z)$ to put poles on top of these zeros, but this would make $IC(z)$ unstable and thereby make the entire system unstable. A remedy that would work for both approaches would be to cascade with the plant a digital filter having poles matching the unit circle zeros of the plant and to envelop the cascade within a feedback loop. If, for example, the plant had a single zero on the unit circle at $z = 1$, the cascade filter would be a digital integrator. The loop around the cascade would need to be designed to be stable for adaptive inverse control.

When the plant is nonminimum-phase with zeros outside the unit circle, we have seen in Chapters 5, 6 and 7 how adaptive inverse control can readily cope with such a plant. Dealing with this kind of plant with a self-tuning regulator is difficult, much more difficult than dealing with a minimum-phase plant. The literature is not clear on how this can be done.

Comparing adaptive inverse control with the self-tuning regulator reveals cases where one approach is advantageous, and cases where the other approach is advantageous. Also, there are many cases where both approaches give equivalent performance although the system configurations and methods of adaptation are totally different.

Bibliography for Appendix C

[1] K.J. ÅSTRÖM, and B. WITTENMARK, *Adaptive control,* 2nd ed. (Menlo Park, CA: Addison Wesley, 1995).

[2] K.J. ÅSTRÖM, and B. WITTENMARK, "On self-tuning regulators," *Automatica*, Vol. 9, No. 2 (1973).

[1] This situation is not so clear. Refer to simulation examples in Chapter 12.

Appendix D

Adaptive Inverse Control for Unstable Linear SISO Plants

If the plant to be controlled, $P(z)$, is unstable, it is impossible to directly apply adaptive inverse control to realize a stable well-controlled system. The reason for this is that a feed-forward controller driving an unstable plant will leave the plant unstable. The first step in the practical utilization of adaptive inverse control for an unstable plant must therefore be stabilization of the plant. Figure D.1 shows an unstable plant $P(z)$ being stabilized by feedback. The plant disturbance is represented as an additive disturbance at the plant output $N(z)$. A feedback filter used within the feedback loop has the transfer function $FB(z)$. The necessary delay around the feedback loop is z^{-1}.[1]

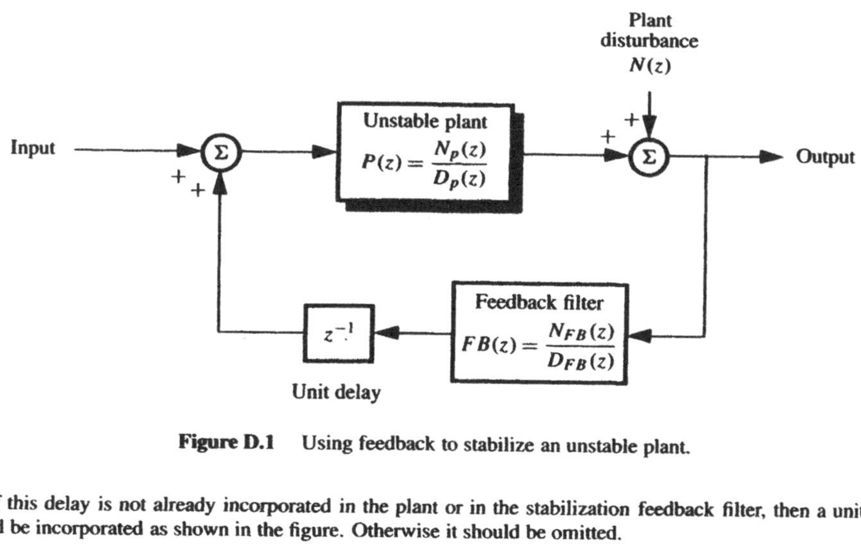

Figure D.1 Using feedback to stabilize an unstable plant.

[1] If this delay is not already incorporated in the plant or in the stabilization feedback filter, then a unit delay should be incorporated as shown in the figure. Otherwise it should be omitted.

It is convenient to describe the plant transfer function as a ratio of a numerator polynomial (responsible for its zeros) and a denominator polynomial (responsible for its poles):

$$P(z) = \frac{N_p(z)}{D_p(z)}. \tag{D.1}$$

The transfer function of the feedback filter can be described as

$$FB(z) = \frac{N_{FB}(z)}{D_{FB}(z)}. \tag{D.2}$$

The input-output transfer function of the feedback system of Fig. D.1 is given by

$$\begin{aligned} H(z) &= \frac{P(z)}{1 - FB(z) \cdot P(z) \cdot z^{-1}} \\ &= \frac{N_p(z) \cdot D_{FB}(z)}{D_p(z) \cdot D_{FB}(z) - N_{FB}(z) \cdot N_p(z) \cdot z^{-1}}. \end{aligned} \tag{D.3}$$

The objective is to choose $FB(z)$ so that $H(z)$ will be stable. Once this is done, then $H(z)$ becomes an equivalent plant whose dynamic response can be controlled by adaptive inverse control and whose disturbance can be canceled by the associated adaptive disturbance canceling techniques.

The roots of the denominator of (D.3) need to be placed inside the unit circle in the z-plane by appropriate choice of $N_{FB}(z)$ and $D_{FB}(z)$. The question is, how does one make this appropriate choice? Is the choice unique? Should the choice be optimal? Or is there a wide latitude in this process? To begin, let the choice be arbitrary as long as $H(z)$ is stable. Furthermore choose the feedback filter $FB(z)$ to be stable.

D.1 DYNAMIC CONTROL OF STABILIZED PLANT

Consider first the issue of dynamic control of $H(z)$. This is the same issue as dynamic control of the plant $P(z)$, since the plant output is really the same as the output of $H(z)$. We need to examine the model-reference inverse of the transfer function $H(z)$. Using Eq. (D.3),

$$(M(z))\left(\frac{1}{H(z)}\right) = \left(\frac{1}{N_p(z)}\right)\left(\frac{D_p(z) \cdot D_{FB}(z) - N_p(z) \cdot N_{FB}(z) \cdot z^{-1}}{D_{FB}(z)}\right)(M(z)). \tag{D.4}$$

For sake of argument, imagine the inverse controller as being a cascade of three FIR filters, each having a transfer function that closely approximates the first, second, and third factors of the right-hand side of equation (D.4) respectively. Since $FB(z)$ has been chosen to be stable, the second factor will be stable since its denominator $D_{FB}(z)$ has all its roots inside the unit circle. A filter whose transfer function closely approximates the second factor is therefore easy to realize as a causal filter without modeling delay. A filter whose transfer function closely approximates the first factor is easy to realize if $N_p(z)$ has all its roots inside the unit circle in the z-plane. If not, realization requires modeling delay. In any event, ease of realization of the first factor depends on plant characteristics and will be the same regardless of the choice of $FB(z)$. Realization of the second factor will always be easy since $FB(z)$ was chosen to be a stable filter and one that stabilizes $H(z)$, but otherwise is completely flexible. Realization of the third factor will be easy because $M(z)$ will always be chosen to be stable.

It may be noted that the cascade of the three FIR filters could in fact be a single FIR filter. This filter is represented as $\widehat{C}(z)$ in Fig. D.2, which is a block diagram of an adaptive inverse control scheme for dynamic control of an unstable plant.

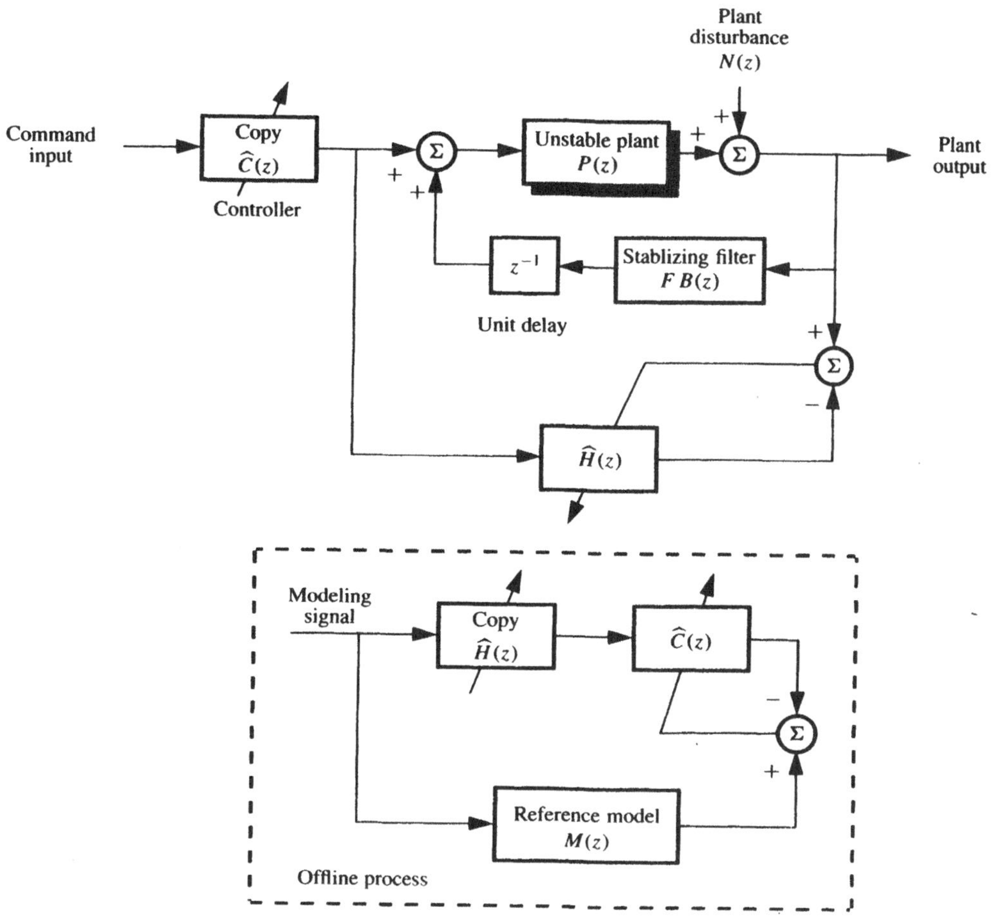

Figure D.2 Inverse control of dynamics of unstable plant.

To design the filter $FB(z)$, some idea of the plant characteristics would be useful. Then root-locus or Nyquist design of this plant stabilizer could be done. The design is not critical, and it could be chosen by experiment. The above arguments show that if the stabilizing filter is itself stable, a wide latitude in its design is acceptable. Variation in this design will result in variation in $\widehat{C}(z)$, but the overall system dynamic response will always be the same. Ease of realization of $\widehat{C}(z)$ will not depend on the choice of $FB(z)$.

The next issue is plant disturbance canceling. Since stabilization of the plant is necessary, a question arises about the optimality of the stabilization approach of Fig. D.1 from the point of view of plant disturbance cancelation. Does $FB(z)$ need to be optimized, or is there wide latitude in its choice for optimal plant disturbance cancelation?

D.2 ADAPTIVE DISTURBANCE CANCELING FOR THE STABILIZED PLANT

Figure D.3 shows a block diagram of an adaptive disturbance canceler for an unstable plant $P(z)$ stabilized by feedback. A feedback filter with transfer function $FB(z)$ provides the stabilization signal. An adaptive process using the plant driving signal $U(z)$ is used to adaptively find $\hat{H}(z)$, an estimate of $H(z)$ given by Eq. (D.3). A copy of $\hat{H}(z)$ is used to obtain the disturbance at the plant output. This disturbance is filtered by $Q(z)$ and z^{-1} and is subtracted from the plant input to cancel the plant disturbance as well as it can in the least squares sense.

We would like to show that adaptive disturbance canceling with an unstable plant will be optimal regardless of the choice of the feedback filter $FB(z)$, as long as it stabilizes the plant. We will prove that this is true if the plant is minimum-phase and the feedback filter is chosen to be stable.

Please refer to Chapter 8, particularly Section 8.2, "Proof of Optimality for the Adaptive Plant Disturbance Canceler," and Fig. 8.2. Many of the mathematical ideas of that section will be useful here. Figure D.4 shows a collection of block diagrams that will help in obtaining an expression for $Q^*(z)$, the optimal transfer function for the disturbance canceling feedback. The filter $Q(z)$ in Fig. D.3 adapts by an offline process toward this Wiener solution. Of interest is the development of expressions for $Q^*(z)$ and for the power of the uncanceled plant disturbance.

Figure D.4(a) represents the unstable plant, $P(z)$, and its additive stationary output disturbance $N(z)$. Figure D.4(b) again shows $P(z)$ with its output disturbance $N(z)$, but here the disturbance is represented as if it originated from a unit-power white disturbance $L(z)$ which has gone through a stable, causal, minimum phase filter $\Gamma(z)$. The situation is analogous to that of Fig. 8.2(b). Figure D.4(c) shows the plant $P(z)$ now being stabilized with feedback. The output disturbance is no longer $N(z)$. It is equivalent to

$$\frac{N(z)}{1 - P(z) \cdot z^{-1} \cdot FB(z)}. \tag{D.5}$$

In Fig. D.4(c), the plant output disturbance is generated by $L(z)$ propagating through the filter $\Gamma'(z)$. The transfer function of this filter is

$$\Gamma'(z) = \Gamma(z) \frac{1}{1 - P(z) \cdot z^{-1} \cdot FB(z)}. \tag{D.6}$$

Figure D.4(d) is a simplified representation of Fig. D.4(c). Figure D.4(e) shows the action of the actual plant disturbance canceler. The equivalent plant disturbance drives the disturbance canceling path consisting of the delay z^{-1}, the filter $-Q^*(z)$, and the transfer function of the stabilized plant $H(z)$. The optimal transfer function $-Q^*(z)$ minimizes the power of the plant output disturbance after cancelation. The block diagram of Fig. D.4(e) works just like that of Fig. 8.2(d).

We need an expression for $Q^*(z)$. The Shannon-Bode method will be used to obtain it. The causal Wiener filter is given by Eq. (2.74).

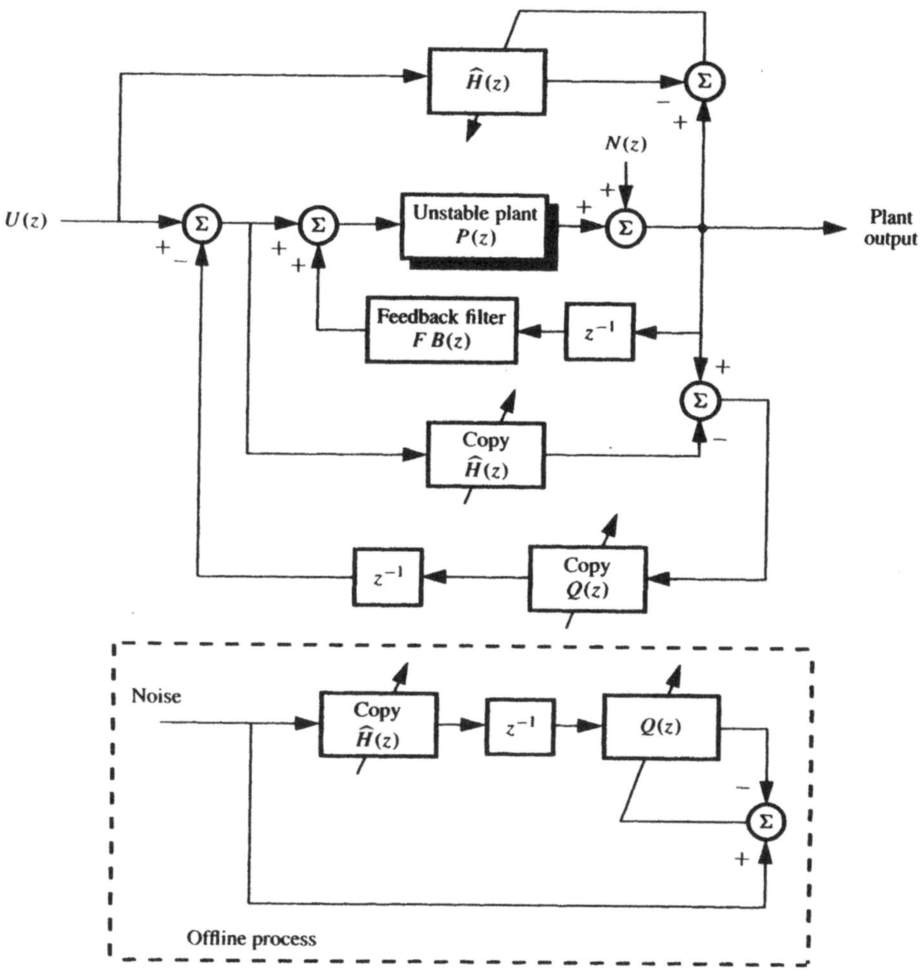

Figure D.3 An adaptive noise-canceler for an unstable plant.

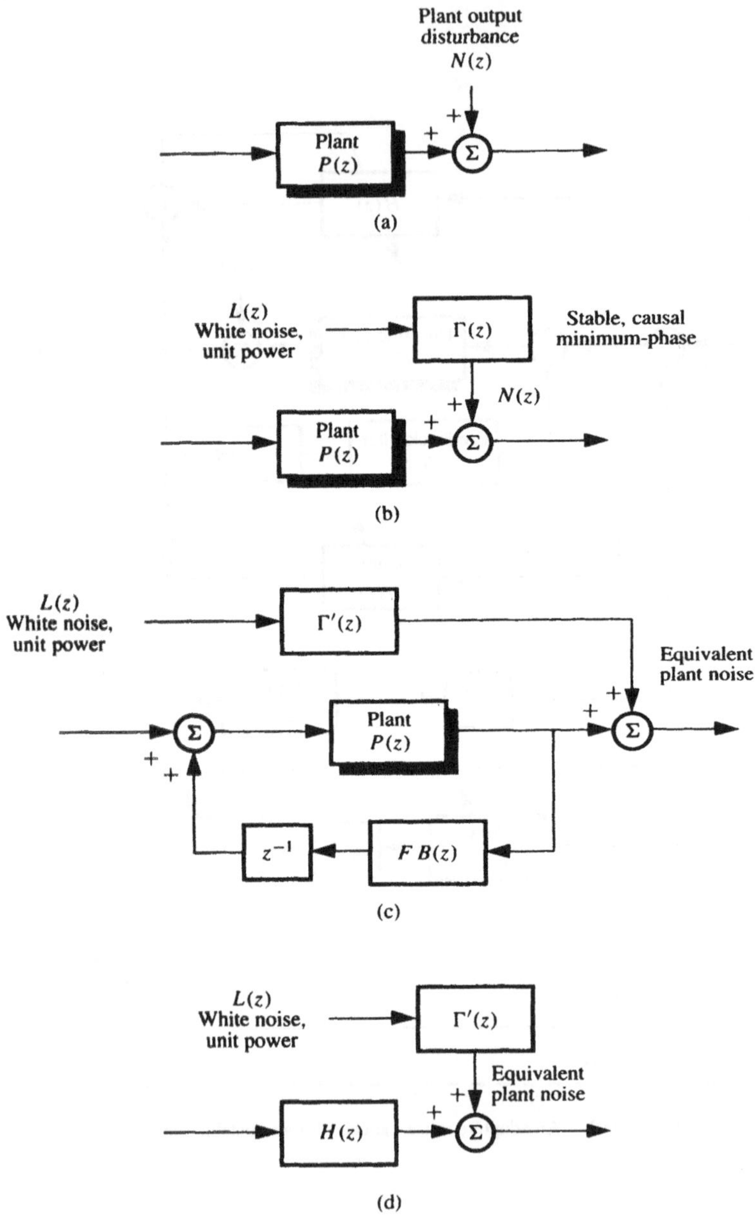

Figure D.4 Calculation of plant output disturbance after cancelation. (continued on next page)

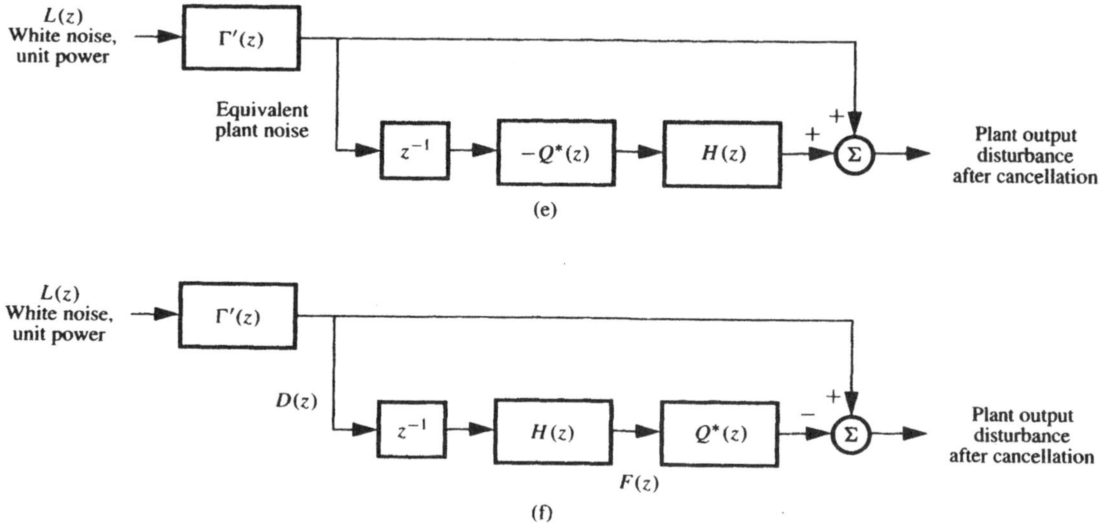

Figure D.4 (continued).

Figure D.4(f) is equivalent to Fig. D.4(e), except that the order of filtering by $H(z)$ and $Q^*(z)$ has been commuted. Since $Q^*(z)$ is the Wiener filter, we make use of our previous Wiener filter notation and designate the input to $Q^*(z)$ as $F(z)$ (f_k in the time domain), and we designate its desired response as $D(z)$ (d_k in the time domain). From Eq. (2.74), the Wiener solution is

$$Q^*(z) = \frac{1}{\Phi^+_{ff}(z)}\left[\frac{\Phi_{fd}(z)}{\Phi^-_{ff}(z)}\right]_+ . \tag{D.7}$$

Now, referring to Fig. D.4(f), we note that

$$\Phi_{ff}(z) = \Gamma'(z)\Gamma'(z^{-1})H(z)H(z^{-1}). \tag{D.8}$$

Also,

$$\Phi_{fd}(z) = \Gamma'(z)\Gamma'(z^{-1})H(z^{-1})z. \tag{D.9}$$

We will need to factor $\Phi_{ff}(z)$ as

$$\Phi_{ff}(z) = \Phi^+_{ff}(z) \cdot \Phi^-_{ff}(z). \tag{D.10}$$

To accomplish this in a useful way, we will factor $\Gamma'(z)$ in the following manner:

$$\Gamma'(z) = \Gamma'_1(z) \cdot \Gamma'_2(z). \tag{D.11}$$

Let the factor $\Gamma'_1(z)$ have all the zeros and poles of $\Gamma'(z)$ that are inside the unit circle. Let the factor $\Gamma'_2(z)$ have all the zeros of $\Gamma'(z)$ that are outside the unit circle. It has no poles because $\Gamma'(z)$ has no poles outside the unit circle, a fact that can be verified by inspection of Eqs. (D.3) and (D.6) as follows. The factor $\Gamma(z)$ of (D.6) is stable and minimum-phase.

The denominator factor of (D.6) has all roots inside the unit circle, since $FB(z)$ was chosen to stabilize the plant $P(z)$. Therefore, $\Gamma'(z)$ is stable, with all poles inside the unit circle.

The factor $\Gamma'_2(z)$ of Eq. (D.11) has zeros only. These zeros comprise all the zeros of $\Gamma'(z)$ which are outside the unit circle. Referring to Eq. (D.6), all such zeros are caused by poles of the denominator. The feedback filter $FB(z)$ was chosen to stabilize the plant. By itself, it could be stable or unstable and accomplish this task. We choose $FB(z)$ to be stable. As such, the zeros of $\Gamma'_2(z)$ will be caused only by the poles of $P(z)$ which are outside the unit circle. Therefore, $\Gamma'_2(z)$ depends only on plant characteristics and does not depend on the choice of the feedback filter $FB(z)$.

Referring to Eq. (D.8), factorization of $\Phi_{ff}(z)$ also requires factorization of $H(z) \cdot H(z^{-1})$. To do this factorization, we first note that $H(z)$ is stable and, from (D.3), it may be deduced that $H(z)$ is minimum-phase. The zeros of $H(z)$ correspond to the zeros of the plant and to the poles of $FB(z)$. We now assume that the plant $P(z)$ has all of its zeros inside the unit circle. Since $FB(z)$ is stable with all of its poles inside the unit circle, $H(z)$ is minimum-phase and

$$H(z) = H^+(z) \text{ and } H(z^{-1}) = H^-(z). \tag{D.12}$$

Factorization of $\Phi_{ff}(z)$ proceeds in the following way. Substituting (D.11) into (D.8),

$$\Phi_{ff}(z) = \Gamma'_1(z)\Gamma'_2(z)\Gamma'_1(z^{-1})\Gamma'_2(z^{-1})H(z)H(z^{-1}). \tag{D.13}$$

Grouping the poles and zeros inside the unit circle and, respectively, those outside the unit circle yields

$$\Phi^+_{ff}(z) = \Gamma'_1(z)\Gamma'_2(z^{-1})H(z), \text{ and} \tag{D.14}$$

$$\Phi^-_{ff}(z) = \Gamma'_2(z)\Gamma'_1(z^{-1})H(z^{-1}). \tag{D.15}$$

To find $Q^*(z)$, we can substitute (D.11) into (D.9) to get

$$\Phi_{fd}(z) = \Gamma'_1(z)\Gamma'_2(z)\Gamma'_1(z^{-1})\Gamma'_2(z^{-1})H(z^{-1})z, \tag{D.16}$$

and then substitute this and (D.14) and (D.15) into (D.7). The result is

$$\begin{aligned} Q^*(z) &= \frac{1}{\Gamma'_1(z)\Gamma'_2(z^{-1})H(z)} \left[\frac{\Gamma'_1(z)\Gamma'_2(z)\Gamma'_1(z^{-1})\Gamma'_2(z^{-1})H(z^{-1})z}{\Gamma'_2(z)\Gamma'_1(z^{-1})H(z^{-1})} \right]_+ \\ &= \frac{1}{\Gamma'_1(z)\Gamma'_2(z^{-1})H(z)} \left[\Gamma'_1(z)\Gamma'_2(z^{-1})z\right]_+ . \end{aligned} \tag{D.17}$$

To evaluate the bracketed factor of (D.17), it is useful to note that $\Gamma'_1(z)$ and $\Gamma'_2(z)$ are both stable and causal and that their product is stable and causal. This product may be expanded therefore as follows.

$$\Gamma'(z) = \Gamma'_1(z)\Gamma'_2(z^{-1}) = \gamma_0 + \gamma_1 z^{-1} + \gamma_2 z^{-2} + \cdots. \tag{D.18}$$

The bracketed factor of (D.17) may now be evaluated as

$$\begin{aligned} [\Gamma'_1(z)\Gamma'_2(z^{-1})z]_+ &= \gamma_1 + \gamma_2 z^{-1} + \gamma_3 z^{-2} + \cdots \\ &= \left(\Gamma'_1(z)\Gamma'_2(z^{-1}) - \gamma_0\right) z. \end{aligned} \tag{D.19}$$

Substituting this into (D.17) gives $Q^*(z)$:

$$Q^*(z) = \frac{\left(\Gamma_1'(z)\Gamma_2'(z^{-1}) - \gamma_0\right) z}{\left(\Gamma_1'(z)\Gamma_2'(z^{-1})H(z)\right)}. \tag{D.20}$$

Now that we have $Q^*(z)$, we can use it to find the power of the uncanceled plant disturbance. Refer to Fig. D.4(f). The plant output disturbance after cancelation is modeled as originating with white disturbance $L(z)$ of unit power which has propagated through a transfer function. We need to find this disturbance transfer function. From Fig. D.4(f), by inspection,

$$\begin{pmatrix} \text{disturbance transfer} \\ \text{function} \end{pmatrix} = \Gamma'(z)\left(1 - Q^*(z) \cdot z^{-1} \cdot H(z)\right). \tag{D.21}$$

Substitution of (D.20) into (D.21) gives

$$\begin{aligned} \begin{pmatrix} \text{disturbance transfer} \\ \text{function} \end{pmatrix} &= \Gamma'(z)\left(1 - \frac{\Gamma_1'(z)\Gamma_2'(z^{-1}) - \gamma_0}{\Gamma_1'(z)\Gamma_2'(z^{-1})}\right) \\ &= \frac{\Gamma'(z)\gamma_0}{\Gamma_1'(z)\Gamma_2'(z^{-1})}. \end{aligned} \tag{D.22}$$

Substitution of (D.11) into the above yields

$$\begin{aligned} \begin{pmatrix} \text{disturbance transfer} \\ \text{function} \end{pmatrix} &= \frac{\Gamma_1'(z)\Gamma_2'(z)\gamma_0}{\Gamma_1'(z)\Gamma_2'(z^{-1})} \\ &= \frac{\Gamma_2'(z)}{\Gamma_2'(z^{-1})}\gamma_0. \end{aligned} \tag{D.23}$$

The uncanceled plant disturbance is therefore equivalent to white noise of unit power propagating through this transfer function.

We are on the verge of a very useful result. We will show next that the disturbance transfer function (D.23) does not change as the stabilization feedback filter $FB(z)$ is changed. This transfer function is a function of the plant characteristic, but it turns out not be a function of $FB(z)$.

Regarding (D.23), the factor $\Gamma_2'(z)$, defined by (D.11) and (D.6), has been previously established as dependent on $P(z)$ but independent of the choice of $FB(z)$. The same is true of course for the denominator factor of (D.23), $\Gamma_2'(z^{-1})$. All that remains is a question about γ_0: Does it depend on $FB(z)$?

Equation (D.18) defines γ_0 as the first term in the expansion of $\Gamma'(z)$ in a power series of z^{-1}. The first term can be found by evaluating expression (D.6) as $z^{-1} \to 0$. The result is

$$\Gamma'(z)|_{z^{-1}\to 0} = \left(\Gamma(z)|_{z^{-1}\to 0}\right)\left[\frac{1}{1 - P(z) \cdot z^{-1} \cdot FB(z)}\right]_{z^{-1}\to 0}, \tag{D.24}$$

or

$$\Gamma'(z)|_{z^{-1}\to 0} = \Gamma(z)|_{z^{-1}\to 0}. \tag{D.25}$$

Therefore, the first term of the expansion of $\Gamma'(z)$ equals the first term of the expansion of $\Gamma(z)$. But $\Gamma(z)$ does not depend at all on $FB(z)$. It relates only to the original plant disturbance with no stabilization feedback and no adaptive disturbance canceling. Therefore, the first term in the expansion of $\Gamma'(z)$ does not depend on $FB(z)$ and therefore γ_0 of Eq. (D.23) does not depend on $FB(z)$. We have now established that the disturbance transfer function does not depend on $FB(z)$. Therefore, the power of the plant output disturbance after cancelation does not depend on the design of the feedback filter $FB(z)$.

In establishing this fact, we have assumed that $FB(z)$ is itself stable and that it stabilizes the plant $P(z)$. We have also assumed that $P(z)$ is minimum-phase. Further analysis has been performed with the minimum-phase restriction on $P(z)$ relaxed. The zeros of $P(z)$ were allowed to be either inside or outside the unit circle or some inside, some outside. The analysis is straightforward, but too lengthy to include here. *The conclusion is that the plant output disturbance power, after adaptive cancelation, is independent of $FB(z)$ as long as $FB(z)$ is itself stable and is able to stabilize the plant $P(z)$. There are no restrictions on $P(z)$ except that it should not have zeros on the unit circle.*[2]

We have not proved that linear feedback stabilization of the plant, exemplified by the diagram in Fig. D.1, is optimal. We simply do not know of any other way to stabilize the unstable plant. Given that the plant is stabilized as in Fig. D.1, minimization of plant disturbance proceeds from there. We have shown that the choice of the feedback $FB(z)$, as long as it stabilizes the plant and it alone is stable, is not critical to the disturbance canceling process. All such feedback filters would perform equivalently from the point of view of disturbance canceling. The minimum plant output disturbance power would not depend upon the choice of $FB(z)$. *Recall that in the first section of this appendix, we showed that the ability of an adaptive inverse controller to control plant dynamics does not also depend on the choice of $FB(z)$, as long as it alone is stable and it is able in a feedback connection to stabilize the plant.*

Adaptive inverse control can be made to work with an unstable plant as well as with a stable one, as long as the unstable plant is first stabilized with feedback. The design of the feedback is not critical. Control of system dynamics is optimized separately from control of plant disturbance. Optimization of system dynamics does not interfere with minimization of plant disturbance.

D.3 A SIMULATION STUDY OF PLANT DISTURBANCE CANCELING: AN UNSTABLE PLANT WITH STABILIZATION FEEDBACK

In order to demonstrate that the choice of $FB(z)$, the stabilization feedback, does not affect the ability of an adaptive canceler to cancel plant disturbance (as long as $FB(z)$ itself is stable), the system of Fig. D.3 was simulated. An unstable plant $P(z)$ was chosen, and three different feedback filters, $FB_i(z)$, $i = 1, 2, 3$, each capable of stabilizing the plant, were chosen.

[2]Even this restriction can be circumvented. If, for example, the unlikely case occurred that the plant had a zero at $z = 1$, the plant could be preceded by an integrator having the transfer function $(1 - z^{-1})^{-1}$.

For each of the stabilization feedback filters, corresponding transfer functions of the equivalent stable plant $H_i(z)$, $i = 1, 2, 3$, were derived analytically. Utilizing these transfer functions in the offline process of Fig. D.3, we were able to compute, for each choice of the stabilization feedback filter, optimal disturbance canceling feedback filters $Q_i(z)$, $i = 1, 2, 3$.

When the disturbance canceling feedback was established and working, we were able to measure the power of the plant disturbance which remained at the system output for all three cases. While making these measurements, the plant input $U(z)$ was set to zero.

For each $i = 1, 2, 3$, we obtained a different transfer function $H_i(z)$ and a different disturbance canceling loop transfer function $Q_i(z)$. However, as expected, the power of the uncanceled plant disturbance remained the same in all three cases.

In all of these simulations, the unstable plant was

$$P(z) = \frac{1}{1 - 2z^{-1}}. \tag{D.26}$$

The plant disturbance $N(z)$ was simulated by filtering a zero-mean white noise through a low-pass 100-weight filter having the transfer function

$$1 + z^{-1} + \cdots + z^{-99}. \tag{D.27}$$

The power output was normalized so that

$$E[n_k^2] = 1. \tag{D.28}$$

The plant model $\widehat{H}(z)$ was obtained from $H(z)$ but truncated to have 100 weights, and the disturbance canceling loop had a $Q(z)$ limited to 50 weights. The three choices of plant stabilization filter were $FB_1(z) = -1.5$, $FB_2(z) = -2$, and

$$FB_3(z) = -\left(1 + \frac{1}{2}z^{-1}\right). \tag{D.29}$$

The result of the simulation experiments are given in Table D.1. In column 2 of this table, the transfer functions of the stabilization filters $FB_i(z)$ are given. In column 3, transfer functions of the equivalent plants $H_i(z)$ (i.e., overall transfer functions of the plant $P(z)$ with corresponding stabilization feedback, as obtained from Eq. (D.3)), are given. For case one, the equivalent plant had one pole, for case two, the equivalent plant had no poles; for case three, the equivalent plant had two poles.

In column 4 of Table D.1, the transform of the output disturbance of the equivalent plant is given for all three cases. This equivalent disturbance appears at the output of the plant stabilization feedback system shown in Fig. D.1. These transforms take into account the impact of the feedback filter $FB(z)$, in accord with Eq. (D.5).

Using the offline adaptive process depicted in Fig. D.3, we have computed the optimal 50-weight disturbance canceling filter $Q(z)$. In each case, we have assumed that $\widehat{H}(z)$ is as given in column 3, except that it is truncated to allow $\widehat{H}(z)$ to have only 100 weights. In finding $Q(z)$, the modeling signal was designed to have the same spectral character as the equivalent plant disturbance given in column 4 of the table. Although $Q(z)$ had 50 weights, only the first 8 weights for each case are listed in column 5 of the table.

Finally, using $Q_i(z)$ as obtained above, we ran the system of Fig. D.3 for $i = 1, 2, 3$. For each case, we measured the power of the uncanceled plant disturbance. Results are pre-

TABLE D.1 RESULTS OF DISTURBANCE CANCELING EXPERIMENTS WITH AN UNSTABLE DISTURBED PLANT STABILIZED BY VARIOUS FEEDBACK FILTERS

Case number 1	Feedback filter $FB(z)$ 2	Equivalent plant $H(z)$ 3	Equivalent plant noise output 4	Noise canceling feedback $Q(z)$ truncated to its first 8 weights 5								Output disturbance power 6
1	-1.5	$\frac{1}{1-\frac{1}{2}z^{-1}}$	$\frac{N(z)(1-2z^{-1})}{1-\frac{1}{2}z^{-1}}$	0.9933	-0.496	0	0	0	0	0	0	0.02825
2	-2	1	$N(z)(1-2z^{-1})$	0.4935	0.2467	0.1234	0.0617	0.0308	0.0154	0.0077	0.0039	0.02819
3	$-1-\frac{1}{2}z^{-1}$	$\frac{1}{1-z^{-1}+\frac{1}{2}z^{-2}}$	$\frac{N(z)(1-2z^{-1}}{1-z^{-1}+\frac{1}{2}z^{-2}}$	1.4935	-2.2402	1.6168	-0.435	0.0308	0.0154	0.0077	0.0039	0.02841

sented in column 6 of Table D.1. One may note that for the three different stabilization feedback filters and their corresponding disturbance cancelation feedback filters, the power of the uncanceled plant disturbance remained virtually the same. In fact, the difference between the lowest value in column 6 (for case two) and the highest (for case three) did not exceed 0.7 percent of the average uncanceled plant disturbance power. It was a delight to experience such close corroboration between theory and experiment!

It should be noted that both the theoretical proof presented above and the results of this simulation experiment depend on the plant $\widehat{H}(z)$ being correct, equal to $H(z)$, or close to it (i.e., $\widehat{H}(z)$ has enough weights to form a good model of $H(z)$, and the adaptation constant μ is small enough so that weight disturbance in the plant model is negligible), and that both depend on the disturbance cancelation feedback filter $Q(z)$ to have enough weights to achieve good cancelation of the plant disturbance. But what happens if these assumptions are incorrect?

Suppose that the length of the model $\widehat{H}(z)$ were limited. If the stabilization feedback filter were chosen so that the impulse response of the stabilized plant $H(z)$ would be relatively short, then limiting the length of $\widehat{H}(z)$ would not cause errors. However, suppose that the stabilization feedback filter were chosen so that the impulse response of $H(z)$ were long. Then $\widehat{H}(z)$ would be too short to adequately model $H(z)$. The result would be that the power of the uncanceled plant disturbance would grow. In this sense, the ability of the disturbance canceler to cancel plant disturbance *could depend on the choice of the stabilization feedback filter*.

In order to demonstrate this possibility, we reduced the number of weights in $\widehat{H}(z)$ to only five. Then we repeated the entire simulation for all three cases. In case two (where $H(z)$ is very simple), the uncanceled plant disturbance was essentially unaltered. Its power had the value 0.03020. In case one, $H(z)$ is more complicated with a longer impulse response, and limiting the length of $\widehat{H}(z)$ had an effect on the uncanceled plant disturbance power, which grew to 0.0653. In case three, $H(z)$ is even more complicated with a longer impulse response, so that limiting $\widehat{H}(z)$ had an even greater effect, causing the uncanceled plant disturbance power to shoot up to 0.403.

Similarly, one can expect that limiting the number of weights in the disturbance canceling feedback filter $Q(z)$ will introduce some sensitivity to the choice of the stabilization feedback filter. To explore this issue, the simulation was rerun with different lengths for the filter $Q(z)$. $\widehat{H}(z)$ was restored to its original length of 100 weights.

With $Q(z)$ set to have 10 weights, the three cases were run again. The spread in the resulting uncanceled plant disturbance powers varied over a range of 1 percent (instead of 0.7 percent for the 50-weight $Q(z)$). This is not a significant variation. Finally, the experiments were rerun with $Q(z)$ limited to only 3 weights. For cases one and two, the uncanceled plant disturbance power remained essentially the same, but for case three with a more complicated $H(z)$, the uncanceled plant disturbance power jumped up by 30 percent. This is significant. Therefore, with limitations placed on the length of the impulse response of $Q(z)$, we note once again that the ability of the disturbance canceler to cancel plant disturbance *could depend on the choice of the stabilization feedback filter*.

D.4 STABILIZATION IN SYSTEMS HAVING BOTH DISCRETE AND CONTINUOUS PARTS

The above analysis is valid for systems that are all discrete or all continuous. Systems that are part discrete and part continuous contain operators (for example, the sampling switch) that are not commutable with other operators. The result is that the above arguments on the noncritical nature of the choice of stabilization feedback may not be strictly valid.

D.5 SUMMARY

It has been demonstrated above that use of the system of Fig. D.2 will allow adaptive inverse control of overall dynamic response when involving an unstable plant $P(z)$ with stabilization feedback, just as is done with a stable plant. $P(z)$ may have some zeros inside the unit circle in the z-plane and may have some zeros outside it. The stabilized equivalent plant $H(z)$ will have the same zeros. Additional zeros of $H(z)$ come from the poles of the stabilization filter $FB(z)$. If $FB(z)$ is stable, then $H(z)$ will have no more zeros outside the unit circle than $P(z)$ had at the outset.

The difficulty of making an inverse filter relates primarily to the number of zeros outside the unit circle of the filter to be inverted. From the above arguments, we may conclude that the choice of $FB(z)$ is not critical to the process of using inverse control to control overall system dynamics, as long as $FB(z)$ is stable by itself, as long as it stabilizes the plant $P(z)$ within the stabilization loop, and doesn't give $H(z)$ any more zeros outside the unit circle than the ones originally had by $P(z)$.

The system of Fig. D.3 can be utilized for canceling the plant disturbance of unstable plants. Performance of the scheme will be optimal and insensitive to the choice of the stabilization feedback filter $FB(z)$ as long as this filter is stable by itself. However, it remains advantageous to design the stabilization feedback, if possible, so that the impulse response of the stabilized plant would not be excessively long, thus making $H(z)$ easy to model. Otherwise, the impulse responses of $\widehat{H}(z)$ and/or $Q(z)$ would need to be long to achieve best performance.

We conclude that systems of Figs. D.2 and D.3 are indeed capable of adaptive inverse control and adaptive disturbance canceling with unstable plants. The procedure to use is the following: Find a stable filter $FB(z)$ that is capable of stabilizing $P(z)$ within the stabilization loop of Fig. D.1. Then treat the stabilized plant $H(z)$ like any other stable plant and apply the standard methods of adaptive inverse control for dynamic response and plant disturbance.

ppendix E

Orthogonalizing Adaptive Algorithms: RLS, DFT/LMS, and DCT/LMS[1]

We saw in Section 3.6 that the mean square error of an adaptive filter trained with the LMS algorithm decreases over time as a sum of exponentials. The time constants of these exponentials are inversely proportional to the eigenvalues of the autocorrelation matrix of the inputs to the filter. This means that small eigenvalues create slow convergence modes in the MSE function. Large eigenvalues, on the other hand, put a limit on the maximum learning rate that can be chosen without encountering stability problems (see Section 3.4). It results from these two counteracting factors that the best convergence properties are obtained when all the eigenvalues are equal, that is, when the input autocorrelation matrix is proportional to the identity matrix. In this case, the inputs are perfectly uncorrelated and have equal power; in other words, they are samples of a white noise process.

As the eigenvalue spread of the input autocorrelation matrix increases, the convergence speed of LMS deteriorates. As a matter of fact, it is not unusual to find practical situations where the inputs are so correlated that LMS becomes too slow to be an acceptable way to go (see Chapter 7 for an example). To solve this problem, a variety of new algorithms have been proposed in the literature. Their actual implementations and properties are all different but the underlying principle remains the same: trying to orthogonalize as much as possible the input autocorrelation matrix and to follow a steepest-descent path on the transformed error function.

These algorithms can be divided in two families: First, the algorithms such as RLS (recursive least squares) and recursive least squares lattice filters that extract information from past data samples to decorrelate present input signals, and second, the algorithms such as DFT/LMS and DCT/LMS (discrete Fourier and discrete cosine transform — least mean squares), which attempt to decorrelate the inputs by preprocessing them with a transformation (the DFT or the DCT) that is independent of the input signal.

[1] By Françoise Beaufays, Department of Electrical Engineering, Stanford University, and SRI International, Menlo Park, CA.

E.1 THE RECURSIVE LEAST SQUARES ALGORITHM (RLS)

E.1.1 Introduction

The RLS algorithm implements recursively an exact least squares solution [6, 10]. The exact Wiener solution of an adaptive filter is given by $\mathbf{W}^* = \mathbf{R}^{-1}\mathbf{P}$ where $\mathbf{R}$ is the autocorrelation matrix of the inputs and $\mathbf{P}$ is the crosscorrelation between inputs and desired output (see Section 3.3). At each time step, RLS estimates $\mathbf{R}$ and $\mathbf{P}$ based on *all* past data, and updates the weight vector using the so-called matrix inversion lemma (see below).

The two main differences with LMS are that an estimate of the autocorrelation matrix is used to decorrelate the current input data, and that, at least for stationary inputs, the quality of the steady-state solution keeps on improving over time, eventually leading to an exact solution with no misadjustment.

The major disadvantages of RLS over LMS are its poor tracking ability in nonstationary environments [4], its complexity (due to matrix by vector multiplications, $\mathcal{O}(n^2)$ operations are required for each weight update whereas only $\mathcal{O}(n)$ operations are necessary with LMS), and its lack of robustness (the algorithm can be unstable under certain input conditions).

The computational cost and robustness issues have been addressed by researchers in developing other exact least squares algorithms. The lattice filter algorithms, for example, require only $\mathcal{O}(n)$ operations per iteration and are far more robust (although it has been observed that finite precision effects can severely degrade the algorithm performances [13]). The reduction in number of operations comes from the fact that Toeplitz autocorrelation matrices have only n distinct elements. By making a one-to-one correspondence between these n elements and n so-called *reflection coefficients*, and by manipulating the algebra, the computational cost can be reduced from $\mathcal{O}(n^2)$ to $\mathcal{O}(n)$. The price for this improvement is the increased complexity of the algorithm in terms of number of equations to be implemented, number of variables to be stored, and general complication of the algebra.

In this appendix, we will limit ourselves to presenting RLS, the simplest and most basic of the exact least squares algorithms. For a detailed description of other algorithms, we refer the reader to Haykin's textbook on adaptive filters [10].

E.1.2 Derivation of the RLS Algorithm

To avoid instability problems, the error function to be minimized in recursive least squares is generally defined as an exponentially weighted sum of squared errors [10]:

$$\xi_k = \sum_{i=1}^{k} \beta^{k-i}\, e_i^2, \tag{E.1}$$

where $\beta \leq 1$ is a positive constant generally chosen close to one, and e_i is the error signal at time i. At time k, the best estimates of $\mathbf{R}$ and $\mathbf{P}$ are given by

$$\mathbf{R}_k = \sum_{i=1}^{k} \beta^{k-i}\, \mathbf{X}_i\mathbf{X}_i^T \;=\; \beta\,\mathbf{R}_{k-1} + \mathbf{X}_k\mathbf{X}_k^T, \tag{E.2}$$

$$\mathbf{P}_k = \sum_{i=1}^{k} \beta^{k-i}\, \mathbf{X}_i d_i \;=\; \beta\,\mathbf{P}_{k-1} + \mathbf{X}_k d_k, \tag{E.3}$$

where T denotes the transpose. The best estimate of the optimum weight vector $\mathbf{W}^*$ is given by

$$\mathbf{W}_k = \mathbf{R}_k^{-1}\mathbf{P}_k. \tag{E.4}$$

Applying the matrix inversion lemma[2] to (E.2), we get

$$\mathbf{R}_k^{-1} = \beta^{-1}\mathbf{R}_{k-1}^{-1} - \frac{\beta^{-2}\mathbf{R}_{k-1}^{-1}\mathbf{X}_k\mathbf{X}_k^T\mathbf{R}_{k-1}^{-1}}{1+\beta^{-1}\mathbf{X}_k^T\mathbf{R}_{k-1}^{-1}\mathbf{X}_k} \tag{E.5}$$

$$= \beta^{-1}\mathbf{R}_{k-1}^{-1} - \beta^{-1}\mathbf{K}_k\mathbf{X}_k^T\mathbf{R}_{k-1}^{-1} \tag{E.6}$$

where $\mathbf{K}_k$, the *gain vector*, is defined by

$$\mathbf{K}_k = \mathbf{G}_k\mathbf{X}_k, \tag{E.7}$$

and

$$\mathbf{G}_k = \frac{\beta^{-1}\mathbf{R}_{k-1}^{-1}}{1+\beta^{-1}\mathbf{X}_k^T\mathbf{R}_{k-1}^{-1}\mathbf{X}_k}. \tag{E.8}$$

Introducing (E.3) and (E.6) in (E.4), and after some algebraic manipulation, we get the weight update formula

$$\mathbf{W}_k = \mathbf{W}_{k-1} + \mathbf{K}_k\alpha_k \tag{E.9}$$

$$= \mathbf{W}_{k-1} + \mathbf{G}_k\alpha_k\mathbf{X}_k, \tag{E.10}$$

where

$$\alpha_k = d_k - \mathbf{W}_{k-1}^T\mathbf{X}_k. \tag{E.11}$$

Note the formal resemblance with LMS. The weight vector is updated proportionally to the product of the current input $\mathbf{X}_k$ and some error signal α_k. The error in RLS is defined a priori in the sense that it is based on the old weight vector $\mathbf{W}_{k-1}$, whereas in LMS the error $e_k = d_k - \mathbf{W}_k^T\mathbf{X}_k$ is computed a posteriori, that is, based on the current weight vector. A more important difference though is that the constant learning rate μ of LMS is replaced by a matrix $\mathbf{G}_k$ that depends on the data and on the iteration. In that respect, RLS can be thought of as a kind of LMS algorithm having a matrix-controlled optimal learning rate. The weight update formula could also be rewritten as

$$\mathbf{W}_k = \mathbf{W}_{k-1} + \mu_k\mathbf{R}_{k-1}^{-1}\alpha_k\mathbf{X}_k, \tag{E.12}$$

where μ_k is a time and data dependent scalar learning rate equal to

$$\mu_k = \frac{\beta^{-1}}{1+\beta^{-1}\mathbf{X}_k^T\mathbf{R}_{k-1}^{-1}\mathbf{X}_k}. \tag{E.13}$$

[2]Let $\mathbf{A}$ and $\mathbf{B}$ be two positive definite $N \times N$ matrices: $\mathbf{C}$ an $N \times M$ matrix, and $\mathbf{D}$ a positive definite $M \times M$ matrix. The lemma says that if $\mathbf{A} = \mathbf{B} + \mathbf{C}\mathbf{D}^{-1}\mathbf{C}^T$, then $\mathbf{A}^{-1} = \mathbf{B}^{-1} - \mathbf{B}^{-1}\mathbf{C}(\mathbf{D} + \mathbf{C}^T\mathbf{B}^{-1}\mathbf{C})^{-1}\mathbf{C}^T\mathbf{B}^{-1}$.

This decomposition places in evidence the decorrelation operation performed by RLS on the input data.

Equations (E.6 – E.8, E.10, E.11) summarize the algorithm. The weights are typically initialized to zero while the $\mathbf{R}_k^{-1}$ matrix is initialized by $\mathbf{R}_0^{-1} = \delta^{-1}\mathbf{I}$, where δ is a small positive constant and $\mathbf{I}$ is the identity matrix.

While the RLS algorithm has the advantages of a fast convergence rate and low sensitivity to input eigenvalue spread, it has the disadvantage of being computationally intensive: The matrix by vector multiplications in (E.6 – E.8) requires $\mathcal{O}(n^2)$ operations per iteration. It also suffers from possible numerical instability and from bad tracking performance in non-stationary environments when compared to LMS. LMS is intrinsically slow because it does not decorrelate its inputs prior to adaptive filtering but preprocesses the inputs by an estimate of the inverse input autocorrelation matrix in the fashion of RLS, and this leads to the problems cited above. The solution we propose in the next section consists of preprocessing the inputs to the LMS filter with a fixed transformation that *does not* depend on the actual input data. The decorrelation will only be approximative, but the computational cost will remain of $\mathcal{O}(n)$ and the robustness and tracking ability of LMS will not be affected.

.2 THE DFT/LMS AND DCT/LMS ALGORITHMS

E.2.1 Introduction

The DFT/LMS and DCT/LMS algorithms (discrete Fourier/cosine transform — least mean squares) are composed of three simple stages (see Fig. E.1.). First, the tap-delayed inputs are preprocessed by a discrete Fourier or cosine transform. The transformed signals are then normalized by the square root of their power. The resulting equal power signals are input to an adaptive linear combiner whose weights are adjusted using the LMS algorithm. With these two algorithms, the orthogonalizing step is data independent; only the power normalization step is data dependent (i.e. the power levels used to normalize the signals are estimated from the actual data). Because of the simplicity of their components, these algorithms retain the robustness and computational low cost of pure LMS while improving its convergence speed.

The algorithm described in Fig. E.1 is summarized in Table E.1.[3]

The speed improvement over LMS clearly depends on the orthogonalizing capabilities of the transform used. No general proof exists that demonstrates the superiority of one transform over the others; many other transforms than the DFT and the DCT could be considered [11] (for example, the discrete Hartley transform [5] or the Walsh-Hadamard transform [15, 7]). The DFT/LMS, first introduced by Narayan [12], is perceived as the simplest algorithm of the family because of the exponential nature of the DFT and because of the strong intuition scientists have developed about the Fourier transform. It is our experience though that in most practical situations DCT/LMS performs much better than DFT/LMS

[3] In practical applications, the power normalization on the inputs is often replaced by a power normalization on the learning rate of the LMS algorithm. The resulting algorithm is more robust when the input power estimates are noisy[1]. In the remainder of this appendix, we will assume that the power estimates are noise free. In this case, the two algorithms are identical.

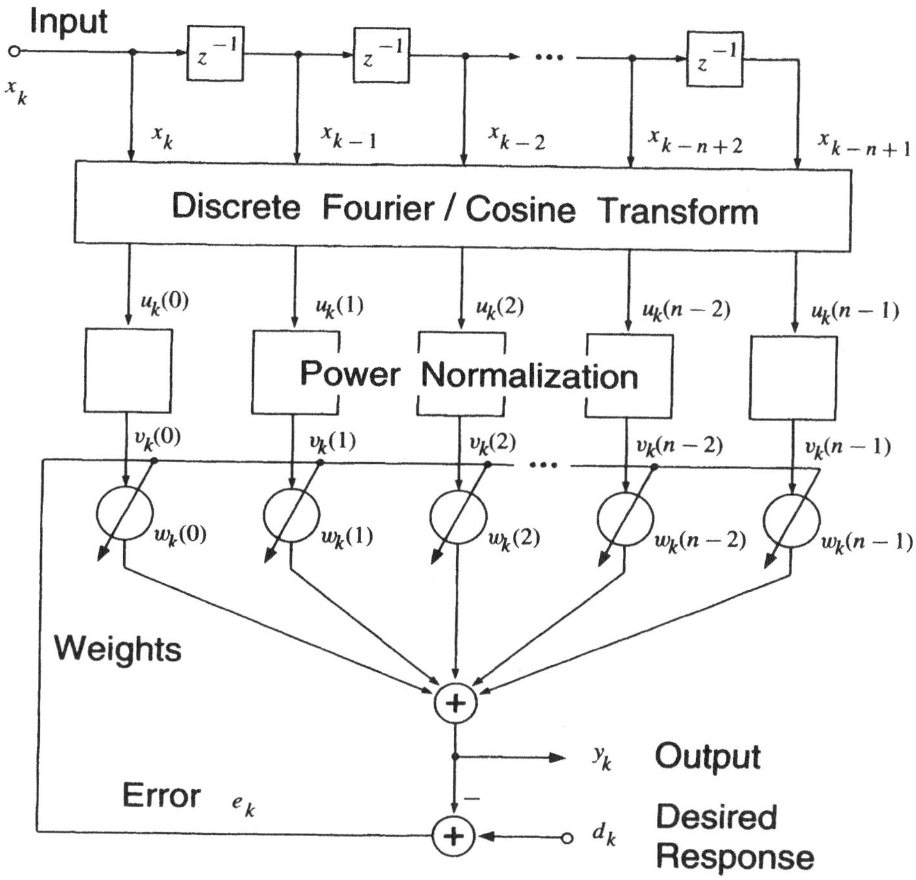

Figure E.1 DFT/LMS and DCT/LMS block diagram.

TABLE E.1 LMS/DFT AND LMS/DCT: SUMMARY OF THE ALGORITHMS

Discrete Fourier/Cosine Transform

$$u_k(i) = \sum_{l=0}^{n-1} \mathbf{T}_n(i,l)\, x_{k-l} \qquad \forall i = 0,\ldots,n-1$$

where

$$\mathbf{T}_n(i,l) = \begin{cases} \mathbf{F}_n(i,l) = \sqrt{1/n}\, e^{j2\pi il/n} & (DFT) \\ \mathbf{C}_n(i,l) = \sqrt{2/n}\, K_i \cos\left(\frac{i(l+1/2)\pi}{n}\right) & (DCT) \end{cases} \qquad \forall i,l = 0,\ldots,n-1$$

with $K_i = 1/\sqrt{2}$ for $i = 0$ and 1 otherwise.

Power Normalization

$v_k(i) = u_k(i)/\sqrt{P_k(i)+\epsilon}$ $\qquad \forall i = 0,\ldots,n-1$

where ϵ is a small constant.

$P_k(i) = \gamma\, P_{k-1}(i) + (1-\gamma)\, u_k^2(i)$ $\qquad \forall i = 0,\ldots,n-1$

$\gamma \in]0,1]$ is generally chosen to be close to 1.

LMS Filtering

$$e_k = d_k - \sum_{i=0}^{n-1} v_k(i) w_k(i)$$

$w_{k+1}(i) = w_k(i) + \mu\, e_k\, v_k^*(i)$ $\qquad \forall i = 0,\ldots,n-1$

where v_k^* is the complex conjugate of v_k.

and other similar algorithms. In addition, it has the advantage over DFT/LMS to be real valued. Nevertheless, we will center our explanations on DFT/LMS for reasons of simplicity, although for practical purposes (cf. for example Sections 7.9 and 7.10) we will always use the DCT/LMS.

In the remainder of this section, we will, through different approaches, explain intuitively and analytically the mechanism of the proposed algorithms. We will then present some theoretical results comparing the performances of LMS, DFT/LMS, and DCT/LMS, and we will conclude with a short discussion of computational cost.

E.2.2 A Filtering Approach

The DFT-DCT performs essentially n linear transformations from the input vector $\mathbf{X}_k = (x_k, x_{k-1}, \ldots, x_{k-n+1})^T$ to the n outputs, $u_k(0), u_k(1), \ldots, u_k(n-1)$. Each transformation is characterized by an impulse response $h_i(l) = \mathbf{T}_n(i,l)$. The corresponding transfer function is given by

$$H_i(\omega) = \sum_{l=0}^{n-1} h_i(l)\, e^{j\omega l}. \tag{E.14}$$

For the DFT, $H_i(\omega)$ is equal to

$$H_i(\omega) = \sqrt{\frac{1}{n}}\, \frac{1 - e^{-j\omega n}}{1 - e^{-j\omega} e^{j\frac{2\pi i}{n}}}, \tag{E.15}$$

which represents a bandpass filter of central frequency $2\pi i/n$. The DFT can thus be seen as a bank of bandpass filters whose central frequencies span the interval $[0, 2\pi]$. Figure E.2 shows the magnitude of a sample transfer function for a 32-point DFT. At each time k, the input signal x_k is decomposed into n signals lying in different frequency bins. If the bandpass filters were perfect, the DFT outputs would be perfectly uncorrelated, but due to the presence of side lobes (see Fig. E.2) there is some leakage from each frequency bin to the others, and thus some correlation between the output signals. As the dimension n of the DFT increases, the amplitudes of the side lobes decrease.

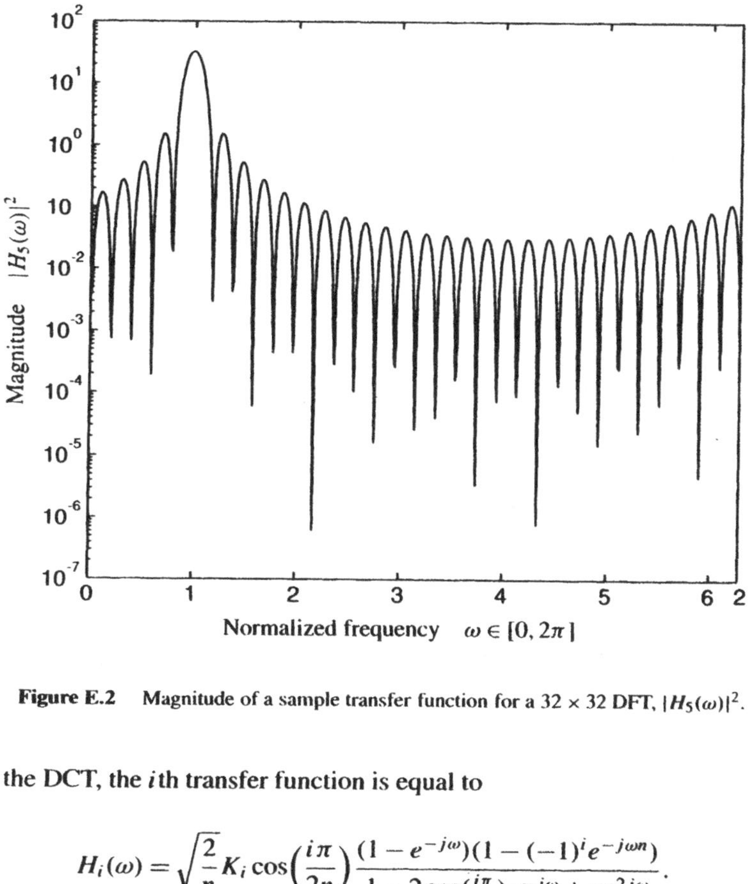

Figure E.2 Magnitude of a sample transfer function for a 32 × 32 DFT, $|H_5(\omega)|^2$.

For the DCT, the ith transfer function is equal to

$$H_i(\omega) = \sqrt{\frac{2}{n}} K_i \cos\left(\frac{i\pi}{2n}\right) \frac{(1 - e^{-j\omega})(1 - (-1)^i e^{-j\omega n})}{1 - 2\cos(\frac{i\pi}{n}) e^{-j\omega} + e^{-2j\omega}}. \tag{E.16}$$

The $H_i(\omega)$'s still represent a bank of bandpass filters but with different central frequencies, different main lobes and side lobes, and different leakage properties. Figure E.3 shows the magnitude of a sample transfer function for a 32-point DCT. The DCT filter looks very different from the DFT one. In particular, the presence of two symmetrical peaks (instead of a single main lobe as in the DFT) comes from the cancelation of the two zeros of the denominator of $H_i(\omega)$ with two zeros of its numerator. More fundamentally, it is a direct consequence of the cosinusoidal nature of the transform. Although the DCT does not separate frequencies the way the DFT does, it is a powerful signal decorrelator as will be demonstrated in the next sections.

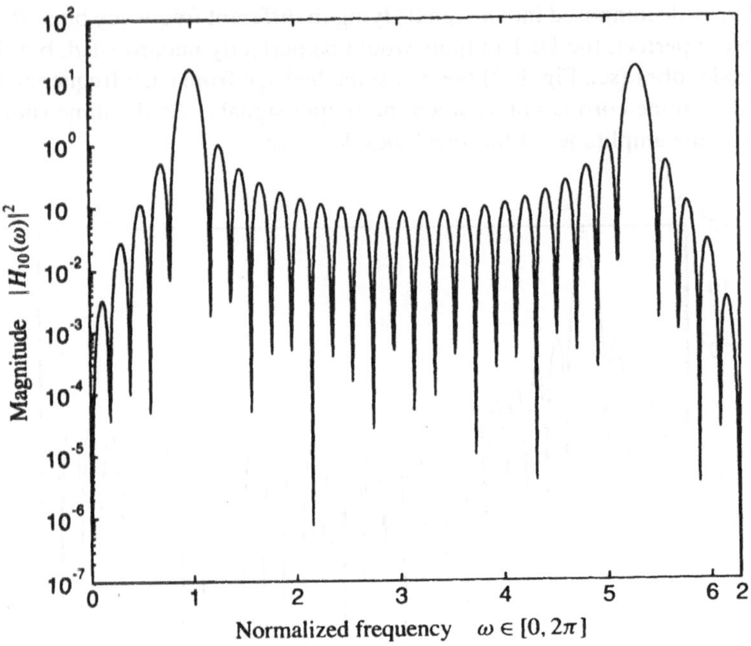

Figure E.3 Magnitude of a sample transfer functions for a 32 × 32 DCT, $|H_{10}(\omega)|^2$.

E.2.3 A Geometrical Approach

The DFT/LMS and DCT/LMS algorithms can also be illustrated geometrically. This approach does not lead to any practical theory but helps understanding conceptually the operations performed by the algorithms block-diagrammed in Fig. E.1. The DFT and DCT matrices defined in Table E.1 are unitary matrices (i.e., their rows are orthogonal to one another and have Euclidean norm one). Unitary transformations perform only rotations and symmetries; they do not modify the shape of the object they transform.

Looking at the MSE formula

$$\xi(\mathbf{W}) = \xi_{\min} + (\mathbf{W} - \mathbf{W}^*)^T \mathbf{R} (\mathbf{W} - \mathbf{W}^*) \tag{E.17}$$

given in Section 3.3, and fixing ξ to some constant value, we see that (E.17) represents a hyperellipsoid in the n-dimensional weight space. A unitary transformation of the inputs rotates the hyperellipsoid and brings it into approximate alignment with the coordinate axes. The slight imperfection in alignment is primarily due to leakage in the transform, DCT or DFT. The idea is illustrated for a simple two-weight case in Fig. E.4. Figure E.4(a) shows the original MSE ellipse, Fig. E.4(b) shows it after transformation by a 2 × 2 DCT matrix. The shape of the ellipse is unchanged and so are the eigenvalues of $\mathbf{R}$.

The power normalization stage (cf., Fig. E.1) can be viewed geometrically as a transformation that, while preserving the elliptical nature of $\xi(\mathbf{W})$, forces it to cross all the coordinate axes at the same distance from the center. This operation is not unitary and it does modify the eigenvalue spread. It almost always improves it. The better the alignment of the hyperellipsoid with the coordinate axes, the more efficient the power normalization will be (a hyperellipsoid perfectly aligned being transformed in an hypersphere). Figure E.4(c) shows the result of power normalization for our example. The new ellipse is more round-shaped and has lower eigenvalue spread.

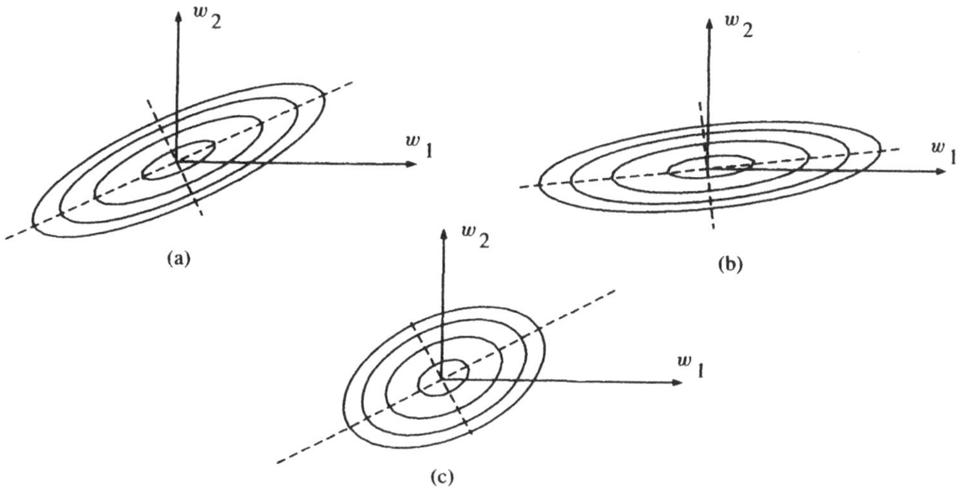

Figure E.4 MSE hyperellipsoid (a) before transformation, (b) after DCT, (c) after power normalization.

E.2.4 An Analytical Approach

In order to find precise information on how well a given transform decorrelates certain classes of input signals, one must set the problem in a more mathematical framework. Transforming a signal $\mathbf{X}_k = (x_k, x_{k-1}, \ldots, x_{k-n+1})^T$ by a matrix $\mathbf{T}_n$ (the DFT or the DCT matrix), transforms its Toeplitz autocorrelation matrix $\mathbf{R} = E[\mathbf{X}_k\mathbf{X}_k^H]$ into a non-Toeplitz matrix $\mathbf{B} = E[\mathbf{T}_n\,\mathbf{X}_k\mathbf{X}_k^H\,\mathbf{T}_n^H] = \mathbf{T}_n\mathbf{R}\mathbf{T}_n^H$ (the superscript H denotes the transpose conjugate). Power normalizing $\mathbf{T}_n\mathbf{X}_k$ transforms its elements $(\mathbf{T}_n\mathbf{X}_k)(i)$ into $(\mathbf{T}_n\mathbf{X}_k)(i)/\sqrt{\text{Power of}(\mathbf{T}_n\mathbf{X}_k)(i)}$, where the power of $(\mathbf{T}_n\mathbf{X}_k)(i)$ can be found on the main diagonal of $\mathbf{B}$. The autocorrelation matrix after transformation and power normalization is thus

$$\mathbf{S} = (\text{diag}\,\mathbf{B})^{-1/2}\,\mathbf{B}\,(\text{diag}\,\mathbf{B})^{-1/2}. \tag{E.18}$$

If $\mathbf{T}$ decorrelated $\mathbf{X}_k$ exactly, $\mathbf{B}$ would be diagonal and $\mathbf{S}$ would be the identity matrix $\mathbf{I}$ and would have all its eigenvalues equal to one, but since the DFT and the DCT are not perfect decorrelators, this does not work out exactly. Some theory has been developed in the past about the decorrelating ability of the DFT and the DCT (see, for example [9, 8, 14]) but the results presented in the literature are in general too weak to allow us to infer anything about the magnitude of the individual eigenvalues of $\mathbf{S}$, which is our main interest. For example, it has been proven that the autocorrelation matrix $\mathbf{B}$ obtained after the DFT or the DCT *asymptotically converges* to a diagonal matrix, *asymptotically* meaning as n, the size of **B**, tends to infinity, and *converges* being understood in a weak norm sense.[4] From this result, we can deduce that **S** will asymptotically converge to identity as n tends to infinity. However, we cannot conclude anything about the possible convergence of the individual eigenvalues of **S** to one, everything depends on *how* and *how fast* **S** converges to **I**. To obtain stronger results, further assumptions are necessary, for example, regarding the class of input signals to be considered.

E.2.5 Eigenvalues and Eigenvalue Spread for Markov-1 Inputs

First-order Markov signals are a very general, practical, and yet simple class of signals. They result from white noise passing through a single-pole lowpass filter. Such a filter has an impulse response that decreases geometrically with a rate ρ given by the filter pole. A Markov-1 input signal $\mathbf{X}_k = (x_k, x_{k-1}, \ldots, x_{k-n+1})^T$ of parameter $\rho \in [0,\ 1]$ has an autocorrelation matrix equal to

$$\mathbf{R} = \begin{pmatrix} 1 & \rho & \rho^2 & \cdots & \rho^{n-1} \\ \rho & 1 & \rho & \cdots & \rho^{n-2} \\ \rho^2 & \rho & 1 & & \\ \vdots & \vdots & & \ddots & \vdots \\ \rho^{n-1} & \rho^{n-2} & & \cdots & 1 \end{pmatrix}. \tag{E.19}$$

For n large (theoretically for n tending to infinity), the minimum and maximum eigenvalues of an autocorrelation matrix **R** are given by the minimum and maximum of the power spectrum of the signal that generated this autocorrelation [9, 8]. This result is a direct consequence of the fact that **R** is Toeplitz. It can easily be checked that in our case the power spectrum of x_k is given by

$$P(\omega) = \sum_{l=-\infty}^{+\infty} \rho^l e^{-j\omega l} = \frac{1}{1 - 2\rho\cos(\omega) + \rho^2}. \tag{E.20}$$

Its maximum and minimum are respectively $1/(1-\rho)^2$ and $1/(1+\rho)^2$. The eigenvalue spread of **R** thus tends to

$$\lim_{n\to\infty}\Big(\text{Eigenvalue spread before transformation}\Big) = \left(\frac{1+\rho}{1-\rho}\right)^2. \tag{E.21}$$

This eigenvalue spread can be extremely large for highly correlated signals (ρ close to 1).

The autocorrelation **S** of the signals obtained after transformation by the DFT or the DCT and after power normalization is not Toeplitz anymore, and the previous theory cannot be applied. The analysis is further complicated by the fact that only asymptotically do the eigenvalues stabilize to fixed magnitudes independent of n, and that power normalization is a nonlinear operation. Successive matrix manipulations and passages to the limit allowed us to prove the following results (see [1, 3] for a complete derivation):

$$\lim_{n\to\infty}\Big(\text{Eigenvalue spread after DFT}\Big) = \frac{1+\rho}{1-\rho}, \tag{E.22}$$

[4] Two matrices converge to one another in a weak norm sense when their weak norms converge to one another. The weak norm of a matrix **A** is defined as the square root of the arithmetic average of the eigenvalues of $\mathbf{A}^H\mathbf{A}$.

$$\lim_{n\to\infty}\left(\text{Eigenvalue spread after DCT}\right) = 1+\rho. \tag{E.23}$$

Note that with the DCT, the eigenvalue spread is never higher than 2!

As a numerical example, let the correlation ρ be equal to 0.95. The eigenvalue spread before transformation is 1,521, after the DFT 39, after the DCT 1.95. In this case, using the DCT/LMS instead of LMS would speed up the filter weight convergence by a factor roughly equal to 750.

These results confirm, for a simple but very practical class of signals, the high quality of the DCT as a signal decorrelator.

E.2.6 Computational Cost of DFT/LMS and DCT/LMS

In addition to their fast convergence and robustness, DFT/LMS and DCT/LMS have the advantage of a very low computational cost. The inputs $x_k, x_{k-1}, \ldots, x_{k-n+1}$ being delayed samples of the same signal, the DFT/DCT can be computed in $\mathcal{O}(n)$ operations. For the DFT, we have

$$u_k(i) = \sum_{l=0}^{n-1} e^{j\frac{2\pi il}{n}} x_{k-l} \tag{E.24}$$

$$= x_k + \sum_{l=1}^{n} e^{j\frac{2\pi il}{n}} x_{k-l} - e^{j\frac{2\pi in}{n}} x_{k-n} \tag{E.25}$$

$$= e^{j\frac{2\pi i}{n}} u_{k-1}(i) + x_k - x_{k-n}. \tag{E.26}$$

The $u_k(i)$'s can thus be found by an $\mathcal{O}(n)$ recursion from the $u_{k-1}(i)$'s. This type of DFT is sometimes called *sliding-DFT*. A similar $\mathcal{O}(n)$ recursion can be derived with more algebra for the DCT:

$$u_k^{DCT}(i) = u_{k-1}^{DCT}(i)\cos(\frac{\pi i}{n}) - u_k^{DST}(i)\sin(\frac{\pi i}{n}) + \sqrt{\frac{2}{n}}\cos(\frac{\pi i}{2n})(x_k - (-1)^i x_{k-n}), \tag{E.27}$$

$$u_k^{DST}(i) = u_{k-1}^{DST}(i)\cos(\frac{\pi i}{n}) + u_k^{DCT}(i)\sin(\frac{\pi i}{n}) + \sqrt{\frac{2}{n}}\sin(\frac{\pi i}{2n})(x_k - (-1)^i x_{k-n}). \tag{E.28}$$

$u_k^{DCT}(i)$ is the ith output of the DCT, $u_k^{DST}(i)$ is the ith output of a DST (discrete sine transform) defined exactly like the DCT but replacing "cos" by "sin" (interlacing two recursions is necessary and comes basically from the fact that $\cos(a+b) = \cos(a)\cos(b) - \sin(a)\sin(b)$). Other $\mathcal{O}(n)$ implementations of the sliding-DFT and the sliding-DCT can be found in the literature. In particular, the so-called *LMS spectrum analyzer* [16] has been shown to implement the sliding-DFT in a very robust way, even when implemented in very precision floating point chips [2].

The power estimates $P_k(i)$ are also computed with simple $\mathcal{O}(n)$ recursions (see Table E.1). Finally, the last step: The LMS adaptation of the variable weights, is $\mathcal{O}(n)$. The overall algorithm is thus $\mathcal{O}(n)$.

For the most part, the DCT/LMS algorithm is superior to the DFT/LMS algorithm. Both are robust algorithms, containing three robust steps: transformation, power normalization (like automatic gain control in a radio or TV), and LMS adaptive filtering. These algorithms are easy to program and understand. They use a minimum of computation, only slightly more than LMS alone. They work almost as well as RLS but don't have robustness problems. The lattice forms of RLS are more robust, but they are much more difficult to program and to understand. All in all, the DFT/LMS and DCT/LMS algorithms should find increased use in practical real-time applications.

Bibliography for Appendix E

[1] F. BEAUFAYS, "Two-layer linear structures for fast adaptive filtering," Ph.D. diss., Information Systems Lab., Stanford University, Stanford, CA, 1994, June 1995.

[2] F. BEAUFAYS, and B. WIDROW, "On the advantages of the LMS spectrum analyzer over non-adaptive implementations of the sliding-DFT," *IEEE Trans. on Circuits and Systems*, Vol. 42, No. 4 (April 1995), pp. 218–220.

[3] F. BEAUFAYS, and B. WIDROW, "Transform domain adaptive filters: An analytical approach," *IEEE Trans. on Signal Processing*, Vol. 43, No. 2 (February 1995), pp. 422–431.

[4] N. BERSHAD, and O. MACCHI, "Comparison of RLS and LMS algorithms for tracking a chirped signal," in *Proc. ICASSP*, Glasgow, Scotland, 1989, pp. 896–899.

[5] R. N. BRACEWELL, *The Hartley transform* (New York: Oxford University Press, 1986).

[6] G. F. FRANKLIN, J. D. POWELL, and M. L. WORKMAN, *Digital control of dynamic systems*, 2nd ed. (Reading, MA: Addison-Wesley, 1990).

[7] R. C. GONZALEZ, and P. WINTZ, *Digital image processing*, 2nd ed. (Reading, MA: Addison-Wesley, 1987).

[8] R. M. GRAY, "Toeplitz and circulant matrices: II," Technical Report 6504-1, Information Systems Lab., Stanford University, Stanford, CA, April 1977.

[9] U. GRENANDER, and G. SZEGO, *Toeplitz forms and their applications*, 2nd ed. (New York: Chelsea Publishing Company, 1984).

[10] S. HAYKIN, *Adaptive filter theory*, 2nd ed. (Englewood Cliffs, NJ: Prentice Hall, 1991).

[11] D. F. MARSHALL, W. K. JENKINS, and J. J. MURPHY, "The use of orthogonal transforms for improving performance of adaptive filters," *IEEE Trans. on Circuits and Systems*, Vol. 36, No. 4 (April 1989), pp. 474–484.

[12] S. S. NARAYAN, A. M. PETERSON, and M. J. NARASIMHA, "Transform domain LMS algorithm," *IEEE Trans. on Acoustics, Speech, and Signal Processing*, Vol. 31, No. 3 (June 1983), pp. 609–615.

[13] R. C. NORTH, J. R. ZEIDLER, W. H. KU, and T. R. ALBERT, "A floating-point arithmetic error analysis of direct and indirect coefficient updating techniques for adaptive lattice filters," *IEEE Trans. on Signal Processing*, Vol. 41, No. 5 (May 1993), pp. 1809–1823.

[14] K. R. RAO, and P. YIP, *Discrete cosine transform*, (San Diego, CA: Academic Press, Inc, 1990).

[15] J. L. WALSH, "A closed set of normal orthogonal functions," *Am. J. Math.*, Vol. 45, No. 1 (1923), pp. 5–24.

[16] B. WIDROW, PH. BAUDRENGHIEN, M. VETTERLI, and P. F. TITCHENER, "Fundamental relations between the LMS algorithm and the DFT," *IEEE Trans. on Circuits and Systems*, Vol. 34, No. 7 (July 1997), pp. 814–819.

Appendix F

A MIMO Application: An Adaptive Noise-Canceling System Used for Beam Control at the Stanford Linear Accelerator Center[1]

F.1 INTRODUCTION

A large number of beam-based feedback loops have recently been implemented at a series of stations along the Stanford Linear Accelerator Center (SLAC) in order to maintain the trajectory of the electron beam down its 3-km length to an accuracy of 20 microns. Since the accelerator parameters vary gradually with time, we have found it necessary to use adaptive methods to measure these parameters to allow the feedback system to perform optimally. Eight beam parameters are being controlled at each station. We therefore have a series of eight-input eight-output MIMO systems.

The next section gives a general description of the accelerator. Sections F.3 and F.4 explain the functioning of the steering feedback loops and the need to have the loops communicate with each other. Sections F.5 and F.6 describe the theoretical background and simulations used in designing the adaptive system and Section F.7 relates our experiences with the system implemented on the actual accelerator.

F.2 A GENERAL DESCRIPTION OF THE ACCELERATOR

SLAC is a complex of particle accelerators located near (and operated by) Stanford University. It is a government facility whose primary goal is the pursuit of elementary particle physics: the study of the smallest particles of matter and the forces between them. SLAC

[1]By Thomas M. Himel, Stanford Linear Accelerator Center, Stanford, CA.

is one of only half a dozen such facilities in the world. It was built in the late 1960s and since then its primary accelerator has been gradually upgraded and extended. Its present state-of-the-art accelerator is known as the SLC (for SLAC Linear Collider) and is depicted in Fig. F.1.

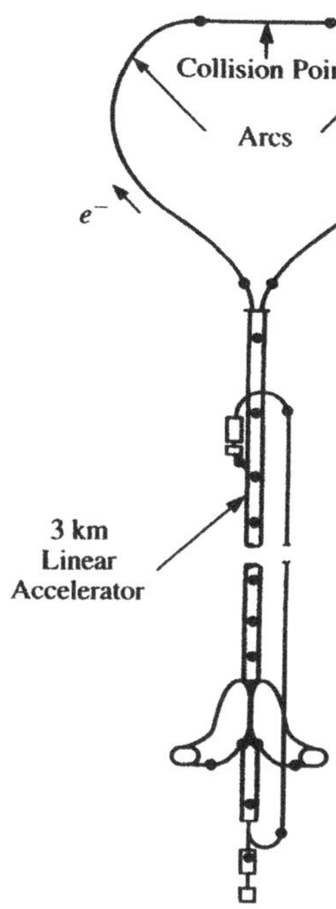

Figure F.1 Diagram of the Stanford Linear Collider with locations of steering feedback loops depicted with large dots.

The SLC is a pulsed accelerator. On each pulse a bunch of 3×10^{10} positrons (anti-electrons) is accelerated down the linear accelerator (Linac) from the bottom to the top of the figure. All those positrons are contained in a cylindrical volume 150 microns in radius and 1 mm long. The positrons are followed by a similar bunch of electrons 20 meters behind. It takes these particles 10 microseconds to reach the end of the Linac. There the two bunches follow separate paths. Magnetically deflected, the electrons go to the left and the positrons go to the right. They head around the arcs, a set of a thousand bending and focusing magnets, which are used to direct the beams toward a head-on collision. Just before collision they are focused by superconducting quadrupole magnets down to a radius of about 2 microns (comparable to the transistor size on integrated circuits). The beams pass through each other. For a small fraction of the pulses, an individual electron and an individual positron will collide head on and convert their energy into matter. This matter is likely to be in the form of the neutral intermediate vector boson, the Z^0. There is a special detector around the beam crossing point which measures the decay fragments of the Z^0 to try to learn more about this elusive particle and the weak force for which it is the intermediary. The study of the Z^0 is one of the major goals of the SLC.

Now, back to the main subject of this section: the operation of the accelerator. All of the action in the above pulse happens in less than 20 microseconds. In the next 8.3 milliseconds nothing much happens. Capacitor banks are recharged and components are given a chance to cool down. The accelerator is then pulsed again giving a repetition rate of 120 pulses-per-second. The important point from the controls standpoint is that the accelerator is pulsed. A beam is not present all the time. One cannot make a 1-MHz bandwidth feedback loop. The best one can do is to sample a beam parameter at 120 Hz and apply the information from that measurement to make a correction, if needed, on the next pulse.

The heart of the SLC is the original 3-km-long linear accelerator which has been heavily refurbished over the years by giving it a new control system, new focusing magnets, and doubling its energy. The Linac is used to simultaneously accelerate a bunch of electrons and a similar bunch of positrons so that each particle has an energy of about 47 GeV (giga or billion electron volts). A few examples will aid in the understanding of this very large energy.

- If you were to put an ordinary 1.5-volt D-cell battery between two parallel metal plates and set an electron near the negative plate, it would be accelerated to the positive plate and end up with an energy of 1.5 eV. To get to 50 GeV, you would need a stack of batteries five times the distance to the moon.

- Einstein's theory of special relativity says matter and energy are different forms of the same thing and relates them with the famous equation $E = mc^2$. At rest, an electron has a mass about 1/2,000 that of a proton. At the end of the SLC, its energy corresponds to a mass of about 50 times that of a proton.

- Special relativity also dictates that nothing can go faster than the speed of light. As particles are given more and more energy their speed just gets closer and closer to that limit. At the end of the SLC an electron has a velocity of 99.999999995% the speed of light. This is so close that if the electron beam were to race a beam of light to the moon, the light would only be 2 cm ahead on arrival.

The key to obtaining such a high energy in such a short distance is that the Linac uses an oscillating electromagnetic field instead of a DC electric field. The field is contained in a cylindrical copper waveguide that has a 9-mm radius hole down its center. The electromagnetic wave has a wavelength of 10.5 cm and the particles "surf" on it. The waveguide is carefully designed so the wave travels at the speed of light; hence, the electrons are constantly accelerated near the crest of the wave.

Creating the electromagnetic wave are 240 klystrons evenly spaced along the 3 km. Each klystron is a vacuum tube about 6 feet high and is similar in concept to tubes used to make radar waves and to power radio transmitters. At SLAC these tubes are very high power. Each one produces 50 megawatts although only for a 3.5-microsecond period. Due to their high power output, these tubes are fairly temperamental, break down fairly often, and need frequent adjustments and repairs. In fact, these klystrons cycling on and off are one of the major sources of disturbances to the beam and drive the need for beam-based feedback systems and adaptive disturbance cancelation.

F.3 TRAJECTORY CONTROL

Having covered how the particles are accelerated, we now come to the matter of trajectory control. As a bunch of electrons is accelerated down the Linac, it must be kept centered in the copper waveguide. Certainly it cannot be allowed to stray more than 9 mm from the center line because it would hit the waveguide. In fact, the requirements are much more stringent than that. The electron bunch is a lot of charge in a small volume and generates a large electromagnetic field between the bunch and the walls of the waveguide. If the bunch is off-center, it is attracted to the wall. Because of these fields, known as wake-fields, the bunch trajectory must be kept stable to better than 20 microns, the diameter of a human hair.

The first line of defense in trajectory control is passive: A set of 300 magnetic focusing lenses (quadrupole electromagnets) spaced along the length of the accelerator focus the beam like optical lenses focus light. Each lens gives a transverse kick to a particle proportional to the distance the particle is from the correct trajectory. Hence if a bunch starts off down the Linac slightly off-center, it will follow a roughly sinelike trajectory about the center line as it speeds down the accelerator. The actual trajectory is a bit more complicated and is known as a betatron oscillation. A particle executes about 30 full oscillations as it goes down the Linac.

In the second line of defense are quasi-DC dipole electromagnets. Near each quadrupole lens there are two dipole magnets: one which kicks the beam vertically and one which kicks it horizontally. Operators can adjust these magnet strengths remotely via the SLC control system. They can be controlled individually by typing in the desired value. They can also be controlled en masse by using an automated steering program which calculates the proper magnet settings to zero the beam orbit. Information for this program is obtained from 300 beam position monitors (BPMs) spaced down the Linac. This steering procedure is typically done several times a day. The beam position monitors are typically made of four strips placed in a cylinder around the nominal beam trajectory. The beam is capacitively coupled to these strips. If the beam is too high, it will be closer to the top strip and will make a larger pulse on that strip. Cables run from each strip to digitizing electronics. By comparing the amplitude on the top and bottom strips (and similarly left and right), we determine the beam position with a typical precision of 20 microns.

In the third line of defense are steering feedback loops. Let's consider the source of disturbances to the beam which make feedback necessary.

- As mentioned above klystrons periodically break and hence are often being turned on and off. This not only causes sudden changes in the beam energy but also changes the trajectory.

- The electrons are accelerated near the crest of an electromagnetic wave. The exact phase of the beam with respect to the wave determines the energy gain. This phase is difficult to keep stable over the 3-km length of the Linac and varies by a few degrees (which corresponds to a few trillionths of a second change in timing). These changes make the energy and trajectory change.

- Operators purposely adjust some magnets to improve the beam quality. Feedback helps stabilize the beam downstream of these adjustments.

In addition to using feedback to control the beam position, adaptive techniques are used to cancel disturbances. We begin our discussion of beam control with a description of beam steering feedback.

F.4 STEERING FEEDBACK

There are presently 20 steering feedback loops running in the SLC [1]. Of these loops, 7 are placed one after the other along the Linac. A typical Linac steering loop (number $n+1$) is illustrated in Fig. F.2. It measures and controls eight states: the position and angle of the electron beam in both the horizontal and vertical directions and the same for positrons. These eight states cannot be directly measured as there is no sensor which can directly determine the angle of the beam. Instead, the piece of the system labeled "transport positions, angles to BPM readings" takes the eight states (labeled "positions, angles at loop $n+1$") to 40 beam position monitor (BPM) readings. These 40 readings correspond to the horizontal and vertical position for both electrons and positrons at each of the 10 BPM locations used by loop $n+1$. The 40 measurements to determine eight states give a factor of 5 redundancy which helps reduce the effect of sensor noise and allows the system to keep running even when a BPM is broken. The boxes labeled "calculate positions, angles from BPM readings" and "LQG feedback controller" represent our feedback system software which uses the BPM readings to estimate the eight states which the controller then uses to calculate settings for the eight actuators that loop $n+1$ controls. These actuators consist of four horizontal dipole magnets and four vertical dipole magnets which bend the beam. The summing junction above the "LQG feedback controller" box represents the effect of these actuators on the states of the beam.

The signal flow lines in Fig. F.2 are labeled to indicate the number of signals present. The blocks are linear memoryless matrices, except for the $z^{-1}I$ block which is a one-beam pulse time delay and the "LQG Feedback Controller" block which contains a low-pass filter in addition to its linear matrices.

A lot of real-life complexity is left out of Fig. F.2 for clarity. All of the components of loop $n+1$ are spread out over a distance of 220 meters along the Linac. The sensors and actuators are controlled by three different microprocessors (because of the large distances and large number of devices, the SLC has a microprocessor for each 100 meters of accelerator) which use simple point-to-point communication links to get all the data to one point. This typical loop takes data and makes corrections at a rate of 20 Hz. It doesn't run at the full accelerator pulse rate of 120 Hz due to speed limitations of the 16 MHz 80386 microprocessors on which the software currently runs.

Each feedback loop is designed using our knowledge of the accelerator optics and the state space formalism of control theory. The linear quadratic Gaussian (LQG) method is used to design the optimum filters to minimize the RMS disturbance seen in the beam. Since there is a fair amount of white noise in the incoming beam disturbance, this filter averages over about six pulses. Hence the typical loop corrects most of a step change in six pulses.

As described so far, there is a problem with the system. We have seven loops in a row in the Linac all looking at the same beam (see Fig. F.1). Figure F.3 depicts the beam trajectory in the region of two of these loops. Figure F.3(a) shows the trajectory on the first pulse after there has been a sudden disturbance (such as an operator adjusting a dipole magnet

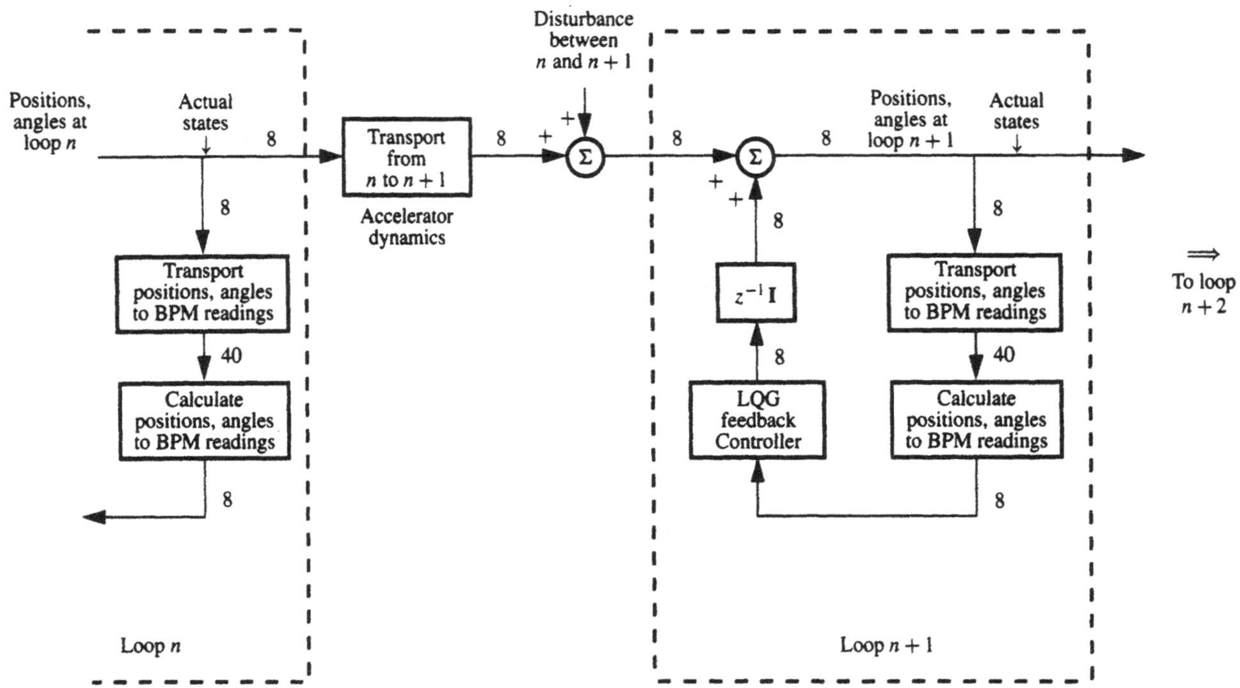

Figure F.2 Diagram of typical feedback loop.

strength) upstream of the two loops. The plot of transverse beam position as a function of distance along the Linac shows the sinelike trajectory caused by the focusing quadrupole lenses. At this time, the loops have not had a chance to make any correction. Figure F.3(b) shows the trajectory on the next pulse. To keep this example simple, the loops were set up to completely fix an error they see in one pulse instead of the six that the actual loops use. The first loop has completely corrected the original disturbance. The second loop has also made a correction *which was unnecessary* because the first loop has corrected for the disturbance. Of course, on the next pulse the second loop would correct its error but the damage has been done: The loops have overshot the mark for a pulse. The problem gets much worse with seven loops in a row. The overshoot can be reduced by having each loop respond more slowly but the system still overshoots and then rings for many pulses. The system is stable and the ringing gradually dies out, but the overall response of the loops is not optimal; hence, the beam positions and angles have a larger RMS than need be.

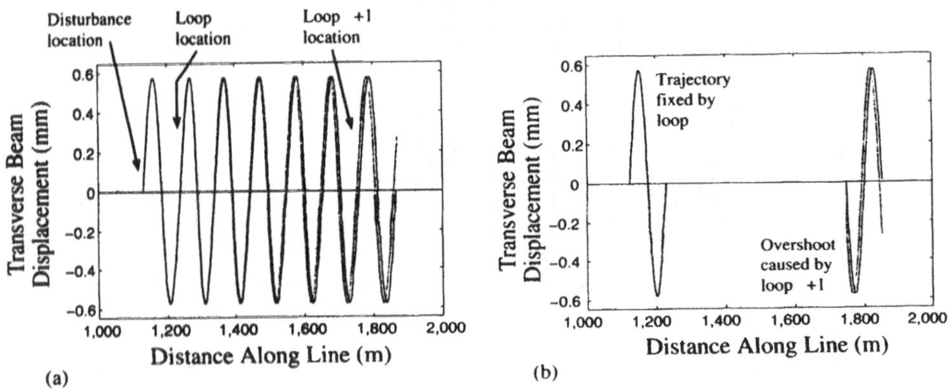

Figure F.3 Feedback's response to a disturbance. Shown is the beam trajectory on the first pulse (a) and second pulse (b) after a sudden disturbance is introduced. The response of the two feedback loops shows the need for the adaptive noise-canceling system.

The proper solution is to have each loop only correct for disturbances which happen between it and the next upstream loop. This would completely eliminate the overshooting caused by multiple loops correcting for the same disturbance.

F.5 ADDITION OF A MIMO ADAPTIVE NOISE CANCELER TO FAST FEEDBACK

An individual loop (say loop $n + 1$) has only a few local beam position monitors to detect disturbances in the beam. It has no way to tell how far upstream the disturbance occurred. Since we want loop $n+1$ to correct for disturbances downstream of loop n, but not upstream, the upstream disturbances can be thought of as noise. With this picture in mind, a noise canceler can be used to solve our problem. For reasons explained in the next section, it must be an adaptive noise canceler. The rest of this section shows how an adaptive noise canceler can be used to solve this problem.

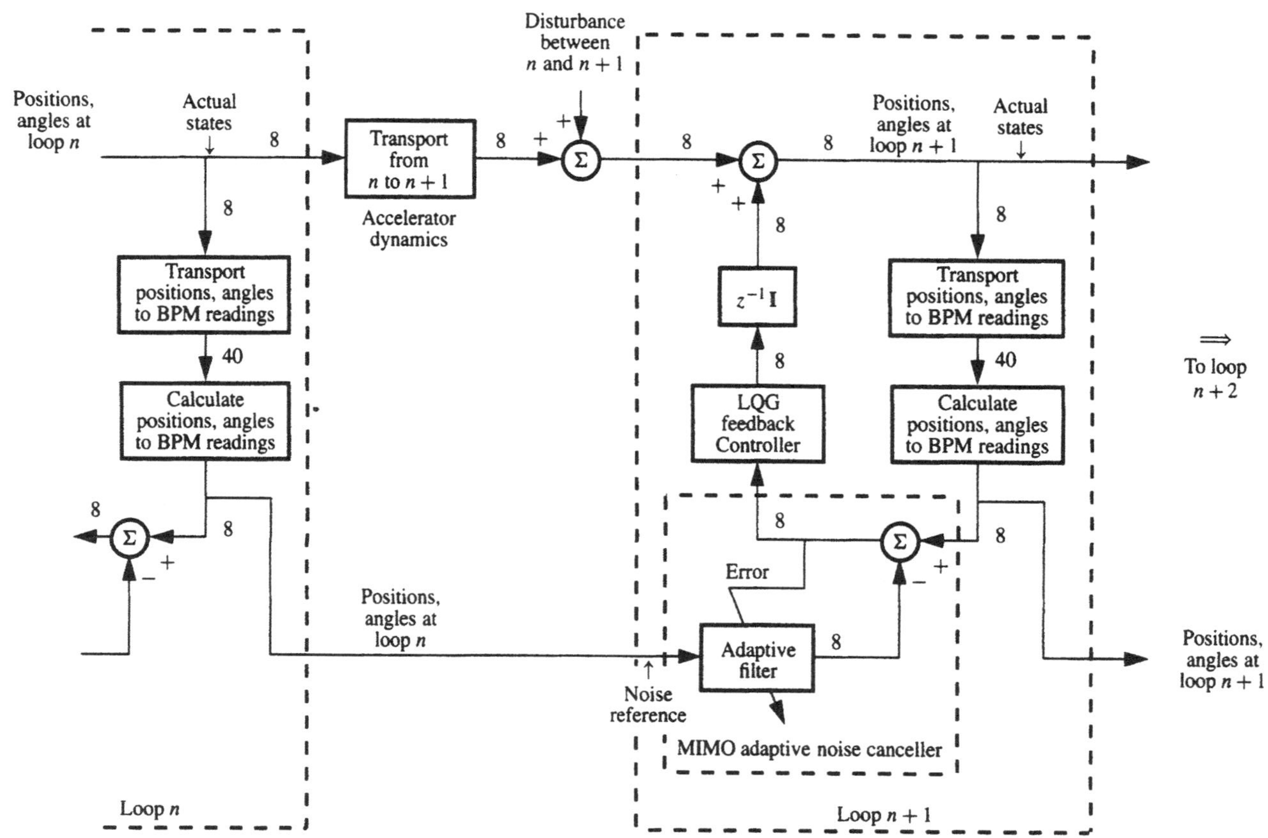

Figure F.4 Adaptive MIMO noise canceler added to the typical feedback loop.

The addition of a MIMO adaptive noise canceler to the typical feedback loop of Fig. F.2 is depicted in Fig. F.4. The bold lines represent information carried by the beam and the bold boxes represent transfer functions which are part of the plant (accelerator). The nonbold items represent things implemented as part of our feedback system.

At a single point in the accelerator it takes eight parameters to characterize the state of the beam. These are the positions and angles of the electron and positron beams in the horizontal and vertical directions. The line in the upper left labeled "Positions, angles at loop n" represents these eight states. Since loop n is responsible for maintaining these states at their desired setpoints (which are typically zero since we want the beam to go in a straight line down the center of the tube), as far as loop $n+1$ is concerned, these states are noise. Loop n reads some BPMs (beam position monitors) and calculates the positions and angles from their readings. It both uses the numbers for its own feedback loop and sends them via a communications link (labeled "Measured positions, angles at loop n") to loop $n+1$ which uses them as its noise reference signal for its adaptive noise canceler.

Similar information is carried to loop $n+1$ by the beam itself. Between the two loops, the beam is following its sinelike trajectory so that positions and angles transform into each other. This is represented by the box labeled "Transport from n to $n+1$" and represents the accelerator dynamics between the two loops. It is very important to note that our problem is static; the transport of this beam pulse does not depend on the positions and angles of the previous beam pulse. Hence, the box can be represented as a simple 8×8 matrix.

In addition to this simple transport of the beam, there may be an additional "Disturbance between n and $n+1$" which gets added in. This disturbance could be due to a klystron tripping off or an operator adjusting a magnet for example. It is this disturbance that loop $n+1$ is intended to correct so it corresponds to the *signal* that we want the noise canceler to extract.

The last box which needs explanation is the "LQG Feedback Controller." This represents the controller of feedback loop $n+1$. As its input, the controller now takes the output of the MIMO adaptive noise canceler which represents our best estimate of the "Disturbance between n and $n+1$." That is precisely what we want loop $n+1$ to correct. The output of the controller controls dipole magnets which steer the beam between n and $n+1$. Hence its output is shown summed into the positions and angles of the beam transported from loop n.

In summary, before the implementation of the adaptive noise canceler, the series of seven feedback loops overcorrected for deviations in the position and angle of the beam. This happened because each feedback loop acted independently and hence they all applied a correction for the same disturbance. The addition of the MIMO adaptive noise cancelers allows each loop to separate disturbances which happen immediately upstream from those which occur upstream of the previous loop. This cures the overcorrection problem.

This completes the description of the overall system. The next section examines the adaptive algorithm used to update the weights.

F.6 ADAPTIVE CALCULATION

Before delving into the details of the adaptive calculation, it is worthwhile to ask why adaptation is necessary at all. What is varying? The box labeled "Transport from n to $n+1$" in Fig. F.4 is what varies. It accounts for the sinelike trajectory, caused by the focusing magnets, that the beam follows as it travels down the accelerator. For example, if loop $n+1$ is 90° of the betatron (sinelike) oscillation downstream of loop n, then a position offset at loop n becomes an angle at loop $n+1$ and an angle transforms into a position. The transformation depends critically on the number of betatron oscillations between the loops. This is parameterized as the *phase advance* where 360 degrees of phase advance corresponds to one full oscillation. Figure F.3 shows two loops separated by $5 \times 360°$ of phase advance which is the average for the loops in the SLC. The dotted line in Figure F.3(a) shows a betatron oscillation where the focusing strength is wrong by 1 percent, an error typical of the real Linac. Note that the position and angle at the second loop are quite different due to the 1 percent error. It is this significant variation of the "Transport from n to $n+1$" which forces the use of an adaptive method for the noise canceler.

The updates of the weights in the adaptive filter are done using the sequential regression (SER) algorithm. The equations used in the SER algorithm are explained in [2], so we won't give them here. Instead it is interesting to examine our experiences which led to the use of this algorithm and to mention minor modifications that were made to ensure it would be robust.

We started out using the least mean square (LMS) algorithm for updating the weights (matrix elements). This is the simplest algorithm, is very fast computationally, and has been successfully used in many applications. In the design phase of the project, simulations were done to check our algorithms. Two problems turned up.

- As explained in [2], the LMS method is stable only if the learning rate is less than the inverse of the largest eigenvalue of the input correlation matrix. For our problem it is the natural jitter of the beam due to magnet supply fluctuations and klystron problems that cause the variations of the positions and angles and hence the information from which the adaptation is done. The amplitude of this jitter can easily change by an order of magnitude in a short time which in turn means the eigenvalues change by that amount. If the jitter increased too much, the LMS method would become unstable and our feedback system would malfunction, making things still worse. To ensure this doesn't happen, we would have to pick a learning rate much smaller than the typical optimum value which would result in a very slow convergence.

- The LMS method has a different convergence rate for each eigenmode which depends on the corresponding eigenvalue. Unless we carefully scaled our inputs, the convergence of the slowest eigenmode would be much less than optimum.

To avoid these problems we went to the SER algorithm. Basically it adaptively estimates the inverse of the input correlation matrix. This is used to scale the inputs so that all the eigenvalues of the correlation matrix of the scaled inputs are equal to 1. This cures the above two problems at the cost of added complexity and CPU time.

Even with the SER method, the calculation of the weights becomes unstable for a short time if the beam jitter suddenly increases. This is because it takes a while for its estimate of the inverse of the input correlation matrix to converge to the new value. During that time the values for the weights have run away. This problem was found in simulation along with the solution. We simply do not update the weights if the inverse correlation matrix is receiving large updates.

Having tested the algorithm with the computer simulation, we went ahead and implemented it in the SLC control system. The software was tested and debugged using a hardware accelerator simulator. This hardware was capable of mimicking the accelerator response to three simple loops, each having one beam position monitor and one dipole corrector. In this environment the proper operation of both the method of cascading to reduce overshoot and the adaptive learning of the beam transport matrix were verified. Finally (about six months into the project) the time had come to use it on the real accelerator.

F.7 EXPERIENCE ON THE REAL ACCELERATOR

The new communications links were installed on the accelerator while the software was being written and tested. With the software ready, we first turned on just the adaptive algorithm. The results were not used to control the beam. After confirming that the matrices had converged to reasonable values, we turned on the use of the noise-canceling system. As shown in Fig. F.5 the response to a step disturbance in the beam trajectory was greatly improved with the turn-on of the adaptive noise-canceling system.

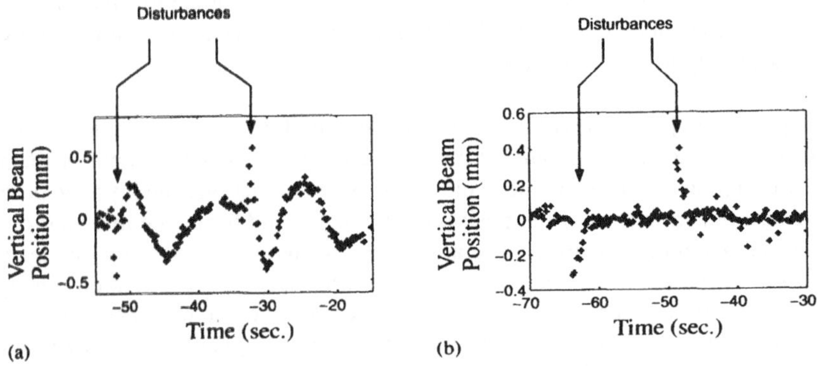

Figure F.5 Response of a chain of six feedback loops to a sudden disturbance in the incoming beam. In (a) adaptive noise-canceling is off so there is a ringing caused by the overcorrection of many loops. In (b) adaptive noise-canceling is on so the whole chain of loops responds just like a single loop. In fact, the first loop did all the work to correct the beam and the downstream loops did virtually nothing.

Over the next few weeks we varied the learning rate to find the optimum value that would allow the adaptation to converge rapidly without having too much noise introduced by the adaptive process. We settled on a learning rate of 0.001 and an adaptive update rate of 10 Hz. This resulted in a convergence time of about 100 seconds. The system was run for several days with learning rates of 0.1 and 0.01 and was completely stable, but with these higher learning rates more random noise showed in the adaptive matrix elements.

The adaptive noise-canceling addition to the fast feedback system has been running stably in seven locations on the SLAC linear collider for over six months. Probably the best measure of its robustness and stability is that operators have made **no** middle of the night phone calls for help to recover from a problem. In fact, there have been no significant problems with the system. Adaptive noise canceling has significantly improved the performance of our feedback systems and helped us achieve our goal of accelerating two beams over a distance of 3 kilometers; pointing them at each other; and then colliding them head-on so they pass through each other even though they have a radius of only 2 microns at the collision point.

In fact, we have even received an unexpected bonus from the adaptive calculation. The adaptive weights can be interpreted as measurements of the beam transport matrix from one loop to the next. These measurements are recorded on disk and can be displayed. Figure F.6 shows an example of such a display. It shows the variation with time of the measured phase advance (which can be simply calculated from the adaptive weights) from one loop to the next. This shows a variation of over 30 degrees which is about 1 percent of the total phase advance between the two loops. We have done many checks and convinced ourselves that this is caused by a real variation in the focusing strengths in the Linac. Accelerator physicists are using plots like Fig. F.6 to try to track down the cause of these changes in focusing strength so they can be fixed. This would make a still more stable beam.

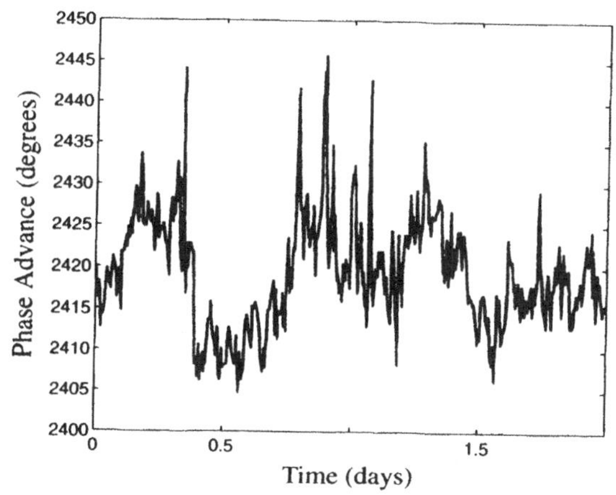

Figure F.6 The adaptively measured phase advance between two loops over a two-day period. The 30-degree variation shows that the cascading of feedback would not have worked without the adaptive measurement of the beam transport matrix.

.8 ACKNOWLEDGEMENTS

The success of this project would not have been possible without the dedicated work of many people in the controls group at SLAC. Particularly important to the project were Stephanie Allison, Phyllis Grossberg, Linda Hendrickson, Dave Nelson, Robert Sass, and Hamid
Shoaee.

Bibliography for Appendix F

[1] F. ROUSE, S. CASTILLO, S. ALLISON, T. GROMME, L. HENDRICKSON, T. HIMEL, K. KRAUTER, R. SASS, and H. SHOAEE, "Database driven fast feedback system for

the Stanford linear collider." in *Nuclear Instruments and Methods in Physics Research,* Sect. A, Vol. 316, No. 2–3 (June 1,1992), pp. 343–350.

[2] B. WIDROW, and S. STERNS, *Adaptive Signal Processing* (Englewood Cliffs, NJ: Prentice Hall, 1985).

Appendix G

Thirty Years of Adaptive Neural Networks: Perceptron, Madaline, and Backpropagation[1]

Fundamental developments in feedforward artificial neural networks from the past 30 years are reviewed. The central theme of this paper is a description of the history, origination, operating characteristics, and basic theory of several supervised neural network training algorithms including the Perceptron rule, the LMS algorithm, three Madaline rules, , , and the backpropagation technique. These methods were developed independently, but with the perspective of history they can all be related to each other. The concept which underlies these algorithms is the "minimal disturbance principle," which suggests that during training it is advisable to inject new information into a network in a manner which disturbs stored information to the smallest extent possible.

G.1 INTRODUCTION

This year marks the thirtieth anniversary of the Perceptron rule and the LMS algorithm, two early rules for training adaptive elements. Both algorithms were first published in 1960. In the years following these discoveries, many new techniques have been developed in the field of neural networks, and the discipline is growing rapidly. One early development was Steinbuch's Learning Matrix [1], a pattern recognition machine based on linear discriminant functions. In the same time frame, Widrow and his students devised Madaline Rule I (MRI), the earliest popular learning rule for neural networks with multiple adaptive elements [2]. Other early work included the "mode-seeking" technique of Stark, Okajima, and Whipple [3]. This was probably the first example of competitive learning in the literature, though it could be argued that earlier work by Rosenblatt on "spontaneous learning" [4, 5] deserves

[1] By Bernard Widrow and Michael A. Lehr. Reprinted, with permission, from *Proceedings of the IEEE*, Vol. 78, No. 9 (September 1990), pp. 1415–1442.

this distinction. Further pioneering work on competitive learning and self-organization was performed in the 1970s by von der Malsburg [6] and Grossberg [7]. Fukushima explored related ideas with his biologically inspired Cognitron and Neocognitron models [8, 9].

In the mid-1960s, Widrow devised a reinforcement learning algorithm called "punish/reward" or "bootstrapping" [10, 11]. This can be used to solve problems when uncertainty about the error signal causes supervised training methods to be impractical. A related reinforcement learning approach was later explored in a classic paper by Barto, Sutton, and Anderson on the "credit assignment" problem [12]. Barto and his colleagues' technique is also somewhat reminiscent of Albus's adaptive CMAC, a distributed table-lookup system based on models of human memory [13, 14]. Yet another approach related to bootstrapping is the associative reward-penalty algorithm of Barto and Anandan [15]. This method solves associative reinforcement learning tasks, providing a link between pattern classification and stochastic learning automata.

In the 1970s Grossberg developed his adaptive resonance theory (ART), a number of novel hypotheses about underlying principles which govern biological neural systems [16]. These ideas served as the basis for later work by Carpenter and Grossberg involving three classes of ART architectures: ART 1 [17], ART 2 [18], and ART 3 [19]. These are self-organizing neural implementations of pattern clustering algorithms. Other important theory on self-organizing systems was pioneered by Kohonen with his work on feature maps [20, 21].

In the early 1980s, Hopfield and others introduced outer product rules as well as equivalent approaches based on the early work of Hebb [22] for training a class of recurrent (signal feedback) networks now called Hopfield models [23, 24]. More recently, Kosko extended some of the ideas of Hopfield and Grossberg to develop his adaptive bidirectional associative memory (BAM) [25], a network model employing differential as well as Hebbian and competitive learning laws. Another model utilizing a differential learning mechanism is Harry Klopf's drive reinforcement theory [26], an extension of Hebbian learning which explains Pavlovian classical conditioning. Other significant models from the past decade include probabilistic ones such as Hinton, Sejnowski, and Ackley's Boltzmann Machine [27, 28], which to oversimplify, is a Hopfield model that settles into solutions by a simulated annealing process governed by Boltzmann statistics. The Boltzmann Machine is trained by a clever two-phase Hebbian-based technique.

While these developments were taking place, adaptive systems research at Stanford traveled an independent path. After devising their Madaline I rule, Widrow and his students developed uses for the Adaline and Madaline. Early applications included, among others, speech and pattern recognition [29], weather forecasting [30], and adaptive controls [31]. Work then switched to adaptive filtering and adaptive signal processing [32] after attempts to develop learning rules for networks with multiple adaptive layers were unsuccessful. Adaptive signal processing proved to be a fruitful avenue for research with applications involving adaptive antennas [33], adaptive inverse controls [34], adaptive noise canceling [35], and seismic signal processing [32]. Outstanding work by R. W. Lucky and others at Bell Laboratories led to major commercial applications of adaptive filters and the LMS algorithm to adaptive equalization in high speed modems [36, 37] and to adaptive echo cancelers for long distance telephone and satellite circuits [38]. After 20 years of research in adaptive signal processing, the work in Widrow's laboratory has once again returned to neural networks.

The first major extension of the feedforward neural network beyond Madaline I took place in 1971 when Werbos developed a backpropagation training algorithm which, in 1974, he first published in his doctoral dissertation [39].[2] Unfortunately, Werbos's work remained almost unknown in the scientific community. In 1982, Parker rediscovered the technique [41] and in 1985, published a report on it at M.I.T. [42]. Not long after Parker published his findings, Rumelhart, Hinton, and Williams [43, 44] also rediscovered the technique and, largely as a result of the clear framework within which they presented their ideas, they finally succeeded in making it widely known.

The elements used by Rumelhart and colleagues in the backpropagation network differ from those used in the earlier Madaline architectures. The adaptive elements in the original Madaline structure used hard-limiting quantizers (signums), while the elements in the backpropagation network use only differentiable nonlinearities, or "sigmoid" functions.[3] In digital implementations, the hard-limiting quantizer is more easily computed than any of the differentiable nonlinearities used in backpropagation networks. In 1987, Widrow and Winter looked back at the original Madaline I algorithm with the goal of developing a new technique that could adapt multiple layers of adaptive elements which use the simpler hard-limiting quantizers. The result was Madaline Rule II [45].

David Andes of the U.S. Naval Weapons Center of China Lake, California, modified Madaline II in 1988 by replacing the hard-limiting quantizers in the Adaline with sigmoid functions, thereby inventing Madaline Rule III (MRIII). Widrow and his students were first to recognize that this rule is mathematically equivalent to backpropagation.

The outline above gives only a partial view of the discipline, and many landmark discoveries have not been mentioned. Needless to say, the field of neural networks is quickly becoming a vast one, and in one short survey we could not hope to cover the entire subject in any detail. Consequently, many significant developments, including some of those mentioned above, will not be discussed in this appendix. The algorithms described will be limited primarily to those developed in our laboratory at Stanford and to related techniques developed elsewhere, the most important of which is the backpropagation algorithm. The section headings indicate the range and coverage of this appendix:

G.1 Introduction

G.2 Fundamental Concepts

G.3 Adaptation — The Minimal Disturbance Principle

G.4 Error Correction Rules — Single Threshold Element

The α-LMS algorithm

[2] We should note, however, that in the field of variational calculus the idea of error backpropagation through nonlinear systems existed centuries before Werbos first thought to apply this concept to neural networks. In the past 25 years, these methods have been used widely in the field of optimal control, as discussed by Le Cun [40].

[3] The term "sigmoid" is usually used in reference to monotonically increasing "S-shaped" functions, such as the hyperbolic tangent. In this appendix, however, we generally use the term to denote any smooth nonlinear functions at the output of a linear adaptive element. In other papers, these nonlinearities go by a variety of names, such as "squashing functions," "activation functions," "transfer characteristics," or "threshold functions."

Information about the neural network paradigms not discussed in this paper can be obtained from a number of other sources, such as the concise survey by Richard Lippmann [46], and the collection of classics by Anderson and Rosenfeld [47]. Much of the early work in the field from the 1960s is carefully reviewed in Nilsson's monograph [48]. A good view of some of the more recent results is presented in Rumelhart and McClelland's popular two-volume set [49]. A paper by Moore [50] presents a clear discussion about ART 1 and some of Grossberg's terminology. Another resource is the DARPA Study report [51] which gives a very comprehensive and readable "snapshot"of the field in 1988.

G.2 FUNDAMENTAL CONCEPTS

Today we can build computers and other machines which perform a variety of well-defined tasks with celerity and reliability unmatched by humans. No human can invert matrices or solve systems of differential equations at speeds which rival modern workstations. Nonetheless, there are still many problems which have yet to be solved to our satisfaction by any man-made machine but are easily disentangled by the perceptual or cognitive powers of humans, and often lower mammals, or even fish and insects. No computer vision system can rival the human ability to recognize visual images formed by objects of all shapes and orientations under a wide range of conditions. Humans effortlessly recognize objects in diverse environments and lighting conditions, even when obscured by dirt, or occluded by other objects. Likewise, the performance of current speech recognition technology pales when compared to the performance of the human adult who easily recognizes words spoken by different people, at different rates, pitches, and volumes, even in the presence of distortion or background noise.

The problems solved more effectively by the brain than by the digital computer typically have two characteristics: They are generally ill-defined, and they usually require an enormous amount of processing. Recognizing the character of an object from its image on television, for instance, involves resolving ambiguities associated with distortion and lighting. It also involves filling in information about a three-dimensional scene which is missing from the two-dimensional image on the screen. There are an infinite number of three-dimensional scenes which can be projected into a two-dimensional image. Nonetheless, the brain deals well with this ambiguity, and using learned cues, usually has little difficulty correctly determining the role played by the missing dimension.

As anyone who has performed even simple filtering operations on images is aware, processing high-resolution images requires a great deal of computation. Our brains accomplish this by utilizing massive parallelism, with millions and even billions of neurons in parts of the brain working together to solve complicated problems. Because solid-state operational amplifiers and logic gates can compute many orders of magnitude faster than current estimates of the effective computational speed of neurons in the brain, we may soon be able to build relatively inexpensive machines with the ability to process as much information as the human brain. This enormous processing power will do little to help us solve problems, however, unless we can utilize it effectively. For instance, coordinating many thousands of processors which must efficiently cooperate to solve a problem is not a simple task. If each processor must be programmed separately, and if all contingencies associated with various ambiguities must be designed into the software, even a relatively simple problem can quickly become unmanageable. The slow progress over the past 25 years or so in machine vision and other areas of artificial intelligence is testament to the difficulties associated with solving ambiguous and computationally intensive problems on von Neumann computers and related architectures.

Thus, there is some reason to consider attacking certain problems by designing naturally parallel computers which process information and learn by principles borrowed from the nervous systems of biological creatures. This does not necessarily mean we should attempt to copy the brain part for part. Although the bird served to inspire development of the airplane, birds do not have propellers, and airplanes do not operate by flapping feathered wings. The primary parallel between biological nervous systems and artificial neural networks is that each typically consists of a large number of simple elements that learn and are able to collectively solve complicated and ambiguous problems.

Today, most artificial neural network research and application are accomplished by simulating networks on serial computers. Speed limitations keep such networks relatively small, but even with small networks some surprisingly difficult problems have been tackled. Networks with fewer than 150 neural elements have been used successfully in vehicular control simulations [52], speech generation [53, 54], and undersea mine detection [51]. Small networks have also been used successfully in airport explosive detection [55], expert systems [56, 57], and scores of other applications. Furthermore, efforts to develop parallel neural network hardware are being met with some success, and such hardware should be available in the future for attacking more difficult problems like speech recognition [58, 59].

Whether implemented in parallel hardware or simulated on a computer, all neural networks consist of a collection of simple elements that work together to solve problems. A basic building block of nearly all artificial neural networks, and most other adaptive systems, is the adaptive linear combiner.

G.2.1 The Adaptive Linear Combiner

The adaptive linear combiner is diagrammed in Fig. G.1. Its output is a linear combination of its inputs. In a digital implementation, this element receives at time k an input signal vector or input pattern vector $\mathbf{X}_k = [x_0, x_{1_k}, x_{2_k}, \ldots x_{n_k}]^T$, and a desired response d_k, a special input used to effect learning. The components of the input vector are weighted by a set of coefficients, the weight vector $\mathbf{W}_k = [w_{0_k}, w_{1_k}, w_{2_k}, \ldots w_{n_k}]^T$. The sum of the weighted inputs is then computed, producing a linear output, the inner product $s_k = \mathbf{X}_k^T \mathbf{W}_k$. The components of $\mathbf{X}_k$ may be either continuous analog values or binary values. The weights are essentially continuously variable and can take on negative as well as positive values.

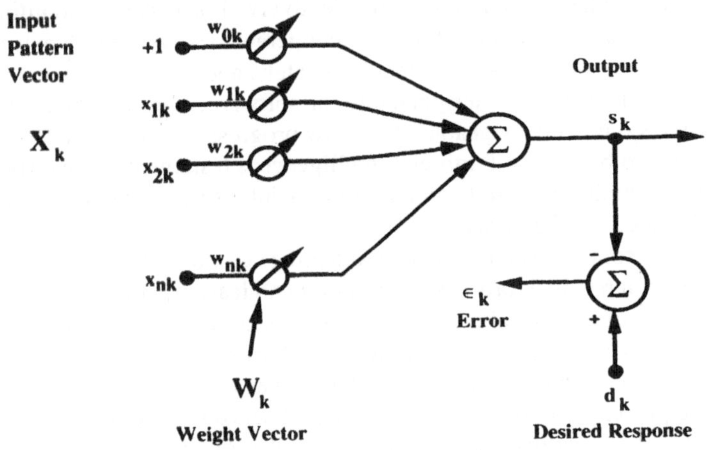

Figure G.1 Adaptive linear combiner.

During the training process, input patterns and corresponding desired responses are presented to the linear combiner. An adaptation algorithm automatically adjusts the weights so that the output responses to the input patterns will be as close as possible to their respective desired responses. In signal processing applications, the most popular method for adapting the weights is the simple LMS (least mean square) algorithm [60, 61], often called the Widrow-Hoff delta rule [44]. This algorithm minimizes the sum of squares of the linear errors over the training set. The linear error ϵ_k is defined to be the difference between the desired response d_k and the linear output s_k, during presentation k. Having this error signal is necessary for adapting the weights. When the adaptive linear combiner is embedded in a multi-element neural network, however, an error signal is often not directly available for each individual linear combiner and more complicated procedures must be devised for adapting the weight vectors. These procedures are the main focus of this appendix.

G.2.2 A Linear Classifier — The Single Threshold Element

The basic building block used in many neural networks is the "adaptive linear element," or Adaline[4] [60], shown in Fig. G.2.

This is an adaptive threshold logic element. It consists of an adaptive linear combiner cascaded with a hard-limiting quantizer which is used to produce a binary ±1 output, $y_k = \text{sgn}(s_k)$. The bias weight w_{0_k} which is connected to a constant input, $x_0 = +1$, effectively controls the threshold level of the quantizer.

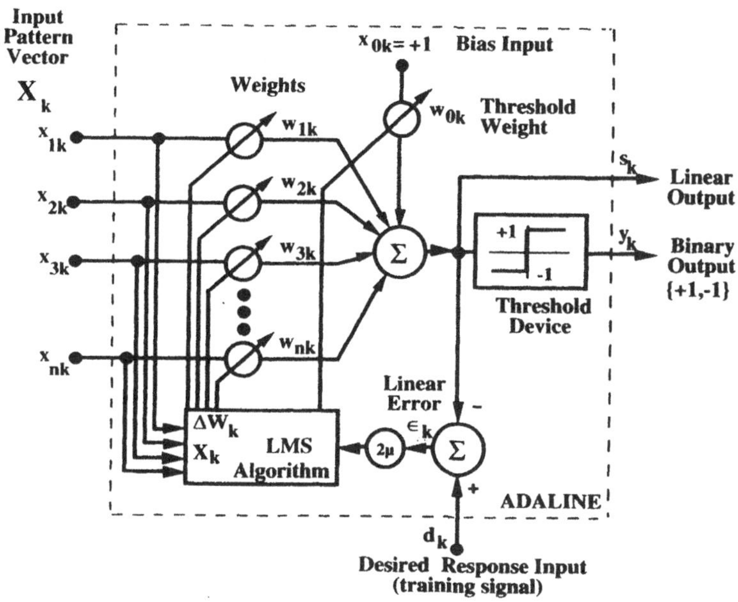

Figure G.2 An adaptive linear element (Adaline).

In single-element neural networks, an adaptive algorithm (such as the LMS algorithm, or the Perceptron rule) is often used to adjust the weights of the Adaline so that it responds correctly to as many patterns as possible in a training set which has binary desired responses. Once the weights are adjusted, the responses of the trained element can be tested by applying various input patterns. If the Adaline responds correctly with high probability to input patterns that were not included in the training set, it is said that generalization has taken place. Learning and generalization are among the most useful attributes of Adalines and neural networks.

[4] In the neural network literature, such elements are often referred to as "adaptive neurons." However, in a conversation between David Hubel of Harvard Medical School and Bernard Widrow, Dr. Hubel pointed out that the Adaline differs from the biological neuron since it contains not only the neural cell body, but also the input synapses and a mechanism for training them.

Linear Separability

With n binary inputs and one binary output, a single Adaline of the type shown in Fig. G.2 is capable of implementing certain logic functions. There are 2^n possible input patterns. A general logic implementation would be capable of classifying each pattern as either $+1$ or -1, in accord with the desired response. Thus, there are 2^{2^n} possible logic functions connecting n inputs to a single binary output. A single Adaline is capable of realizing only a small subset of these functions, known as the linearly separable logic functions or threshold logic functions [62]. These are the set of logic functions that can be obtained with all possible weight variations.

Figure G.3 shows a two-input Adaline element. Figure G.4 represents all possible binary inputs to this element with four large dots in pattern vector space. In this space, the components of the input pattern vector lie along the coordinate axes. The Adaline separates input patterns into two categories, depending on the values of the weights. A critical thresholding condition occurs when the linear output s equals zero:

$$s = x_1 w_1 + x_2 w_2 + w_0 = 0, \tag{G.1}$$

therefore

$$x_2 = -\frac{w_1}{w_2} x_1 - \frac{w_0}{w_2}. \tag{G.2}$$

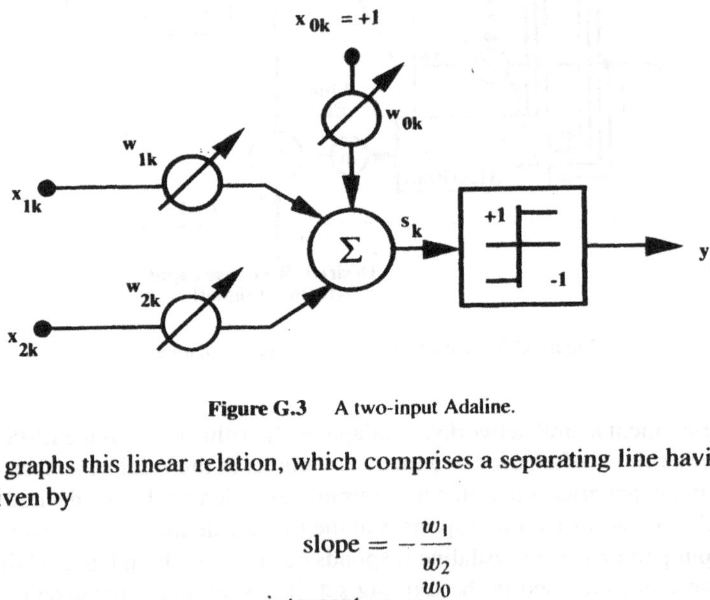

Figure G.3 A two-input Adaline.

Figure G.4 graphs this linear relation, which comprises a separating line having slope and intercept given by

$$\text{slope} = -\frac{w_1}{w_2}$$

$$\text{intercept} = -\frac{w_0}{w_2}. \tag{G.3}$$

The three weights determine slope, intercept, and the side of the separating line that corresponds to a positive output. The opposite side of the separating line corresponds to a negative output. For Adalines with four weights, the separating boundary is a plane; with more than four weights, the boundary is a hyperplane. Note that if the bias weight is zero, the separating hyperplane will be homogeneous — it will pass through the origin in pattern space.

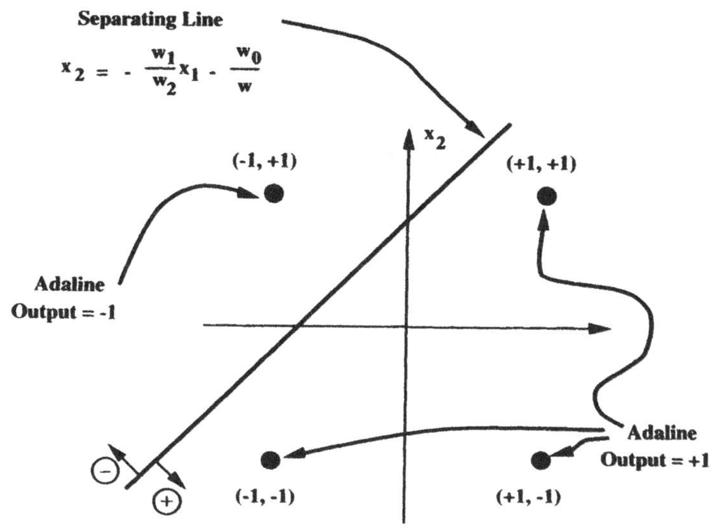

Figure G.4 Separating line in pattern space.

As sketched in Fig. G.4, the binary input patterns are classified as follows:

$$\begin{aligned}(+1,+1) &\rightarrow +1\\(+1,-1) &\rightarrow +1\\(-1,-1) &\rightarrow +1\\(-1,+1) &\rightarrow -1.\end{aligned} \tag{G.4}$$

This is an example of a linearly separable function. An example of a function which is not linearly separable is the two-input exclusive NOR function:

$$\begin{aligned}(+1,+1) &\rightarrow +1\\(+1,-1) &\rightarrow -1\\(-1,-1) &\rightarrow +1\\(-1,+1) &\rightarrow -1.\end{aligned} \tag{G.5}$$

No single straight line exists that can achieve this separation of the input patterns; thus, without preprocessing, no single Adaline can implement the exclusive NOR function.

With two inputs, a single Adaline can realize 14 of the 16 possible logic functions. With many inputs, however, only a small fraction of all possible logic functions are realizable, that is, linearly separable. Combinations of elements or networks of elements can be used to realize functions which are not linearly separable.

Capacity of Linear Classifiers

The number of training patterns or stimuli that an Adaline can learn to correctly classify is an important issue. Each pattern and desired output combination represents an inequality constraint on the weights. It is possible to have inconsistencies in sets of simultaneous inequalities just as with simultaneous equalities. When the inequalities (i.e., the patterns) are determined at random, the number which can be picked before an inconsistency arises is a matter of chance.

In their 1964 dissertations [63, 64], T. M. Cover and R. J. Brown both showed that the average number of random patterns with random binary desired responses that can be absorbed by an Adaline is approximately equal to twice the number of weights.[5] This is the *statistical pattern capacity* C_s of the Adaline. As reviewed by Nilsson [48], both theses included an analytic formula describing the probability that such a training set can be separated by an Adaline (i.e. is linearly separable). The probability is a function of N_p, the number of input patterns in the training set, and N_w, the number of weights in the Adaline, including the threshold weight, if used:

$$P_{\text{Separable}} = \begin{cases} 2^{-(N_p-1)} \sum_{i=0}^{N_w-1} \binom{N_p - 1}{i} & \text{for } N_p > N_w \\ 1 & \text{for } N_p \leq N_w \end{cases} . \tag{G.6}$$

In Fig. G.5 this formula was used to plot a set of analytical curves which show the probability that a set of N_p random patterns can be trained into an Adaline as a function of the ratio N_p/N_w. Notice from these curves that as the number of weights increases, the statistical pattern capacity of the Adaline $C_s = 2N_w$ becomes an accurate estimate of the number of responses it can learn.

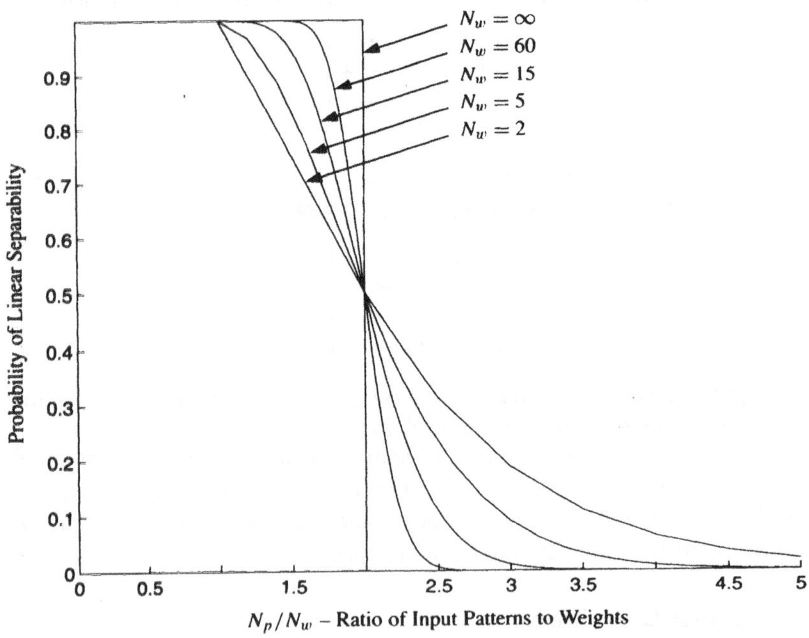

Figure G.5 Probability that an Adaline can separate a training pattern set as a function of the ratio N_p/N_w.

[5] Underlying theory for this result was discovered independently by a number of researchers including, among others, R. O. Winder [65], S. H. Cameron [66], and R. D. Joseph [67].

Another fact that can be observed from Fig. G.5 is that a problem is guaranteed to have a solution if the number of patterns is equal to (or less than) half the statistical pattern capacity; that is, if the number of patterns is equal to the number of weights. We will refer to this as the *deterministic pattern capacity* C_d of the Adaline. An Adaline can learn *any* two-category pattern classification task involving no more patterns than that represented by its deterministic capacity, $C_d = N_w$.

Both the statistical and deterministic capacity results depend upon a mild condition on the positions of the input patterns: The patterns must be in general position with respect to the Adaline.[6] If the input patterns to an Adaline are continuous valued and smoothly distributed (i.e., pattern positions are generated by a distribution function containing no impulses), general position is assured. The general position assumption is often invalid if the pattern vectors are binary. Nonetheless, even when the points are not in general position, the capacity results represent useful upper bounds.

The capacity results apply to randomly selected training patterns. In most problems of interest, the patterns in the training set are not random but exhibit some statistical regularities. These regularities are what make generalization possible. The number of patterns that an Adaline can learn in a practical problem often far exceeds its statistical capacity because the Adaline is able to generalize within the training set and learns many of the training patterns before they are even presented.

G.2.3 Nonlinear Classifiers

The linear classifier is limited in its capacity, and of course is limited to only linearly separable forms of pattern discrimination. More sophisticated classifiers with higher capacities are nonlinear. Two types of nonlinear classifiers are described here. The first is a fixed preprocessing network connected to a single adaptive element, and the other is the multi-element feedforward neural network.

Polynomial Discriminant Functions

Nonlinear functions of the inputs applied to the single Adaline can yield nonlinear decision boundaries. Useful nonlinearities include the polynomial functions. Consider the system illustrated in Fig. G.6 which contains only linear and quadratic input functions. The critical thresholding condition for this system is

$$\begin{aligned} s = {} & w_0 + x_1 w_1 + x_1^2 w_{11} + x_1 x_2 w_{12} + \\ & x_2^2 w_{22} + x_2 w_2 = 0. \end{aligned} \tag{G.7}$$

With proper choice of the weights, the separating boundary in pattern space can be established as shown, for example, in Fig. G.7. This represents a solution for the exclusive NOR function of (G.5). Of course, all of the linearly separable functions are also realizable. The use of such nonlinearities can be generalized for more inputs than two and for higher

[6]Patterns are in general position with respect to an Adaline with no threshold weight if any subset of pattern vectors containing no more than N_w members forms a linearly independent set, or equivalently, if no set of N_w or more input points in the N_w-dimensional pattern space lie on a homogeneous hyperplane. For the more common case involving an Adaline with a threshold weight, general position means that no set of N_w or more patterns in the $(N_w - 1)$-dimensional pattern space lie on a hyperplane not constrained to pass through the origin [63, 48].

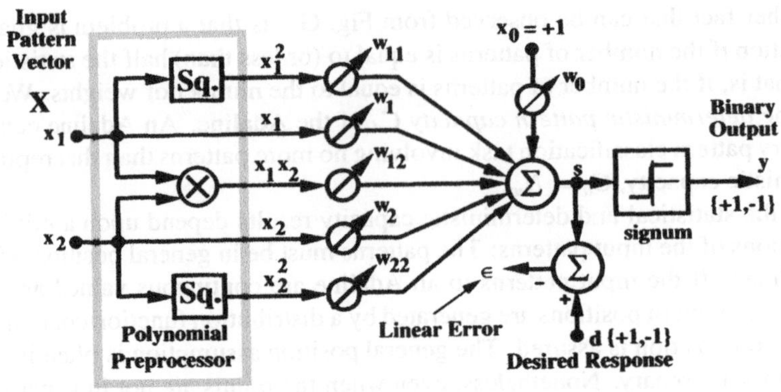

Figure G.6 An Adaline with inputs mapped through nonlinearities.

degree polynomial functions of the inputs. Some of the first work in this area was done by D. F. Specht [68, 69, 70] at Stanford in the 1960s when he successfully applied polynomial discriminants to the classification and analysis of electrocardiographic signals. Work on this topic has also been done by Barron [71, 72, 73] and by A. G. Ivakhnenko [74] in the former Soviet Union.

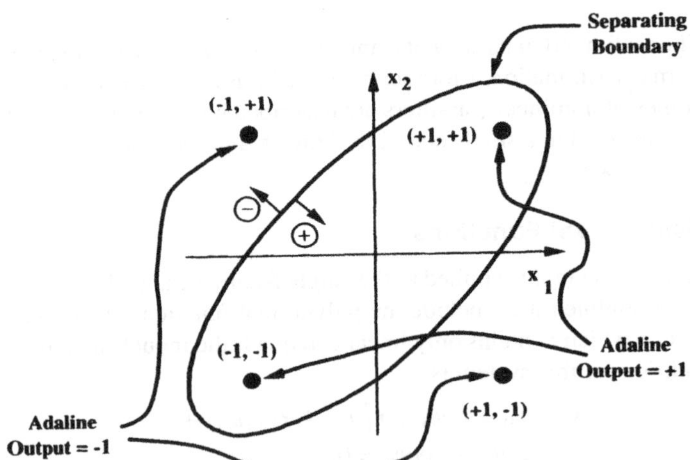

Figure G.7 An elliptical separating boundary for realizing a function which is not linearly separable.

The polynomial approach offers great simplicity and beauty. Through it one can realize a wide variety of adaptive nonlinear discriminant functions by adapting only a single Adaline element. Several methods have been developed for training the polynomial discriminant function. Specht developed a very efficient noniterative (i.e., single pass through the training set) training procedure, the polynomial discriminant method (PDM), which allows the polynomial discriminant function to implement a nonparametric classifier based on the Bayes decision rule. Other methods for training the system include iterative error correction rules such as the Perceptron and α-LMS rules, and iterative gradient descent procedures such as the μ-LMS and SER (also called RLS) algorithms [32]. Gradient descent with a single adaptive element is typically much faster than with a layered neural network. Furthermore, as we shall see, when the single Adaline is trained by a gradient descent procedure, it will converge to a unique global solution.

After the polynomial discriminant function has been trained by a gradient descent procedure, the weights of the Adaline will represent an approximation to the coefficients in a multidimensional Taylor series expansion of the desired response function. Likewise, if appropriate trigonometric terms are used in place of the polynomial preprocessor, the Adaline's weight solution will approximate the terms in the (truncated) multidimensional Fourier series decomposition of a periodic version of the desired response function. The choice of preprocessing functions determines how well a network will generalize for patterns outside the training set. Determining "good" functions remains a focus of current research [75, 76]. Experience seems to indicate that unless the nonlinearities are chosen with care to suit the problem at hand, often better generalization can be obtained from networks with more than one adaptive layer. In fact, one can view multilayer networks as single-layer networks with trainable preprocessors which are essentially self-optimizing.

Madaline I

One of the earliest trainable layered neural networks with multiple adaptive elements was the Madaline I structure of Widrow and Hoff [2, 77]. Mathematical analyses of Madaline I were developed in the Ph.D. dissertations of Ridgway [78], Hoff [77], and Glanz [79]. In the early 1960s, a 1,000-weight Madaline I was built out of hardware [80] and used in pattern recognition research. The weights in this machine were memistors, electrically variable resistors developed by Widrow and Hoff which are adjusted by electroplating a resistive link [81].

Madaline I was configured in the following way. Retinal inputs were connected to a layer of adaptive Adaline elements, the outputs of which were connected to a fixed logic device that generated the system output. Methods for adapting such systems were developed at that time. An example of this kind of network is shown in Fig. G.8. Two Adalines are connected to an AND logic device to provide an output.

With weights suitably chosen, the separating boundary in pattern space for the system of Fig. G.8 would be as shown in Fig. G.9. This separating boundary implements the exclusive NOR function of (G.5).

Madalines were constructed with many more inputs, with many more Adaline elements in the first layer, and with various fixed logic devices such as AND, OR, and majority vote-taker elements in the second layer. Those three functions, illustrated in Fig. G.10, are all threshold logic functions. The given weight values will implement these three functions, but the weight choices are not unique.

Feedforward Networks

The Madalines of the 1960s had adaptive first layers and fixed threshold functions in the second (output) layers [78, 48]. The feedforward neural networks of today often have many layers, and usually all layers are adaptive. The backpropagation networks of Rumelhart and

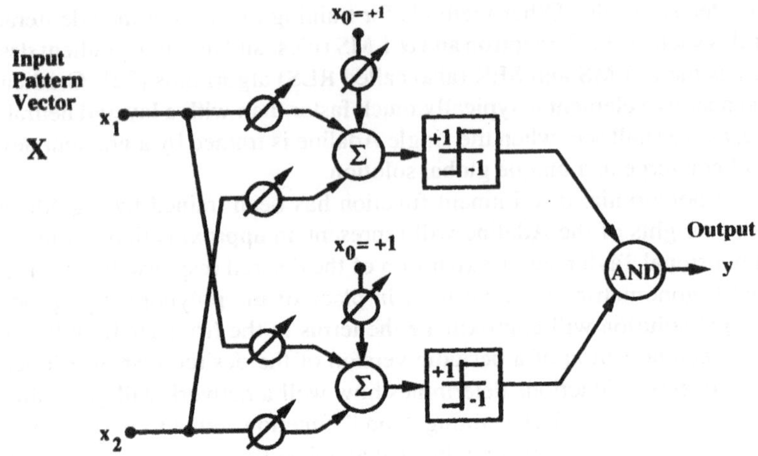

Figure G.8 A two-Adaline form of Madaline.

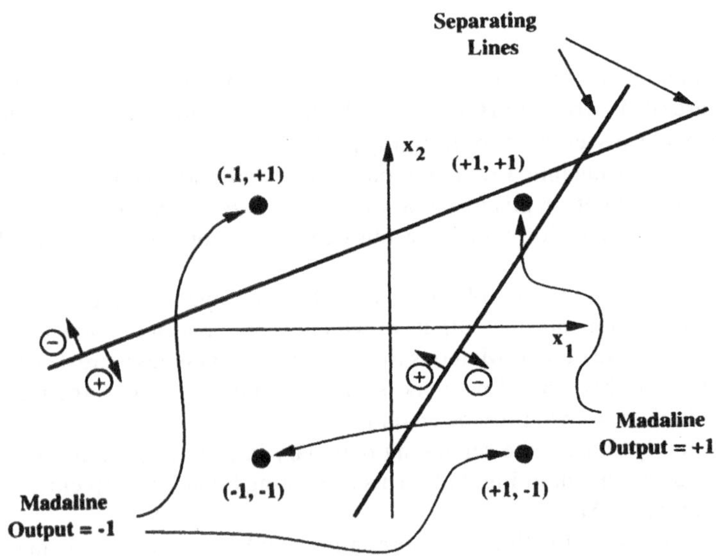

Figure G.9 Separating lines for Madaline of Fig. G.8

his colleagues [49] are perhaps the best-known examples of multilayer networks. A fully-connected three-layer[7] feedforward adaptive network is illustrated in Fig. G.11. In a fully

[7] In Rumelhart and colleagues terminology, this would be called a four-layer network, following Rosenblatt's convention of counting layers of signals, including the input layer. For our purposes, we find it more useful to count only layers of computing elements. We do not count as a layer the set of input terminal points.

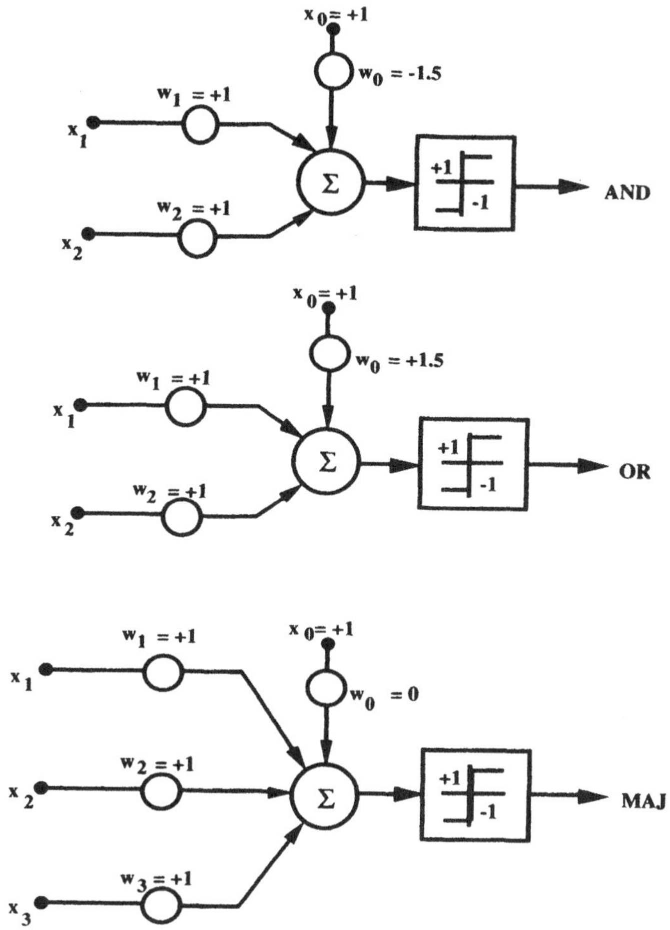

Figure G.10 Fixed-weight Adaline implementations of AND, OR, and MAJ logic functions.

connected layered network, each Adaline receives inputs from every output in the preceding layer.

During training, the response of each output element in the network is compared with a corresponding desired response. Error signals associated with the output elements are readily computed, so adaptation of the output layer is straightforward. The fundamental difficulty associated with adapting a layered network lies in obtaining "error signals" for hidden layer Adalines, that is, for Adalines in layers other than the output layer. The backpropagation and Madaline III algorithms contain methods for establishing these error signals.

There is no reason why a feedforward network must have the layered structure of Fig. G.11. In Werbos's development of the backpropagation algorithm [39], in fact, the Adalines are ordered and each receives signals directly from each input component and from the

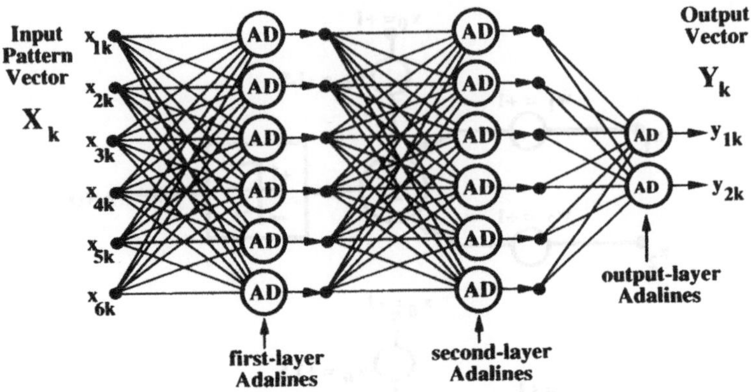

Figure G.11 A three-layer adaptive neural network.

output of each preceding Adaline. Many other variations of the feedforward network are possible. An interesting area of current research involves a generalized backpropagation method which can be used to train "high order" or "sigma-pi" networks that incorporate a polynomial preprocessor for each Adaline [49, 82].

One characteristic that is often desired in pattern recognition problems is invariance of the network output to changes in the position and size of the input pattern or image. Various techniques have been used to achieve translation, rotation, scale, and time invariance. One method involves including in the training set several examples of each exemplar transformed in size, angle, and position, but with a desired response which depends only on the original exemplar [80]. Other research has dealt with various Fourier and Mellin transform preprocessors [83, 84], as well as neural preprocessors [85]. Giles and colleagues have developed a clever averaging approach which removes unwanted dependencies from the polynomial terms in high-order threshold logic units (polynomial discriminant functions) [76] and high-order neural networks [82]. Other approaches have considered Zernike moments [86], graph matching [87], spatially repeated feature detectors [9], and time-averaged outputs [88].

Capacity of Nonlinear Classifiers

An important consideration that should be addressed when comparing various network topologies concerns the amount of information they can store.[8] Of the nonlinear classifiers mentioned above, the pattern capacity of the Adaline driven by a fixed preprocessor composed of smooth nonlinearities is the simplest to determine. If the inputs to the system are smoothly distributed in position, the outputs of the preprocessing network will be in general position with respect to the Adaline. Thus, the inputs to the Adaline will satisfy the condition required in Cover's Adaline capacity theory. Accordingly, the deterministic and statistical pattern capacities of the system are essentially equal to those of the Adaline.

[8] We should emphasize that the information referred to here corresponds to the maximum number of binary input/output mappings a network achieves with properly adjusted weights, not the number of bits of information that can be stored directly into the network's weights.

The capacities of Madaline I structures which utilize both the majority element and the "OR" element were experimentally estimated by J. S. Koford in the early 1960s. Although the logic functions that can be realized with these output elements are quite different, both types of elements yield essentially the same statistical storage capacity. The average number of patterns that a Madaline I network can learn to classify was found to be equal to the capacity per Adaline multiplied by the number of Adalines in the structure. The statistical capacity C_s is therefore approximately equal to twice the number of adaptive weights. Although the Madaline and the Adaline have roughly the same capacity per adaptive weight, without preprocessing the Adaline can separate only linearly separable sets, while the Madaline has no such limitation.

A great deal of theoretical and experimental work has been directed toward determining the capacity of both Adalines and Hopfield networks [89, 90, 91, 92]. Somewhat less theoretical work has been focused on the pattern capacity of multilayer feedforward networks, though some knowledge exists about the capacity of two-layer networks. Such results are of particular interest because the two-layer network is surprisingly powerful. With a sufficient number of hidden elements, a signum network with two layers can implement any Boolean function.[9] Equally impressive is the power of the two-layer sigmoid network. Given a sufficient number of hidden Adaline elements, such networks can implement any continuous input-output mapping to arbitrary accuracy [94, 95, 96]. Although two-layer networks are quite powerful, it is likely that some problems can be solved more efficiently by networks with more than two layers. Nonfinite-order predicate mappings (e.g., the connectedness problem [97]) can often be computed by small networks which use signal feedback [98].

In the mid-1960s, Cover studied the capacity of a feedforward signum network with an arbitrary number of layers[10] and a single output element [63, 99]. He determined a lower bound on the minimum number of weights N_w needed to enable such a network to realize any Boolean function defined over an arbitrary set of N_p patterns in general position. Recently, Baum extended Cover's result to multioutput networks and also used a construction argument to find corresponding upper bounds for the special case of the two-layer signum network [100]. Consider a two-layer fully connected feedforward network of signum Adalines which has N_x input components (excluding the bias inputs) and N_y output components. If this network is required to learn to map any set containing N_p patterns which are in general position to any set of binary desired response vectors (with N_y components), it follows from Baum's results[11] that the minimum requisite number of weights N_w can be bounded by

$$\frac{N_y N_p}{1+\log_2(N_p)} \leq N_w < N_y\left(\frac{N_p}{N_x}+1\right)(N_x+N_y+1)+N_y. \tag{G.8}$$

From Eq. (G.8), it can be shown that for a two-layer feedforward network with several times as many inputs and hidden elements as outputs (say, at least five times as many), the deter-

[9]This can be seen by noting that any Boolean function can be written in the sum of products form [93], and that such an expression can be realized with a two-layer network by using the first-layer Adalines to implement AND gates, while using the second-layer Adalines to implement OR gates.

[10]Actually, the network can be an arbitrary feedforward structure and need not be layered.

[11]The upper bound used here is Baum's loose bound: min #hidden nodes $\leq N_y\lceil N_p/N_x\rceil < N_y(N_p/N_x+1)$.

ministic pattern capacity is bounded below by something slightly smaller than N_w/N_y. It also follows from Eq. (G.8) that the pattern capacity of any feedforward network with a large ratio of weights to outputs (i.e., N_w/N_y at least several thousand) can be bounded above by a number somewhat larger than $(N_w/N_y)\log_2(N_w/N_y)$. Thus, the deterministic pattern capacity C_d of a two-layer network can be bounded by

$$\frac{N_w}{N_y} - K_1 \le C_d \le \frac{N_w}{N_y}\log_2\left(\frac{N_w}{N_y}\right) + K_2 \tag{G.9}$$

where K_1 and K_2 are positive numbers which are small terms if the network is large with few outputs relative to the number of inputs and hidden elements.

It is easy to show that Eq. (G.8) also bounds the number of weights needed to ensure that N_p patterns can be learned with probability 1/2, except in this case the lower bound on N_w becomes $(N_y N_p - 1)/(1+\log_2(N_p))$. It follows that Eq. (G.9) also serves to bound the statistical capacity C_s of a two-layer signum network.

It is interesting to note that the capacity bounds (G.9) encompass the deterministic capacity for the single-layer network comprised of a bank of N_y Adalines. In this case each Adaline would have N_w/N_y weights, so the system would have a deterministic pattern capacity of N_w/N_y. As N_y becomes large, the statistical capacity also approaches N_w/N_y (for N_x finite). Until further theory on feedforward network capacity is developed, it seems reasonable to use the capacity results from the single-layer network to estimate that of multilayer networks.

Little is known about the number of binary patterns that layered sigmoid networks can learn to classify correctly. The pattern capacity of sigmoid networks cannot be smaller than that of signum networks of equal size, however, because as the weights of a sigmoid network grow toward infinity, it becomes equivalent to a signum network with a weight vector in the same direction. Insight relating to the capabilities and operating principles of sigmoid networks can be winnowed from the literature [101, 102, 103].

A network's capacity is of little utility unless it is accompanied by useful generalizations to patterns not presented during training. In fact, if generalization is not needed, we can simply store the associations in a lookup table and will have little need for a neural network. The relationship between generalization and pattern capacity represents a fundamental trade-off in neural network applications: The Adaline's inability to realize all functions is in a sense a strength rather than the fatal flaw envisioned by some critics of neural networks [97], because it helps limit the capacity of the device and thereby improves its ability to generalize.

For good generalization, the training set should contain a number of patterns at least several times larger than the network's capacity (i.e., $N_p >> N_w/N_y$). This can be understood intuitively by noting that if the number of degrees of freedom in a network (i.e., N_w) is larger than the number of constraints associated with the desired response function (i.e., $N_y N_p$), the training procedure will be unable to completely constrain the weights in the network. Apparently, this allows effects of initial weight conditions to interfere with learned information and degrade the trained network's ability to generalize. A detailed analysis of generalization performance of signum networks as a function of training set size is described in [104].

A Nonlinear Classifier Application

Neural networks have been used successfully in a wide range of applications. To gain some insight about how neural networks are trained and what they can be used to compute, it is instructive to consider Sejnowski and Rosenberg's 1986 NETtalk demonstration [53, 54]. With the exception of work on the traveling salesman problem with Hopfield networks [105], this was the first neural network application since the 1960s to draw widespread attention. NETtalk is a two-layer feedforward sigmoid network with 80 Adalines in the first layer and 26 Adalines in the second layer. The network is trained to convert text into phonetically correct speech, a task well suited to neural implementation. The pronunciation of most words follows general rules based upon spelling and word context, but there are many exceptions and special cases. Rather than programming a system to respond properly to each case, the network can learn the general rules and special cases by example.

One of the more remarkable characteristics of NETtalk is that it learns to pronounce words in stages suggestive of the learning process in children. When the output of NETtalk is connected to a voice synthesizer, the system makes babbling noises during the early stages of the training process. As the network learns, it next conquers the general rules, and like a child, tends to make a lot of errors by using these rules even when not appropriate. As the training continues, however, the network eventually abstracts the exceptions and special cases and is able to produce intelligible speech with few errors.

The operation of NETtalk is surprisingly simple. Its input is a vector of seven characters (including spaces) from a transcript of text, and its output is phonetic information corresponding to the pronunciation of the center (fourth) character in the seven-character input field. The other six characters provide context which helps determine the desired phoneme. To read text, the seven character window is scanned across a document in computer memory and the network generates a sequence of phonetic symbols which can be used to control a speech synthesizer. Each of the seven characters at the network's input is a 29-component binary vector, with each component representing a different alphabetic character or punctuation mark. A one is placed in the component associated with the represented character while all other components are set to zero.[12]

The system's 26 outputs correspond to 23 articulatory features and 3 additional features which encode stress and syllable boundaries. When training the network, the desired response vector has zeros in all components except those which correspond to the phonetic features associated with the center character in the input field. In one experiment, Sejnowski and Rosenberg had the system scan a 1,024-word transcript of phonetically transcribed continuous speech. With the presentation of each seven-character window, the system's weights were trained by the backpropagation algorithm in response to the network's output error. After roughly 50 presentations of the entire training set, the network was able to produce accurate speech from data the network had not been exposed to during training.

[12]The input representation often has a considerable impact on the success of a network. In NETtalk, the inputs are sparsely coded in 29 components. One might consider instead choosing a 5-bit binary representation or the 7-bit ASCII code. It should be clear, however, that in this case the sparse representation helps simplify the network's job of interpreting input characters as 29 distinct symbols. Usually the appropriate input encoding is not difficult to decide. When intuition fails, however, one sometimes must experiment with different encodings to find one that works well.

Backpropagation is not the only technique that might be used to train NETtalk. In other experiments, the slower Boltzmann learning method was used, and, in fact, Madaline Rule III could be used as well. Likewise, if the sigmoid network were replaced by a similar signum network, Madaline Rule II would also work, although more first-layer Adalines would likely be needed for comparable performance.

The remainder of this appendix develops and compares various adaptive algorithms for training Adalines and artificial neural networks to solve classification problems such as NETtalk. These same algorithms can be used to train networks for other problems such as those involving nonlinear control [52], system identification [52, 106], signal processing [32], or decision making [57].

G.3 ADAPTATION — THE MINIMAL DISTURBANCE PRINCIPLE

The iterative algorithms described in this appendix are all designed in accord with a single underlying principle. These techniques — the two LMS algorithms, Mays's rules, and the Perceptron procedure for training a single Adaline, the MRI rule for training the simple Madaline, as well as MRII, MRIII, and backpropagation techniques for training multilayer Madalines — all rely upon the principle of minimal disturbance: *Adapt to reduce the output error for the current training pattern, with minimal disturbance to responses already learned.* Unless this principle is practiced, it is difficult to simultaneously store the required pattern responses. The minimal disturbance principle is intuitive. It was the motivating idea that led to the discovery of the LMS algorithm and the Madaline rules. In fact, the LMS algorithm had existed for several months as an error reduction rule before it was discovered that the algorithm uses an instantaneous gradient to follow the path of steepest descent and minimize the mean square error of the training set. It was then given the name "LMS" (least mean square) algorithm.

G.4 ERROR CORRECTION RULES — SINGLE THRESHOLD ELEMENT

As adaptive algorithms evolved, principally two kinds of online rules have come to exist. One kind, *error correction rules*, alter the weights of a network to correct a certain proportion of the error in the output response to the present input pattern. The other kind, *gradient rules*, alter the weights of a network during each pattern presentation by gradient descent with the objective of reducing mean square error, averaged over all training patterns. Both types of rules invoke similar training procedures. Because they are based upon different objectives, however, they can have significantly different learning characteristics.

Error correction rules, of necessity, often tend to be ad hoc. They are most often used when training objectives are not easily quantified, or when a problem does not lend itself to tractable analysis. A common application, for instance, concerns training neural networks that contain discontinuous functions. An exception is the α-LMS algorithm, an error correction rule which has proven to be an extremely useful technique for finding solutions to well-defined and tractable linear problems.

We begin with error correction rules applied initially to single Adaline elements, and then to networks of Adalines.

G.4.1 Linear Rules

Linear error correction rules alter the weights of the adaptive threshold element with each pattern presentation to make an error correction which is proportional to the error itself. The one linear rule, α-LMS, is described next.

The α-LMS algorithm

The α-LMS algorithm or Widrow-Hoff delta rule applied to the adaptation of a single Adaline (Fig. G.2) embodies the *minimal disturbance principle*. The weight update equation for the original form of the algorithm can be written as

$$\mathbf{W}_{k+1} = \mathbf{W}_k + \alpha \frac{\epsilon_k \mathbf{X}_k}{|\mathbf{X}_k|^2}. \tag{G.10}$$

The time index or adaptation cycle number is k. $\mathbf{W}_{k+1}$ is the next value of the weight vector, $\mathbf{W}_k$ is the present value of the weight vector, and $\mathbf{X}_k$ is the present input pattern vector. The present linear error ϵ_k is defined to be the difference between the desired response d_k and the linear output $s_k = \mathbf{W}_k^T \mathbf{X}_k$ before adaptation:

$$\epsilon_k \triangleq d_k - \mathbf{W}_k^T \mathbf{X}_k. \tag{G.11}$$

Changing the weights yields a corresponding change in the error:

$$\Delta \epsilon_k = \Delta(d_k - \mathbf{W}_k^T \mathbf{X}_k) = -\mathbf{X}_k^T \Delta \mathbf{W}_k. \tag{G.12}$$

In accordance with the α-LMS rule of Eq. (G.10), the weight change is as follows:

$$\Delta \mathbf{W}_k = \mathbf{W}_{k+1} - \mathbf{W}_k = \alpha \frac{\epsilon_k \mathbf{X}_k}{|\mathbf{X}_k|^2}. \tag{G.13}$$

Combining Eqs. (G.12) and (G.13), we obtain

$$\Delta \epsilon_k = -\alpha \frac{\epsilon_k \mathbf{X}_k^T \mathbf{X}_k}{|\mathbf{X}_k|^2} = -\alpha \epsilon_k. \tag{G.14}$$

Therefore, the error is reduced by a factor of α as the weights are changed while holding the input pattern fixed. Presenting a new input pattern starts the next adaptation cycle. The next error is then reduced by a factor of α, and the process continues. The initial weight vector is usually chosen to be zero and is adapted until convergence. In nonstationary environments, the weights are generally adapted continually.

The choice of α controls stability and speed of convergence [32]. For input pattern vectors independent over time, stability is ensured for most practical purposes if

$$0 < \alpha < 2. \tag{G.15}$$

Making α greater than 1 generally does not make sense, since the error would be overcorrected. Total error correction comes with $\alpha = 1$. A practical range for α is

$$0.1 < \alpha < 1.0. \tag{G.16}$$

This algorithm is self-normalizing in the sense that the choice of α does not depend on the magnitude of the input signals. The weight update is collinear with the input pattern and of a magnitude inversely proportional to $|\mathbf{X}_k|^2$. With binary ± 1 inputs, $|\mathbf{X}_k|^2$ is equal to the number of weights and does not vary from pattern to pattern. If the binary inputs are the usual 1 and 0, no adaptation occurs for weights with 0 inputs, while with ± 1 inputs, all weights are adapted each cycle and convergence tends to be faster. For this reason, the symmetric inputs $+1$ and -1 are generally preferred.

Figure G.12 provides a geometrical picture of how the α-LMS rule works. In accord with Eq. (G.13), $\mathbf{W}_{k+1}$ equals $\mathbf{W}_k$ added to $\Delta\mathbf{W}_k$, and $\Delta\mathbf{W}_k$ is parallel with the input pattern vector $\mathbf{X}_k$. From Eq. (G.12), the change in error is equal to the negative dot product of $\mathbf{X}_k$ and $\Delta\mathbf{W}_k$. Since the α-LMS algorithm selects $\Delta\mathbf{W}_k$ to be collinear with $\mathbf{X}_k$, the desired error correction is achieved with a weight change of the smallest possible magnitude. When adapting to respond properly to a new input pattern, the responses to previous training patterns are therefore minimally disturbed, on the average.

The α-LMS algorithm corrects error, and if all input patterns are of equal length, it minimizes mean square error [32]. The algorithm is best known for this property.

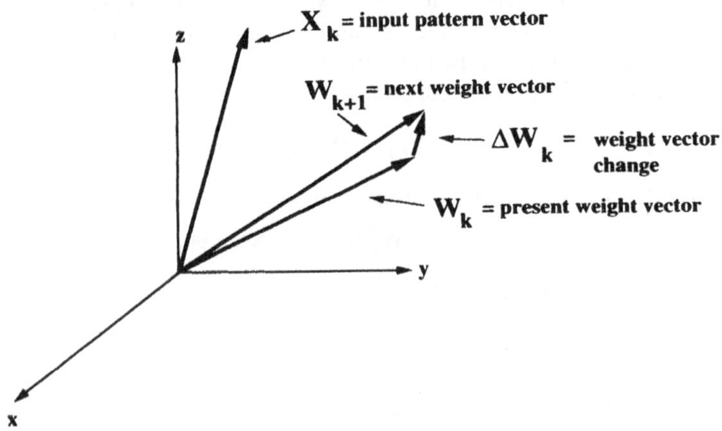

Figure G.12 Weight correction by the LMS rule.

G.4.2 Nonlinear Rules

The α-LMS algorithm is a linear rule which makes error corrections that are proportional to the error. It is known [107] that in some cases this linear rule may fail to separate training patterns that are linearly separable. Where this creates difficulties, nonlinear rules may be used. In the following paragraphs, we describe early nonlinear rules which were devised by Rosenblatt [108, 5] and Mays[107]. These nonlinear rules also make weight vector changes collinear with the input pattern vector (the direction which causes minimal disturbance), changes which are based on the linear error but are not directly proportional to it.

The Perceptron Learning Rule

The Rosenblatt α-Perceptron [108, 5], diagrammed in Fig. G.13, processed input patterns with a first layer of sparse, randomly connected, fixed-logic devices. The outputs of the fixed first layer fed a second layer which consisted of a single adaptive linear threshold element. Other than the convention that its input signals were $\{1,0\}$ binary, and that no bias weight was included, this element is equivalent to the Adaline element. The learning rule for the α-Perceptron is very similar to LMS, but its behavior is in fact quite different.

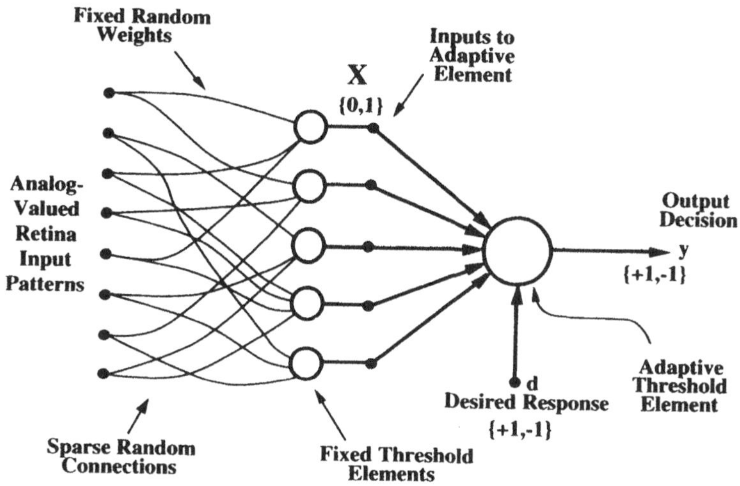

Figure G.13 Rosenblatt's α-Perceptron.

It is interesting to note that Rosenblatt's Perceptron learning rule was first presented in 1960 [108], and Widrow and Hoff's LMS rule was first presented the same year, a few months later [61]. These rules were developed independently in 1959.

The adaptive threshold element of the α-Perceptron is shown in Fig. G.14. Adapting with the Perceptron rule makes use of the "quantizer error" $\tilde{\tilde{\epsilon}}_k$, defined to be the difference between the desired response and the output of the quantizer

$$\tilde{\tilde{\epsilon}}_k \stackrel{\Delta}{=} d_k - y_k. \tag{G.17}$$

The Perceptron rule, sometimes called the Perceptron convergence procedure, does not adapt the weights if the output decision y_k is correct, that is, if $\tilde{\tilde{\epsilon}}_k = 0$. If the output decision disagrees with the binary desired response d_k, however, adaptation is effected by adding the input vector to the weight vector when the error $\tilde{\tilde{\epsilon}}_k$ is positive, or subtracting the input vector from the weight vector when the error $\tilde{\tilde{\epsilon}}_k$ is negative. Note that the quantizer error $\tilde{\tilde{\epsilon}}_k$ is always equal to either $+2$, -2, or 0. Thus, half the product of the input vector and the quantizer error $\tilde{\tilde{\epsilon}}_k$ is added to the weight vector. The Perceptron rule is identical to the α-LMS algorithm, except that with the Perceptron rule, half of the quantizer error, $\tilde{\tilde{\epsilon}}_k/2$, is used in place of the normalized linear error $\epsilon_k/|\mathbf{X}_k|^2$ of the α-LMS rule. The Perceptron

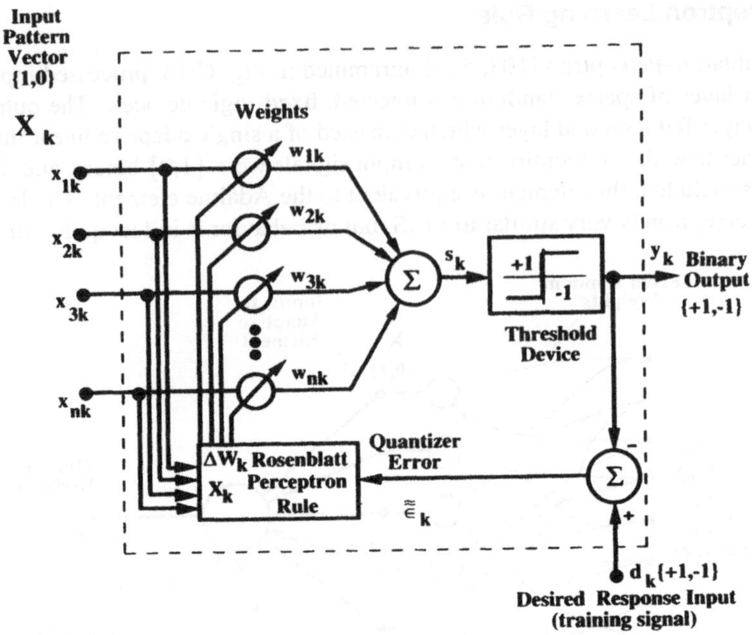

Figure G.14 The adaptive threshold element of the Perceptron.

rule is nonlinear in contrast to the LMS rule which is linear (compare Figs. G.2 and G.14). Nonetheless, the Perceptron rule can be written in a form which is very similar to the α-LMS rule of Eq. (G.10):

$$\mathbf{W}_{k+1} = \mathbf{W}_k + \alpha \frac{\tilde{\tilde{\epsilon}}_k}{2} \mathbf{X}_k. \tag{G.18}$$

Rosenblatt normally set α to one. In contrast to α-LMS, the choice of α does not affect the stability of the Perceptron algorithm, and it affects convergence time only if the initial weight vector is nonzero. Also, while α-LMS can be used with either analog or binary desired responses, Rosenblatt's rule can be used only with binary desired responses.

The Perceptron rule stops adapting when the training patterns are correctly separated. There is no restraining force controlling the magnitude of the weights, however. The direction of the weight vector, not its magnitude, determines the decision function. The Perceptron rule has been proven to be capable of separating any linearly separable set of training patterns [5, 109, 48, 107]. If the training patterns are not linearly separable, the Perceptron algorithm goes on forever, and often does not yield a low-error solution, even if one exists. In most cases, if the training set is not separable, the weight vector tends to gravitate toward zero[13] so that even if α is very small, each adaptation can dramatically affect the switching function implemented by the Perceptron.

[13]This results because the length of the weight vector decreases with each adaptation that does not cause the linear output s_k to change sign and assume a magnitude greater than that before adaptation. Although there are

This behavior is very different from that of the α-LMS algorithm. Continued use of α-LMS does not lead to an unreasonable weight solution if the pattern set is not linearly separable. Nor, however, is this algorithm guaranteed to separate any linearly separable pattern set. α-LMS typically comes close to achieving such separation, but its objective is different, that is, error reduction at the linear output of the adaptive element.

Rosenblatt also introduced variants of the fixed-increment rule that we have discussed thus far. A popular one was the absolute-correction version of the Perceptron rule.[14] This rule is identical to that stated in Eq. (G.18) except the increment size α is chosen with each presentation to be the smallest integer which corrects the output error in one presentation. If the training set is separable, this variant has all the characteristics of the fixed-increment version with α set to 1, except that it usually reaches a solution in fewer presentations.

Mays's Algorithms

In his Ph.D. dissertation, [107], C. H. Mays described an "increment adaptation" rule[15] and a "modified relaxation adaptation" rule. The fixed-increment version of the Perceptron rule is a special case of the increment adaptation rule.

Increment adaptation in its general form involves the use of a "dead zone" for the linear output s_k, equal to $\pm\gamma$ about zero. All desired responses are ± 1 (refer to Fig. G.14). If the linear output s_k falls outside the dead zone ($|s_k| \geq \gamma$), adaptation follows a normalized variant of the fixed-increment Perceptron rule (with $\alpha/|\mathbf{X}_k|^2$ used in place of α). If the linear output falls within the dead zone, whether or not the output response y_k is correct, the weights are adapted by the normalized variant of the Perceptron rule as though the output response y_k had been incorrect. The weight update rule for Mays's increment adaptation algorithm can be written mathematically as

$$\mathbf{W}_{k+1} = \begin{cases} \mathbf{W}_k + \alpha \, \tilde{\tilde{\epsilon}}_k \dfrac{\mathbf{X}_k}{2|\mathbf{X}_k|^2} & \text{if } |s_k| \geq \gamma \\ \mathbf{W}_k + \alpha d_k \dfrac{\mathbf{X}_k}{|\mathbf{X}_k|^2} & \text{if } |s_k| < \gamma \end{cases}, \tag{G.19}$$

where $\tilde{\tilde{\epsilon}}_k$ is the quantizer error of Eq. (G.17).

With the dead zone $\gamma = 0$, Mays's increment adaptation algorithm reduces to a normalized version of the Perceptron rule (G.18). Mays proved that if the training patterns are linearly separable, increment adaptation will always converge and separate the patterns in a finite number of steps. He also showed that use of the dead zone reduces sensitivity to weight errors. If the training set is not linearly separable, Mays's increment adaptation rule typically performs much better than the Perceptron rule because a sufficiently large dead zone tends to cause the weight vector to adapt away from zero when any reasonably good solution exists. In such cases, the weight vector may sometimes appear to meander rather aimlessly, but it will typically remain in a region associated with relatively low average error.

exceptions, for most problems this situation occurs only rarely if the weight vector is much longer than the weight increment vector.

[14]The terms "fixed-increment" and "absolute correction" are due to Nilsson [48] Rosenblatt referred to methods of these types, respectively, as quantized and nonquantized learning rules.

[15]The increment adaptation rule was proposed by others before Mays, though from a different perspective [109].

The increment adaptation rule changes the weights with increments that generally are not proportional to the linear error, ϵ_k. The other Mays rule, modified relaxation, is closer to α-LMS in its use of the linear error ϵ_k (refer to Fig. G.2). The desired response and the quantizer output levels are binary ± 1. If the quantizer output y_k is wrong or if the linear output s_k falls within the dead zone $\pm\gamma$, adaptation follows α-LMS to reduce the linear error. If the quantizer output y_k is correct and the linear output s_k falls outside the dead zone, the weights are not adapted. The weight update rule for this algorithm can be written as

$$\mathbf{W}_{k+1} = \begin{cases} \mathbf{W}_k & \text{if } \tilde{\tilde{\epsilon}}_k = 0 \text{ and } |s_k| \geq \gamma \\ \mathbf{W}_k + \alpha\epsilon_k \frac{\mathbf{X}_k}{|\mathbf{X}_k|^2} & \text{otherwise} \end{cases}, \tag{G.20}$$

where $\tilde{\tilde{\epsilon}}_k$ is the quantizer error of Eq. (G.17).

If the dead zone γ is set to ∞, this algorithm reduces to the α-LMS algorithm (G.10). Mays showed that, for dead zone $0 < \gamma < 1$, and learning rate $0 < \alpha \leq 2$, this algorithm will converge and separate any linearly separable input set in a finite number of steps. If the training set is not linearly separable, this algorithm performs much like Mays's increment adaptation rule.

Mays's two algorithms achieve similar pattern separation results. The choice of α does not affect stability, although it does affect convergence time. The two rules differ in their convergence properties, but there is no consensus on which is the better algorithm. Algorithms like these can be quite useful, and we feel that there are many more to be invented and analyzed.

The α-LMS algorithm, the Perceptron procedure, and Mays's algorithms can all be used for adapting the single Adaline element or they can be incorporated into procedures for adapting networks of such elements. Multilayer network adaption procedures which use some of these algorithms will be discussed below.

G.5 ERROR CORRECTION RULES — MULTI-ELEMENT NETWORKS

The algorithms discussed next are the Widrow-Hoff Madaline rule from the early 1960s, now called Madaline Rule I (MRI) and Madaline Rule II (MRII), developed by Widrow and Winter in 1987.

G.5.1 Madaline Rule I

The MRI rule allows the adaptation of a first layer of hard-limited (signum) Adaline elements whose outputs provide inputs to a second layer, consisting of a single fixed threshold logic element which may be, for example, the OR gate, AND gate, or Majority Vote Taker discussed previously. The weights of the Adalines are initially set to small random values.

Figure G.15 shows a Madaline I architecture with five fully connected first-layer Adalines. The second layer is a majority element (MAJ). Because the second-layer logic element is fixed and known, it is possible to determine which first-layer Adalines can be

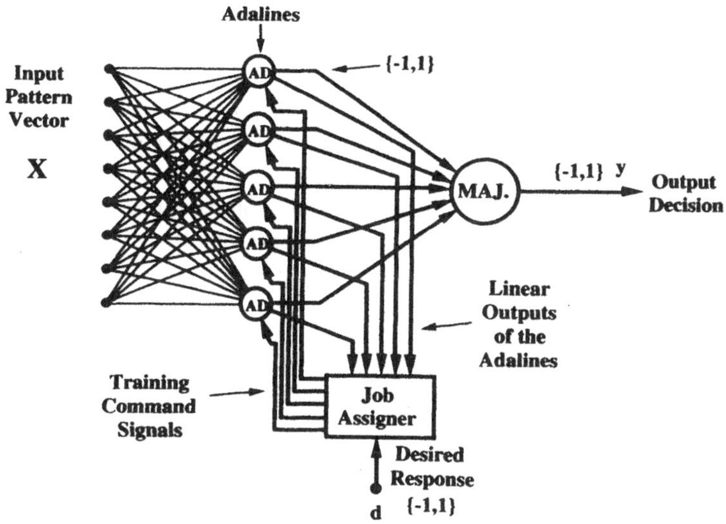

Figure G.15 A five-Adaline example of the Madaline I architecture.

adapted to correct an output error. The Adalines in the first layer assist each other in solving problems by automatic load-sharing.

One procedure for training the network in Fig. G.15 follows. A pattern is presented, and if the output response of the majority element matches the desired response, no adaptation takes place. However, if, for instance, the desired response is $+1$ and three of the five Adalines read -1 for a given input pattern, one of the latter three must be adapted to the $+1$ state. The element that is adapted by MRI is the one whose linear output s_k is closest to zero, that is, the one whose analog response is closest to the desired response. If more of the Adalines were originally in the -1 state, enough of them are adapted to the $+1$ state to make the majority decision equal $+1$. The elements adapted are those whose linear outputs are closest to zero. A similar procedure is followed when the desired response is -1. When adapting a given element, the weight vector can be moved in the LMS direction far enough to reverse the Adaline's output (absolute correction or "fast" learning), or it can be adapted by the small increment determined by the α-LMS algorithm (statistical or "slow" learning). The one desired response, d_k, is used for all Adalines that are adapted. The procedure can also be modified to allow one of Mays's rules to be used. In that event, for the case we have considered (majority output element), adaptations take place if at least half of the Adalines either have outputs which differ from the desired response or have analog outputs which are in the dead zone. By setting the dead zone of Mays's increment adaptation rule to zero, the weights can also be adapted by Rosenblatt's Perceptron rule.

Differences in initial conditions and the results of subsequent adaptation cause the various elements to take "responsibility" for certain parts of the training problem. The basic principle of load sharing is summarized thus: *Assign responsibility to the Adaline or Adalines that can most easily assume it.*

In Fig. G.15, the "job assigner," a purely mechanized process, assigns responsibility during training by transferring the appropriate adapt commands and desired response signals to the selected Adalines. The job assigner utilizes linear-output information. Load sharing is important, since it results in the various adaptive elements developing individual weight vectors. If all the weight vectors were the same, there would be no point in having more than one element in the first layer.

When training the Madaline, the pattern presentation sequence should be random. Experimenting with this, Ridgway [78] found that cyclic presentation of the patterns could lead to cycles of adaptation. These cycles would cause the weights of the entire Madaline to cycle, preventing convergence.

The adaptive system of Fig. G.15 was suggested by common sense and was found to work well in simulations. Ridgway found that the probability that a given Adaline will be adapted in response to an input pattern is greatest if that element had taken such responsibility during the previous adapt cycle when the pattern was most recently presented. The division of responsibility stabilizes at the same time that the responses of individual elements stabilize to their share of the load. When the training problem is not perfectly separable by this system, the adaptation process tends to minimize error probability, although it is possible for the algorithm to "hang up" on local optima.

The Madaline structure of Fig. G.15 has two layers — the first layer consists of adaptive logic elements, the second of fixed logic. There are a variety of fixed-logic devices that could be used for the second layer. A variety of MRI adaptation rules were devised by Hoff and described in his doctoral dissertation [77] which can be used with all possible fixed-logic output elements. An easily described training procedure results when the output element is an OR gate. During training, if the desired output for a given input pattern is $+1$, only the one Adaline whose linear output is closest to zero would be adapted if any adaptation is needed, that is, if all Adalines give -1 outputs. If the desired output is -1, all elements must give -1 outputs, and any giving $+1$ outputs must be adapted.

The MRI rule obeys the "minimal disturbance principle" in the following sense. No more Adaline elements are adapted than necessary to correct the output decision and any dead-zone constraint. The elements whose linear outputs are nearest to zero are adapted because they require the smallest weight changes to reverse their output responses. Furthermore, whenever an Adaline is adapted, the weights are changed in the direction of its input vector, providing the requisite error correction with minimal weight change.

G.5.2 Madaline Rule II

The MRI rule was recently extended to allow the adaptation of multilayer binary networks by Winter and Widrow with the introduction of Madaline Rule II (MRII) [45, 85, 110]. A typical two-layer MRII network is shown in Fig. G.16. The weights in both layers are adaptive.

Training with the MRII rule is similar to training with the MRI algorithm. The weights are initially set to small random values. Training patterns are presented in a random sequence. If the network produces an error during a training presentation, we begin by adapting first-layer Adalines. By the "minimal disturbance principle," we select the first-layer Adaline with the smallest linear output magnitude and perform a "trial adaptation" by inverting its binary output. This can be done without adaptation by adding a perturbation Δs

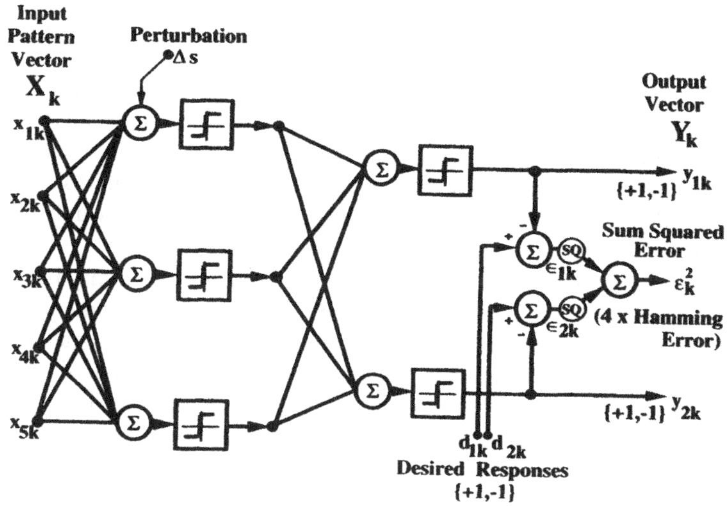

Figure G.16 Typical two-layer Madaline II architecture.

of suitable amplitude and polarity to the Adaline's sum (refer to Fig. G.16). If the output Hamming error is reduced by this bit inversion, that is, if the number of output errors is reduced, the perturbation Δs is removed and the weights of the selected Adaline element are changed by α-LMS in a direction collinear with the corresponding input vector — the direction which reinforces the bit reversal with minimal disturbance to the weights. Conversely, if the trial adaptation does not improve the network response, no weight adaptation is performed.

After finishing with the first element, we perturb and update other Adalines in the first layer which have "sufficiently small" linear-output magnitudes. Further error reductions can be achieved, if desired, by reversing pairs, triples, and so forth, up to some predetermined limit. After exhausting possibilities with the first layer, we move on to the next layer and proceed in a like manner. When the final layer is reached, each of the output elements is adapted by α-LMS. At this point, a new training pattern is selected at random and the procedure is repeated. The goal is to reduce Hamming error with each presentation, thereby hopefully minimizing the average Hamming error over the training set. Like MRI, the procedure can be modified so that adaptations follow an absolute correction rule or one of Mays's rules rather than α-LMS. Like MRI, MRII can "hang up" on local optima.

G.6 STEEPEST-DESCENT RULES — SINGLE THRESHOLD ELEMENT

Thus far, we have described a variety of adaptation rules that act to reduce a given proportion of the error with the presentation of each training pattern. Often, the objective of adaptation is to reduce error averaged in some way over the training set. The most common error function is mean square error (MSE), although in some situations other error criteria may be more appropriate [111, 112, 113]. The most popular approaches to mean square error reduction in both single-element and multi-element networks are based upon the method of steepest descent. More sophisticated gradient approaches such as quasi-Newton [32, 114, 115, 116] and conjugate gradient [116, 117] techniques often have better convergence properties, but the conditions under which the additional complexity is warranted are not generally known. The discussion that follows is restricted to minimization of MSE by the method of steepest descent [118, 119]. More sophisticated learning procedures usually require many of the same computations used in the basic steepest descent procedure.

Adaptation of a network by steepest descent starts with an arbitrary initial value $\mathbf{W}_0$ for the system's weight vector. The gradient of the mean square error function is measured and the weight vector is altered in the direction corresponding to the negative of the measured gradient. This procedure is repeated, causing the MSE to be successively reduced on average and causing the weight vector to approach a locally optimal value.

The method of steepest descent can be described by the relation

$$\mathbf{W}_{k+1} = \mathbf{W}_k + \mu(-\nabla_k), \tag{G.21}$$

where μ is a parameter that controls stability and rate of convergence, and ∇_k is the value of the gradient at a point on the MSE surface corresponding to $\mathbf{W} = \mathbf{W}_k$.

To begin, we derive rules for steepest descent minimization of the MSE associated with a single Adaline element. These rules are then generalized to apply to full-blown neural networks. Like error correction rules, the most practical and efficient steepest descent rules typically work with one pattern at a time. They minimize mean square error, approximately, averaged over the entire set of training patterns.

G.6.1 Linear Rules

Steepest-descent rules for the single threshold element are said to be linear if weight changes are proportional to the linear error, the difference between the desired response d_k and the linear output of the element, s_k.

Mean Square Error Surface of the Linear Combiner

In this section we demonstrate that the MSE surface of the linear combiner of Fig. G.1 is a quadratic function of the weights and is thus easily traversed by gradient descent.

Let the input pattern $\mathbf{X}_k$ and the associated desired response d_k be drawn from a statistically stationary population. During adaptation, the weight vector varies so that even with stationary inputs, the output s_k and error ϵ_k will generally be nonstationary. Care must be taken in defining the mean square error since it is time varying. The only possibility is an ensemble average, defined below.

At the kth iteration, let the weight vector be $\mathbf{W}_k$. Squaring and expanding Eq. (G.11) yields

$$\epsilon_k^2 = \left(d_k - \mathbf{X}_k^T\mathbf{W}_k\right)^2 \tag{G.22}$$

$$= d_k^2 - 2d_k\mathbf{X}_k^T\mathbf{W}_k + \mathbf{W}_k^T\mathbf{X}_k\mathbf{X}_k^T\mathbf{W}_k. \tag{G.23}$$

Now assume an ensemble of identical adaptive linear combiners, each having the same weight vector $\mathbf{W}_k$ at the kth iteration. Let each combiner have individual inputs $\mathbf{X}_k$ and d_k derived from stationary ergodic ensembles. Each combiner will produce an individual error ϵ_k represented by Eq. (G.23). Averaging Eq. (G.23) over the ensemble yields

$$E\left[\epsilon_k^2\right]_{\mathbf{W}=\mathbf{W}_k} = E\left[d_k^2\right] - 2E\left[d_k \mathbf{X}_k^T\right]\mathbf{W}_k + \mathbf{W}_k^T E\left[\mathbf{X}_k \mathbf{X}_k^T\right]\mathbf{W}_k. \tag{G.24}$$

Defining the vector $\mathbf{P}$ as the crosscorrelation between the desired response (a scalar) and the $\mathbf{X}$-vector[16] then yields

$$\mathbf{P}^T \triangleq E\left[d_k \mathbf{X}_k^T\right] = E\left[d_k, d_k x_{1k}, \ldots d_k x_{nk}\right]^T. \tag{G.25}$$

The input correlation matrix $\mathbf{R}$ is defined in terms of the ensemble average

$$\mathbf{R} \triangleq E\left[\mathbf{X}_k \mathbf{X}_k^T\right] = E\begin{bmatrix} 1 & x_{1k} & \cdots & x_{nk} \\ x_{1k} & x_{1k}x_{1k} & \cdots & x_{1k}x_{nk} \\ \vdots & \vdots & & \vdots \\ x_{nk} & x_{nk}x_{1k} & \cdots & x_{nk}x_{nk} \end{bmatrix}. \tag{G.26}$$

This matrix is real, symmetric, and positive definite, or in rare cases, positive semidefinite. The mean square error ξ_k can thus be expressed as

$$\xi_k \triangleq E\left[\epsilon_k^2\right]_{\mathbf{W}=\mathbf{W}_k} = E\left[d_k^2\right] - 2\mathbf{P}^T\mathbf{W}_k + \mathbf{W}_k^T \mathbf{R} \mathbf{W}_k. \tag{G.27}$$

Note that the mean square error is a quadratic function of the weights. It is a convex hyperparaboloidal surface, a function that never goes negative. Figure G.17 shows a typical mean square error surface for a linear combiner with two weights. The position of a point on the grid in this figure represents the value of the Adaline's two weights. The height of the surface at each point represents the mean square error over the training set when the Adaline's weights are fixed at the values associated with the grid point. Adjusting the weights involves descending along this surface toward the unique minimum point ("the bottom of the bowl") by the method of steepest descent.

The gradient ∇_k of the mean square error function with $\mathbf{W} = \mathbf{W}_k$ is obtained by differentiating Eq. (G.27):

$$\nabla_k \triangleq \left\{ \begin{matrix} \frac{\partial E\left[\epsilon_k^2\right]}{\partial w_{0k}} \\ \vdots \\ \frac{\partial E\left[\epsilon_k^2\right]}{\partial w_{nk}} \end{matrix} \right\}_{\mathbf{W}=\mathbf{W}_k} = -2\mathbf{P} + 2\mathbf{R}\mathbf{W}_k. \tag{G.28}$$

This is a linear function of the weights. The optimal weight vector $\mathbf{W}^*$, generally called the Wiener weight vector, is obtained from Eq. (G.28) by setting the gradient to zero:

$$\mathbf{W}^* = \mathbf{R}^{-1}\mathbf{P}. \tag{G.29}$$

[16]We assume here that $\mathbf{X}$ includes a bias component $x_{0k} = +1$.

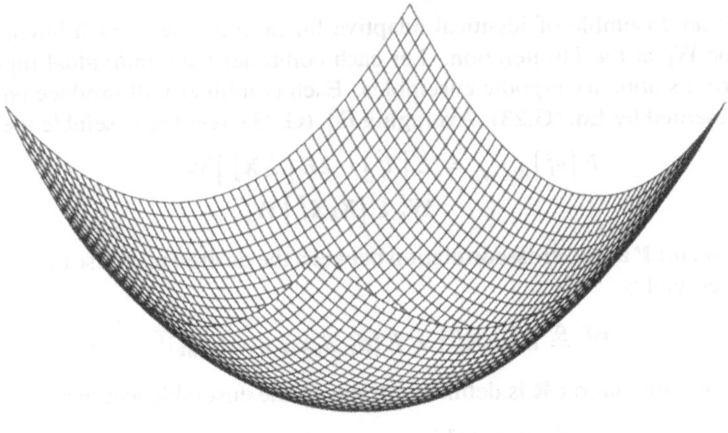

Figure G.17 Typical mean square error surface of a linear combiner.

This is a matrix form of the Wiener-Hopf equation [120, 121, 122]. In the next section we examine μ-LMS, an algorithm which enables us to obtain an accurate estimate of $\mathbf{W}^*$ without first computing $\mathbf{R}^{-1}$ and $\mathbf{P}$.

The μ-LMS Algorithm

The μ-LMS algorithm works by performing approximate steepest descent on the mean square error surface in weight space. Because it is a quadratic function of the weights, this surface is convex and has a unique (global) minimum.[17] An instantaneous gradient based upon the square of the instantaneous linear error is

$$\hat{\nabla}_k = \frac{\partial \epsilon_k^2}{\partial \mathbf{W}_k} = \left\{ \begin{array}{c} \frac{\partial \epsilon_k^2}{\partial w_{0k}} \\ \vdots \\ \frac{\epsilon_k^2}{\partial w_{nk}} \end{array} \right\}. \tag{G.30}$$

LMS works by using this crude gradient estimate in place of the true gradient ∇_k of Eq. (G.28). Making this replacement into Eq. (G.21) yields

$$\mathbf{W}_{k+1} = \mathbf{W}_k + \mu\left(-\hat{\nabla}_k\right) = \mathbf{W}_k - \mu\frac{\partial \epsilon_k^2}{\partial \mathbf{W}_k}. \tag{G.31}$$

The instantaneous gradient is used because it is readily available from a single data sample. The true gradient is generally difficult to obtain. Computing it would involve averaging the instantaneous gradients associated with all patterns in the training set. This is usually impractical and almost always inefficient.

[17]Unless the autocorrelation matrix of the pattern vector set has m zero eigenvalues, in which case the minimum MSE solution will be an m dimensional subspace in weight space [32].

Performing the differentiation in Eq. (G.31) and replacing the linear error by definition (G.11) gives

$$\begin{aligned} \mathbf{W}_{k+1} &= \mathbf{W}_k - 2\mu\epsilon_k \frac{\partial \epsilon_k}{\partial \mathbf{W}_k} \\ &= \mathbf{W}_k - 2\mu\epsilon_k \frac{\partial \left(d_k - \mathbf{W}_k^T \mathbf{X}_k\right)}{\partial \mathbf{W}_k}. \end{aligned} \tag{G.32}$$

Noting that d_k and $\mathbf{X}_k$ are independent of $\mathbf{W}_k$, yields

$$\mathbf{W}_{k+1} = \mathbf{W}_k + 2\mu\epsilon_k \mathbf{X}_k. \tag{G.33}$$

This is the μ-LMS algorithm. The learning constant μ determines stability and convergence rate. For input patterns independent over time, convergence of the mean and variance of the weight vector is ensured [32] for most practical purposes if

$$0 < \mu < \frac{1}{\operatorname{tr}[\mathbf{R}]}, \tag{G.34}$$

where $\operatorname{tr}[\mathbf{R}] = \sum(\text{diagonal elements of } \mathbf{R})$ is the average signal power of the $\mathbf{X}$-vectors, that is, $E(\mathbf{X}^T\mathbf{X})$. With μ set within this range,[18] the μ-LMS algorithm converges in the mean to $\mathbf{W}^*$, the optimal Wiener solution discussed above. A proof of this can be found in [32].

In the μ-LMS algorithm, and other iterative steepest descent procedures, use of the instantaneous gradient is perfectly justified if the step size is small. For small μ, $\mathbf{W}$ will remain essentially constant over a relatively small number of training presentations, K. The total weight change during this period will be proportional to

$$\begin{aligned} -\sum_{\ell=0}^{K-1} \frac{\partial \epsilon_{k+\ell}^2}{\partial \mathbf{W}_{k+\ell}} &\simeq -K\left(\frac{1}{K}\sum_{\ell=0}^{K-1} \frac{\partial \epsilon_{k+\ell}^2}{\partial \mathbf{W}_k}\right) \\ &= -K\frac{\partial}{\partial \mathbf{W}_k}\left(\frac{1}{K}\sum_{\ell=0}^{K-1} \epsilon_{k+\ell}^2\right) \\ &\simeq -K\frac{\partial \xi}{\partial \mathbf{W}_k}, \end{aligned} \tag{G.35}$$

where ξ denotes the mean square error function. Thus, on average the weights follow the true gradient. It is shown in [32] that the instantaneous gradient is an unbiased estimate of the true gradient.

Comparison of μ-LMS and α-LMS

We have now presented two forms of the LMS algorithm, μ-LMS (G.33) above, and α-LMS (G.10) in Section G.4.1. They are very similar algorithms, both using the LMS instantaneous gradient. α-LMS is self-normalizing, with the parameter α determining the fraction of the instantaneous error to be corrected with each adaptation. μ-LMS is a constant-coefficient linear algorithm which is considerably easier to analyze than α-LMS. Comparing

[18] Horowitz and Senne [123] have proven that Eq. (G.34) is not sufficient in general to guarantee convergence of the weight vector's variance. For input patterns generated by a zero-mean Gaussian process independent over time, instability can occur in the worst case if μ is greater than $1/(3\operatorname{tr}[\mathbf{R}])$.

the two, the α-LMS algorithm is like the μ-LMS algorithm with a continually variable learning constant. Although α-LMS is somewhat more difficult to implement and analyze, it has been demonstrated experimentally to be a better algorithm than μ-LMS when the eigenvalues of the input autocorrelation matrix, $\mathbf{R}$, are highly disparate, giving faster convergence for a given level of gradient noise[19] propagated into the weights. It will be shown next that μ-LMS has the advantage that it will always converge in the mean to the minimum mean square error solution, while α-LMS may converge to a somewhat biased solution.

We begin with α-LMS of Eq. (G.10):

$$\mathbf{W}_{k+1} = \mathbf{W}_k + \alpha \frac{\epsilon_k \mathbf{X}_k}{|\mathbf{X}_k|^2}. \tag{G.36}$$

Replacing the error with its definition (G.11) and rearranging terms yields

$$\mathbf{W}_{k+1} = \mathbf{W}_k + \alpha \frac{(d_k - \mathbf{W}_k^T \mathbf{X}_k)\mathbf{X}_k}{|\mathbf{X}_k|^2} \tag{G.37}$$

$$= \mathbf{W}_k + \alpha \left(\frac{d_k}{|\mathbf{X}_k|} - \mathbf{W}_k^T \frac{\mathbf{X}_k}{|\mathbf{X}_k|} \right) \frac{\mathbf{X}_k}{|\mathbf{X}_k|}. \tag{G.38}$$

We define a new training set of pattern vectors and desired responses, $\{\tilde{\tilde{\mathbf{X}}}_k, \tilde{\tilde{d}}_k\}$, by normalizing elements of the original training set as follows:[20]

$$\tilde{\tilde{\mathbf{X}}}_k \triangleq \frac{\mathbf{X}_k}{|\mathbf{X}_k|} \qquad \tilde{\tilde{d}}_k \triangleq \frac{d_k}{|\mathbf{X}_k|}. \tag{G.39}$$

Equation (G.38) then becomes

$$\mathbf{W}_{k+1} = \mathbf{W}_k + \alpha \left(\tilde{\tilde{d}}_k - \mathbf{W}_k^T \tilde{\tilde{\mathbf{X}}}_k \right) \tilde{\tilde{\mathbf{X}}}_k . \tag{G.40}$$

This is the μ-LMS rule of Eq. (G.33) with 2μ replaced by α. The weight adaptations chosen by the α-LMS rule are equivalent to those of the μ-LMS algorithm presented with a different training set—the normalized training set defined by (G.39). The solution that will be reached by the μ-LMS algorithm is the Wiener solution of this training set,

$$\tilde{\tilde{\mathbf{W}}}^* = (\tilde{\tilde{\mathbf{R}}})^{-1} \tilde{\tilde{\mathbf{P}}}, \tag{G.41}$$

where

$$\tilde{\tilde{\mathbf{R}}} = E[\tilde{\tilde{\mathbf{X}}}_k \tilde{\tilde{\mathbf{X}}}_k^T] \tag{G.42}$$

is the input correlation matrix of the normalized training set and the vector

$$\tilde{\tilde{\mathbf{P}}} = E[\tilde{\tilde{d}}_k \tilde{\tilde{\mathbf{X}}}_k] \tag{G.43}$$

[19] Gradient noise is the difference between the gradient estimate and the true gradient.

[20] The idea of a normalized training set was suggested by Derrick Nguyen.

is the crosscorrelation between the normalized input and the normalized desired response. Therefore α-LMS converges in the mean to the Wiener solution of the normalized training set. When the input vectors are binary with ± 1 components, all input vectors have the same magnitude and the two algorithms are equivalent. For nonbinary training patterns, however, the Wiener solution of the normalized training set generally is no longer equal to that of the original problem, so α-LMS converges in the mean to a somewhat biased version of the optimal least squares solution.

The idea of a normalized training set can also be used to relate the stable ranges for the learning constants α and μ in the two algorithms. The stable range for α in the α-LMS algorithm given in Eq. (G.15) can be computed from the corresponding range for μ given in Eq. (G.34) by replacing $\mathbf{R}$ and μ in Eq. (G.34) by $\tilde{\tilde{\mathbf{R}}}$ and $\alpha/2$, respectively, and then noting that $\mathrm{tr}[\tilde{\tilde{\mathbf{R}}}]$ is equal to one:

$$0 < \alpha < \frac{2}{\mathrm{tr}[\tilde{\tilde{\mathbf{R}}}]}, \quad \text{or}$$
$$0 < \alpha < 2. \tag{G.44}$$

G.6.2 Nonlinear Rules

The Adaline elements considered thus far use at their outputs either hard-limiting quantizers (signums), or no nonlinearity at all. The input-output mapping of the hard-limiting quantizer is $y_k = \mathrm{sgn}(s_k)$. Other forms of nonlinearity have come into use in the past two decades, primarily of the sigmoid type. These nonlinearities provide saturation for decision making, yet they have differentiable input-output characteristics that facilitate adaptivity. We generalize the definition of the Adaline element to include the possible use of a sigmoid in place of the signum, and then determine suitable adaptation algorithms.

Figure G.18 shows a "sigmoid Adaline" element which incorporates a sigmoidal nonlinearity. The input-output relation of the sigmoid can be denoted by $y_k = \mathrm{sgm}(s_k)$. A typical sigmoid function is the hyperbolic tangent:

$$y_k = \tanh(s_k) = \left(\frac{1 - e^{-2s_k}}{1 + e^{-2s_k}}\right). \tag{G.45}$$

We shall adapt this Adaline with the objective of minimizing the mean square of the sigmoid error $\tilde{\epsilon}_k$, defined as

$$\tilde{\epsilon}_k \triangleq d_k - y_k = d_k - \mathrm{sgm}(s_k). \tag{G.46}$$

Backpropagation for the Sigmoid Adaline

Our objective is to minimize $E[(\tilde{\epsilon}_k)^2]$, averaged over the set of training patterns by proper choice of the weight vector. To accomplish this, we shall derive a backpropagation algorithm for the sigmoid Adaline element. An instantaneous gradient is obtained with each input vector presentation, and the method of steepest descent is used to minimize error as was done with the μ-LMS algorithm of Eq. (G.33).

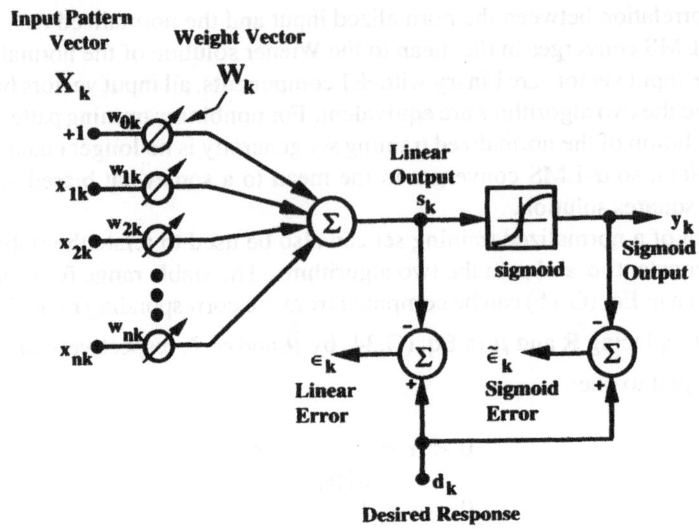

Figure G.18 Adaline with sigmoidal nonlinearity.

Referring to Fig. G.18, the instantaneous gradient estimate obtained during presentation of the kth input vector $\mathbf{X}_k$ is given by

$$\hat{\nabla}_k = \frac{\partial(\tilde{\epsilon}_k)^2}{\partial \mathbf{W}_k} = 2\tilde{\epsilon}_k \frac{\partial \tilde{\epsilon}_k}{\partial \mathbf{W}_k}. \tag{G.47}$$

Differentiating Eq. (G.46) yields

$$\frac{\partial \tilde{\epsilon}_k}{\partial \mathbf{W}_k} = -\frac{\partial \mathrm{sgm}(s_k)}{\partial \mathbf{W}_k} = -\mathrm{sgm}'(s_k)\frac{\partial s_k}{\partial \mathbf{W}_k}. \tag{G.48}$$

We may note that

$$s_k = \mathbf{X}_k^T \mathbf{W}_k. \tag{G.49}$$

Therefore,

$$\frac{\partial s_k}{\partial \mathbf{W}_k} = \mathbf{X}_k. \tag{G.50}$$

Substituting into Eq. (G.48) gives

$$\frac{\partial \tilde{\epsilon}_k}{\partial \mathbf{W}_k} = -\mathrm{sgm}'(s_k)\mathbf{X}_k. \tag{G.51}$$

Inserting this into Eq. (G.47) yields

$$\hat{\nabla}_k = -2\tilde{\epsilon}_k \mathrm{sgm}'(s_k)\mathbf{X}_k. \tag{G.52}$$

Using this gradient estimate with the method of steepest descent provides a means for minimizing the mean square error even after the summed signal s_k goes through the nonlinear sigmoid. The algorithm is

$$\mathbf{W}_{k+1} = \mathbf{W}_k + \mu(-\hat{\nabla}_k) \tag{G.53}$$

$$= \mathbf{W}_k + 2\mu\tilde{\epsilon}_k \text{sgm}'(s_k)\mathbf{X}_k. \tag{G.54}$$

Equation (G.54) is the backpropagation algorithm for the sigmoid Adaline element. The backpropagation name makes more sense when the algorithm is utilized in a layered network, which will be studied below. Implementation of algorithm (G.54) is illustrated in Fig. G.19.

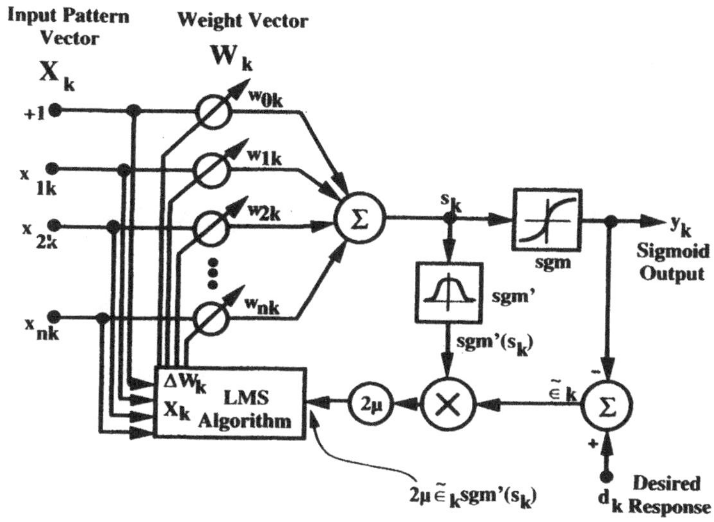

Figure G.19 Implementation of backpropagation for the sigmoid Adaline element.

If the sigmoid is chosen to be the hyperbolic tangent function (G.45), then the derivative $\text{sgm}'(s_k)$ is given by

$$\text{sgm}'(s_k) = \frac{\partial\,(\tanh(s_k))}{\partial s_k} = 1 - (\tanh(s_k))^2 = 1 - y_k^2. \tag{G.55}$$

Accordingly Eq. (G.54) becomes

$$\mathbf{W}_{k+1} = \mathbf{W}_k + 2\mu\tilde{\epsilon}_k(1 - y_k^2)\mathbf{X}_k. \tag{G.56}$$

Madaline Rule III for the Sigmoid Adaline

The implementation of algorithm (G.54), illustrated in Fig. G.19, requires accurate realization of the sigmoid function and its derivative function. These functions may not be realized accurately when implemented with analog hardware. Indeed, in an analog network, each Adaline will have its own individual nonlinearities. Difficulties in adaptation have been encountered in practice with the backpropagation algorithm because of imperfections in the nonlinear functions.

To circumvent these problems, a new algorithm has been devised by David Andes for adapting networks of sigmoid Adalines. This is the Madaline Rule III (MRIII) algorithm.

The idea of MRIII for a sigmoid Adaline is illustrated in Fig. G.20. The derivative of the sigmoid function is not used here. Instead, a small perturbation signal Δs is added to the sum s_k, and the effect of this perturbation upon output y_k and error $\tilde{\epsilon}_k$ is noted.

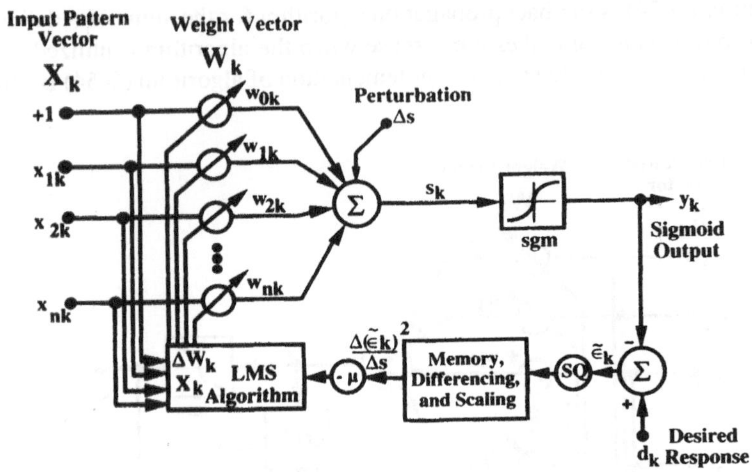

Figure G.20 Implementation of the MRIII algorithm for the sigmoid Adaline element.

An instantaneous estimated gradient can be obtained as follows:

$$\hat{\nabla}_k = \frac{\partial(\tilde{\epsilon}_k)^2}{\partial \mathbf{W}_k} = \frac{\partial(\tilde{\epsilon}_k)^2}{\partial s_k}\frac{\partial s_k}{\partial \mathbf{W}_k} = \frac{\partial(\tilde{\epsilon}_k)^2}{\partial s_k}\mathbf{X}_k. \tag{G.57}$$

Since Δs is small,

$$\hat{\nabla}_k \simeq \left(\frac{\Delta(\tilde{\epsilon}_k)^2}{\Delta s}\right)\mathbf{X}_k. \tag{G.58}$$

Another way to obtain an approximate instantaneous gradient by measuring the effects of the perturbation Δs can be obtained from Eq. (G.57).

$$\hat{\nabla}_k = \frac{\partial(\tilde{\epsilon}_k)^2}{\partial s_k}\mathbf{X}_k = 2\tilde{\epsilon}_k\frac{\partial\tilde{\epsilon}_k}{\partial s_k}\mathbf{X}_k \simeq 2\tilde{\epsilon}_k\left(\frac{\Delta\tilde{\epsilon}_k}{\Delta s}\right)\mathbf{X}_k. \tag{G.59}$$

Accordingly, there are two forms of the MRIII algorithm for the sigmoid Adaline. They are based on the method of steepest descent, using the estimated instantaneous gradients:

$$\mathbf{W}_{k+1} = \mathbf{W}_k - \mu\left(\frac{\Delta(\tilde{\epsilon}_k)^2}{\Delta s}\right)\mathbf{X}_k \tag{G.60}$$

or,

$$\mathbf{W}_{k+1} = \mathbf{W}_k - 2\mu\tilde{\epsilon}_k \left(\frac{\Delta\tilde{\epsilon}_k}{\Delta s}\right)\mathbf{X}_k. \tag{G.61}$$

For small perturbations, these two forms are essentially identical. Neither one requires a priori knowledge of the sigmoid's derivative, and both are robust with respect to natural variations, biases, and drift in the analog hardware. Which form to use is a matter of implementational convenience. The algorithm of Eq. (G.60) is illustrated in Fig. G.20.

Regarding algorithm (G.61), some changes can be made to establish a point of interest. Note that, in accord with Eq. (G.46),

$$\tilde{\epsilon}_k = d_k - y_k. \tag{G.62}$$

Adding the perturbation Δs causes a change in ϵ_k equal to

$$\Delta\tilde{\epsilon}_k = -\Delta y_k. \tag{G.63}$$

Now, Eq. (G.61) may be rewritten as

$$\mathbf{W}_{k+1} = \mathbf{W}_k + 2\mu\tilde{\epsilon}_k \left(\frac{\Delta y_k}{\Delta s}\right)\mathbf{X}_k. \tag{G.64}$$

Since Δs is small, the ratio of increments may be replaced by a ratio of differentials finally giving

$$\mathbf{W}_{k+1} \simeq \mathbf{W}_k + 2\mu\tilde{\epsilon}_k \frac{\partial y_k}{\partial s_k}\mathbf{X}_k \tag{G.65}$$

$$= \mathbf{W}_k + 2\mu\tilde{\epsilon}_k \operatorname{sgm}'(s_k)\mathbf{X}_k. \tag{G.66}$$

This is identical to the backpropagation algorithm (G.54) for the sigmoid Adaline. Thus, backpropagation and MRIII are mathematically equivalent if the perturbation Δs is small, but MRIII is robust, even with analog implementations.

MSE Surfaces of the Adaline

Figure G.21 shows a linear combiner connected to both sigmoid and signum devices. Three errors, ϵ, $\tilde{\epsilon}_k$, and $\tilde{\tilde{\epsilon}}$ are designated in this figure. They are

$$\begin{aligned} \text{linear error} &= \epsilon = d - s \\ \text{sigmoid error} &= \tilde{\epsilon} = d - \operatorname{sgm}(s) \\ \text{signum error} &= \tilde{\tilde{\epsilon}} = d - \operatorname{sgn}(\operatorname{sgm}(s)) = d - \operatorname{sgn}(s). \end{aligned} \tag{G.67}$$

To demonstrate the nature of the square error surfaces associated with these three types of error, a simple experiment with a two-input Adaline was performed. The Adaline was driven by a typical set of input patterns and their associated binary $\{+1, -1\}$ desired responses. The sigmoid function used was the hyperbolic tangent. The weights could have been adapted to minimize the mean square error of ϵ,$\tilde{\epsilon}$, or $\tilde{\tilde{\epsilon}}$. The mean square error surfaces of $E[(\epsilon)^2]$, $E[(\tilde{\epsilon})^2]$, $E[(\tilde{\tilde{\epsilon}})^2]$ plotted as functions of the two weight values, are shown in Figs. G.22, G.23, and G.24, respectively.

Although the above experiment is not all encompassing, we can infer from it that minimizing the mean square of the linear error is easy, minimizing the mean square of the sigmoid error is more difficult, but typically much easier than minimizing the mean square of the signum error. Only the linear error is guaranteed to have an MSE surface with a unique global minimum (assuming invertible **R**-matrix). The other MSE surfaces can have local optima [124, 125].

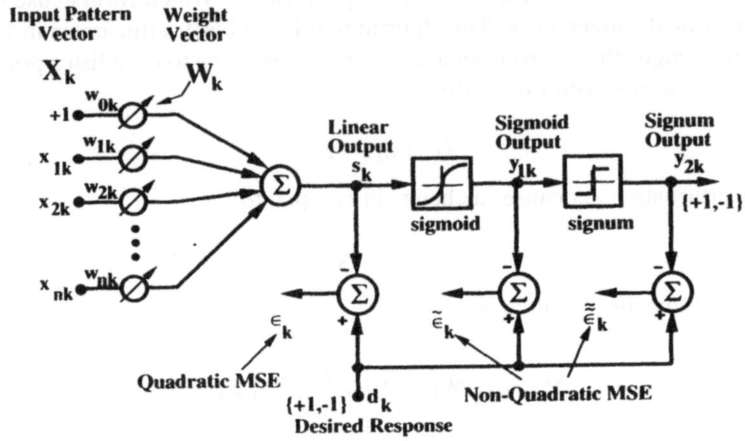

Figure G.21 The linear, sigmoid, and signum errors of the Adaline.

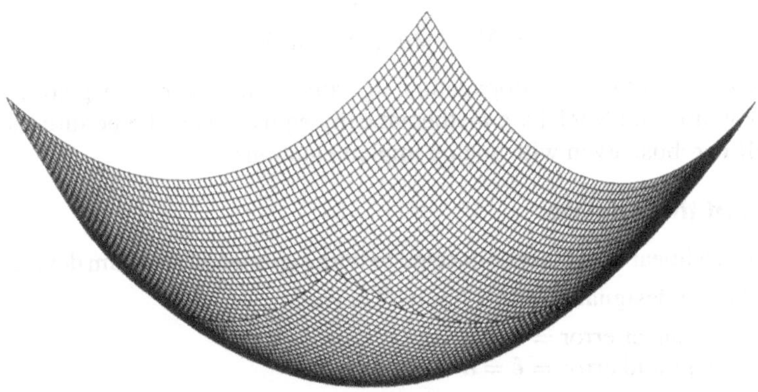

Figure G.22 Example MSE surface of linear error.

In nonlinear neural networks, gradient methods generally work better with sigmoid rather than signum nonlinearities. Smooth nonlinearities are required by the MRIII and backpropagation techniques. Moreover, sigmoid networks are capable of forming internal representations which are more complex than simple binary codes, and thus, these networks can often form decision regions which are more sophisticated than those associated with similar signum networks. In fact, if a noiseless infinite-precision sigmoid Adaline could be

Figure G.23 Example MSE surface of sigmoid error.

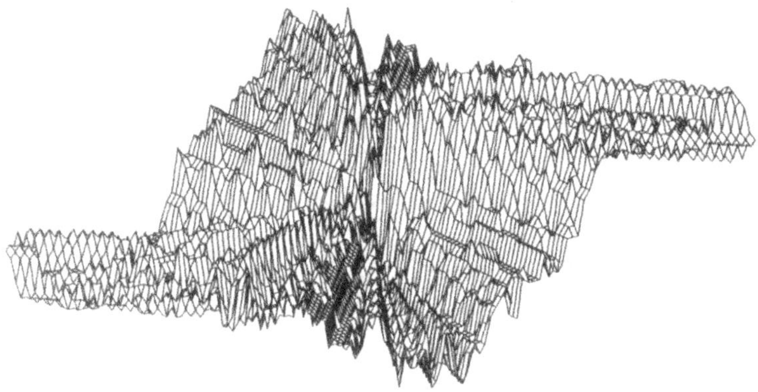

Figure G.24 Example MSE surface of signum error.

constructed, it would be able to convey an infinite amount of information at each time step. This is in contrast to the maximum Shannon information capacity of one bit associated with each binary element.

The signum does have some advantages over the sigmoid in that it is easier to implement in hardware and much simpler to compute on a digital computer. Furthermore, the outputs of signums are binary signals which can be efficiently manipulated by digital computers. In a signum network with binary inputs, for instance, the output of each linear combiner can be computed without performing weight multiplications. This involves simply adding together the values of weights with $+1$ inputs and subtracting from this the values of all weights that are connected to -1 inputs.

Sometimes a signum is used in an Adaline to produce decisive output decisions. The error probability is then proportional to the mean square of the output error $\tilde{\tilde{\epsilon}}$. To minimize this error probability approximately, one can easily minimize $E[(\epsilon)^2]$ instead of directly minimizing $E[(\tilde{\tilde{\epsilon}})^2]$ [60]. However, with only a little more computation, one could

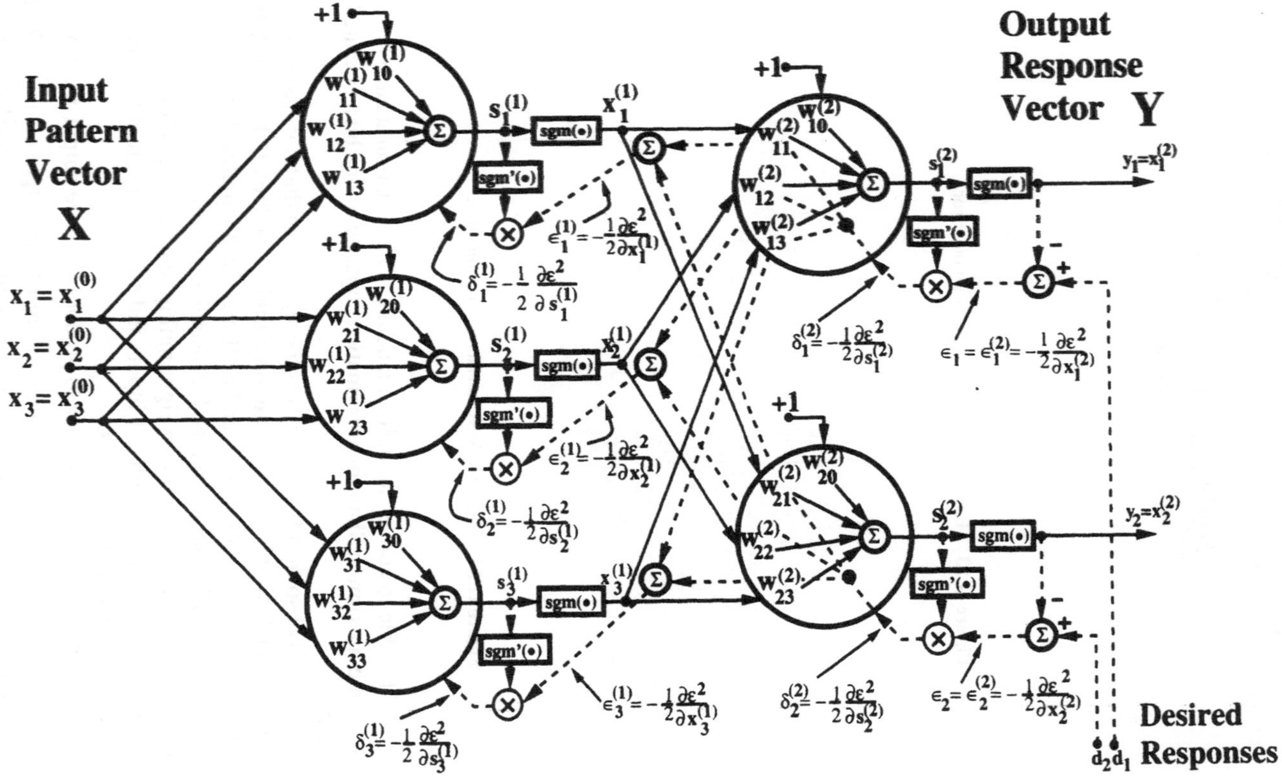

Figure G.25 Example two-layer backpropagation network architecture.

minimize $E[(\tilde{\epsilon})^2]$ and typically come much closer to the objective of minimizing $E[(\tilde{\tilde{\epsilon}})^2]$. The sigmoid can therefore be used in training the weights even when the signum is used to form the Adaline output, as in Fig. G.21.

7 STEEPEST-DESCENT RULES — MULTI-ELEMENT NETWORKS

We now study rules for steepest descent minimization of the MSE associated with entire networks of sigmoid Adaline elements. Like their single-element counterparts, the most practical and efficient steepest descent rules for multi-element networks typically work with one pattern presentation at a time. We will describe two steepest descent rules for multi-element sigmoid networks: backpropagation and Madaline Rule III.

G.7.1 Backpropagation for Networks

The publication of the backpropagation technique by Rumelhart and colleagues [44] has unquestionably been the most influential development in the field of neural networks during the past decade. In retrospect, the technique seems simple. Nonetheless, largely because early neural network research dealt almost exclusively with hard-limiting nonlinearities, the idea never occurred to neural network researchers throughout the 1960s.

The basic concepts of backpropagation are easily grasped. Unfortunately, these simple ideas are often obscured by relatively intricate notation, so formal derivations of the backpropagation rule are often tedious. We present an informal derivation of the algorithm and illustrate how it works for the simple network shown in Fig. G.25. A more formal derivation would require the use of ordered derivatives to make precise which weights are treated as variables in each of the partial derivatives used below [126].

The backpropagation technique is a nontrivial generalization of the single sigmoid Adaline case of Section G.6.2. When applied to multi-element networks, the backpropagation technique adjusts the weights in the direction opposite the instantaneous error gradient:

$$\hat{\nabla}_k = \frac{\partial \varepsilon_k^2}{\partial \mathbf{W}_k} = \begin{Bmatrix} \frac{\partial \varepsilon_k^2}{\partial w_{1k}} \\ \vdots \\ \frac{\partial \varepsilon_k^2}{\partial w_{nk}} \end{Bmatrix}. \tag{G.68}$$

Now, however, $\mathbf{W}_k$ is a long n-component vector of all weights in the entire network. The instantaneous sum squared error ε_k^2 is the sum of the squares of the errors at each of the N_y outputs of the network. Thus,

$$\varepsilon_k^2 = \sum_{i=1}^{N_y} \epsilon_{ik}^2. \tag{G.69}$$

In the network example shown in Fig. G.25, the sum square error is given by

$$\varepsilon^2 = (d_1 - y_1)^2 + (d_2 - y_2)^2, \tag{G.70}$$

where we now suppress the time index k for convenience.

In its simplest form, backpropagation training begins by presenting an input pattern vector $\mathbf{X}$ to the network, sweeping forward through the system to generate an output response vector $\mathbf{Y}$, and computing the errors at each output. The next step involves sweeping the effects of the errors backward through the network to associate a "square error derivative" δ with each Adaline, computing a gradient from each δ, and finally updating the weights of each Adaline based upon the corresponding gradient. A new pattern is then presented and the process is repeated. The initial weight values are normally set to small random numbers. The algorithm will not work properly with multilayer networks if the initial weights are either zero or poorly chosen nonzero values.[21].

We can get some idea about what is involved in the calculations associated with the backpropagation algorithm by examining the network of Fig. G.25. Each of the five large circles represents a linear combiner, as well as some associated signal paths for error backpropagation, and the corresponding adaptive machinery for updating the weights. This detail is shown in Fig. G.26. The solid lines in these diagrams represent forward signal paths through the network, and the dotted lines represent the separate backward paths that are used in association with calculations of the square error derivatives δ. From Fig. G.25, we see that the calculations associated with the backward sweep are of a complexity which is roughly equal to that represented by the forward pass through the network. The backward sweep requires the same number of function calculations as the forward sweep, but no weight multiplications in the first layer.

As stated above, after a pattern has been presented to the network, and the response error of each output has been calculated, the next step of the backpropagation algorithm involves finding the instantaneous square error derivative δ associated with each summing junction in the network. The square error derivative associated with the jth Adaline in layer ℓ is defined as[22]

$$\delta_j^{(\ell)} \triangleq -\frac{1}{2}\frac{\partial \varepsilon^2}{\partial s_j^{(\ell)}}. \tag{G.71}$$

Each of these derivatives in essence tells us how sensitive the sum square output error of the network is to changes in the linear output of the associated Adaline element.

The instantaneous square error derivatives are first computed for each element in the output layer. The calculation is simple. As an example, below we derive the required ex-

[21] Recently, Nguyen has discovered that a more sophisticated choice of initial weight values in hidden layers can lead to reduced problems with local optima and dramatic increases in network training speed [102]. Experimental evidence suggests that it is advisable to choose the initial weights of each hidden layer in a quasi-random manner which ensures that at each position in a layer's input space the outputs of all but a few of its Adalines will be saturated, while ensuring that each Adaline in the layer is unsaturated in some region of its input space. When this method is used, the weights in the output layer are set to small random values.

[22] In Fig. G.25, all notation follows the convention that superscripts within parentheses indicate the layer number of the associated Adaline or input node, while subscripts identify the associated Adaline(s) within a layer.

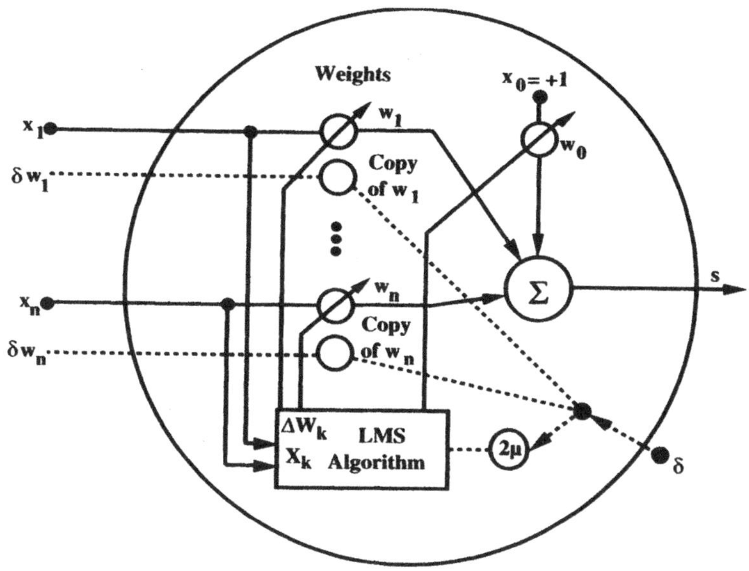

Figure G.26 Detail of linear combiner and associated circuitry in backpropagation network.

pression for $\delta_1^{(2)}$, the derivative associated with the top Adaline element in the output layer of Fig. G.25. We begin with the definition of $\delta_1^{(2)}$ from Eq. (G.71),

$$\delta_1^{(2)} \triangleq -\frac{1}{2}\frac{\partial \varepsilon^2}{\partial s_1^{(2)}}. \tag{G.72}$$

Expanding the squared error term ε^2 by Eq. (G.70) yields

$$\delta_1^{(2)} = -\frac{1}{2}\frac{\partial\left((d_1 - y_1)^2 + (d_2 - y_2)^2\right)}{\partial s_1^{(2)}} \tag{G.73}$$

$$= -\frac{1}{2}\frac{\partial\left(d_1 - \text{sgm}(s_1^{(2)})\right)^2}{\partial s_1^{(2)}}$$

$$-\frac{1}{2}\frac{\partial\left(d_2 - \text{sgm}(s_2^{(2)})\right)^2}{\partial s_1^{(2)}}. \tag{G.74}$$

We note that the second term is zero. Accordingly,

$$\delta_1^{(2)} = -\frac{1}{2}\frac{\partial\left(d_1 - \text{sgm}(s_1^{(2)})\right)^2}{\partial s_1^{(2)}}. \tag{G.75}$$

Observing that d_1 and $s_1^{(2)}$ are independent yields

$$\delta_1^{(2)} = -\left(d_1 - \text{sgm}(s_1^{(2)})\right)\frac{\partial\left(-\text{sgm}(s_1^{(2)})\right)}{\partial s_1^{(2)}} \tag{G.76}$$

$$= \left(d_1 - \text{sgm}(s_1^{(2)})\right)\text{sgm}'(s_1^{(2)}). \tag{G.77}$$

We denote the error $d_1 - \text{sgm}(s_1^{(2)})$, by $\epsilon_1^{(2)}$. Therefore,

$$\delta_1^{(2)} = \epsilon_1^{(2)}\text{sgm}'(s_1^{(2)}). \tag{G.78}$$

Note that this corresponds to the computation of $\delta_1^{(2)}$ as illustrated in Fig. G.25. The value of δ associated with the other output element in the figure can be expressed in an analogous fashion. Thus each square error derivative δ in the output layer is computed by multiplying the output error associated with that element by the derivative of the associated sigmoidal nonlinearity. Note from Eq. (G.55) that if the sigmoid function is the hyperbolic tangent, Eq. (G.78) becomes simply

$$\delta_1^{(2)} = \epsilon_1^{(2)}(1 - (y_1)^2). \tag{G.79}$$

Developing expressions for the square error derivatives associated with hidden layers is not much more difficult (refer to Fig. G.25). We need an expression for $\delta_1^{(1)}$, the square error derivative associated with the top element in the first layer of Fig. G.25. The derivative $\delta_1^{(1)}$ is defined by

$$\delta_1^{(1)} \triangleq -\frac{1}{2}\frac{\partial \varepsilon^2}{\partial s_1^{(1)}}. \tag{G.80}$$

Expanding this by the chain rule, noting that ε^2 is determined entirely by the values of $s_1^{(2)}$ and $s_2^{(2)}$, yields

$$\delta_1^{(1)} = -\frac{1}{2}\left(\frac{\partial \varepsilon^2}{\partial s_1^{(2)}}\frac{\partial s_1^{(2)}}{\partial s_1^{(1)}} + \frac{\partial \varepsilon^2}{\partial s_2^{(2)}}\frac{\partial s_2^{(2)}}{\partial s_1^{(1)}}\right). \tag{G.81}$$

Using the definitions of $\delta_1^{(2)}$ and $\delta_2^{(2)}$, and then substituting expanded versions of Adaline linear outputs $s_1^{(2)}$ and $s_2^{(2)}$ gives

$$\delta_1^{(1)} = \delta_1^{(2)}\frac{\partial s_1^{(2)}}{\partial s_1^{(1)}} + \delta_2^{(2)}\frac{\partial s_2^{(2)}}{\partial s_1^{(1)}} \tag{G.82}$$

$$= \delta_1^{(2)}\frac{\partial}{\partial s_1^{(1)}}\left(w_{10}^{(2)} + \sum_{i=1}^{3} w_{1i}^{(2)}\text{sgm}(s_i^{(1)})\right)$$

$$+\delta_2^{(2)}\frac{\partial}{\partial s_1^{(1)}}\left(w_{20}^{(2)} + \sum_{i=1}^{3} w_{2i}^{(2)}\text{sgm}(s_i^{(1)})\right). \tag{G.83}$$

Noting that $\partial[\text{sgm}(s_i^{(\ell)})]/\partial s_j^{(\ell)} = 0, i \neq j$, leaves

$$\delta_1^{(1)} = \delta_1^{(2)} w_{11}^{(2)} \text{sgm}'(s_1^{(1)}) + \delta_2^{(2)} w_{21}^{(2)} \text{sgm}'(s_1^{(1)}) \tag{G.84}$$

$$= \left[\delta_1^{(2)} w_{11}^{(2)} + \delta_2^{(2)} w_{21}^{(2)}\right] \text{sgm}'(s_1^{(1)}). \tag{G.85}$$

Now, we make the following definition:

$$\epsilon_1^{(1)} \triangleq \delta_1^{(2)} w_{11}^{(2)} + \delta_2^{(2)} w_{21}^{(2)}. \tag{G.86}$$

Accordingly,

$$\delta_1^{(1)} = \epsilon_1^{(1)} \text{sgm}'(s_1^{(1)}). \tag{G.87}$$

Referring to Fig. G.25, we can trace through the circuit to verify that $\delta_1^{(1)}$ is computed in accord with Eqs. (G.86) and (G.87). The easiest way to find values of δ for all the Adaline elements in the network is to follow the schematic diagram of Fig. G.25.

Thus, the procedure for finding $\delta^{(\ell)}$, the square error derivative associated with a given Adaline in hidden layer ℓ, involves respectively multiplying each derivative $\delta^{(\ell+1)}$ associated with each element in the layer immediately downstream from a given Adaline by the weight which connects it to the given Adaline. These weighted square error derivatives are then added together, producing an error term $\epsilon^{(\ell)}$, which, in turn, is multiplied by $\text{sgm}'(s^{(\ell)})$, the derivative of the given Adaline's sigmoid function at its current operating point. If a network has more than two layers, this process of backpropagating the instantaneous square error derivatives from one layer to the immediately preceding layer is successively repeated until a square error derivative δ is computed for each Adaline in the network. This is easily shown at each layer by repeating the chain rule argument associated with Eq. (G.81).

We now have a general method for finding a derivative δ for each Adaline element in the network. The next step is to use these δ's to obtain the corresponding gradients. Consider an Adaline somewhere in the network which, during presentation k, has a weight vector $\mathbf{W}_k$, an input vector $\mathbf{X}_k$, and a linear output $s_k = \mathbf{W}_k^T \mathbf{X}_k$.

The instantaneous gradient for this Adaline element is

$$\hat{\nabla}_k = \frac{\partial \varepsilon_k^2}{\partial \mathbf{W}_k}. \tag{G.88}$$

This can be written as

$$\hat{\nabla}_k = \frac{\partial \varepsilon_k^2}{\partial \mathbf{W}_k} = \frac{\partial \varepsilon_k^2}{\partial s_k} \frac{\partial s_k}{\partial \mathbf{W}_k}. \tag{G.89}$$

Note that $\mathbf{W}_k$ and $\mathbf{X}_k$ are independent so

$$\frac{\partial s_k}{\partial \mathbf{W}_k} = \frac{\partial \mathbf{W}_k^T \mathbf{X}_k}{\partial \mathbf{W}_k} = \mathbf{X}_k. \tag{G.90}$$

Therefore,

$$\hat{\nabla}_k = \frac{\partial \varepsilon_k^2}{\partial s_k} \mathbf{X}_k. \tag{G.91}$$

For this element,

$$\delta_k = -\frac{1}{2} \frac{\partial \varepsilon_k^2}{\partial s_k}. \tag{G.92}$$

Accordingly,

$$\hat{\nabla}_k = -2\delta_k \mathbf{X}_k. \tag{G.93}$$

Updating the weights of the Adaline element using the method of steepest descent with the instantaneous gradient is a process represented by

$$\mathbf{W}_{k+1} = \mathbf{W}_k + \mu(-\hat{\nabla}_k) = \mathbf{W}_k + 2\mu\delta_k \mathbf{X}_k. \tag{G.94}$$

Thus, after backpropagating all square error derivatives, we complete a backpropagation iteration by adding to each weight vector the corresponding input vector scaled by the associated square error derivative. Equation (G.94) and the means for finding δ_k comprise the general weight update rule of the backpropagation algorithm.

There is a great similarity between Eq. (G.94) and the μ-LMS algorithm (G.33), but one should view this similarity with caution. The quantity δ_k, defined as a squared error derivative, might appear to play the same role in backpropagation as that played by the error in the μ-LMS algorithm. However, δ_k is not an error. Adaptation of the given Adaline is effected to reduce the squared output error ε_k^2, not δ_k of the given Adaline or of any other Adaline in the network. The objective is not to reduce the δ_k's of the network, but to reduce ε_k^2 at the network output.

It is interesting to examine the weight updates that backpropagation imposes on the Adaline elements in the output layer. Substituting Eq. (G.77) into Eq. (G.94) reveals that the Adaline which provides output y_1 in Fig. G.25 is updated by the rule

$$\mathbf{W}_{k+1} = \mathbf{W}_k + 2\mu\epsilon_1^{(2)} \text{sgm}'(s_1^{(2)})\mathbf{X}_k. \tag{G.95}$$

This rule turns out to be identical to the single Adaline version (G.54) of the backpropagation rule. This is not surprising since the output Adaline is provided with both input signals and desired responses, so its training circumstance is the same as that experienced by an Adaline trained in isolation.

There are many variants of the backpropagation algorithm. Sometimes, the size of μ is reduced during training to diminish the effects of gradient noise in the weights. Another extension is the momentum technique [44] which involves including in the weight change vector $\Delta\mathbf{W}_k$ of each Adaline a term proportional to the corresponding weight change from the previous iteration. That is, Eq. (G.94) is replaced by a pair of equations

$$\Delta\mathbf{W}_k = 2\mu(1-\eta)\delta_k\mathbf{X}_k + \eta\Delta\mathbf{W}_{k-1} \tag{G.96}$$

$$\mathbf{W}_{k+1} = \mathbf{W}_k + \Delta\mathbf{W}_k, \tag{G.97}$$

where the momentum constant $0 \leq \eta < 1$ is in practice usually set to something around 0.8 or 0.9.

The momentum technique low-pass filters the weight updates and thereby tends to resist erratic weight changes due either to gradient noise or to high spatial frequencies in the mean square error surface. The factor $(1-\eta)$ in Eq. (G.96) is included to give the filter a DC gain of unity so that the learning rate μ does not need to be stepped down as the momentum constant η is increased. A momentum term can also be added to the update equations of other algorithms discussed in this appendix. A detailed analysis of stability issues associated with momentum updating for the μ-LMS algorithm, for instance, has been described by Shynk and Roy [127].

In our experience, the momentum technique used alone is usually of little value. We have found, however, that it is often useful to apply the technique in situations that require relatively "clean"[23] gradient estimates. One case is a normalized weight update equation which makes the network's weight vector move the same Euclidean distance with each iteration. This can be accomplished by replacing Eqs. (G.96) and (G.97) with

$$\Delta_k = \delta_k \mathbf{X}_k + \eta \Delta_{k-1} \tag{G.98}$$

$$\mathbf{W}_{k+1} = \mathbf{W}_k + \frac{\mu \Delta_k}{\sqrt{\sum_{\text{all Adalines}} |\Delta_k|^2}}, \tag{G.99}$$

where again $0 < \eta < 1$. The weight updates determined by Eqs. (G.98) and (G.99) can help a network find a solution when a relatively flat local region in the mean square error surface is encountered. The weights move by the same amount whether the surface is flat or inclined. It is reminiscent of α-LMS because the gradient term in the weight update equation is normalized by a time-varying factor. The weight update rule could be further modified by including terms from both techniques associated with Eqs. (G.96) through (G.99). Other methods for speeding up backpropagation training include Fahlman's popular quickprop method [128], as well as the delta-bar-delta approach reported in an excellent paper by Jacobs [129].[24]

One of the most promising new areas of neural network research involves backpropagation variants for training various recurrent (signal feedback) networks. Recently, backpropagation rules have been devised for training recurrent networks to learn static associations [130, 131]. More interesting is the online technique of Williams and Zipser [132] which allows a wide class of recurrent networks to learn dynamic associations and trajectories. A more general and computationally viable variant of this technique has been advanced by Narendra and Parthasarathy [106]. These online methods are generalizations of a well-known steepest descent algorithm for training linear IIR filters [133, 32].

An equivalent technique which is usually far less computationally intensive, but best suited for offline computation also exists [39, 44, 134]. This approach, called "backpropagation through time," has been used by Nguyen and Widrow [52] to enable a neural network to learn without a teacher how to back up a computer-simulated trailer truck to a loading dock (Fig. G.27). This is a complicated and highly nonlinear steering task. Nevertheless, with just six inputs providing information about the current position of the truck, a two-layer neural network with only 26 Adalines was able to learn of its own accord to solve this problem. Once trained, the network could successfully back up the truck from any initial position and orientation in front of the loading dock.

[23]"Clean" gradient estimates are those with little gradient noise.

[24]Jacobs's paper, like many other papers in the literature, assumes for analysis that the true gradients rather than instantaneous gradients are used to update the weights, that is, that weights are changed periodically, only after all training patterns are presented. This eliminates gradient noise but can slow down training enormously if the training set is large. The delta-bar-delta procedure in Jacobs's paper involves monitoring changes of the true gradients in response to weight changes. It should be possible to avoid the expense of computing the true gradients explicitly in this case by instead monitoring changes in the outputs of, say, two momentum filters with different time constants.

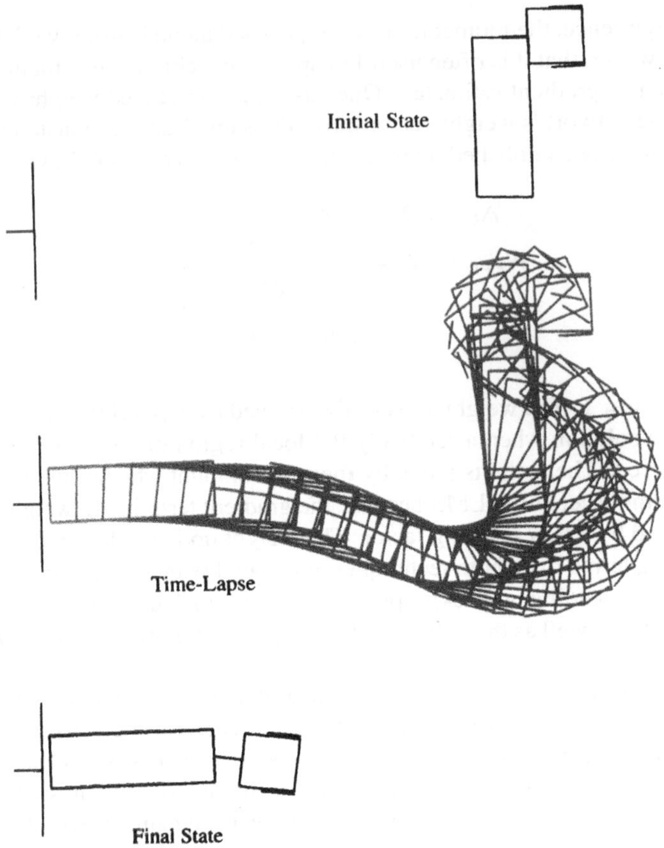

Figure G.27 Example truck backup sequence.

G.7.2 Madaline Rule III for Networks

It is difficult to build neural networks with analog hardware which can be trained effectively by the popular backpropagation technique. Attempts to overcome this difficulty have led to the development of the MRIII algorithm. A commercial analog neurocomputing chip based primarily on this algorithm has already been devised [135]. The method described in this section is a generalization of the single Adaline MRIII technique (G.60). The multi-element generalization of the other single element MRIII rule (G.61) is described in [136].

The MRIII algorithm can be readily described by referring to Fig. G.28. Although this figure shows a simple two-layer feedforward architecture, the procedure to be developed will work for neural networks with any number of Adaline elements in any feedforward structure. In [136], we discuss variants of the basic MRIII approach that allow steepest descent training to be applied to more general network topologies, even those with signal feedback.

Assume that an input pattern **X** and its associated desired output responses d_1 and d_2 are presented to the network of Fig. G.28. At this point, we measure the sum squared output

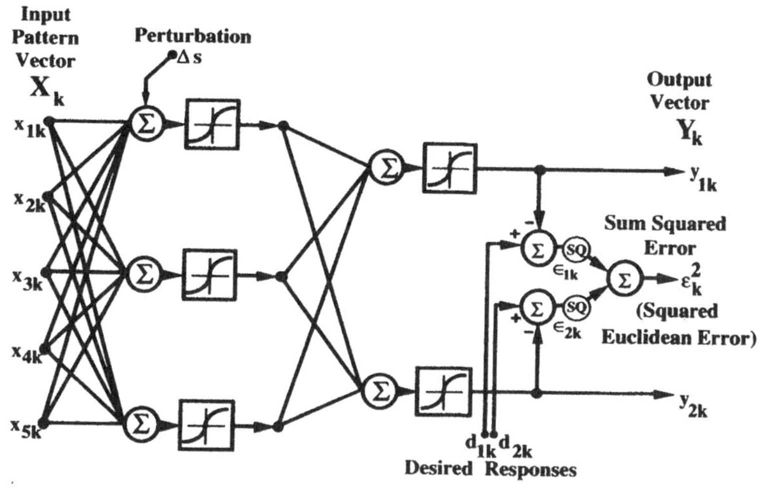

Figure G.28 Example two-layer Madaline III architecture.

response error $\varepsilon^2 = (d_1 - y_1)^2 + (d_2 - y_2)^2 = \epsilon_1^2 + \epsilon_2^2$. We then add a small quantity Δs to a selected Adaline in the network, providing a perturbation to the element's linear sum. This perturbation propagates through the network, and causes a change in the sum of the squares of the errors, $\Delta(\varepsilon^2) = \Delta(\epsilon_1^2 + \epsilon_2^2)$. An easily measured ratio is

$$\frac{\Delta\left(\varepsilon^2\right)}{\Delta s} = \frac{\Delta\left(\epsilon_1^2 + \epsilon_2^2\right)}{\Delta s} \simeq \frac{\partial\left(\varepsilon^2\right)}{\partial s}. \tag{G.100}$$

Below we use this to obtain the instantaneous gradient of ε_k^2 with respect to the weight vector of the selected Adaline. For the kth presentation, the instantaneous gradient is

$$\hat{\nabla}_k = \frac{\partial\left(\varepsilon_k^2\right)}{\partial \mathbf{W}_k} = \frac{\partial\left(\varepsilon_k^2\right)}{\partial s_k}\frac{\partial s_k}{\partial \mathbf{W}_k} = \frac{\partial\left(\varepsilon_k^2\right)}{\partial s_k}\mathbf{X}_k. \tag{G.101}$$

Replacing the derivative with a ratio of differences yields

$$\hat{\nabla}_k \simeq \frac{\Delta\left(\varepsilon_k^2\right)}{\Delta s}\mathbf{X}_k. \tag{G.102}$$

The idea of obtaining a derivative by perturbing the linear output of the selected Adaline element is the same as that expressed for the single element in Section G.6.2, except that here the error is obtained from the output of a multi-element network rather than from the output of a single element.

The gradient (G.102) can be used to optimize the weight vector in accord with the method of steepest descent:

$$\mathbf{W}_{k+1} = \mathbf{W}_k - \mu\frac{\Delta(\varepsilon_k^2)}{\Delta s}\mathbf{X}_k. \tag{G.103}$$

Maintaining the same input pattern, one could either perturb all the elements in the network in sequence, adapting after each gradient calculation, or else the derivatives could be computed and stored to allow all Adalines to be adapted at once. These two MRIII approaches both involve the same weight update equation (G.103), and if μ is small, both lead to equivalent solutions. With large μ, experience indicates that adapting one element at a time results in convergence after fewer iterations, especially in large networks. Storing the gradients, however, has the advantage that after the initial unperturbed error is measured during a given training presentation, each gradient estimate requires only the perturbed error measurement. If adaptations take place after each error measurement, however, both perturbed and unperturbed errors must be measured for each gradient calculation. This is because each weight update changes the associated unperturbed error.

G.7.3 Comparison of MRIII with MRII

MRIII was derived from MRII by replacing the signum nonlinearities with sigmoids. The similarity of these algorithms becomes evident when comparing Fig. G.28, representing MRIII, with Fig. G.16, representing MRII.

The MRII network is highly discontinuous and nonlinear. Using an instantaneous gradient to adjust the weights is not possible. In fact, from the MSE surface for the signum Adaline that we presented in Section G.6.2, it is clear that even gradient descent techniques which use the true gradient could run into severe problems with local minima. The idea of adding a perturbation to the linear sum of a selected Adaline element is workable, however. If the Hamming error has been reduced by the perturbation, the Adaline is adapted to reverse its output decision. This weight change is in the LMS direction, along its **X**-vector. If adapting the Adaline would not reduce network output error, it is not adapted. This is in accord with the minimal disturbance principle. The Adalines selected for possible adaptation are those whose analog sums are closest to zero, that is, the Adalines which can be adapted to give opposite responses with the smallest weight changes. It is useful to note that with binary ± 1 desired responses, the Hamming error is equal to $1/4$ the sum square error. Minimizing the output Hamming error is therefore equivalent to minimizing the output sum square error.

The MRIII algorithm works in a similar manner. All the Adalines in the MRIII network are adapted, but those whose analog sums are closest to zero will usually be adapted most strongly, because the sigmoid has its maximum slope at zero, contributing to high gradient values. As with MRII, the objective is to change the weights for the given input presentation to reduce the sum square error at the network output. In accord with the minimal disturbance principle, the weight vectors of the Adaline elements are adapted in the LMS direction, along their **X**-vectors, and are adapted in proportion to their capabilities for reducing the sum square error (the square of the Euclidean error) at the output.

G.7.4 Comparison of MRIII with Backpropagation

In Section G.6.2, we argued that, for the sigmoid Adaline element, the MRIII algorithm (G.61) is essentially equivalent to the backpropagation algorithm (G.54). The same argument can be extended to the network of Adaline elements, demonstrating that if Δs is small and adaptation is applied to all elements in the network at once, then MRIII is essentially equivalent to backpropagation. That is, to the extent that the sample derivative $\Delta\varepsilon_k^2/\Delta s$

from Eq. (G.103) is equal to the analytical derivative $\partial\varepsilon_k^2/\partial s_k$ from Eq. (G.91), the two rules follow identical instantaneous gradients, and thus perform identical weight updates.

The backpropagation algorithm requires fewer operations than MRIII to calculate gradients, since it is able to take advantage of a priori knowledge of the sigmoid nonlinearities and their derivative functions. Conversely, the MRIII algorithm uses no prior knowledge about the characteristics of the sigmoid functions. Rather, it acquires instantaneous gradients from perturbation measurements. Using MRIII, tolerances on the sigmoid implementations can be greatly relaxed compared to acceptable tolerances for successful backpropagation.

Steepest-descent training of multi-layer networks implemented by computer simulation or by precise parallel digital hardware is usually best carried out by backpropagation. During each training presentation, the backpropagation method requires only one forward computation through the network followed by one backward computation in order to adapt all the weights of an entire network. To accomplish the same effect with the form of MRIII that updates all weights at once, one measures the unperturbed error followed by a number of perturbed error measurements equal to the number of elements in the network. This process can represent a significant amount of computation.

If a network is to be implemented in analog hardware, however, experience has shown that MRIII offers strong advantages over backpropagation. Comparison of Fig. G.25 with Fig. G.28 demonstrates the relative simplicity of MRIII: All the apparatus for backward propagation of error-related signals is eliminated, and the weights do not need to carry signals in both directions (see Fig. G.26). MRIII is a much simpler algorithm to build and to understand, and in principle it produces the same instantaneous gradient as the backpropagation algorithm. The momentum technique and most other common variants of the backpropagation algorithm can be applied to MRIII training.

G.7.5 MSE Surfaces of Neural Networks

In Section G.6.2, "typical" mean square error surfaces of sigmoid and signum Adalines were shown indicating that sigmoid Adalines are much more conducive to gradient approaches than signum Adalines. The same phenomena result when Adalines are incorporated into multi-element networks. The MSE surfaces of MRII networks are reasonably chaotic and will not be explored here. In this section we examine only MSE surfaces from a typical backpropagation training problem with a sigmoidal neural network.

In a network with more than two weights, the MSE surface is high-dimensional and difficult to visualize. It is possible, however, to look at slices of this surface by plotting the mean square error surface created by varying two of the weights while holding all others constant. The surfaces plotted in Figs. G.29 and G.30 show two such slices of the MSE surface from a typical learning problem involving, respectively, an untrained sigmoidal network and a trained one. The first surface resulted from varying two first-layer weights of an untrained network. The second surface resulted from varying the same two weights after the network was fully trained. The two surfaces are similar, but the second one has a deeper minimum which was carved out by the backpropagation learning process. Figures G.31 and G.32 resulted from varying a different set of two weights in the same network. Fig. G.31 is the result from varying a first-layer weight and third-layer weight in the untrained net-

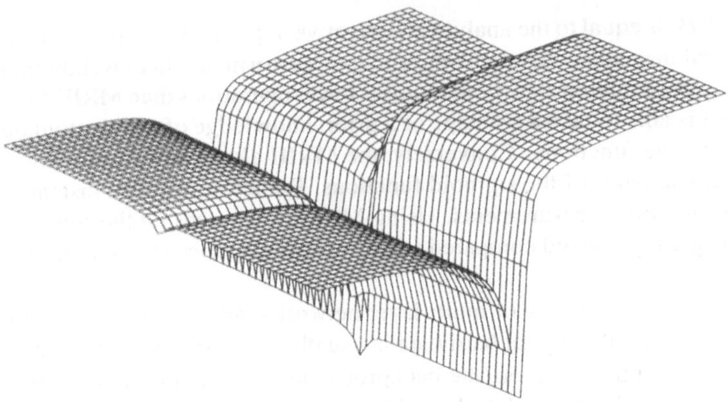

Figure G.29 Example MSE surface of an untrained sigmoidal network as a function of two first-layer weights.

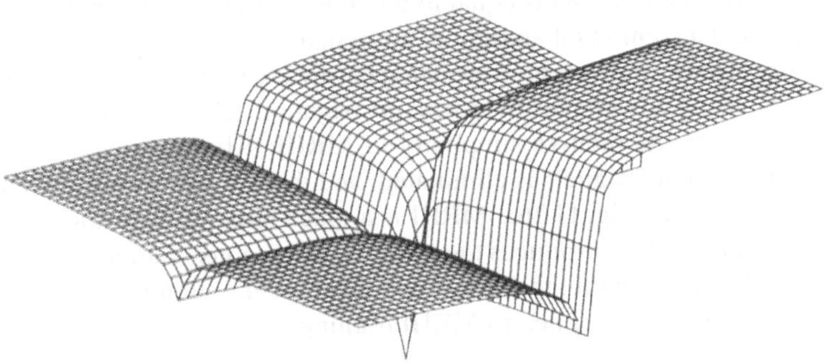

Figure G.30 Example MSE surface of a trained sigmoidal network as a function of two first-layer weights.

work while Fig. G.32 is the surface that resulted from varying the same two weights after the netwcrk was trained.

By studying many plots, it becomes clear that backpropagation and MRIII will be subject to convergence on local optima. The same is true for MRII. The most common remedy for this is the sporadic addition of noise to the weights or gradients. Some of the "simulated annealing" methods [137] do this. Another method involves retraining the network several times using different random initial weight values until a satisfactory solution is found.

Solutions found by people in everyday life are usually not optimal, but many of them are useful. If a local optimum yields satisfactory performance, often there is simply no need to search for a better solution.

G.8 SUMMARY

This year (1990) is the thirtieth anniversary of the publication of the Perceptron rule by Rosenblatt and the LMS algorithm by Widrow and Hoff. It has also been 16 years since Werbos first published the backpropagation algorithm. These learning rules and several others

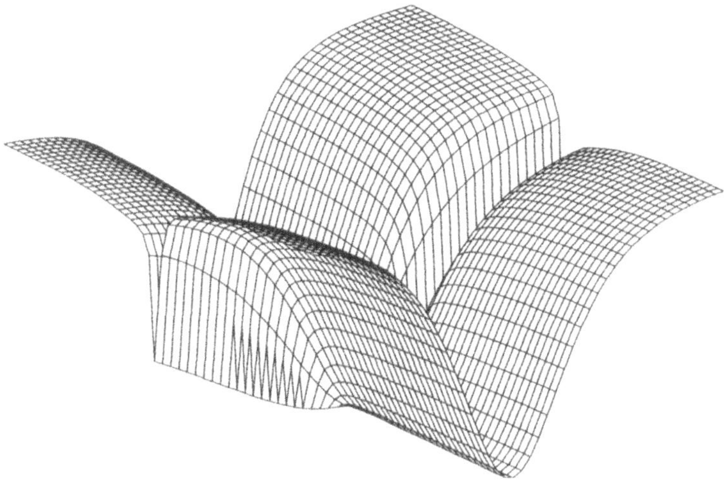

Figure G.31 Example MSE surface of an untrained sigmoidal network as a function of a first-layer weight and a third-layer weight.

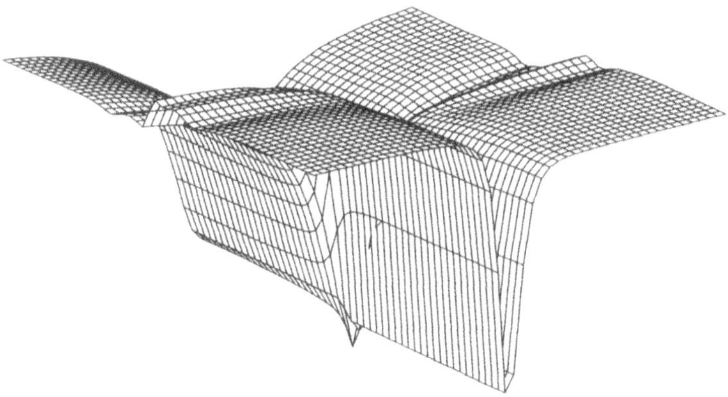

Figure G.32 Example MSE surface of a trained sigmoidal network as a function of a first-layer weight and a third-layer weight.

have been studied and compared. Although they differ significantly from each other, they all belong to the same "family."

A distinction was drawn between error correction rules and steepest descent rules. The former includes the Perceptron rule, Mays's rules, the α-LMS algorithm, the original Madaline I rule of 1962, and the Madaline II rule. The latter includes the μ-LMS algorithm, the Madaline III rule, and the backpropagation algorithm. The chart in Fig. G.33 categorizes the learning rules that have been studied.

Although these algorithms have been presented as established learning rules, one should not gain the impression that they are perfect and frozen for all time. Variations are possible for every one of them. They should be regarded as substrates upon which to build

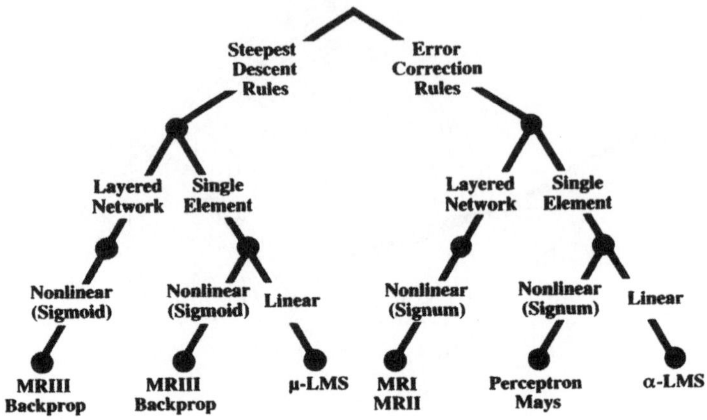

Figure G.33 Learning rules.

new and better rules. There is a tremendous amount of invention waiting "in the wings." We look forward to the next 30 years.

Acknowledgments

This work was sponsored by SDIO Innovative Science and Technology Office and managed by ONR under contract #N00014-86-K-0718, by the Department of the Army Belvoir R D & E Center under contracts #DAAK70-87-P-3134 and #DAAK70-89-K-0001, by a grant from the Lockheed Missiles and Space Company, by NASA under contract #NCA2-389, and by Rome Air Development Center under contract #F30602-88-D-0025, subcontract E-21-T22-S1.

Bibliography for Appendix G

[1] K. STEINBUCH and V.A.W. PISKE, "Learning matrices and their applications," *IEEE Trans. on Elect. Computers*, Vol. EC-12, No. 5 (December 1963), pp. 846–862.

[2] B. WIDROW, "Generalization and information storage in networks of Adaline 'neurons'," in M. C. Yovitz, G. T. Jacobi, and G. Goldstein, eds., *Self-Organizing Systems 1962* (Washington, DC: Spartan Books, 1962), pp. 435–461.

[3] L. STARK, M. OKAJIMA, and G.H. WHIPPLE, "Computer pattern recognition techniques: Electrocardiographic diagnosis," *Communications of the ACM*, Vol. 5 (October 1962), pp. 527–532.

[4] F. ROSENBLATT, "Two theorems of statistical separability in the perceptron," in *Mechanization of Thought Processes*, Proceedings of a Symposium held at the National Physical Laboratory, November 1958, Vol 1 (London: HM Stationery Office, 1959), pp. 421–456.

[5] F. ROSENBLATT, *Principles of neurodynamics: Perceptrons and the theory of brain mechanisms* (Washington, DC: Spartan Books, 1962).

[6] C. VON DER MALSBURG, "Self-organizing of orientation sensitive cells in the striate cortex," *Kybernetik*, Vol. 14 (1973), pp. 85–100.

[7] S. GROSSBERG, "Adaptive pattern classification and universal recoding, I: Parallel development and coding of neural feature detectors," *Biological Cybernetics*, Vol. 23 (1976), pp. 121–134.

[8] K. FUKUSHIMA, "Cognitron: A self-organizing multilayered neural network," *Biological Cybernetics*, Vol. 20 (1975), pp. 121–136.

[9] K. FUKUSHIMA, "Neocognitron: A self-organizing neural network model for a mechanism of pattern recognition unaffected by shift in position," *Biological Cybernetics*, Vol. 36, No. 4 (1980), pp. 193–202.

[10] B. WIDROW, "Bootstrap learning in threshold logic systems," presented at the American Automatic Control Council (Theory Committee), IFAC Meeting, London, England, June 1966.

[11] B. WIDROW, N.K. GUPTA, and S. MAITRA, "Punish/reward: Learning with a critic in adaptive threshold systems," *IEEE Trans. Syst. , Man, Cybern.*, Vol. SMC-3, No. 5 (September 1973), pp. 455–465.

[12] A.G. BARTO, R.S. SUTTON, and C.W. ANDERSON, "Neuronlike adaptive elements that can solve difficult learning control problems," *IEEE Trans. Syst., Man, Cybern.*, Vol. SMC-13 (1983), pp. 834–846.

[13] J.S. ALBUS, "A new approach to manipulator control: The cerebellar model articulation controller (cmac)," *Journal of Dynamic Systems, Measurement, and Control*, Vol. 97 (1975), pp. 220–227.

[14] W.T. MILLER III, "Sensor-based control of robotic manipulators using a general learning algorithm," *IEEE Journal of Robotics and Automation*, Vol. RA-3, No. 2 (April 1987), pp. 157–165.

[15] A.G. BARTO, and P. ANANDAN, "Pattern recognizing stochastic learning automata," *IEEE Transactions on Systems, Man, and Cybernetics*, Vol. 15 (1985), pp. 360–375.

[16] S. GROSSBERG, "Adaptive pattern classification and universal recoding, II: Feedback, expectation, olfaction, and illusions," *Biological Cybernetics*, Vol. 23 (1976) pp. 187–202.

[17] G.A. CARPENTER, and S. GROSSBERG, "A massively parallel architecture for a self-organizing neural pattern recognition machine," *Computer Vision, Graphics, and Image Processing*, Vol. 37 (1983), pp. 54–115.

[18] G.A. CARPENTER, and S. GROSSBERG, "Art 2: Self-organization of stable category recognition codes for analog output patterns," *Applied Optics*, Vol. 26, No. 23 (December 1, 1987), pp. 4919–4930.

[19] G.A. CARPENTER, and S. GROSSBERG, "Art 3 hierarchical search: Chemical transmitters in self-organizing pattern recognition architectures," in *Proceedings of the International Joint Conference on Neural Networks*, Vol. II (Washington, DC: Lawrence Erlbaum, January 1990), pp. 30–33.

[20] T. KOHONEN, "Self-organized formation of topologically correct feature maps," *Biological Cybernetics*, Vol. 43 (1982), pp. 59–69.

[21] T. KOHONEN, *Self-organization and associative memory*, 2nd ed. (New York: Springer-Verlag, 1988).

[22] D.O. HEBB, *The organization of behavior* (New York: John Wiley, 1949).

[23] J.J. HOPFIELD, "Neural networks and physical systems with emergent collective computational abilities," *Proc. Natl. Acad. Sci.*, Vol. 79 (April 1982), pp. 2554–2558.

[24] J.J. HOPFIELD, "Neurons with graded response have collective computational properties like those of two-state neurons," *Proc. Natl. Acad. Sci.*, Vol. 81 (May 1984), pp. 3088–3092.

[25] B.KOSKO, "Adaptive bidirectional associative memories," *Applied Optics*, Vol. 26, No. 23 (December 1, 1987), pp. 4947–4960.

[26] H.A. KLOPF. "Drive-reinforcement learning: A real-time learning mechanism for unsupervised learning," in *Proceedings of the IEEE First International Conference on Neural Networks*, Vol. II, San Diego, CA, June 21–24, 1987, pp. 441–445.

[27] G.E. HINTON, T.J. SEJNOWSKI, and D.H. ACKLEY, "Boltzmann machines: Constraint satisfaction networks that learn," Technical Report CMU-CS-84-119, Carnegie-Mellon University, Department of Computer Science, 1984.

[28] G.E. HINTON and T.J. SEJNOWSKI, "Learning and relearning in boltzmann machines," in D. E. Rumelhart and J. L. McClelland, eds., *Parallel Distributed Processing*, Vol. 1 (Cambridge, MA: M.I.T. Press, 1986), chap. 7.

[29] L.R. TALBERT, G.F. GRONER, J.S. KOFORD, R.J. BROWN, P.R. LOW, and C.H. MAYS, "A real-time adaptive speech-recognition system," Technical report, Stanford University, 1963.

[30] M.J.C. HU, "Application of the Adaline System to Weather Forecasting," E. E. Degree diss., Technical Report 6775-1, Stanford Electron. Labs., Stanford, CA, June 1964.

[31] B. WIDROW, "The original adaptive neural net broom-balancer," in *Proceedings of the IEEE International Symposium on Circuits and Systems*, Philadelphia, PA, May 4–7, 1987, pp. 351–357.

[32] B. WIDROW and S.D. STEARNS, *Adaptive signal processing* (Englewood Cliffs, NJ: Prentice Hall, 1985).

[33] B. WIDROW, P. MANTEY, L. GRIFFITHS, and B. GOODE, "Adaptive antenna systems," *Proceedings of the IEEE*, Vol. 55 (December 1967), pp. 2143–2159.

[34] B. WIDROW, "Adaptive inverse control," in *Proceedings of the 2nd International Federation of Automatic Control Workshop*, Lund, Sweden, July 1–3, 1986, pp. 1–5.

[35] B. WIDROW, J.M. MCCOOL, J.R. GLOVER, JR., J. KAUNITZ, C. WILLIAMS, R.H. HEARN, J.R. ZEIDLER, E. DONG, JR., and R.C. GOODLIN, "Adaptive noise cancelling: Principles and applications," *Proceedings of the IEEE*, Vol. 63 (December 1975), pp. 1692–1716.

[36] R.W. LUCKY, "Automatic equalization for digital communication," *Bell System Technical Journal*, Vol. 44 (April 1965), pp. 547–588.

[37] R.W. LUCKY, J. SALZ, and E.J. WELDON, *Principles of data communication* (New York: McGraw-Hill, 1968).

[38] M.M. SONDHI, "An adaptive echo canceller," *Bell System Technical Journal*, Vol. 46 (March 1967), pp. 497–511.

[39] P. WERBOS, "Beyond regression: New tools for prediction and analysis in the behavioral sciences," Ph.D. diss., Harvard University, August 1974.

[40] Y. LE CUN, "A theoretical framework for back-propagation," in D. Touretzky, G. Hinton, and T. Sejnowski, eds., *Proceedings of the 1988 Connectionist Models Summer School* (San Mateo, CA: Morgan Kaufmann), June 17–26, 1988, pp. 21–28.

[41] D.B. PARKER, "Learning-logic," Invention Report S81-64, File 1, Office of Technology Licensing, Stanford University, Stanford, CA, October 1982.

[42] D.B. PARKER, "Learning-logic," Technical Report TR-47, Center for Computational Research in Economics and Management Science, M.I.T., April 1985.

[43] D.E. RUMELHART, G.E. HINTON, and R.J. WILLIAMS, "Learning internal representations by error propagation," ICS Report 8506, Institute for Cognitive Science, University of California at San Diego, La Jolla, CA, September 1985.

[44] D.E. RUMELHART, G.E. HINTON, and R.J. WILLIAMS, "Learning internal representations by error propagation," in D.E. Rumelhart and J.L. McClelland, eds., *Parallel Distributed Processing*, Vol. 1 (Cambridge, MA: M.I.T. Press, 1986), chap. 8.

[45] B. WIDROW, R.G. WINTER, and R. BAXTER, "Learning phenomena in layered neural networks," in *Proceedings of the IEEE First International Conference on Neural Networks*, Vol. II, San Diego, CA, June 1987, pp. 411–429.

[46] R.P. LIPPMANN, "An introduction to computing with neural nets," *IEEE ASSP Magazine*, April 1987.

[47] J.A. ANDERSON, and E. ROSENFELD, eds., *Neurocomputing: Foundations of research* (Cambridge, MA: M.I.T. Press, 1988).

[48] N. NILSSON, *Learning machines* (New York: McGraw-Hill, 1965).

[49] D.E. RUMELHART, and J.L. MCCLELLAND, eds., *Parallel distributed processing*, Vols. 1 and 2 (Cambridge, MA: M.I.T. Press, 1986).

[50] B. MOORE, "Art 1 and pattern clustering," in D. Touretzky, G. Hinton, and T. Sejnowski, eds., *Proceedings of the 1988 Connectionist Models Summer School* (San Mateo, CA: Morgan Kaufmann), June 17–26, 1988, pp. 174–185.

[51] A.B. GSCHWENDTER, ed., *DARPA Neural Network Study* (Fairfax, VA: AFCEA International Press, 1988).

[52] D. NGUYEN, and B. WIDROW, "The truck backer-upper: An example of self-learning in neural networks," in *Proceedings of the International Joint Conference on Neural Networks*, Vol. II, Washington, DC, June 1989, pp. 357–363.

[53] T.J. SEJNOWSKI, and C.R. ROSENBERG, "NETtalk: A parallel network that learns to read aloud," Technical Report JHU/EECS-86/01, Johns Hopkins University, 1986.

[54] T.J. SEJNOWSKI, and C.R. ROSENBERG, "Parallel networks that learn to pronounce English text," *Complex Systems*, Vol. 1 (1987), pp. 145–168.

[55] P.M. SHEA, and V. LIN, "Detection of explosives in checked airline baggage using an artificial neural system," in *Proceedings of the International Joint Conference on Neural Networks*, Vol. II, Washington, DC, June 1989, pp. 31–34.

[56] D.G. BOUNDS, P.J. LLOYD, B. MATHEW, and G. WADDELL, "A multilayer perceptron network for the diagnosis of low back pain," in *Proceedings of the IEEE Second International Conference on Neural Networks*, Vol. II, San Diego, CA, July 1988, pp. 481–489.

[57] G. BRADSHAW, R. FOZZARD, and L. CECI, "A connectionist expert system that actually works," in D. S. Touretzky, ed., *Advances in Neural Information Processing Systems I* (San Mateo, CA: Morgan Kaufmann, 1989), pp. 248–255.

[58] N. MOKHOFF, "Neural nets making the leap out of lab," *Electronic Engineering Times*, 22 January 1990, p. 1.

[59] C.A. MEAD, *Analog VLSI and neural systems* (Reading, MA: Addison-Wesley, 1989).

[60] B. WIDROW, and M.E. HOFF, JR., "Adaptive switching circuits," in *1960 IRE Western Electric Show and Convention Record*, P. 4, August 23, 1960, pp. 96–104.

[61] B. WIDROW, and M.E. HOFF, JR., "Adaptive switching circuits," Technical Report 1553–1, Stanford Electron. Labs., Stanford, CA, June 30, 1960.

[62] P.M. LEWIS II, and C.L. COATES, *Threshold logic* (New York: John Wiley, 1967).

[63] T.M. COVER, "Geometrical and statistical properties of linear threshold devices," Ph.D. diss., Technical Report 6107-1, Stanford Electron. Labs., Stanford, CA, May 1964.

[64] R.J. BROWN, "Adaptive multiple-output threshold systems and their storage capacities," Ph.D. diss., Technical Report 6771-1, Stanford Electron. Labs., Stanford, CA, June 1964.

[65] R.O. WINDER, "Threshold logic," Ph.D. diss., Princeton University, 1962.

[66] S.H. CAMERON, "An estimate of the complexity requisite in a universal decision network," in *Proceedings of 1960 Bionics Symposium*, Dayton, OH, December 1960, Wright Air Development Division Technical Report 60-600, pp. 197–211.

[67] R.D. JOSEPH, "The number of orthants in n-space intersected by an s-dimensional subspace," Tech. Memorandum 8, Project PARA, Cornell Aeronautical Laboratory, Buffalo, New York 1960.

[68] D.F. SPECHT, "Generation of polynomial discriminant functions for pattern recognition," Ph.D. diss., Technical Report 6764-5, Stanford Electron. Labs., Stanford, CA, May 1966.

[69] D.F. SPECHT, "Vectorcardiographic diagnosis using the polynomial discriminant method of pattern recognition," *IEEE Trans. Biomed. Eng.*, Vol. BME-14 (April 1967), pp. 90–95.

[70] D.F. SPECHT, "Generation of polynomial discriminant functions for pattern recognition," *IEEE Trans. Electron. Comput.*, Vol. EC-16 (June 1967), pp. 308–319.

[71] A.R. BARRON, "Adaptive learning networks: Development and application in the united states of algorithms related to gmdh," in S. J. Farlow, ed., *Self-Organizing Methods in Modeling* (New York: Marcel Dekker Inc., 1984), pp. 25–65.

[72] A.R. BARRON, "Predicted squared error: A criterion for automatic model selection," in S. J. Farlow, ed., *Self-Organizing Methods in Modeling* (New York: Marcel Dekker Inc., 1984), pp. 87–103.

[73] A.R. BARRON, and R.L. BARRON, "Statistical learning networks: A unifying view," in *1988 Symposium on the Interface: Statistics and Computing Science*, Reston, VA, April 21–23, 1988, pp. 192–203.

[74] A.G. IVAKHNENKO, "Polynomial theory of complex systems," *IEEE Trans. Syst., Man, Cybern.*, Vol. SMC-1 (October 1971), pp. 364–378.

[75] YOH-HAN PAO, "Functional link nets: Removing hidden layers," *AI Expert* (April 1989), pp. 60–68.

[76] C.L. GILES, and T. MAXWELL, "Learning, invariance, and generalization in high-order neural networks," *Applied Optics*, Vol. 26, No. 23 (1 December 1987), pp. 4972–4978.

[77] M.E. HOFF, JR., "Learning phenomena in networks of adaptive switching circuits," Ph.D. diss., Technical Report 1554-1, Stanford Electron. Labs., Stanford, CA, July 1962.

[78] W.C. RIDGWAY III, "An adaptive logic system with generalizing properties," Ph.D. diss., Technical Report 1556-1, Stanford Electron. Labs., Stanford, CA, April 1962.

[79] F.H. GLANZ, "Statistical extrapolation in certain adaptive pattern-recognition systems," Ph.D. diss., Technical Report 6767-1, Stanford Electron. Labs., Stanford, CA, May 1965.

[80] B. WIDROW, "Adaline and madaline"—1963, plenary speech, in *Proceedings of the IEEE First International Conference on Neural Networks*, Vol. I, San Diego, CA, June 23, 1987, pp. 145–158.

[81] B. WIDROW, "An adaptive 'Adaline' neuron using chemical 'memistors'," Technical Report 1553-2, Stanford Electron. Labs., Stanford, CA, October 17, 1960.

[82] C.L. GILES, R.D. GRIFFIN, and T. MAXWELL, "Encoding geometric invariances in higher order neural networks," in D. Z. Anderson, ed., *Neural Information Processing Systems* (American Institute of Physics, 1988), pp. 301–309.

[83] D. CASASENT, and D. PSALTIS, "Position, rotation, and scale invariant optical correlation," *Applied Optics*, Vol. 15, No. 7 (July 1976), pp. 1795–1799.

[84] W.L. REBER, and J. LYMAN, "An artificial neural system design for rotation and scale invariant pattern recognition," in *Proceedings of the IEEE First International Conference on Neural Networks*, Vol. IV, San Diego, CA, June 1987, pp. 277–283.

[85] B. WIDROW, and R.G. WINTER, "Neural nets for adaptive filtering and adaptive pattern recognition," *IEEE Computer*, March 1988, pp. 25–39.

[86] A. KHOTANZAD, and YAW HUA HONG, "Rotation invariant pattern recognition using Zernike moments," in *Proceedings of the 9th International Conference on Pattern Recognition*, Vol. I, 1988, pp. 326–328.

[87] C. VON DER MALSBURG, "Pattern recognition by labeled graph matching," *Neural Networks*, Vol. 1, No. 2 (1988), pp. 141–148.

[88] A. WAIBEL, T. HANAZAWA, G. HINTON, K. SHIKANO, and K.J. LANG, "Phoneme recognition using time delay neural networks," *IEEE Trans. Acoust., Speech, and Signal Processing*, Vol. ASSP-37, No. 3 (March 1989), pp. 328–339.

[89] C.M. NEWMAN, "Memory capacity in neural network models: Rigorous lower bounds," *Neural Networks*, Vol. 1, No. 3 (1988), pp. 223–238.

[90] Y.S. ABU-MOSTAFA, and J. ST. JACQUES, "Information capacity of the Hopfield model," *IEEE Transactions on Information Theory*, Vol. IT-31, No. 4 (1985), pp. 461–464.

[91] Y.S. ABU-MOSTAFA, "Neural networks for computing?," J.S. Denker, ed., *Neural Networks for Computing, American Institute of Physics Conference Proceedings No. 151* (New York: American Institute of Physics, 1986), pp. 1–6.

[92] S.S. VENKATESH, "Epsilon capacity of neural networks," J.S. Denker, ed., *Neural Networks for Computing, American Institute of Physics Conference Proceedings No. 151* (New York: American Institute of Physics, 1986), pp. 440–445.

[93] J.D. GREENFIELD, *Practical digital design using IC's*, 2nd ed. (New York: John Wiley, 1983).

[94] M. STINCHCOMBE, and H. WHITE, "Universal approximation using feedforward networks with non-sigmoid hidden layer activation functions," in *Proceedings of the International Joint Conference on Neural Networks*, Vol. I, Washington, DC, June 1989, pp. 613–617.

[95] G. CYBENKO, "Continuous valued neural networks with two hidden layers are sufficient," Technical report, Department of Computer Science, Tufts University, March 1988.

[96] B. IRIE, and S. MIYAKE, "Capabilities of three-layered perceptrons," in *Proceedings of the IEEE Second International Conference on Neural Networks*, Vol. I, San Diego, CA, July 1988, pp. 641–647.

[97] M.L. MINSKY, and S.A. PAPERT, *Perceptrons: An introduction to computational geometry*, expanded ed. (Cambridge, MA: M.I.T. Press, 1988).

[98] M.W. ROTH, "Survey of neural network technology for automatic target recognition," *IEEE Transactions on Neural Networks*, Vol. 1,No. 1 (March 1990), pp. 28–43.

[99] T. M. COVER, "Capacity problems for linear machines," in L. N. Kanal, ed., *Pattern recognition*, P. 3 (Washington, DC: Thompson Book Co., 1968), pp. 283–289.

[100] E.B. BAUM, "On the capabilities of multilayer perceptrons," *Journal of Complexity*, Vol. 4, No. 3 (September 1988), pp. 193–215.

[101] A. LAPEDES, and R. FARBER, "How neural networks work," Technical Report LA-UR-88-418, Los Alamos National Laboratory, Los Alamos, NM, 1987.

[102] D. NGUYEN, and B. WIDROW, "Improving the learning speed of 2-layer neural networks by choosing initial values of the adaptive weights," in *Proceedings of the International Joint Conference on Neural Networks*, San Diego, CA, June 1990.

[103] G. CYBENKO, "Approximation by superpositions of a sigmoidal function," *Mathematics of Control, Signals, and Systems*, Vol. 2, No. 4 (1989).

[104] E.B. BAUM, and D. HAUSSLER, "What size net gives valid generalization?," *Neural Computation*, Vol. 1, No. 1 (1989), pp. 151–160.

[105] J.J. HOPFIELD, and D.W. TANK, "Neural computations of decisions in optimization problems," *Biological Cybernetics*, Vol. 52 (1985), pp. 141–152.

[106] K.S. NARENDRA, and K. PARTHASARATHY, "Identification and control of dynamical systems using neural networks," *IEEE Transactions on Neural Networks*, Vol. 1, No. 1 (March 1990), pp. 4–27.

[107] C.H. MAYS, "Adaptive threshold logic," Ph.D. diss., Technical Report 1557-1, Stanford Electron. Labs., Stanford, CA, April 1963.

[108] F. ROSENBLATT, "On the convergence of reinforcement procedures in simple perceptrons," Cornell Aeronautical Laboratory Report VG-1196-G-4, Buffalo, February 1960.

[109] H. BLOCK, "The perceptron: A model for brain functioning, I," *Reviews of Modern Physics*, Vol. 34 (January 1962), pp. 123–135.

[110] R. G. WINTER, "Madaline rule II: A new method for training networks of adalines," Ph.D. diss., Stanford University, Stanford, CA, January 1989.

[111] E. WALACH, and B. WIDROW, "The least mean fourth (lmf) adaptive algorithm and its family," *IEEE Transactions on Information Theory*, Vol. IT-30, No. 2 (March 1984), pp. 275–283.

[112] E.B. BAUM, and F. WILCZEK, "Supervised learning of probability distributions by neural networks," in D.Z. Anderson, ed., *Neural Information Processing Systems* (New York: American Institute of Physics, 1988), pp. 52–61.

[113] S.A. SOLLA, E. LEVIN, and M. FLEISHER, "Accelerated learning in layered neural networks," *Complex Systems*, Vol. 2 (1988), pp. 625–640.

[114] D.B. PARKER, "Optimal algorithms for adaptive neural networks: Second order back propagation, second order direct propagation, and second order Hebbian learning," in *Proceedings of the IEEE First International Conference on Neural Networks*, Vol. II, San Diego, CA, June 1987, pp. 593–600.

[115] A.J. OWENS, and D.L. FILKIN, "Efficient training of the back propagation network by solving a system of stiff ordinary differential equations," in *Proceedings of the International Joint Conference on Neural Networks*, Vol. II, Washington, DC, June 1989, pp. 381–386.

[116] D.G. LUENBERGER, *Linear and nonlinear programming*, 2nd ed. (Reading, MA: Addison-Wesley, 1984).

[117] A. KRAMER, and A. SANGIOVANNI-VINCENTELLI, "Efficient parallel learning algorithms for neural networks," in D. S. Touretzky, ed., *Advances in Neural Information Processing Systems I* (San Mateo, CA: Morgan Kaufmann, 1989), pp. 40–48.

[118] R.V. SOUTHWELL, *Relaxation methods in engineering science* (New York: Oxford, 1940).

[119] D.J. WILDE, *Optimum seeking methods* (Englewood Cliffs, NJ: Prentice Hall, 1964).

[120] N. WIENER, *Extrapolation, interpolation, and smoothing of stationary time series, with engineering applications* (New York: John Wiley, 1949).

[121] T. KAILATH, "A view of three decades of linear filtering theory," *IEEE Trans. Inform. Theory*, Vol. IT-20 (March 1974), pp. 145–181.

[122] H. BODE, and C. SHANNON, "A simplified derivation of linear least squares smoothing and prediction theory," *Proc. IRE*, Vol. 38 (April 1950), pp. 417–425.

[123] L.L. HOROWITZ, and K.D. SENNE, "Performance advantage of complex LMS for controlling narrow-band adaptive arrays," *IEEE Transactions on Circuits and Systems*, Vol. CAS-28, No. 6 (June 1981), pp. 562–576.

[124] E.D. SONTAG, and H.J. SUSSMANN, "Backpropagation separates when perceptrons do," in *Proceedings of the International Joint Conference on Neural Networks*, Vol. I, Washington, DC, June 1989, pp. 639–642.

[125] E.D. SONTAG, and H.J. SUSSMANN, "Backpropagation can give rise to spurious local minima even for networks without hidden layers," *Complex Systems*, Vol. 3 (1989), pp. 91–106.

[126] P. WERBOS, "Generalization of backpropagation with application to a recurrent gas market model," *Neural Networks*, Vol. 1 (1988), pp. 339–356.

[127] J.J. SHYNK, and S. ROY, "The LMS algorithm with momentum updating," in *ISCAS 88*, Espoo, Finland, June 1988.

[128] S.E. FAHLMAN, "Faster learning variations on backpropagation: An empirical study," in D. Touretzky, G. Hinton, and T. Sejnowski, eds., *Proceedings of the 1988 Connectionist Models Summer School* (San Mateo, CA: Morgan Kaufmann), June 17–26, 1988, pages 38–51.

[129] R.A. JACOBS, "Increased rates of convergence through learning rate adaptation," *Neural Networks*, Vol. 1, No. 4 (1988), pp. 295–307.

[130] F.J. PINEDA, "Generalization of backpropagation to recurrent neural networks," *Physical Review Letters*, Vol. 18, No. 59 (1987), pp. 2229–2232.

[131] L.B. ALMEIDA, "A learning rule for asynchronous perceptrons with feedback in a combinatorial environment," in *Proceedings of the IEEE First International Conference on Neural Networks*, Vol. II, San Diego, CA, June 1987, pp. 609–618.

[132] R.J. WILLIAMS, and D. ZIPSER, "A learning algorithm for continually running fully recurrent neural networks," ICS Report 8805, Institute for Cognitive Science, University of California at San Diego, La Jolla, CA, October 1988.

[133] S.A. WHITE, "An adaptive recursive digital filter," in *Proc. 9th Asilomar Conf. Circuits Syst. Comput.* (November 1975), p. 21.

[134] B. PEARLMUTTER, "Learning state space trajectories in recurrent neural networks," in D. Touretzky, G. Hinton, and T. Sejnowski, eds., *Proceedings of the 1988 Connectionist Models Summer School* (San Mateo, CA: Morgan Kaufmann, 1988), June 17–26, pp. 113–117.

[135] M. HOLLER, S. TAM, H. CASTRO, and R. BENSON, "An electrically trainable artificial neural network (ETANN) with 10240 'floating gate' synapses, in *Proceedings of the International Joint Conference on Neural Networks*, Vol. II, Washington, DC, June 1989, pp. 191–196.

[136] D. ANDES, B. WIDROW, M.A. LEHR, and E.A. WAN, "MRIII: A robust algorithm for training analog neural networks," in *Proceedings of the International Joint Conference on Neural Networks*, Vol. I (Washington, DC: Lawrence Erlbaum, January 1990), pp. 533–536.

[137] S. KIRKPATRICK, C.D. GELATT, JR., and M.P. VECCHI, "Optimization by simulated annealing," *Science*, Vol. 220 (May 1983), pp. 671–680.

ppendix H

Neural Control Systems

A variety of methods for control for nonlinear systems were presented in Chapter 11. This appendix describes additional methods for nonlinear control which are based on neural networks. The networks of interest are multilayered, and adaptation is done with the backpropagation algorithm. For a description of these networks and the backpropagation learning algorithm, refer to Appendix G.

1 A NONLINEAR ADAPTIVE FILTER BASED ON NEURAL NETWORKS

Figure H.1 shows the architecture of a nonlinear filter made by connecting a multilayer neural network to the taps of a tapped delay line. The connections in the network are variable weights, as indicated in the figure. Each of the neural elements contains a sigmoidal nonlinearity. The error at the filter output is the difference between the desired response and the network response. The error is used by the backpropagation algorithm to adapt the weights.

The same filter of Fig. H.1 can be represented more simply by the diagram of Fig. H.2. Furthermore, with even less detail, a nonlinear adaptive filter can be drawn in a general way as in Fig. H.3. It is understood that the filter is of the form of a tapped delay line with a multilayered neural network attached. The error signal is used with backpropagation to adapt the weights. The number of weights, the number of neurons, and the number of layers would need to be specified in a given case.

2 A MIMO NONLINEAR ADAPTIVE FILTER

A MIMO filter is shown in Fig. H.4. Once again, this is a transversal (tapped delay-line) filter with neural networks (connected to the delay line taps). The weights are adapted by backpropagation using both error signals. This MIMO filter can be represented by the diagram of Fig. H.3. As such, the input, output, error, and desired response signals would be vector rather than scalar.

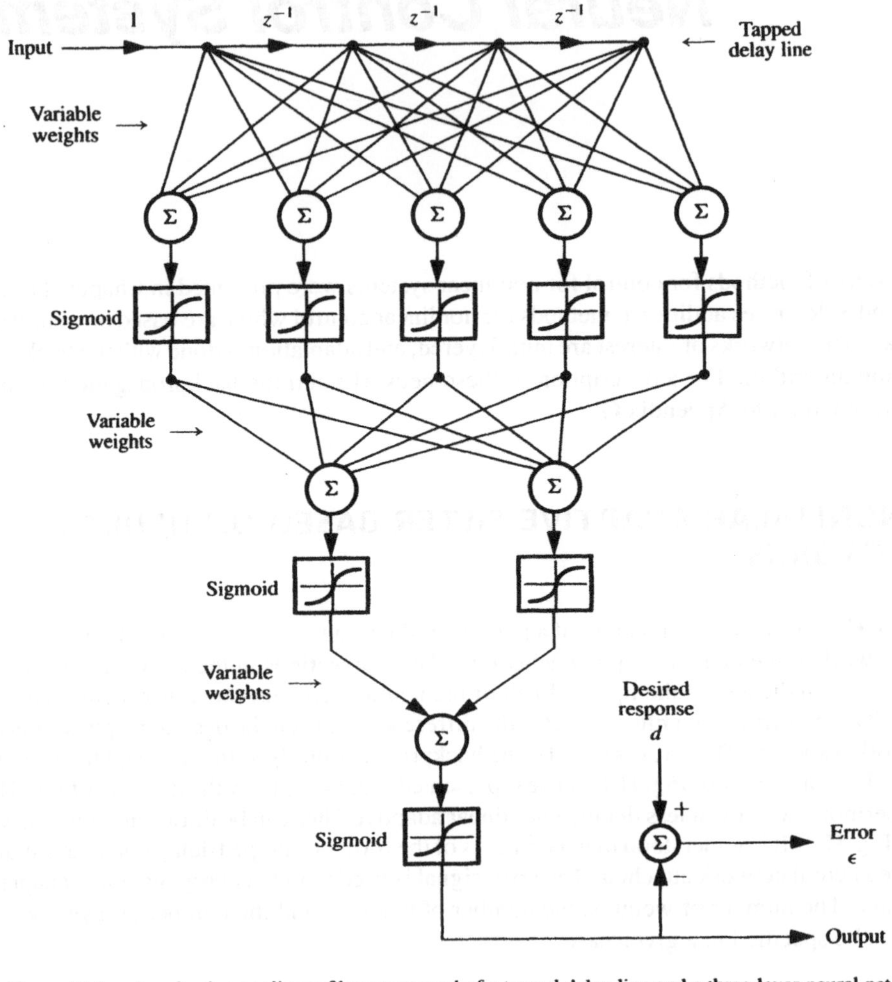

Figure H.1 An adaptive nonlinear filter composed of a tapped delay line and a three-layer neural net.

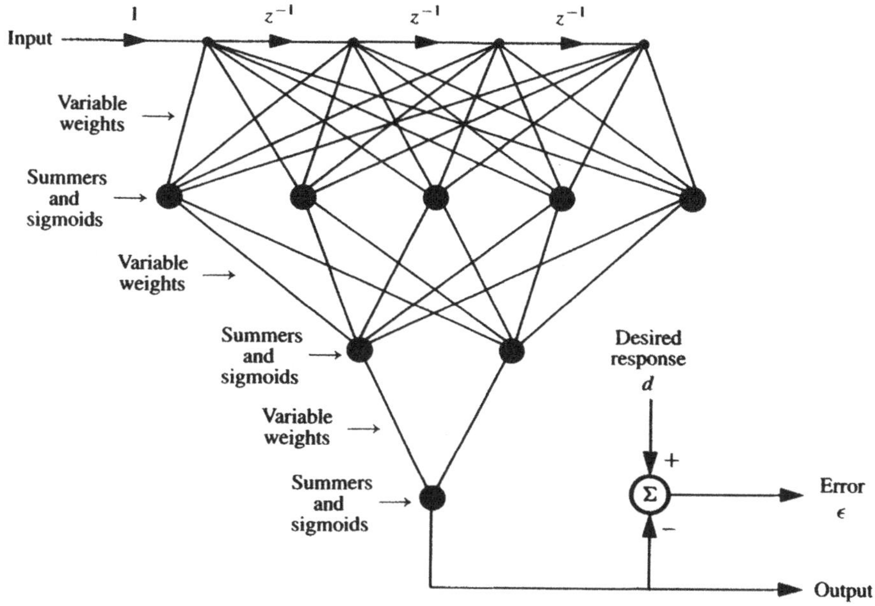

Figure H.2 A simpler diagram for the adaptive filter of Fig. H.1.

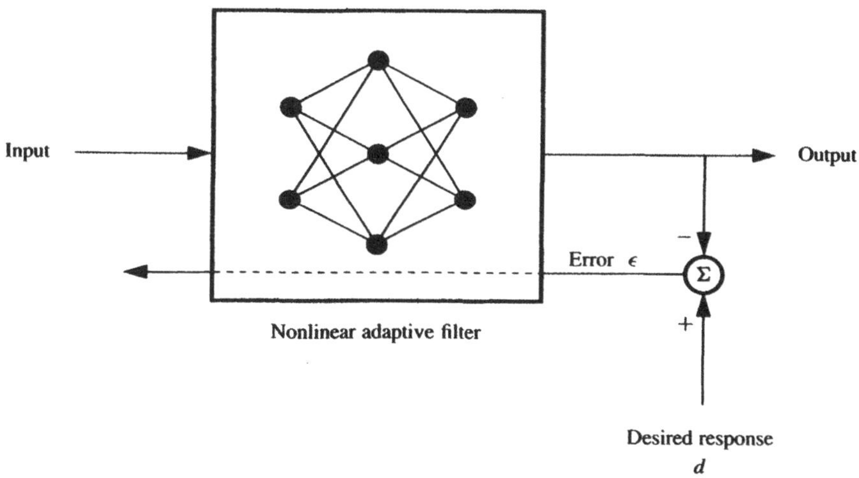

Figure H.3 A general representation for an adaptive transversal filter based on neural networks.

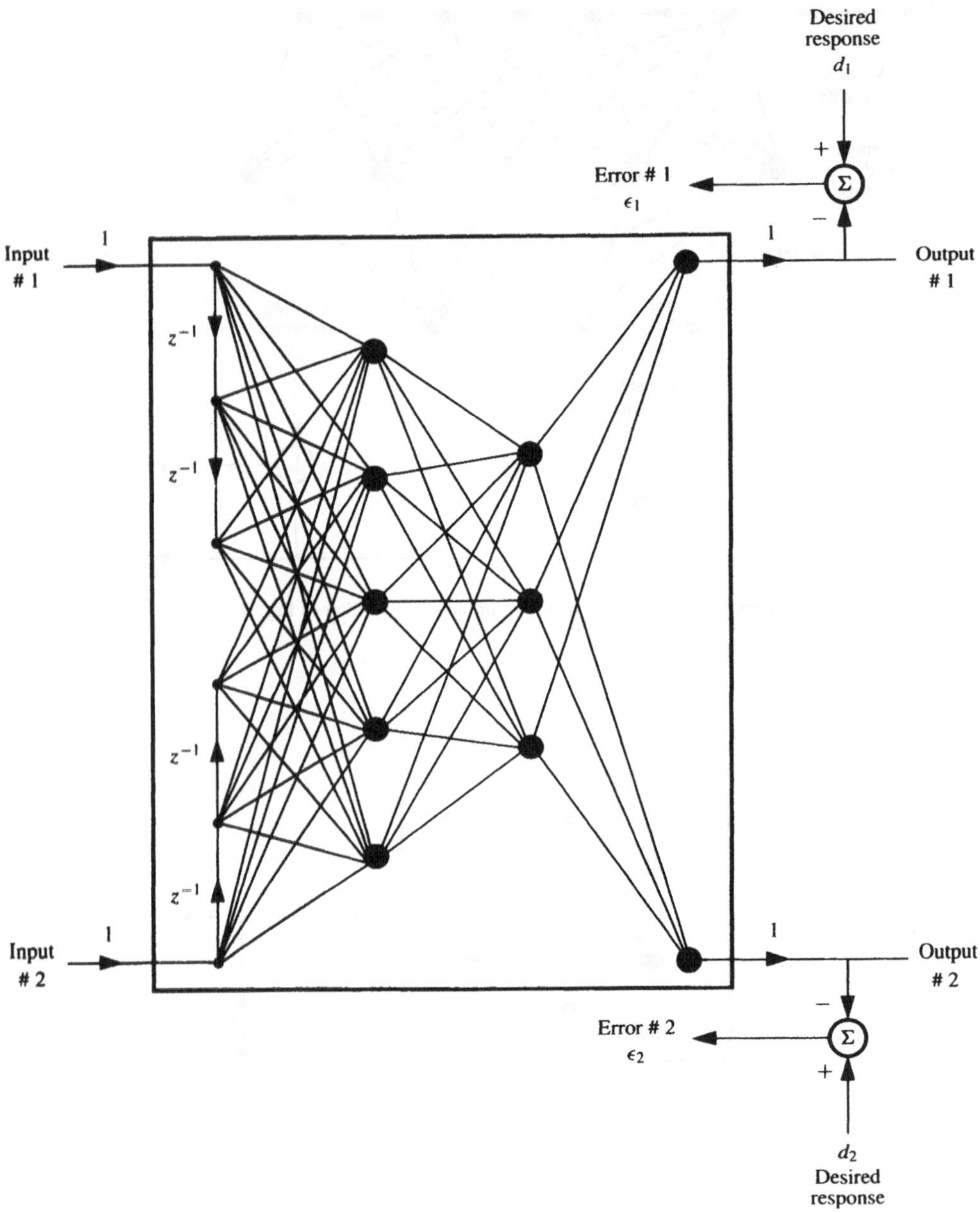

Figure H.4 A two-input two-output MIMO adaptive filter.

H.3 A CASCADE OF LINEAR ADAPTIVE FILTERS

The capability of achieving learning in a cascade of adaptive filters is of fundamental importance to neural control systems. To explore this, we begin with a cascade of linear adaptive filters and show how to train them by using the filtered-ϵ LMS algorithm developed in Section 7.2.

Figure H.5 shows a cascade of two adaptive filters, $A(z)$ and $B(z)$. The input to the cascade is applied to $A(z)$. The output of the cascade is the output of $B(z)$. The output is compared with the desired response, and the error is used to adapt $B(z)$ with conventional LMS. $A(z)$ is adapted using the same error filtered by $B^{-1}(z)$. This works well as long as $B(z)$ is adapted slowly and is always minimum-phase.

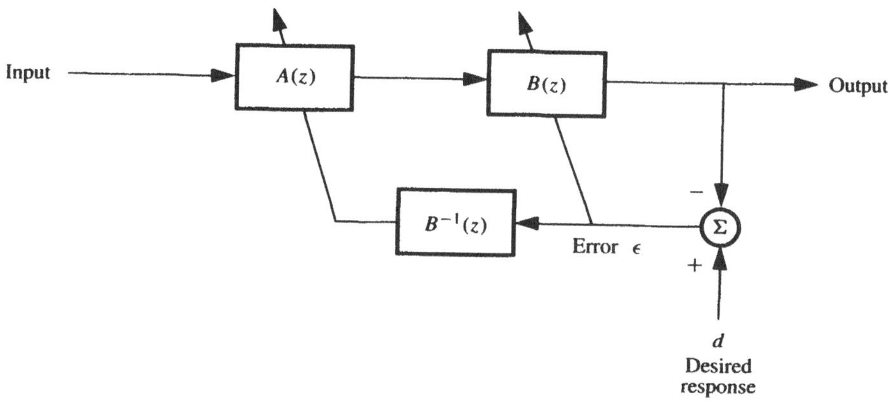

Figure H.5 Training a cascade of adaptive filters by means of the filtered-ϵ LMS algorithm.

If $B(z)$ is nonminimum-phase, then a good inverse would be a delayed inverse. Assume that an offline process is used to get $B_\Delta^{-1}(z)$. Figure H.6 shows how it could be used to adapt the cascade of $A(z)$ and $B(z)$.

H.4 A CASCADE OF NONLINEAR ADAPTIVE FILTERS

We next explore the question of how to train a cascade of adaptive nonlinear filters based on neural networks. The training method will be a form of backpropagation which will play a similar role to that played by filtered-ϵ LMS for the linear case.

Consider the cascade of nonlinear filters shown in Fig. H.7. In training this cascade, the objective is to adapt all the weights to minimize the mean square of the error. How to do this becomes clear when the diagram of Fig. H.7 is redrawn as in Fig. H.8. The two diagrams appear to be different but they function identically.

The system of Fig. H.7 is a cascade of two nonlinear filters, each a tapped delay line connected to a two-layer neural network. The system of Fig. H.8, its equivalent, is a single nonlinear filter consisting of a tapped delay line connected to a four-layer neural network. The delays are z^{-1} in the flow paths which bring signals to the inputs of the neural network.

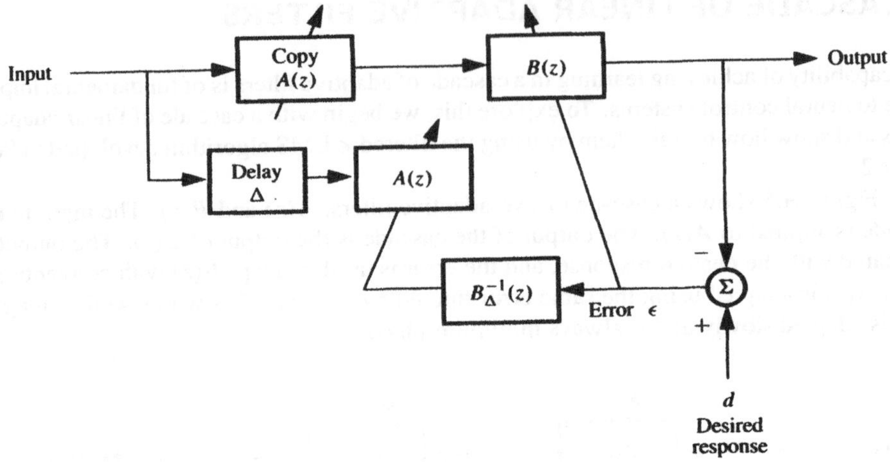

Figure H.6 Training a cascade of $A(z)$ and $B(z)$ where $B(z)$ is nonminimum-phase.

Since there are no delays within the network itself, the four layers can be adapted with the backpropagation algorithm.

Standard backpropagation will not be the correct algorithm, however. In order for the system of Fig. H.8 to be equivalent to that of Fig. H.7, it is necessary that the values of the weights of the first filter of Fig. H.7, which are labeled $a, b, \ldots, i$, be translated into the weights of the first two layers of the network of Fig. H.8. This must be done exactly as shown. Accordingly, all the weights of Fig. H.8 labeled a must have identical values. All the b weights must have identical values, and so forth. When adapting these weights, the initial conditions must satisfy this restriction, and this restriction must be satisfied at the completion of each network adaptation cycle. A change in the a weights should be computed as the average charge that would be computed by conventional backpropagation if there were no restrictions in the weights. A change in the b weights would be computed in like manner, and so forth.

To adapt the cascade of Fig. H.7, one would compute weight changes with the diagram of Fig. H.8 and copy the corresponding weight values into the system of Fig. H.7. To represent symbolically the adaptation of a cascade of nonlinear filters like that of Fig. H.7 by proper application of error backpropagation, the diagram of Fig. H.9 is used. The same diagram could represent both SISO and MIMO systems.

H.5 NONLINEAR INVERSE CONTROL SYSTEMS BASED ON NEURAL NETWORKS

Adaptation of cascaded nonlinear filters can be used in nonlinear inverse control. An adaptive model-reference control system based on these principles is shown in Fig. H.10. The principle of operation is related to that of adaptive inverse control with the filtered-ϵ algorithm. It is useful to compare the nonlinear system of Fig. H.10 with the linear system of Fig. 7.4. In Fig. 7.4, the overall system error is filtered through the plant inverse and then

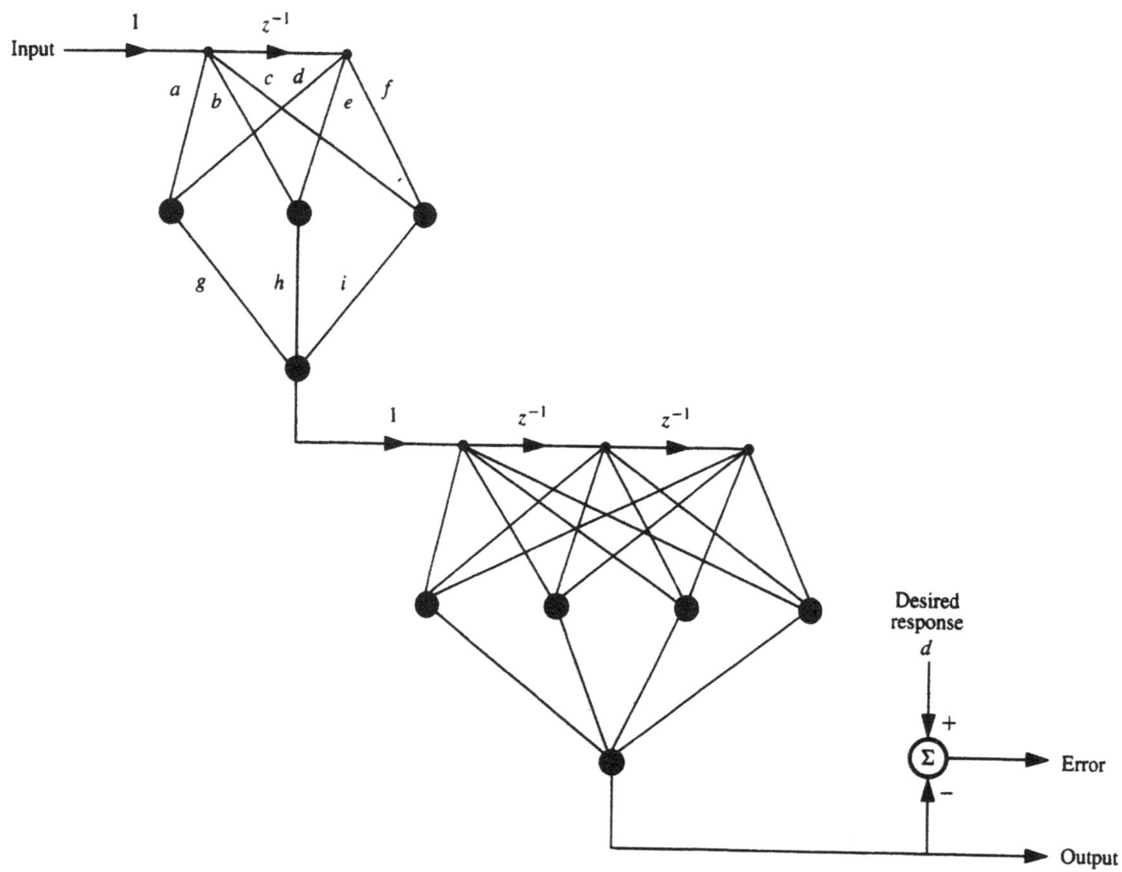

Figure H.7 A cascade of two nonlinear filters, each a tapped delay line connected to a two-layer neural network.

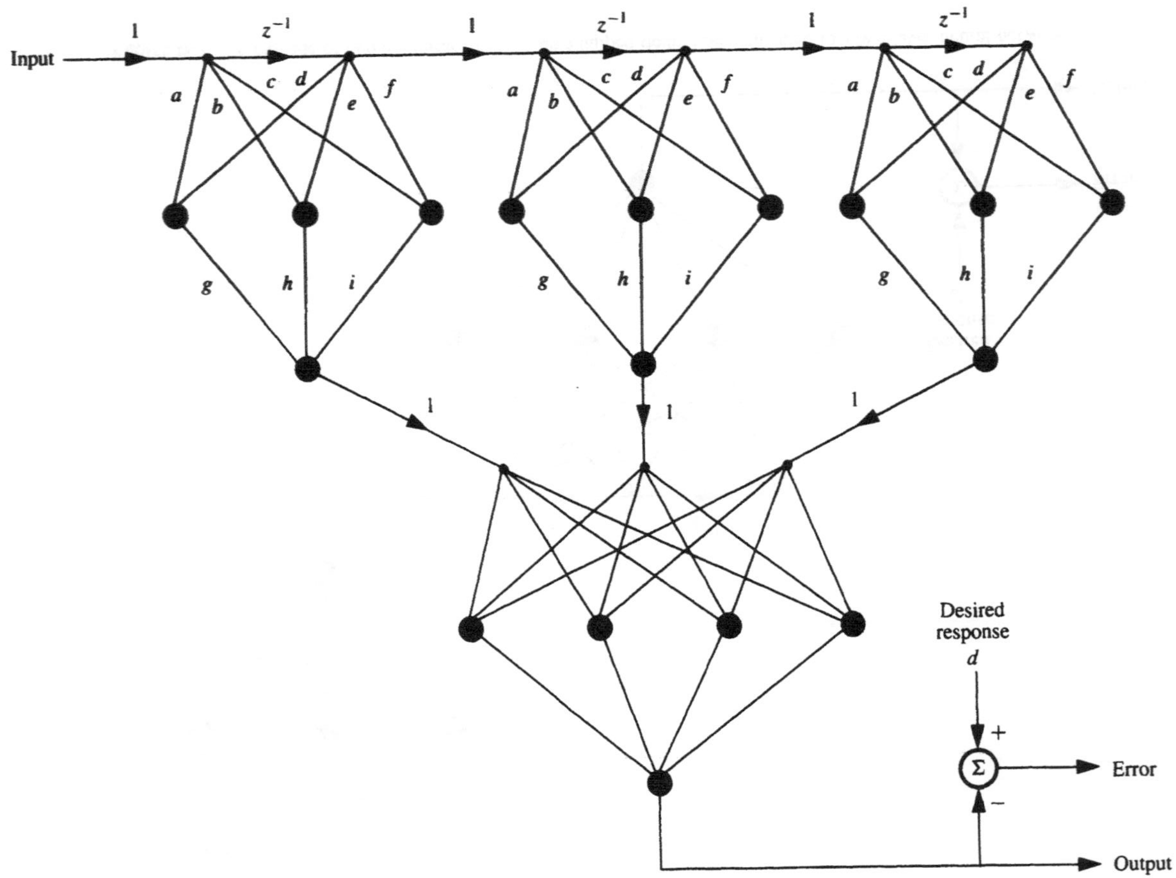

Figure H.8 A system that is equivalent to that of Fig. H.7. It is a tapped delay line connected to a four-layer neural network.

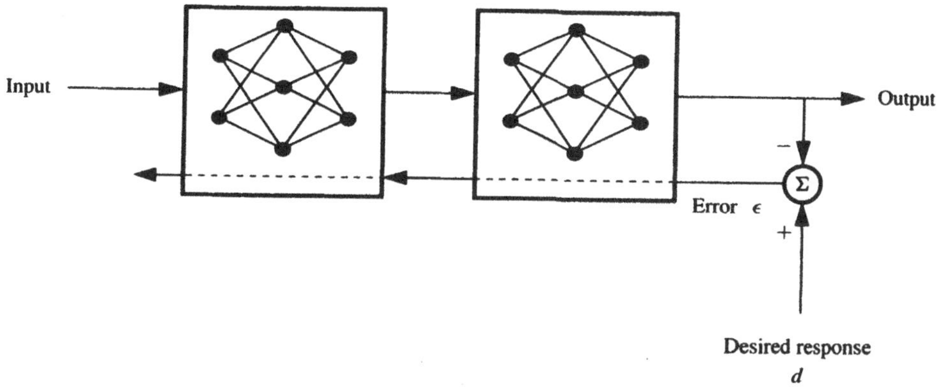

Figure H.9 A trainable cascade of nonlinear adaptive filters.

used to adapt the controller. Minimizing the mean square of this error optimizes the controller. Having a precise plant inverse is not critical for the minimization of this error.

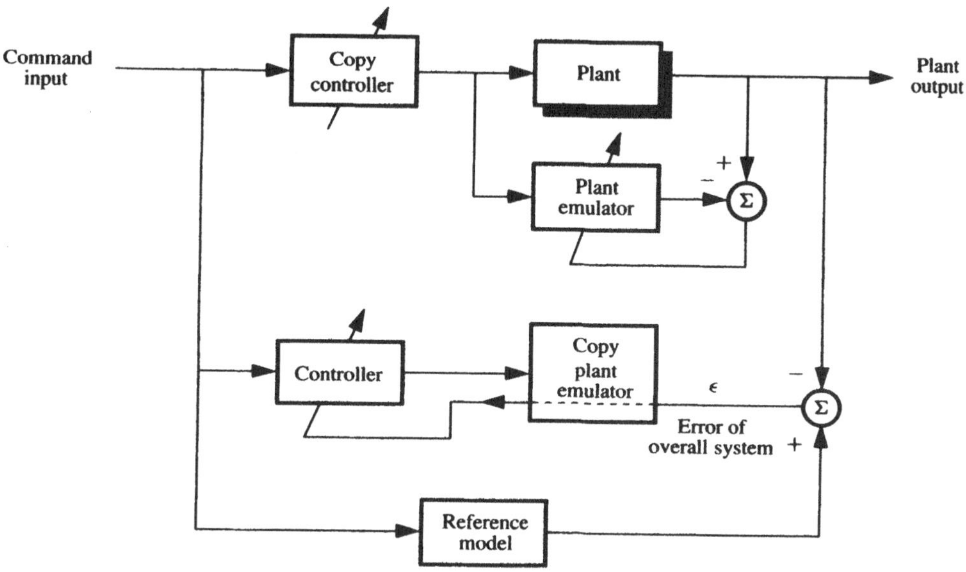

Figure H.10 A model-reference adaptive inverse control system for a nonlinear plant. The plant emulator and the controller are tapped delay lines with neural networks connected to the taps.

In Fig. H.10, the overall system error is backpropagated through a direct model of the plant and then used to adapt the controller. The plant model or "emulator" adapted by the backpropagation algorithm, is illustrated in Fig. H.10. The controller weights are used to train the cascade of Fig. H.9. This process is not used to train the "emulator copy," however. Its weights are simply copied from those of the emulator itself. Having a precise plant emulator is not critical for the minimization of the overall system error.

The following sections report on current laboratory and commercial applications of these neural network principles.

H.6 THE TRUCK BACKER-UPPER

Vehicular control by artificial neural networks is a topic that has generated widespread interest. At Purdue University, tests have been performed using neural networks to control a model helicopter [1]. In a much larger project, a full-sized self-driving van named ALVINN (Autonomous Land Vehicle In a Neural Network) complete with video camera "eyes," and an onboard "brain" made from four workstations, has been developed and built at Carnegie-Mellon University [2]. ALVINN learned to drive by watching humans drive, and can drive long distances at normal highway speeds, negotiating through traffic without human intervention. The system is not yet perfect, of course, so when ALVINN drives, a human is always present to take over the controls if something goes wrong.

In this section, we consider a system less complicated and more easily described than ALVINN — that of a neural network which has learned to steer a computer-simulated truck and trailer while backing to a loading platform. A solution to this highly nonlinear control problem was obtained by self-learning. The inputs to the two-layer network are "state" variables: the angle and position of the rear of the trailer and the angle of the cab (see Fig. H.11). The output of the neural network is the angle of the steering wheel. The work was done by Nguyen and Widrow [3,4,5]. The learning algorithm they used, which is based on the backpropagation algorithm, is called "backpropagation-through-time."

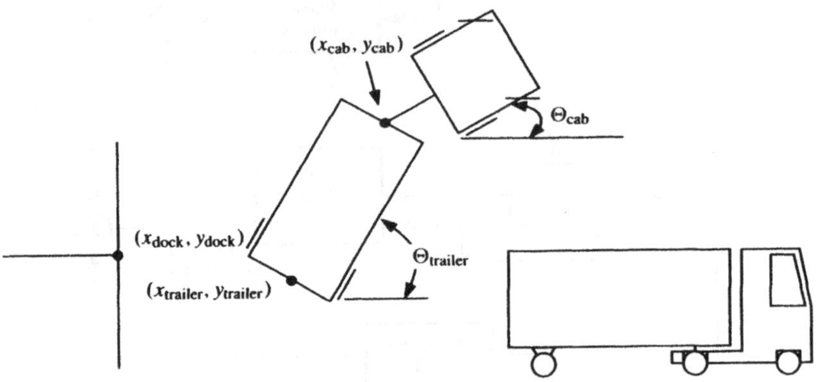

Figure H.11 Truck, trailer, and loading dock.

The truck was only allowed to back up. Backing was done as a sequence of small steps. On the scale of a real 18-wheeler, each step would be a distance of approximately 1 meter. The truck backs from its initial position until it hits something and stops. The desired final state of the system involves having the rear of the trailer parallel to the loading platform and positioned at its center. The actual final state is compared with the desired final state and the difference is a state error vector. After each backing-up sequence is completed, the final error vector is used to modify the controller weights so that if the truck were placed in the same initial position and allowed to retry the backup sequence, the new final state error would have a smaller magnitude than before.

Figure H.12 is a diagram of the neural net controller steering the truck — a controller governing a "plant" represented by the truck kinematics. To train the controller, an emulator of the truck kinematics is needed. This is designed as a two-layer neural network. It is trained by backpropagation (as shown in Fig. H.13) to produce the same output states as the plant when both the emulator and plant have the same driving function.

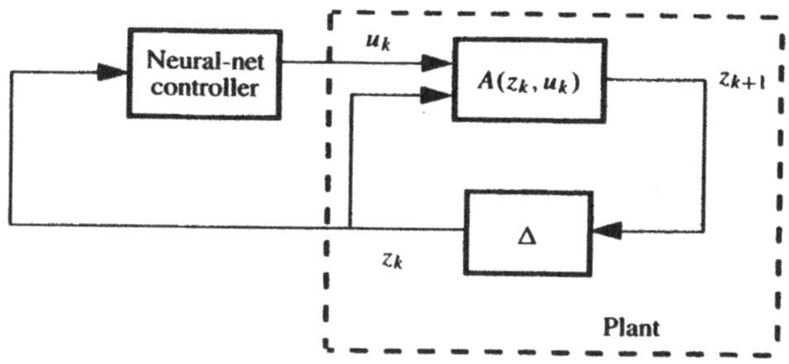

Figure H.12 Plant and controller.

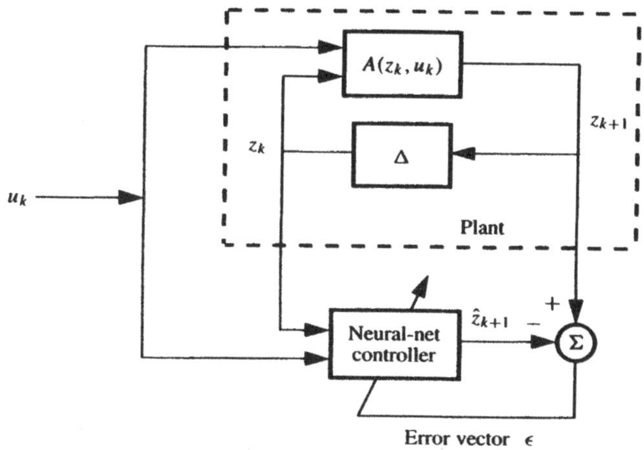

Figure H.13 Training the neural-net plant emulator.

The controller is also a two-layer neural network, trained as shown in Fig. H.14. The initial position or state of the truck, z_0 is applied to the controller which generates a single output, the steering wheel angle. Using this initial steering signal, u_0, the truck backs up a step. The truck has now gone from the initial state z_0 to the next state z_1. The process of using the controller to set the steering angle and then backing a step is repeated until either the truck hits something or the number of time steps exceeds a predetermined constant. The final state of the run is z_K.

The controller and emulator are each composed of two layers of adaptive neurons. Every backing step is analogous to signals going through four layers. Backing from state to

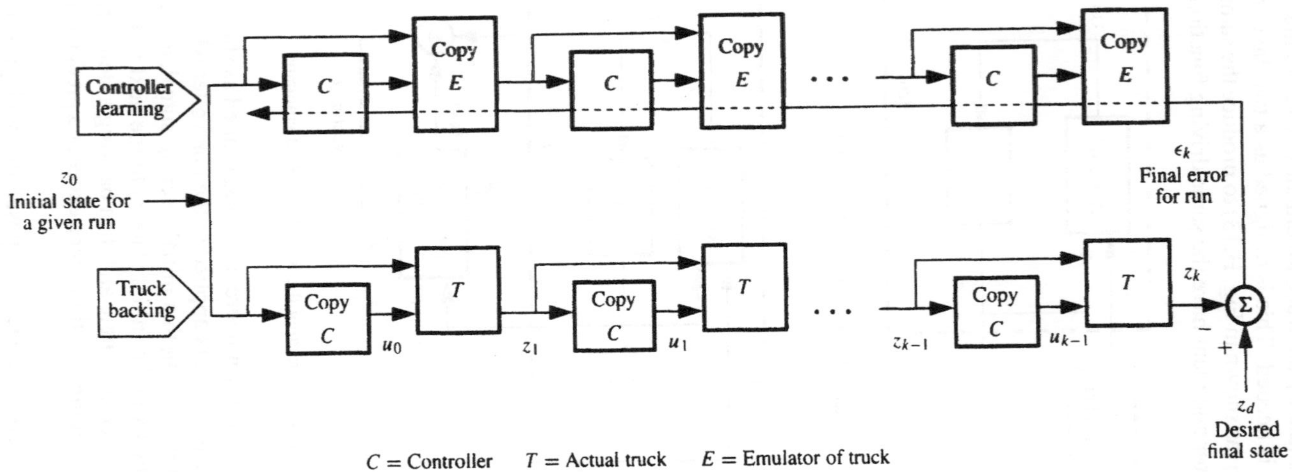

C = Controller T = Actual truck E = Emulator of truck

Figure H.14 Training the controller with backpropagation.

state is represented by signals going through the layers of neural net. The whole backup sequence can thus be represented as the forward propagation of the state vector through a giant feedforward neural network containing a number of layers equal to four times the number of backing steps. In a process called backpropagation-through-time, the final error vector is backpropagated through all the layers of this composite network. Only the controller learns by this, at the end of a run. The backpropagation algorithm dictates weight changes at the various steps. Actual changes in the controller weights are made by averaging the corresponding changes over the steps. The changes dictated for the emulator weights are ignored. Adaptation of the emulator weights is done in accord with the scheme shown in Fig. H.13.

After each backup sequence, the backpropagation-through-time algorithm finds a gradient of the squared positional error of the truck's final state with respect to the weights of the controller. This gradient is used to update the controller's weights by stochastic gradient descent.

Once learning is complete, the truck is able to back up satisfactorily from almost any initial position, even "jackknifed," and even from initial positions that were not previously encountered during training. The controller's ability to react and respond reasonably to new positions is an example of generalization. An illustration of the functioning of an already-trained system is shown in Fig. H.15. This is a laboratory exercise that could, in the future, have implications for vehicle control. More importantly, however, it serves as a visual demonstration of the capabilities of nonlinear networks. This demonstration helped development of the *Intelligent Arc Furnace™* controller described next.

7 APPLICATIONS TO STEEL MAKING

An electric arc furnace is used to melt and process scrap steel. The heat energy comes from a three-phase power line of rather massive capacity (often 30 megawatts or more — enough electrical power for a city of 30,000 people). The three-phase line connects to a bank of step-down transformers to supply current for three electrodes that stick down into the furnace. The electrodes made of graphite, are about 1 foot in diameter, and are about 20 feet long. Three independent servos control the depth of the electrodes into the furnace.

When starting a new "heat," scrap steel is loaded into the furnace and the servos are activated to drive the electrodes down toward the scrap pile. When an arc is first struck, sparks fly and the noise is deafening. One's first impression is that it's like Dante's Inferno.

Because the cost of installing and operating a large arc furnace is so great, even small changes in efficiency have a tremendous impact on economics. The motivation for the development of "intelligent control" is clear. In this section we describe the *Intelligent Arc Furnace™* controller, invented by a recent Stanford graduate, Bill Staib of Neural Applications Corporation [6]. The figures in this section were supplied by the inventor.

Figure H.16 shows an arc furnace, its three-phase power system, and instrumentation that provides signals useful for the control of the electrode servos. Currents and voltages in the system are sensed, digitized, and fed to a 486 PC that implements the neural control system. Numerical processing is performed by an 80 megaflop Intel i860 microprocessor. A microphone placed near the furnace provides the computer with the sounds of Dante's Inferno. From all the sensed variables, a state vector is obtained.

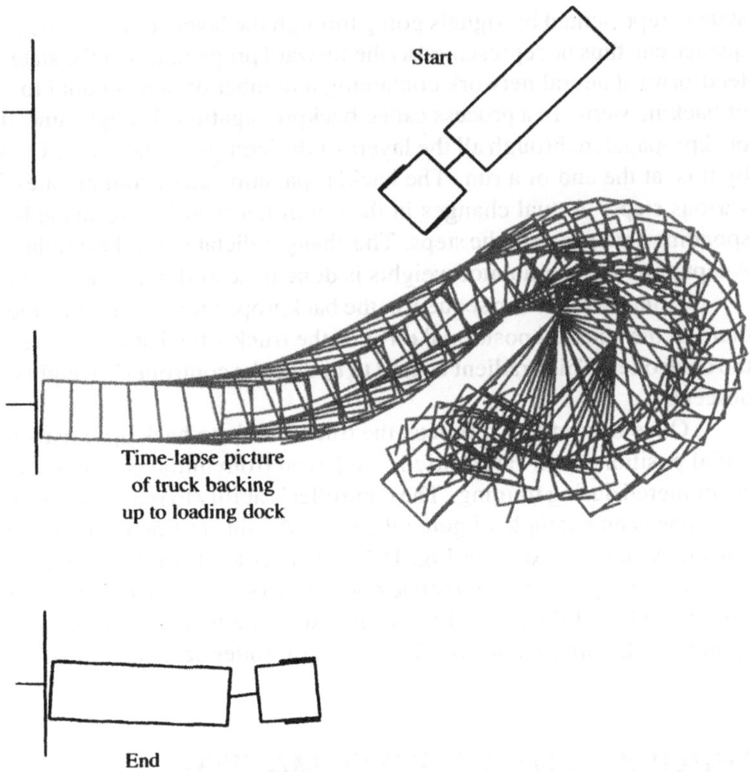

Figure H.15 Example of a truck backup sequence.

Figure H.17 shows the training of a neural network emulator of the furnace. The emulator is used in the training of the controller or regulator, another neural network. Figure H.18 shows the training of the regulator. The learning algorithm is a variant of the backpropagation algorithm. It works in a similar way to the training process for a single stage of Fig. H.14 of the truck-backer.

The results with neural control thus far have been excellent compared with the control systems that commonly exist for arc furnaces. Consumption of electric power is reduced by 5–8 percent, wear and tear on the furnace and the electrodes is reduced by about 20 percent, the power factor on the input power lines is brought closer to 1, and the daily throughput of steel is increased by 10 percent. The neural controllers are being installed by Neural Applications Corporation just as quickly as they can be produced. These improvements are reportedly worth millions of dollars per year per furnace.

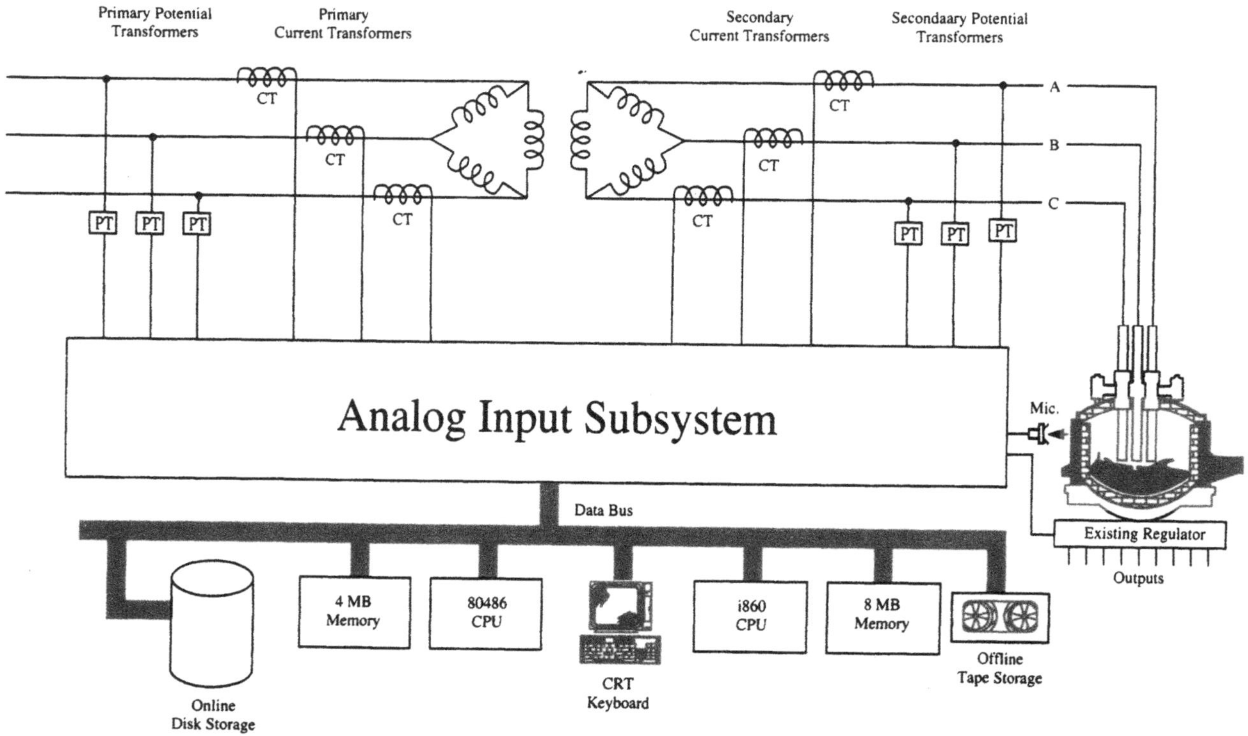

Figure H.16 Electric arc furnace and data acquisition system.

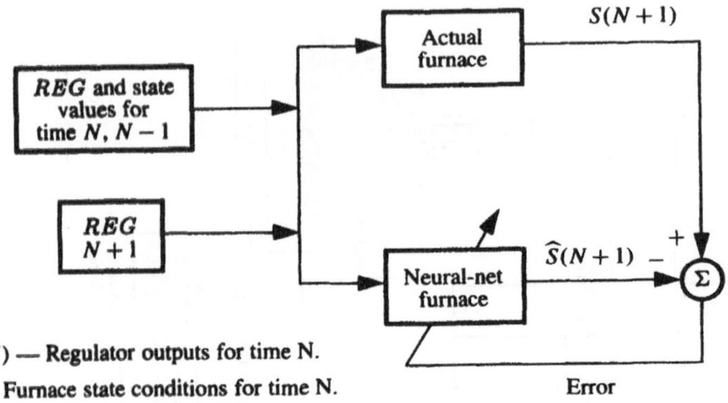

REG(*N*) — Regulator outputs for time N.
S(*N*) — Furnace state conditions for time N.

Figure H.17 Arc furnace emulator training.

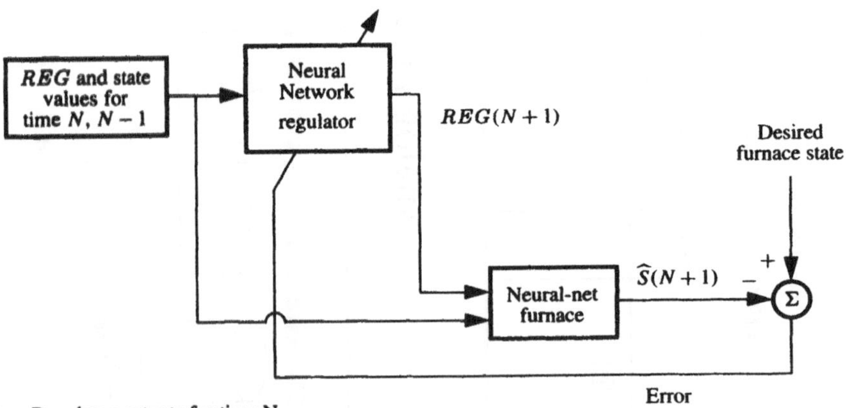

REG(*N*) — Regulator outputs for time N.
S(*N*) — Furnace state conditions for time N.

Figure H.18 Arc furnace regulator training.

.8 APPLICATIONS OF NEURAL NETWORKS IN THE CHEMICAL PROCESS INDUSTRY

Pavilion Technologies, Inc., of Austin, Texas has embedded neural networks and fuzzy logic into their Process Insights package for chemical manufacturing and control applications [7]. With this package, the user takes historical process data and uses it to build a predictive model of plant behavior. The model is then used to change the control setpoints in the plant to optimize behavior. Pavilion Technologies is a spin-off of MCC (Microelectronics and Computer Technology Corporation), Austin, Texas, where the original work was done in 1989–1990 by John Havener of Texas Eastman and Jim Keeler of MCC/Pavilion Technologies. In the original application conducted at the Texas Eastman Facility, Longview, Texas, the neural network produced setpoint changes that reduced by one-third the requirement of an expensive chemical additive needed to remove byproduct impurities during production. The facility produces plastics and chemical intermediates such as aldehydes and olefins. Since that work was complete, the technology and Pavilion's Process Insights package has been used in nearly 200 real-world applications, including modeling and optimization of distillation columns, modeling and control of plastics production, modeling and control of impurity levels in boilers, and so forth. These applications have generated tremendous paybacks, with savings of some applications totaling millions of dollars per year in a single unit production facility. Texas Eastman, a division of Eastman Kodak, has been so satisfied with the results that they are currently encouraging the use of neural networks throughout their Longview plant [8].

In making these applications, the first step is plant modeling or plant emulation. Typically, the plant has many inputs (such as pressures, temperatures, flow rates, feed-stock characteristics.) and one or more output parameters (such as yield, impurity levels, variance). The figures in this section were supplied by Pavilion. In Fig. H.19, an adaptive neural network is used to model an unknown plant, that is, to learn the plant's dynamics from historical data.

Once the plant emulator converges, it can be used to train the neural-net controller. Figure H.20 shows how this is done. The error vector is the difference between the plant output vector and the desired state vector. This error is backpropagated through the neural plant model to provide error signals for the adaptation of the weights of the controller. The controller weights are adapted by the backpropagation algorithm to minimize the sum of squares of the components of the error vector. Pavilion uses fuzzy logic in its Process Insights package to establish constraints on some of the controlled variables.

It is interesting to compare Figs. H.14, H.17, and H.20. Very similar things are going on in the vehicle control system (the truck-backer), in the arc furnace control system, and in the chemical process control system. An emulator is made of the process to be controlled, and the controller is adapted by backpropagating the system error through the emulator. This is a very powerful idea and it leads to useful applications. The reader should be aware however that this is not the only means of neural control. For instance, one method used by the Lord Corporation in the process of developing a new adhesive product [9] involved training a neural network to model the product's adhesive properties as a function of the ingredient proportions. After the network was trained, a search was performed over the neural network's input space to find the formulation believed by the network to be optimum. This formulation was then fabricated and its adhesive properties were tested. This created a new

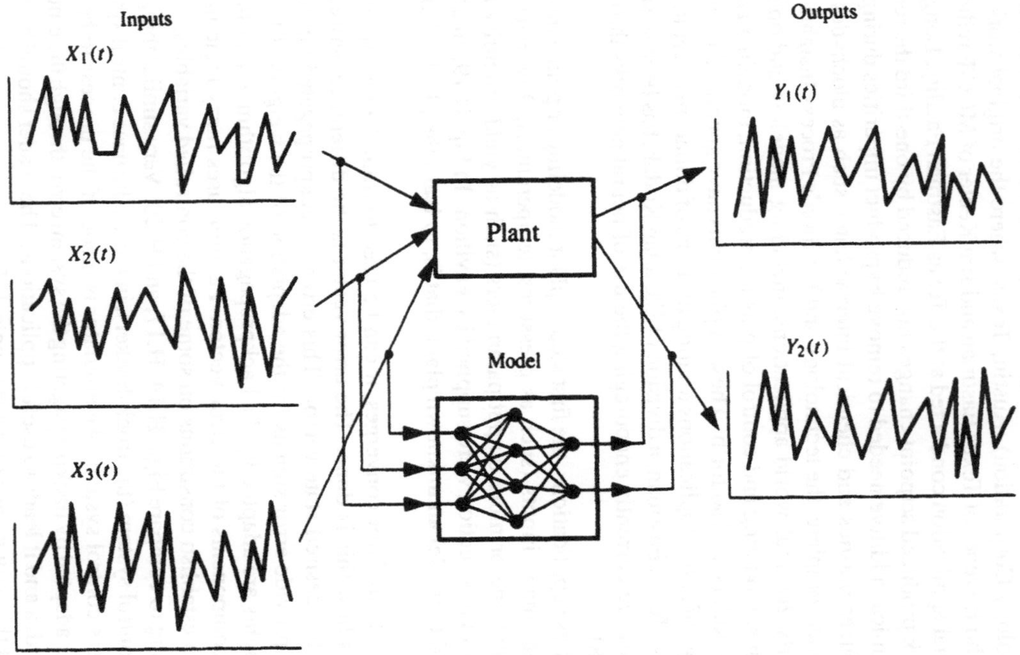

Figure H.19 Adaptive plant emulation.

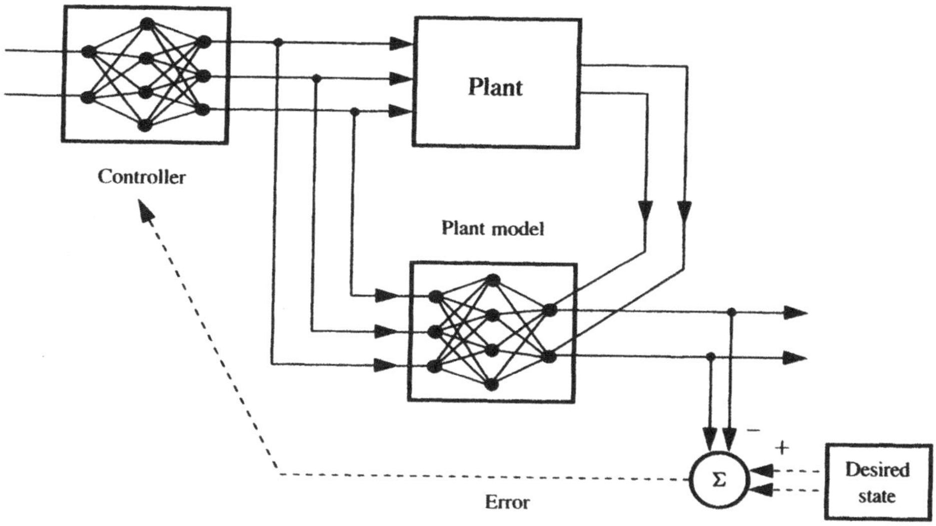

Figure H.20 Using the plant model for training the controller by error backpropagation.

data point and the network was retrained to create a more accurate model. The network's input space was again searched, and this process was repeated until adequate results were obtained.

This section has described only a small fraction of the commercial and industrial applications of neural networks that exist today. The list is long and impressive and growing rapidly. Before the turn of the century, we can reasonably expect to see neural networks become a household word and a part of everyday life. In Japan, fuzzy logic has already achieved this status.

Bibliography for Appendix H

[1] T.J. PALLETT, and S. Ahmad, "Real-time neural network control of a miniature helicopter in vertical flight," in *Proceedings of the 17th International Conference on Applications of Artificial Intelligence in Engineering — AIENG/92*, Waterloo, Ontario, Canada, 1992, pp. 143–160.

[2] O. PORT, "Sure, it can drive, but how is it at changing tires?" *Business Week*, March 2, 1992, pp. 98–99.

[3] D.H. NGUYEN, and B. Widrow, "Neural networks for self-learning control systems," *IEEE Control Systems Magazine*, Vol. 10, No. 3 (April 1990), pp. 18–23.

[4] D.H. NGUYEN, and B. WIDROW, "Neural networks for self-learning control systems," *Int'l. J. Control*, Vol. 54, No. 6 (1991), pp. 1439–1451.

[5] W.E. STAIB, and R.B. STAIB, "The intelligent arc furnace™ controller: A neural network electrode position optimization system for the electric arc furnace," in *Int'l. Joint Conference on Neural Networks*, IEEE, 1992.

[6] R.B. FERGUSON, "Chemical process optimization utilizing neural network systems," in *SICHEM '92*, Seoul, Korea (Erlbaum, 1992).

[7] "Use of neural networks spreads plantwide," *Texas Eastman News*, Vol. 29, No. 9 (April 29, 1993), p. 1.

Glossary

Text Conventions

1. Boldfaced uppercase symbols denote column vectors or matrices.
2. The estimate of a scalar, vector, or matrix is designated by the use of *hat* ($\hat{}$) placed over the pertinent symbol.
3. The symbol $|\ \ |$ denotes the magnitude or absolute value of the scalar enclosed within.
4. The symbol $\|\ \ \|$ stands for the norm of the vector enclosed within.
5. The inverse of a nonsingular matrix or the inverse of the transfer function is denoted by the superscript $^{-1}$.
6. Transposition of a vector or a matrix is denoted by superscript ^{T}.
7. Superscript *plus* ($^+$) on a rational function of z denotes creation of a new function, which has all the poles and zeros of the original function which are located inside the unit circle in the z-plane.
8. Superscript *minus* ($^-$) on a rational function of z denotes creation of a new function, which has all the poles and zeroes of the original function which are located outside the unit circle on the z-plane.
9. Superscript *asterisk* (*) denotes the optimal (or Wiener) solution.
10. The *asterisk* ($*$) between two sequences indicates performing convolution.
11. Transformation of a weight vector to the coordinates defined by the principal axes of the quadratic surface is denoted by superscript *prime* ($'$).
12. Subscript $_{causal}$ denotes that given impulse response is a causal one.
13. Subscript ($_\Delta$) denotes the delayed version (e.g., delayed inverse).
14. Subscript $_{COPY}$ indicates that the weights of the given filter are copied from the filter computed by an adaptive algorithm at another part of the system.
15. Subscript $_{OPT}$ means the optimal value of the parameter.
16. Time reversal may be denoted by putting ($\tilde{}$) over the pertinent symbol.
17. Symbols *trace* ( ) and *tr* stand for taking trace of the matrix within the parenthesis or immediately to the right.
18. Taking gradient is denoted by (∇).
19. Summation operator is denoted by ($\sum$).

20. The symbol [$]_+$ denotes an operation on an impulse response, where only the causal part is taken, with all the noncausal impulses being discarded.
21. The expectation operator is denoted by $E[\ \ ]$, where the quantity enclosed is a random variable or random vector. Similarly, cov [ ] denotes covariance.
22. Operator ($)_{\text{ave}}$ denotes taking average of all the values enclosed within.
23. z^{-1} denotes unit sample delay used in defining the z-transform of a sequence.
24. When an expression or a symbol of an expression is used as a subscript of N, then the numerator of the expression is taken.
25. When an expression or a symbol of an expression is used as a subscript of D, the denominator of the expression is taken.
26. The symbol diag $[\lambda_1, \ldots, \lambda_p, \ldots, \lambda_n]$ denotes a diagonal matrix whose elements on the main diagonal are $\lambda_1, \ldots, \lambda_p, \ldots, \lambda_n$.
27. The symbol $\ll$ denotes that the expression on the left of the symbol is much smaller than the one on the right.
28. Signal vectors and transfer function matrices of MIMO systems are designated by [ ].
29. Symbol $\triangleq$ denotes definition.
30. Symbol $\approx$ denotes approximate equality.
31. In constructing block diagrams, the following symbols are used. The symbol

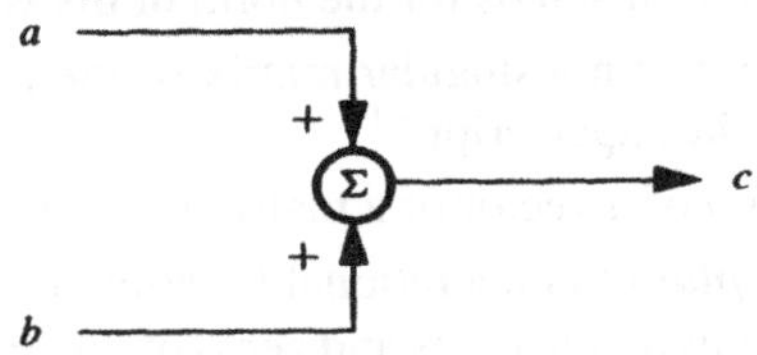

denotes an adder with $c = a + b$. The same symbol with algebraic signs as follows

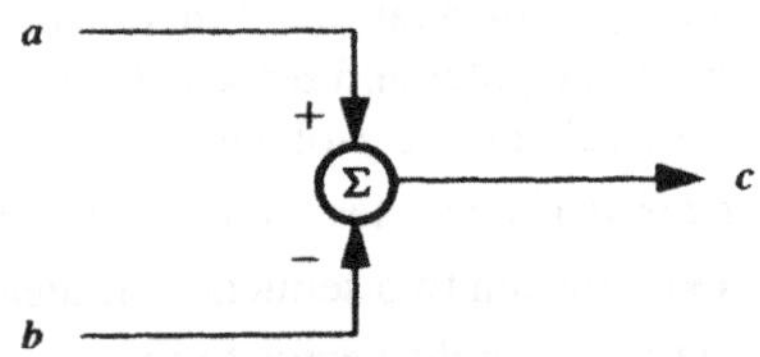

denotes a subtractor with $c = ab$. The symbol

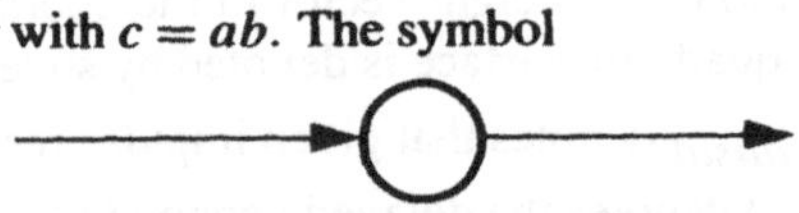

stands for multiplicative weight. Number inside the circle (if present) indicates the value of the weight. Arrow through the weight denotes that we are dealing with a variable weight. The symbol

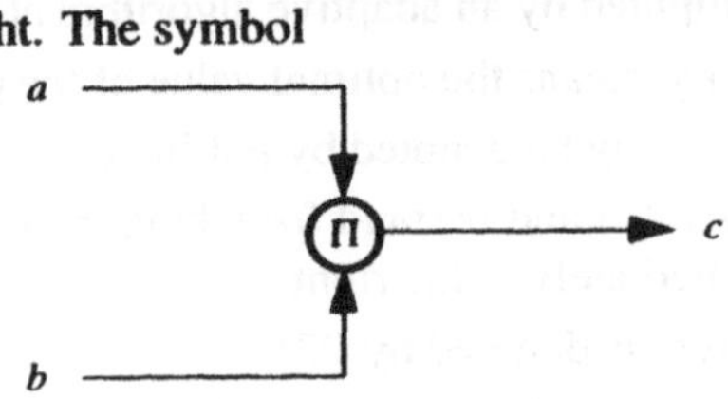

means that $c = a \cdot b$. A box with shadowing

represents a physical plant. While a simple box

denotes an analog filter or a digital filter. An arrow through such a box denotes that the impulse response is being adapted. The arrow can be connected to a signal or not connected at all. In the first case, the signal is used as an error input used to adapt the filter. In the second case, actual adaptation is performed in another part of the system. Then the impulse response is copied into the filter under consideration.

32. Operator z denotes z-transform.

Abbreviations

Adaline	Adaptive Linear Neuron
ADC	Analog-to-Ditigal Conversion
AIS	Audio Input Signal
BPM	Beam Position Monitor
DAC	Digital-to-Analog Conversion
DCT	Digital Cosine Transform
DFT	Digital Fourier Transform
DSD	Differential Steepest Descent
EPS	Earphone Signal
FIR	Finite Impulse Response
GEN	(Noise) Generator
Hg	Mercury
IIR	Infinite Impulse Response
IV	Intravenous
LMS	Least Mean Square
LRS	Linear Random Search
MAX	Maximum
MIMO	Multiple-Input Multiple-Output
MIN	Minimum
MSE	Mean Square Error
RLS	Recursive Least Squares Algorithm
SAN	Strong Ambient Noise
SISO	Single-Input Single-Output
SLAC	Stanford Linear Accelerator Center
TLU	Threshold Logic Unit

Principal Symbols

a	auxiliary scalar variable
a	location of a zero of a rational function
AD	neural element
b	location of a zero of a rational function
c_k	kth impulse response of a controller
$C(z)$	transfer function of a controller
$\mathbf{C}$	controller weight vector
d_k	desired response at instant k
f_k	digital signal at instant k
$F(z)$	transfer function of f_k
$FB(z)$, $FC(z)$	feedback controllers
g_k	output of a digital filter at time k
$G(z)$	z-transform of a sequence g_k
h_k, h_l, h_μ	impulses of impulse responses
$H(z)$	transfer function of a plant or a digital filter
$H_{AB}(z)$	transfer function from point A to point B
$H_{ij}(z)$	transfer function of a MIMO system from input i to output j
i_k	command input at time instant k
$\mathbf{I}$	identity matrix
$I(z)$	z-transform of command input
$IC(z)$	input controller
j	square root of -1
j	index
k	time index
K	number of inputs and outputs of a MIMO system
l	index
l_k	"source" noise at time instant k
$L(z)$	z-transform of noise l_k
m	number of weights in the controller
m	index
M	misadjustment
$M(z)$	model reference
M_1, M_2	microphones
n	number of weights in the adaptive plant model
n_k	noise at instant k
$\tilde{n}_k$	equivalent plant noise at sample k
N	data length
$\mathbf{N}_k$	noise vector
$O(\ )$	the order of
p_k	kth impulse of plant impulse response
$\mathbf{P}$	plant weight vector
$\mathbf{P}$	crosscorrelation vector

$P(z)$	plant transfer function
P	plant impulse response
q_k	kth impulse response of plant noise canceling filter
Q	orthonormal model matrix of the autocorrelation matrix R
$Q(z)$	transfer function of plant noise canceling filter
r_p	geometric ratio
r_k	noise signal at time instant k
R	autocorrelation matrix
s_k	digital signal at sample k
SQ	squaring device
u_k	plant input at time k
u'_k	controller output at time k
$U(z)$, $U'(z)$	transfer function of u_k and u'_k
$\mathbf{U}_k$, $\mathbf{U}'_k$	vectors of u_k and u'_k
v_k	variance of weight noise vector
$\mathbf{V}$	vector of weight errors
$\mathbf{V_0}$	initial weight error vector
w_l	weight number l
w_{lk}	weight number l at time instant k
$\mathbf{W}$	weight vector
x_k	input signal at time instant k
x_{lk}	input number l at time instant k
$\mathbf{X}_k$	input vector
$X(z)$	transform of signal x_k
$\mathcal{X}$	matrix of vectors X_k
y_k	plant dynamic response at sample k
$Y(z)$	z-transform of signal y_k, digital filter transfer function
z_k	plant output at time instant k
α	location of a pole of a rational function
β	individual weight noise variance
β	location of a pole of a rational function
$\gamma_0, \gamma_1, \gamma_2$	coefficients of a polynomial.
δ_k	dither signal at sample k
$\delta_{(m)}$	Kronecher delta
Δ	delay
Δ_k	vector of dither inputs δ_k
$\Delta\mathbf{P}$, $\Delta\mathbf{C}$	errors in plant model and controller
ϵ_k	error
$\underline{\epsilon}_k$	dynamic system error
ϵ'_k	ideal error
$\Gamma(z)$	noise transform
λ_1, λ_p	eigenvalues
Λ	diagonal matrix of eigenvalues

μ, μ_1, μ_2	adaptation constants
μ	index
$\xi, \bar{\xi}$	mean square error
ω	normalized angular frequency
$\phi_{ss}(m), \phi_{nn}(m), \phi_{ii}(m),$ $\phi_{ff}(m), \phi_{uu}(m)$	autocorrelation functions
$\phi_{fg}(m), \phi_{yd}(m),$ $\phi_{uy}(m), \phi_{uz}(m)$	crosscorrelation functions
$\Phi_{ss}(z), \Phi_{nn}(z), \Phi_{ii}(z),$ $\Phi_{ff}(z), \Phi_{uu}(z)$	z-transforms of autocorrelation functions
$\Phi_{fg}(z), \Phi_{yd}(z),$ $\Phi_{uy}(z), \Phi_{uz}(z)$	z-transforms of crosscorrelation functions
τ, τ_p	time constants
$\tau_{p_{mse}}$	time constant of mean square error
∞	infinity

Index

D

E

F

Printed and bound by CPI Group (UK) Ltd, Croydon, CR0 4YY
16/09/2026
14963463-0004